W0254196

Progress in Gene Therapy - basic and clinical frontiers

PROGRESS IN GENE THERAPY - *basic and clinical frontiers*

EDITORS:

R. Bertolotti, S.H. Parvez and T. Nagatsu

Utrecht • Boston • Tokyo
2000

VSP BV
P.O. Box 346
3700 AH Zeist
The Netherlands

Tel: +31 30 692 5790
Fax: +31 30 693 2081
vsppub@compuserve.com
www.vsppub.com

First published in 2000

ISBN 90-6764-328-9

Printed in The Netherlands by Ridderprint bv, Ridderkerk.

CONTENTS

PREFACE

Gene Therapy is expected to revolutionize the practice of Medicine at the turn of the third Millenium. Therapeutic/prophylactic benefits should arise both from gene transfer and gene repair/inactivation protocols devised for patient's somatic cells. Gene expression cassettes, designed for the production of therapeutic proteins and non-coding RNA, are thus experimented together with emerging gene repair/inactivation techniques, on a variety of inherited, acquired and infectious/parasitic diseases which also include complex neuro-degenerative processes. From monogenic diseases to cancer, limb/heart ischemia, DNA vaccines, AIDS, arthritis, wound healing, ..., the limit is likely to be our imagination and safety concerns. Moreover, in not so far a future, we may shift from the basic field of disease prevention /treatment to functional enhancement and ageing prevention. Long-term safety and ethical concerns are still to be solved. Yet, Gene Therapy has already been proposed for tantalizing non-disease applications.

The present book represents a collection of chapters on the main aspects of Gene Therapy, some of those we have already treated in the past. The book updates and further develops the current survey. It is aimed at understanding why Human Gene Therapy is likely to be a medical break-through although definitive clinical success still needs time to accomplish. Highlights are focused both on technical/biomedical concepts and on experimental/clinical data that appear to confer potential universality to Gene Therapy. Current and emerging approaches are thus presented together with problems and temptative issues. Yet, this book is not intended to provide an exhaustive review of the pathologies that are currently approached with Gene Therapy. It should rather be perceived as a booster for new applications in every field of medicine.

R. Bertolotti

H. Parvez

T. Nagatsu

Progress in Gene Therapy: Basic and Clinical Frontiers, pp. 1-34
R. Bertolotti *et al.* (Eds)

Gene therapy: transient or long-term minigene expression and gene repair/inactivation

Roger Bertolotti*

CNRS, Molecular Genetics, Department of Hepato-Gastroenterology, Faculty of Medicine, University of Nice Sophia Antipolis, 06107 Nice, France

Table of Contents

---

* E-mail: Roger.Bertolotti@unice.fr

1. INTRODUCTION

a) Current gene therapy relies on minigene transfer/expression

In its original design, gene therapy was devoted to inherited diseases. It was aimed at compensating the deficiency resulting from a dysfunctional gene upon introducing an active counterpart (transgene) into appropriate somatic cells (Anderson, 1984; Blaese *et al.*, 1995). Such an approach relies on gene transfer (transfection). It involves either direct transfection techniques with naked DNA and non-viral vectors (Graham and van der Eb, 1973; Neumann *et al.*, 1982; Wolff *et al.*, 1990; Felgner *et al.*, 1996; Bertolotti *et al.*, 1998; and chapters from this book) or transduction with viral vectors (Mann *et al.*, 1983; Miller *et al.*, 1993; Palese and Roizman, 1996; Bertolotti *et al.*, 1998; and chapters from this book).

Such a gene compensating approach to inherited diseases is now overflown by other potential applications of transgene expression aimed at treating/preventing acquired disorders and infectious/parasitic diseases (reviews: Anderson, 1998; Bertolotti *et al.*, 1998 and 1999). This unexpected explosion of gene therapy essentially results 1) from the observation that therapeutic benefits can result from transient transgene expression (review: Bertolotti, 1998c), and 2) from the huge inherent potential of minigenes. Indeed, current transgenes are minigenes, *i.e.* constructs where DNA corresponding to messenger RNA (cDNA) or non-coding RNA are placed under the control of various mammalian gene expression cassettes. Such artificial genes are usually very small as compared to normal human/mammalian ones, and can thus be easily handled in viral vectors or bacterial plasmids. In addition, they provide an unlimited combination of expression cassettes and expressible sequences, and are thus flexible enough to fit a very broad field of applications.

b) Gene repair and targeted integration of minigenes

The most obvious approach to tackle an inherited disease is to repair the dysfunctional gene in appropriate somatic cell targets. Reestablishing the wild type sequence of the mutant gene before or after the onset of the disease would prevent or reverse the genetic disorder, respectively. Prevention is of course the best way since complete phenotypic reversion is not likely to be possible in many cases. Although ideal for inherited diseases, this method of gene therapy requires efficient techniques of *in situ* gene repair that are not yet available (see below). However, this approach is all the more important as it should also provide for targeted integration of therapeutic minigenes (Bertolotti, 1998c and 1999); unlike current gene transfer technology, non-random integration of therapeutic minigenes into host genomic DNA will prevent insertional mutagenesis hazards and provide for optimized minigene expression (see below).

2. MINIGENE EXPRESSION THERAPY

a) Minigenes as short synthetic counterparts of human genes

As illustrated with the human gene encoding hypoxanthine-guanine phosphoribosyl transferase (HPRT; Fig. 1B), human/mammalian genes are usually large (20-100 kilobases); their coding sequence is interrupted by non-coding intervening sequences (introns) that are removed upon splicing of nascent mRNA (Breathnach and Chambon, 1981; Leff *et al.*, 1986). Like for human HPRT, mature mRNA are usually small: depending on tissues, they have an average size of about 1.5 kilobases (kb), ranging from about 0.2 to 14 kb. Interestingly, most large mRNA have long non-coding ends that can be eventually deleted in their cDNA counterparts. Therefore, by contrast to most human/mammalian genes, full-length coding sequences are usually small enough to be readilly cloned upon reverse-transcription of mature mRNA and inserted into appropriate constructs for expression. Such mammalian expression cassettes were first generated from mammalian cell viruses (Mulligan and Berg, 1980) and readilly used to construct full-length cDNA expression libraries (Okayama and Berg, 1983).

The structure of a minigene is illustrated with the human HPRT gene in figure 1A where a cDNA encoding full-length HPRT protein is inserted in one of the original simian virus expression cassette. Other minigenes are shown in figure 2. Typical minigenes comprise thus 1) a viral or mammalian promoter for transcription initiation, 2) a cDNA encoding the protein to be

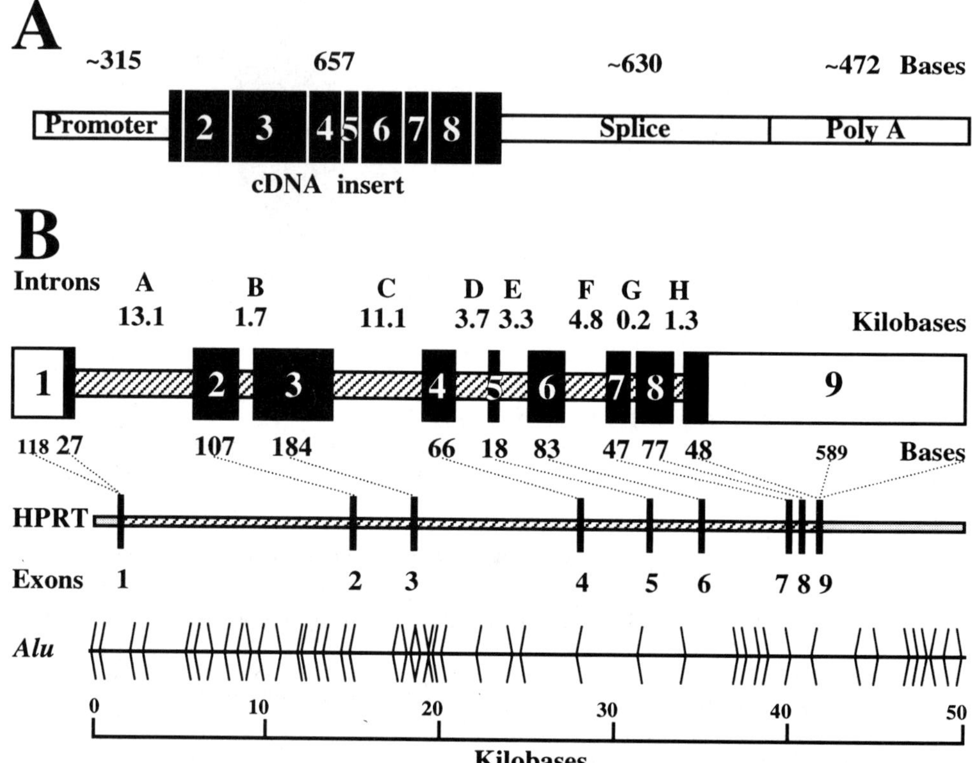

Figure 1. HPRT: structure of a minigene (A) and of the human gene (B; based on Edwards *et al.*, 1990, and EMBL/GeneBank M26434) (Bertolotti, 1998a and 1999). **A)** The shortest full-length HPRT cDNA corresponding to complete coding sequence (657 base-long) is inserted in an SV40 expression cassette (Mulligan and Berg, 1981) comprising 1) the early promoter-enhancer sequence (Promoter) of the SV40 virus on the 5' end of the cDNA for transcription initiation, and 2) on the 3' end, the small t-antigen intervening sequence for mRNA splicing (Splice), and the early transcript termination/polyadenylation sequence (Poly A) for efficient termination of transcription and subsequent polyadenylation of the mRNA. Such a synthetic gene of about 2 kilobases (kb) is very small as compared to its normal human counterpart (~ 40 kb). **B)** The HPRT gene comprises 9 exons (boxes) and 8 introns (hatched areas). Introns and flanking sequences (stippled areas) contain repetitive elements such as *Alu* (46 repeats, positioned with orientation) and LINE (not shown). Enlarged at the upper part of the figure, exons, not in scale with introns, span about 1364 bases out of a ~40 kb genomic sequence. The coding sequence is 657 base-long (filled boxes). Exons 1 and 9 comprise 5' and 3' non-coding regions (empty boxes), respectively.

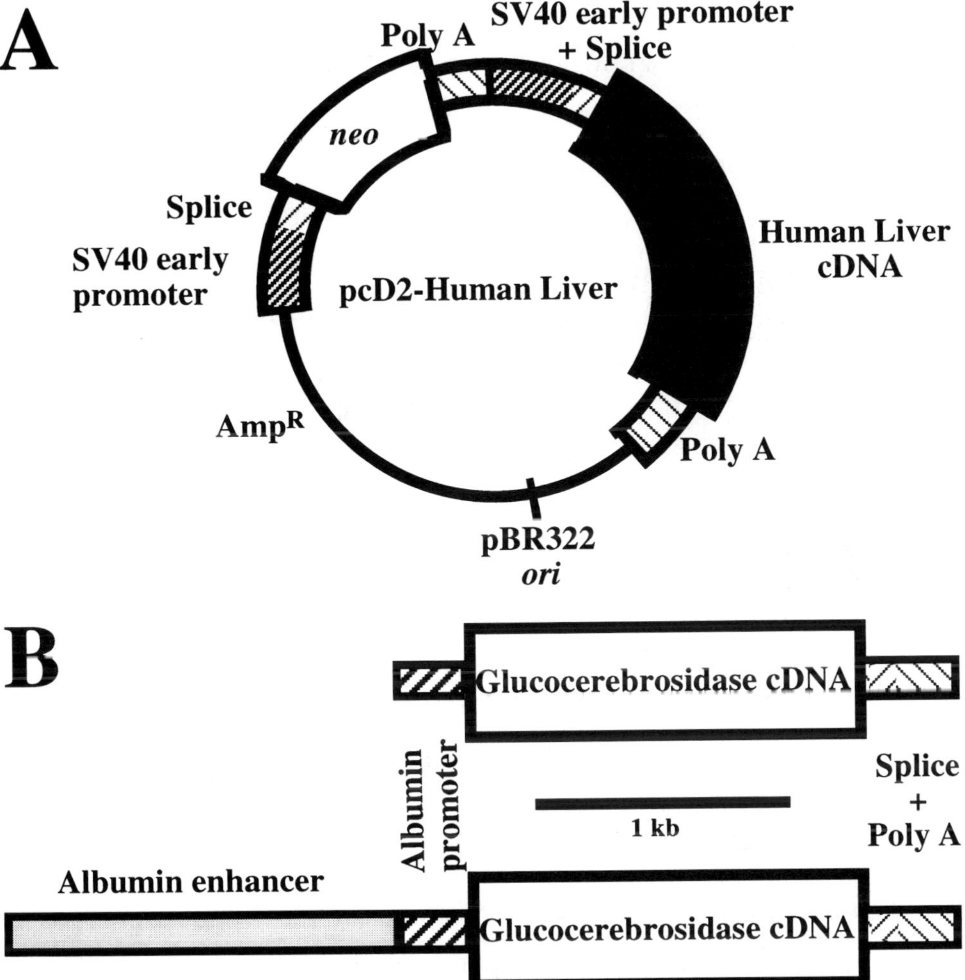

Figure 2. Minigenes from our liver expression programs (Bertolotti, 1998a). **A.** The viral SV40 early promoter ( ) is active both in mammalian cells (Mulligan and Berg, 1980) including differentiated hepatoma lines (Bertolotti and Lutfalla, 1983; Lutfalla *et al.*, 1985) and in the fission yeast (Lee and Nurse, 1987). It is used to drive transcription of full-length cDNA in pcD-libraries (Okayama and Berg, 1983). The pcD2 vector (Chen and Okayama, 1987) we have used to construct human liver cDNA expression libraries (Bertolotti *et al.*, 1995) comprises two SV40-based expression cassettes driving 1) the bacterial *neo* "gene" for transfectant selection (left) and 2) library cDNA encoding human liver proteins (right), respectively. In addition to the SV40 promoter for transcription initiation, both cassettes comprise SV40 DNA sequences providing for efficient transcript termination and maturation (Poly A, Splice; Fig. 1). On the other hand, plasmid pcD contains the pBR322 sequences ( ) required for self-replication (*ori*) and selection (Amp^R) in bacteria (library amplification). **B.** Unlike its SV40 counterpart, the strong albumin promoter ( ) is tissue-specific (Gorski *et al.*, 1986) and is used for liver-restricted expression. However, high-level transcription is achieved only when the *cis*-acting albumin enhancer is present (Pinkert *et al.*, 1987). Both cassettes have been used for the construction of glucocerebrosidase minigenes (McKinley C., Bertolotti R. and Ginns E., 1990-91, unpublished).

produced (or a DNA corresponding to a non-coding RNA), 3) a short intervening sequence for RNA splicing (Splice) either on the 3' (Fig. 1A and 2B) or 5' end (Fig. 2A) of the cDNA, and 4) a DNA sequence for efficient transcript termination and polyadenylation (Poly A).

Additional *cis*-acting elements can be introduced that enhance promoter strength (Fig. 2B; Banerji *et al.*, 1981), confer tissue-specific expression (Fig. 2B; Walker *et al.*, 1983) or implement unique expression features (see below and Bertolotti, 1998c). Unlike in the minigene from Fig. 2B, the albumin enhancer is located 8.5-10.4 kb upstream of its promoter in the albumin gene (Pinkert *et al.*, 1987). In fact, as illustrated with the albumin gene (Pinkert *et al.*, 1987) or the β-globin gene locus (Forrester *et al.*, 1987 and 1989; Grosveld *et al.*, 1987; Festenstein *et al.*, 1996; Bulger *et al.*, 1999; Prioleau *et al.*, 1999), *cis*-acting elements can be fairly far from the promoter and are therefore also responsible for the large size of mammalian/human genes.

b) Minigenes as an unlimited combination of expression cassettes and expressible sequences

Another important feature of minigenes is to provide artificial expression patterns such as synthesis of foreign proteins, ectopic production of tissue-specific proteins or transcription of designed non-coding RNA. Non-coding RNA include basic antisense RNA that impair target gene expression (Izant and Weintraub, 1985), RNA decoys that inhibit HIV replication (Sullenger *et al.*, 1990), and catalytic or splice-altering RNA that can inactivate or repair target RNA (Cech, 1988; Ely *et al.*, 1999; Sullenger and Cech, 1994; Li *et al.*, 1997; Gorman *et al.*, 1998; Puttaraju *et al.*, 1999).

Production of active protein or non-coding RNA is determined by the nature of the expression cassette. Early mammalian expression cassettes were thus shown to express bacterial xanthine-guanine phosphoribosyl transferase (XGPRT) in mammalian cells (Mulligan and Berg, 1980 and 1981). XGPRT is the bacterial homolog of mammalian HPRT; however, unlike its mammalian counterpart, XGPRT phosphoribosylates both xanthine and hypoxanthine. XGPRT was therefore used as a universal bacterial selective marker for mammalian cells either with xanthine selective medium for wild-type cells (Mulligan and Berg, 1981) or with HAT selective medium for HPRT-deficient mutants (Mulligan and Berg, 1980). We have used the same XGPRT minigenes in our early transfection experiments with well-differentiated hepatoma cells of human, rat and mouse origin (Bertolotti and Lutfalla, 1983; Lutfalla *et al.*, 1985). Bacterial XGPRT was used as a selective growth marker for transfectant hepatoma cells; as illustrated in figure 3, growth of XGPRT transfectants in HAT selective medium was directly correlated to the production of active bacterial XGPRT enzyme in hepatoma cells (Lutfalla *et al.*, 1985).

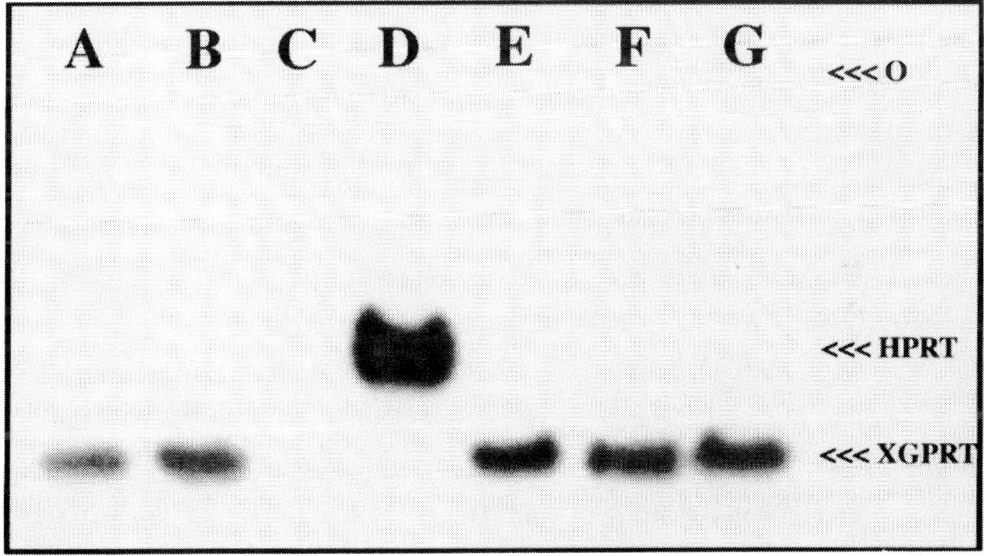

Figure 3. XGPRT and HPRT zymograms of transfected and control cells. Rat and mouse HPRT-deficient hepatoma cells were transfected with plasmids comprising a XGPRT minigene (Bertolotti and Lutfalla, 1983; Lutfalla *et al.*, 1985). Transfectants were selected in HAT medium (Szybalska and Szybalski, 1962; Littlefield, 1964) and their XGPRT-HPRT content analyzed upon electrophoresis of cellular protein extracts on cellulose acetate gels and subsequent fluorography of purine phosphoribosyltransferase activity (Lutfalla *et al.*, 1985). Unlike HPRT-deficient (lane C) and wild-type (lane D) rat controls, transfectant cells from rat (lanes A, B and F) and mouse (lanes E and G) hepatoma lines expressed active bacterial XGPRT enzyme as expected (Lutfalla *et al.*, 1985).

In addition to the expression of heterologous proteins, minigenes can drive the ectopic production of various proteins, *i.e.* production of a protein in a cell that is not programmed to do it. Ectopic production of a single protein can produce dramatic effects such as the conversion of fibroblasts to myoblasts (Davis *et al.*, 1987). Such a myogenic conversion of fibroblasts provides now a potential source of stem cells for gene therapy of Duchenne muscular dystrophy and other myopathies (Lattanzi *et al.*, 1998; Huard *et al.*, 1998). On the other hand, ectopic minigene expression in skeletal muscle is currently envisioned for the treatment of a variety of serum protein deficiencies as illustrated for coagulation factor IX (Herzog *et al.*, 1999; Monahan *et al.*, 1998).

Early expression cassettes based on SV40 viral elements (Fig. 1A and 2A; Mulligan and Berg, 1980 and 1981) are active in most mammalian cell types. Such "universal" expression cassettes driven by the SV40 early promoter or by other viral or ubiquitous mammalian promoters have been used to express proteins and non-coding RNA in a variety of mammalian cells. Importantly enough, they have been instrumental in the development of polyvalent full-length cDNA expression libraries (Fig. 2A; Okayama and Berg, 1983; Aruffo and Seed, 1987; Bertolotti *et al.*, 1995). By contrast to these universal expression cassettes, tissue-specific constructs can provide for transcriptional targeting, *i.e.* they are used to restrict transgene expression to the appropriate type of target cells *in vivo* (Fig. 2B; see Ferrari *et al.*, 1998). Indeed, from early universal expression cassettes based on SV40 viral elements (Mulligan and Berg, 1980) to actual sophisticated constructs, twenty years of gene transfer experiments have unraveled many genomic elements that govern mammalian gene expression. *Cis*-acting genomic elements have thus been identified that enhance gene expression (Banerji *et al.*, 1981), that confer tissue-specificity (Walker *et al.*, 1983; Pinkert *et al.*, 1987), that provide position-independent expression (Stief *et al.*, 1989) relying, possibly, on insulators (review: Bell and Felsenfeld, 1999), ... Such elements, that may include non-mammalian ones, are combined in order to implement unique expression features to a broad spectrum of artificial genes, *e.g.* aforedescribed transcriptional targeting, or regulatable expression with cassettes that are under the control of orally bioavailable drugs (Gossen and Bujard, 1992; Ye *et al.*, 1999; Saez *et al.*, 1997; Wang *et al.*, 1997). However, although sophisticated, current minigene expression cassettes are still far from the very complex genomic *cis*-regulatory logic of a "smart" sea-urchin developmental gene (Yuh *et al.*, 1998).

Even with additional *cis*-acting elements, minigenes usually remain small enough to be amplified in bacterial plasmids (Fig. 2) or viral vectors, and then delivered to target cells. Unlike their normal counterparts, minigenes provide for an unlimited combination of expression cassettes and expressible sequences. They are thus very flexible and can be designed for a broad range of therapeutic/prophylactic applications (Anderson, 1998; Bertolotti, 1998c).

c) Minigene expression therapy has a broad range of applications

Therapeutic minigenes are thus involved in a variety of experimental protocols (see Table 1). In particular, they are used to supply 1) an active protein to mutant deficient cells (inherited diseases; Blaese *et al.*, 1995), 2) an angiogenic factor to endothelial cells *via* ectopic production in skeletal muscles (Baumgartner *et al.*, 1998), 3) a foreign protein to host immune cells (DNA vaccines, Ulmer *et al.*, 1993), 4) a conditional suicide viral enzyme to cancer cells (Culver *et al.*, 1992) or to heterologous grafted lymphocytes (Bonini *et al.*, 1997), and 5) designed non-coding RNA to various kind of recipient cells (see below and Bertolotti, 1998c). Such RNA include anti-HIV ribozymes that are currently challenged in clinical trials on AIDS patients (Ely *et al.*, 1999; Fanning *et al.*, this book).

d) Transient or long-term minigene expression therapy

Except for some viral vectors (*e.g.* alphaviruses or poxviruses: Berglund *et al.*, 1998; Moss, 1996), expression is effective when minigenes

Table 1. GENE THERAPY: a schematic overview (Bertolotti, 1998c)

I. MINIGENE EXPRESSION THERAPY

A. VECTORS
- 1) Viral vectors
  - 1-1) Retroviruses: a) cell-division dependent, b) Lentiviruses
  - 1-2) Adenovirus
  - 1-3) Adeno-Associated virus
  - 1-4) Herpes simplex
  - 1-5) Other viruses: alphaviruses, poxviruses,...
- 2) Pioneer vectors for non-viral receptor-mediated delivery
  - 2-1) Asialoglycoprotein receptor targeting (liver specific)
  - 2-2) Transferrin receptor targeting (general)
- 3) Lipoplexes
- 4) Polyplexes
- 5) Fusigenic Sendai-virus liposomes
- 6) Naked DNA: direct or particle-mediated transfer
- 7) Encapsulated cells

B. INHERITED & DEGENERATIVE DISEASES
- 1) ADA-deficiency
- 2) Cystic fibrosis
- 3) Muscular disorders
- 4) TH-deficiency
- 5) Parkinson's disease
- 6) Others & Stem cell targeting

C. CANCER
- 1) Suicide therapy and bystander effect
- 2) HSVtk-lymphocytes for allogenic grafts
- 3) Immunotherapy (dendritic cell priming, cytokine transfer,...)
- 4) Apoptosis (antiangiogenic and apoptotic genes)

D. CARDIOVASCULAR DISEASES

E. DNA VACCINES

F. PROSPECTS & EMERGING TECHNOLOGIES
- 1) Regulatable expression cassettes
- 2) DNA shuffling to improve gene therapy constructs
- 3) dsDNA-cored recombinase-DNA nucleoprotein filaments

II. RNA INACTIVATION or REPAIR

A. INACTIVATION (Ribozymes)

B. REPAIR (Ribozymes, modified snRNA, pre-*trans*-splicing RNA, tRNA)

III. GENE REPAIR/INACTIVATION

A. CHIMERIC RNA-DNA OLIGONUCLEOTIDES (point mutations)

B. DNA-RECOMBINASE NUCLEOPROTEIN FILAMENTS

IV. ARTIFICIAL CHROMOSOMES

are in the nucleus of transfected cells. Minigenes are either free (transient expression) or integrated into host chromosomes (permanent expression). Yet, transient expression may also result from occasional unstable expression of integrated minigenes and from elimination of transfected cells. On the other hand, stable expression is eventually achieved with some virus-based shuttle vectors that are maintained as self-replicating nuclear episomes (DiMaio *et al.*, 1982; Yates *et al*, 1985; Lutfalla *et al.*, 1989).

In most inherited diseases, the missing gene product is usually a permanent need. Therefore, these diseases essentially request long-lasting minigene expression therapies where the ideal targets are stem cells (see below). By contrast, unexpectedly, transient minigene expression appears to be effective either as a therapeutic or a prophylactic approach for an increasing number of diseases, *e.g.* cancer, limb and heart ischaemia/ stenosis, wound healing, bone repair, infectious and parasitic diseases. This is the great surprise of gene therapy and, most likely, its largest field of application. Such a potential breakthrough in molecular medicine relies on transient expression of a growing spectrum of therapeutic minigenes (Bertolotti, 1998c).

e) Transient expression as a gene therapy breakthrough

Welcome to the exploding field of transient minigene expression therapy, in particular to the ongoing revolution of DNA vaccines. Indeed, experimental DNA vaccines have been devised that provide efficient immunization upon intramuscular or intrasplenic injection of minigenes that encode relevant antigens (Donnelly *et al.*, 1997; Taubes, 1997). Unlike most protein vaccines, DNA vaccines raise both cytolytic T cells and antibodies (Ulmer *et al.*, 1993); they appear to be active against a broad range of infectious and parasitic diseases, in particular against tuberculosis (Huygen *et al.*, 1996; Tascon *et al.*, 1996) and, possibly, malaria (Doolan *et al.*, 1998; Wang *et al.*, 1998). In addition to DNA vaccines that fill entire scientific meetings and a specialized website on Internet (http://www.

genweb.com/dnavax/dnavax.html), transient minigene expression opens new therapeutic/prophylactic avenues for many diseases: cancer (Itaya *et al.*, 1987; Tepper *et al*, 1989; Culver *et al.*, 1992; Roth and Cristiano, 1997; Pardoll, 1998; Folkman, 1998), limb/heart ischemia (Isner *et al.*, 1996; Baumgartner *et al.*, 1998; Losordo *et al.*, 1998; Symes *et al.*, 1999; Rosengart *et al.*, 1999), AIDS (see antisens RNA, ribozymes, RNA decoys, DNA vaccines), arthritis (Ghivizzani *et al.*, 1998), liver cirrhosis (Ueki *et al.*, 1999), wound healing (Sun *et al.*, 1997; Eming *et al.*, 1998; Yamasaki *et al.*, 1998), bone repair (Bonadio *et al.*, 1999), insulin-dependent diabetes (topical induction of pancreatic islet neogenesis; see Edlund, 1999), ... Therapeutic angiogenesis has thus been achieved in patients with critical limb ischemia upon intramuscular injection and ensuing transient local expression (less than 30 days) of a minigene encoding human vascular endothelial growth factor (VEGF; Baumgartner *et al.*, 1998); the same approach is currently applied to selected patients with myocardial ischemia upon direct myocardial transfer of VEGF minigenes (Losordo *et al.*, 1998; Symes *et al.*, 1999; Rosengart *et al.*, 1999). Moreover, the first phase III clinical trial, *i.e.* full-scale evaluation of treatment effect, is a transient anticancer one: gliostoma cells are targeted with a retroviral vector and destroyed upon transient expression and ensuing "bystander" effect of a suicide gene (Ram *et al.*, 1997; Dr. Marie, USA).

Importantly enough, transient expression may request strict inforcement when residual expression is a potential problem to therapeutic efficiency such as in current experimental DNA vaccines where occasional remnant antigen synthesis could result in tolerance rather than immunity. Securing transient expression may involve lytic expression such as in emerging suicidal DNA vaccines (Berglund *et al.*, 1998; Smerdou and Liljeström, 2000 and this book) or regulatable expression cassettes that are under the control of orally bioavailable drugs (Gossen and Bujard, 1992; Ye *et al.*, 1999; Saez *et al.*, 1997; Wang *et al.*, 1997).

f) Long-term minigene expression: inherited/acquired disorders

Long-lasting minigene expression therapy has been quickly extended from classic inherited diseases (Anderson, 1984; Blaese *et al.*, 1995) to complex neuro-degererative processes (Rosenberg *et al.*, 1988; chapters from this book) where transgenic supply of neurotrophic factors or neurotransmitters can antagonize the pathologic processes. In addition, it is currently involved in other types of approaches, *e.g.* life-long protection of the immune system against AIDS upon transduction of hematopoietic progenitor cells with minigenes encoding anti-HIV ribozymes (Ely *et al.*, 1999; Fanning *et al.*, this book). However, although simple in principle, long-term minigene expression therapy is still hampered by many problems (Verma and Somia, 1997; Anderson, 1998).

Current non-viral vectors can provide for efficient transfection and transient expression (see above and Suzuki *et al.*, 2000), but barely promote transgene integration (see below and Kaneda *et al.*, this book). One possible way to solve the problem is 1) either to use shuttle plasmid vectors that are maintained as self-replicating nuclear episomes (DiMaio *et al.*, 1982; Yates *et al*, 1985; Lutfalla *et al.*, 1989) as illustrated by the recent approach of Kaneda and collaborators (Saeki *et al.*, 1998; Kaneda *et al.*, this book), or 2) to develop efficient gene targeting vectors (Bertolotti, 1998c, 1999 and below). Therefore, current clinical trials aimed at long-term transgene expression are based on vectors derived from integrating viruses (Anderson, 1998; Verma and Somia, 1997). However, poor sustained transgene expression is still obtained in current clinical trials that are essentially based on murine onco-retroviral vectors (Anderson, 1998; Verma and Somia, 1997).

Transcriptional shut-off of integrated transgenes have been described for retroviral vectors (Verma and Somia, 1997) that may be inherent to viral vectors[1]; anti-virus/anti-retroelement protection of mammalian cells includes

[1] Conversely, transient transgene expression occasionally relies on infection-dependent production of a transcription factor (Loser *et al.*, 1998).

inactivation of integrated foreign promoters and control elements upon mechanisms such as DNA methylation (Anderson, 1998; Waugh O'Neill *et al.*, 1998; Yoder *et al.*, 1997). Such a viral promoter-specific transcriptional shut-off has been reported for adeno-associated virus (AAV) vectors carrying a cytomegalovirus (CMV) promoter (review: Summerford *et al.*, 2000 and this book). Indeed, comparative studies with AAV vectors showed that minigene expression driven by a CMV promoter declines over time where expression from a mammalian promoter is persistent (Klein *et al.*, 1998; McCown *et al.*, 1996). However, AAV-transduction of this viral promoter provides for sustained expression in several cell types, in particular in skeletal muscle (Xiao *et al.*, 1996; Kessler *et al.*, 1996; Herzog *et al.*, 1999). Gradual inactivation of viral promoters appears thus to be cell-type dependent. Therefore, other expression problems may result from fluctuations in recipient cell differentiation.

Unlike the aforementioned prototypical retroviral vectors based on murine onco-retroviruses, lentiviral vectors are not suject to transcriptional "shut off" (Verma and Somia, 1997). These newly devised retroviral vectors provide for efficient and sustained minigene expression in many different cell types (Naldini *et al.*, 1996a; Miyoshi *et al.*, 1997; Blomer *et al.*, 1997; Follenzi and Naldini, this book). The same holds true for AAV vectors, in particular when expression of the minigene is driven by an internal mammalian promoter (reviews: Summerford *et al.*, 2000 and this book).

Another essential feature of both lentiviral and AAV vectors is their ability to transduce non-dividing cells (Naldini *et al.*, 1996b and 1996a; Follenzi and Naldini, this book; Xiao *et al.*, 1996; Kessler *et al.*, 1996; Summerford *et al.*, 2000 and this book). Such an ability is missing to prototyped onco-retroviral vectors and is indispensable for efficient *in vivo* gene therapy of adult patients whose target cells are mainly quiescent ones. Importantly enough, in addition to long-lived terminally differentiated cells from many tissues and organs, many stem cells such as the hematopoietic progenitors are quiescent-like targets since they have a low proliferating rate in adults. Indeed, HIV-based lentiviral vectors stably transduce quiescent human hematopoietic progenitor cells (Uchida *et al.*, 1998; Case *et al.*, 1999; Evans *et al.*, 1999) that can repopulate immune-deficient NOD-SCID mice (Miyoshi *et al.*, 1999).

Current integrating viral vectors derived both from AAV (Herzog *et al.*, 1999; Snyder *et al.*, 1999; Summerford *et al.*, 2000 and this book) and lentiviruses (Dull *et al.*, 1998; Miyoshi *et al.*, 1998; Zufferey *et al.*, 1998; Follenzi and Naldini, this book) are promissing tools for long-term minigene expression therapy. AAV vectors can accomodate only 3.5-4 kilobases of transgenic sequence(s) while lentiviral vectors have a larger gene delivery capacity (7-7.5 kb). AAV vectors are currently being tested in approved clinical protocols (review: Summerford *et al.*, 2000 and this book) while HIV-based self-inactivating lentiviral vectors with a third generation packaging system (Dull *et al.*, 1998; Miyoshi *et al.*, 1998; Zufferey *et al.*, 1998) may meet the safety requirements for future clinical experimentation (Bukovsky *et al.*, 1999; Follenzi and Naldini, this book).

3. TOWARD GENE REPAIR AND TARGETED INTEGRATION OF MINIGENES

a) Gene repair as the ideal therapy for inherited diseases

Either transient or long-term, minigene expression therapy holds great promises for many acquired disorders and inherited diseases. Such a potential medical breakthrough has however a major limitation for inherited diseases: current minigenes cannot fully substitute for wild-type genes. Indeed, minigenes lack most non-coding genomic sequences and may therefore miss important structural regulatory features specific to their normal counterparts (see, for example, the complex structural features of the β-globin locus; Bulger *et al.*, 1999 and Prioleau *et al.*, 1999). In addition, they are either free or "random-integrated" into host chromosomes. Integration occurs mainly at non physiological locations upon "illegitimate recombination" (Merrihew *et al.*, 1996; Critchlow and Jackson, 1998) or viral DNA processing (*e.g.* retroviruses and adeno-associated virus). Consequently, the main drawback is that minigenes are usually not under complete or normal physiological genetic regulations, an essential concern when dealing with tightly regulated functions such as globin expression or insulin production. In addition, random integration of transgenic DNA into host chromosomes is a potential carcinogenic hazard since random insertional mutagenesis may hit a cancer-prone gene (*e.g.* oncogene, tumor suppressor or DNA repair gene). This is why, in december 1991, we devised a new approach to gene therapy aimed at repairing mutant genes. It is based on premade DNA-human recombinase nucleoprotein filaments and targeted integration of transfecting DNA upon homologous recombination with host chromosomal DNA (Bertolotti R., Grant application to French MRT, 1991; Bertolotti, 1996a, 1996b and 1998a).

b) Recombinase-mediated gene therapy

In vitro, the key reaction of homologous recombination, *i.e.* ATP-dependent homologous DNA pairing-strand exchange with double-stranded DNA (dsDNA), is mediated by a stochiometric complex formed upon polymerisation of recombinase RAD51 onto single-stranded DNA (ssDNA) (Sung, 1994; Sung and Robberson, 1995; Baumann *et al.*, 1996). The resulting helicoidal RAD51-ssDNA filament is the active presynaptic complex within which the reaction occurs upon invasion and subsequent unwinding/stretching of naked dsDNA. The final product of the reaction is an heteroduplex; *in vivo*, it is subjected to mismatch repair mechanisms that result in gene conversion when the incoming bases are maintained, *i.e.* the sequence of target dsDNA is changed. Such a conversion of host chromosomal DNA (gene repair/inactivation) is very poor when transfection is performed with current technology (review: Yanez and Porter, 1998) although there is a recent potential exception based on small PCR genomic fragments (Goncz *et al.*, 1998). The initial aim of our approach was to increase this efficiency to a level compatible with gene therapy protocols upon transfecting somatic cells with active ssDNA-human recombinase filaments. Homologous recombination between transfecting DNA and cognate chromosomal DNA is strongly inhibited by low levels of sequence heterology (Deng and Capecchi, 1992; Te Riele *et al.*, 1992). Therefore, upon publication of the detailed enzymatic properties of eukaryotic recombinase RAD51 (Sung and Robberson, 1995), we devised a more sophisticated approach and invented dsDNA-cored filaments (Bertolotti, 1996b). The idea is to eliminate sequence heterology problems by utilizing a dsDNA core as a recombination-locked therapeutic cassette and drive its targeted integration by using recombination-activated ssDNA tails that are 100% identical to cognate chromosomal DNA. We thus moved from gene conversion to "true" gene targeting, *i.e.* integration of transfecting dsDNA into host chromosomal DNA upon homologous recombination (review: Capecchi, 1989). However, by contrast to classic low-efficiency DNA targeting technology (reviews: Capecchi, 1989; Yanez and Porter, 1998), we provide the physiological substrate for homologous recombination (3' ssDNA-tailed dsDNA molecules) and use it as a premade complex comprising recombination-activated ssDNA tails (recombinase-DNA nucleoprotein filaments). Importantly enough, dsDNA-cored nucleoprotein filaments offer broad DNA exchange potentialities and may thus accomodate minigenes. Therefore, in addition to gene repair/inactivation approaches, they may be instrumental for targeted (non-random) integration of therapeutic minigenes (Bertolotti, 1998c, 1999 and this book).

c) Oligonucleotide-mediated correction of point mutations

Interestingly enough, chimeric RNA-DNA oligonucleotides were recently devised that appear to provide efficient single-base conversion of targeted chromosomal DNA. Indeed, RNA-DNA oligonucleotides were first designed to inactivate key viral proteins upon point-mutating viral episomes (Yoon *et al.*, 1996) and have been shown to be also effective at genomic targeting *in vitro* (Cole-Strauss *et al.*, 1996) and *in vivo* (Kren *et al.,* 1998). As expected, the resulting single-base conversion is stable through clonal propagation (Alexeev and Yoon, 1998). Importantly enough, these RNA-DNA oligonucleotides appear to be active both on proliferating and resting cells (Kren *et al.*, 1998). However, there are still questions about the universality of the method (genes and cell-types; Yoon, 1999 and this book). In addition, its suggested high efficiency await independent confirmation (Strauss, 1998). On the other hand, single-stranded bifunctional oligonucleotides have been recently described that provide attractive correction of chromosomal point mutations in human cultured cells (Culver *et al.*, 1999). They comprise a third strand-forming domain for synergistic targeting of a single-base conversion repair domain.

d) Gene repair and targeted integration of minigenes

From single-base conversion to broad DNA exchange potentialities, gene repair technology is now emerging as a promising approach for gene therapy of inherited diseases. In addition, our recombinase-mediated DNA targeting approach may succeed in efficient non-random chromosomal integration of therapeutic minigenes and may thus be instrumental for perfecting long-term minigene expression therapy.

4. DISCUSSION

a) Genomic DNA transfer

Gene targeting is based on genomic DNA transfer. However, we use only small gene portions (Capecchi, 1989; Bertolotti, 1996b, 1998c and 1999). Minigene expression therapy has been primarily developed because mammalian genes are usually too large to be handle in viral vectors[2] or bacterial plasmids. A new approach is now focussed on artificial human micro/mini-chromosomes (Harrington *et al.*, 1997; Ikeno *et al.*, 1998;

---

[2] Emerging "gut-less" adenoviral vector have higher insert capacity and have recently been used to transfer a small human gene achieving improved *in vivo* transgene expression (Schiedner *et al.*, 1998).

Henning *et al.*, 1999; Mills *et al.*, 1999; Shen *et al.*, 2000) that could accomodate human full-length genes. Such "synthetic" chromosomes may become an attractive alternate to the present gene compensating strategy based on minigene expression therapy.

b) RNA inactivation or repair

Targeted inactivation or repair of RNA is indeed mediated by specific non-coding RNA. In order to achieve long-term therapy, these therapeutic RNA have to be produced in target cells upon standard minigene expression therapy protocols (Ely *et al.*, 1999; Li *et al.*, 1997; Gorman *et al.*, 1998; Puttaraju *et al.*, 1999). However, in this case, the final minigene product is not a protein or a basic non-coding RNA; it is a catalytic RNA that will either specifically cut or modify a target RNA (Zaug *et al.*, 1986; Haseloff and Gerlach, 1988; Sullenger and Cech, 1994; Gorman *et al.*, 1998; Puttaraju *et al.*, 1999) or suppress an mRNA nonsense mutation (Temple *et al.*, 1982; Li *et al.*, 1997). Supression of nonsense mutations (Temple *et al.*, 1982; therapeutic suppressor tRNA) and inactivation of target RNA (Cech, 1988; Sarver *et al.*, 1990; therapeutic ribozymes) are early approaches. Recent developments focus on repair of mutant RNA with therapeutic *trans*-splicing ribozymes (Sullenger and Cech, 1994; Jones *et al.*, 1996; Lan *et al.*, 1998), splice-altering snRNAs (Gorman *et al.*, 1998 and this book) and RNA *trans*-splicing molecules (Puttaraju *et al.*, 1999). Importantly enough, by emending mutant mRNA, such catalytic transgenic RNA should restore the regulated production of a wild-type gene product since, in this case, physiological gene regulation relies on the endogenous mutant gene rather than on the therapeutic minigene. Such a gene therapy approach to inherited diseases is all the more innovative as current minigenes cannot properly substitute for their normal counterparts when the latters are subject to complex genetic regulation (*e.g.* globin genes: Gorman *et al.*, 1998 and this book). In addition, like gene repair, it can be directed at dominant mutations where elimination of the mutant transcript is required for clinical benefit (*e.g.*

sickle cell anemia). For many mutations, RNA correction stands thus as an improved approach for minigene expression therapy of inherited diseases; however, it cannot compare to gene repair.

5. PROSPECTS AND FUTURE ORIENTATIONS

a) Toward the ideal vectors

As discussed above and outlined in table 1, vectors and RNA/DNA delivery are still the Achilles heel of gene therapy. The ideal vectors should be safe, injectable, able to target specific types of cells and to efficiently deliver undamaged therapeutic RNA/DNA to the cell nucleus either for transient expression (minigenes) or for targeted integration into appropriate chromosomal sites (minigenes or gene repair DNA). Toward such an achievement, both viral and non-viral vectors are under intensive investigations and improvements.

Clinical trials based on transient minigene expression involve both types of vectors. However, long-term minigene expression therapy trials are currently restricted to viral vectors (see above) with which safety and immunogenicity are of major concern. Such an immunogenicity problem is most likely responsible for the recent fulgurant death of a teenager patient enrolled in an adenoviral vector-mediated trial for ornithine transcarbamylase deficiency. Emerging helper-dependent adenoviral vectors appear to be safer (Parks *et al.*, 1996; Morsy and Caskey, this book). Low immunogenicity and experimental safety have been instrumental in the development of prototypical retroviral vectors (Verma and Somia, 1997; Anderson, 1998) and of emerging AAV vectors (Summerford *et al.*, 2000 and this book). However, unlike current retroviral vectors, AAV vectors transduce quiescent cells in addition to proliferating ones and are therefore more appropriate for efficient gene therapy protocols. The same holds true for newly devised retroviral vectors based on HIV and other lentiviral viruses; however, safety requirements are of course much more stringent for vectors based on pathogenic HIV and non-human homologs (Follenzi and Naldini, this book). In addition to potential safety and immunogenicity hazards, viral vectors have another limitation: they cannot accomodate large transgenic elements (see above and Verma and Somia, 1997). This is why, although viral vectors are expected to be therapeutically very important in the coming years, they will most likely be outworn by non-viral vectors in the not too distant future. Importantly enough, large scale production is usually easier and cheaper for synthetic delivery systems than for clinical grade viral reagents.

b) Tissue-specific/stem cell delivery

The ability to deliver therapeutic RNA/DNA to the correct cells is an important feature for gene therapy. In this respect, one of the main problems for long-lasting gene therapy of many tissues and organs is to transfect stem cells (Verma and Somia, 1997; Anonymous, 1997). Such a problem is all the more crucial as cell replacement is fairly fast in newborns and children, and in some adult tissues (Kohn and Parkman, 1997; Bertolotti, 1998b). Therefore, as illustrated for hematopoietic, cord blood and myogenic cells, an important preliminary to efficient gene correction or long-term therapeutic minigene expression may be the identification (Anonymous, 1997; Ferrari *et al.*, 1998; Bittner *et al.*, 1999; Gussoni *et al.*, 1999), the purification (Anonymous, 1997), the *in vitro* multiplication/differentiation (Anonymous, 1997) or the conversion-mediated production (Lattanzi *et al.*, 1998; Huard *et al.*, 1998) of appropriate stem/progenitor cells.

The *in vitro* manipulation and multiplication of stem/progenitor cells provide an *ex vivo* strategy for gene therapy. In this case, cells are transfected/transduced *in vitro* and then returned to the patient. Such an *ex vivo* approach has been used in the first approved clinical protocol for gene therapy, *i.e.* gene therapy for adenosine deaminase (ADA)-deficient severe combined immunodeficiency that started trials in 1990 (Blaese *et al.*, 1995). Such trials started with peripheral blood T lymphocytes (Blaese *et al.*, 1995) that are supposed to have a finite life-span and were subsequently shifted to bone marrow hematopoietic stem cells (Bordignon *et al.*, 1995; Hoogerbrugge *et al.*, 1996) and then to cord blood hematopoietic progenitor cells (Kohn *et al.*, 1995 and 1998). As detailed below (Kohn, this book), the goal is to reconstitute hematopoietic lineages with transduced CD34+ stem cells thanks to a selective advantage of ADA-producing T-cell progeny. The same *ex vivo* strategy applies to anti-infectious disease therapy: it is currently being used in an anti-HIV clinical trial where autologous pluripotent CD34+ stem cells are transduced with a minigene encoding an anti-HIV ribozyme (Ely *et al.*, 1999; Fanning *et al.*, this book). In this case, minigene expression acts both on viral suppression and immune restoration. Such an *ex vivo* gene therapy procedure is perfectly suited for hematopoietic cells because they are easy to remove from and return to the body. However, it may prove instrumental in non-hematopoietic cell diseases where the delivery of therapeutic transgenic stem cells to sites throughout the body is critical and can be achieved upon systemic infusion. In this respect, the recent identification of bone marrow-derived myogenic progenitors holds great promise for such an *ex vivo* gene therapy approach of Duchenne muscular dystrophy and other myopathies (Ferrari *et al.*, 1998; Bittner *et al.*, 1999; Gussoni *et al.*, 1999).

The converse approach, *i.e.* tissue-specific delivery of therapeutic RNA/DNA *in vivo*, is still a matter of challenge. This approach was pioneered by Wu and Wu (1988) on a now well-established hepatic model system (Fukuma *et al.,* this book). The principle is to use a tissue-specific ligand to target the gene therapy vector to the relevant differentiated/stem cells. Experimental liver-specific delivery of transfectant DNA was thus achieved by Wu and Wu upon endocytosis mediated by the liver-specific asialoglycoprotein receptor. Such an approach has been recently used to target chimeric RNA/DNA oligonucleotides for single-base conversion of an hepatic gene in rats (Kren *et al.*, 1998). On the other hand, the restricted natural tropism of certain viruses is also being used with standard or pseudotyped vectors (see Stitz *et al.*, this book) and may provide cell/tissue-targeting ligands for non-viral vectors (Wickelgren, 1998). However, the repertoire for cell-specific delivery is currently very poor. Systematic attempts to generate cell- or tissue-targeting molecules include derivatisation of antibodies (see Jiang and Dornburg, 1999) and screening of random peptide phage libraries (Barry *et al.*, 1996; Koivunen *et al.*, 1999). On the other hand, the characterization of a broad variety of stem cells may identify highly specific lineage markers that could be used for cell-targeting.

Cell-targeting is aimed at systemic infusion of the gene therapy vector. Another way to focus on the target cells *in vivo* is to use topical applications where tissue-specific targeting is replaced by restricted diffusion of the vectors. Current *in vivo* protocols for clinical trials are based on this *in situ* approach (Anderson, 1998). Importantly enough, direct injection of gene therapy vectors into the blood-stream may request concealment by PEG or other molecules in order to avoid unspecific adsorption on non-target cells and on matrix elements (see Kircheis and Wagner, 2000).

c) Human genome project and long-term prospects

The Human Genome Project is progressing well. Therefore, the identification and the molecular characterization of many new human genes will be an important point for the development of gene therapy. The Human Genome Projet stands thus as a fantastic booster for gene therapy. It will broaden the field of applications and strengthern our trust in molecular medicine. Ultimately, the limit is likely to be our imagination and ethical concerns. From the basic field of disease prevention and treatment, we might shift to tantalizing functional enhancement and ageing prevention. Indeed, long-term safety concerns are still to be solved; yet, gene therapy has already been proposed for the protection of injury prone athletes (Gerich *et al.*, 1996) and for alopecia and loss of hair color (Li and Hoffman, 1995).

6. REFERENCES

Alexeev, V. and Yoon, K. (1998). Stable and inheritable changes in genotype and phenotype of albino melanocytes induced by an RNA-DNA oligonucleotide. *Nat. Biotechnol.*, 16: 1343-1346.

Anderson, F.W. (1984). Prospects for human gene therapy. *Science*, 226: 401-409.

Anderson, F.W. (1998). Human gene therapy. *Nature*, 392 suppl.: 25-30.

Anonymous (1997). Proceedings of the Hematopoietic Stem Cells International Symposium and Workshop. *Stem Cells*, 15 Suppl. 1: 1-291.

Aruffo, A. and Seed, B. (1987). Molecular cloning of a CD28 cDNA by a high-efficiency COS cell expression system. *Proc. Natl. Acad. Sci. USA*, 84: 8573-8577.

Banerji, J., Rusconi, S. and Schaffner, W. (1981). Expression of a beta-globin gene is enhanced by remote SV40 DNA sequences. *Cell*, 27: 299-308.

Barry, M.A., Dower, W.J. and Johnston, S.A. (1996). Toward cell-targeting gene therapy vectors: selection of cell-binding peptides from random peptide-presenting phage libraries. *Nature Med.*, 2: 299-305.

Baumann, P., Benson, F. and West, S.C. (1996). Human Rad51 protein promotes ATP-dependent homologous pairing and strand transfer reactions *in vitro*. *Cell*, 87: 757-766.

Baumgartner, I., Pieczek, A., Manor, O., Blair, R., Kearney, M., Walsh, K. and Isner, J.M. (1998). Constitutive expression of phVEGF165 after intramuscular gene transfer promotes collateral vessel development in patients with critical limb ischemia. *Circulation*, 97: 1114-1123.

Bell, A.C. and Felsenfeld, G. (1999). Stopped at the border: boundaries and insulators. *Curr. Opin. Genet. Dev.*, 9: 191-198.

Berglund, P., Smerdou, C., Fleeton, M.N., Tubulekas, I. and Liljestrom, P. (1998). Enhancing immune responses using suicidal DNA vaccines. *Nat. Biotechnol.*, 16: 562-565.

Bertolotti, R. (1996a). Recombinase-mediated gene therapy: targeting single stranded DNA to chromosomal DNA with human RAD51 presynaptic fibers. *Hepatology*, 24: 484A.

Bertolotti, R. (1996b). Recombinase-mediated gene therapy: strategies based on Lesch-Nyhan mutants for gene repair/inactivation using human RAD51 nucleoprotein filaments. *Biogenic Amines*, 12: 487-498.

Bertolotti, R. (1998a). Recombinase-DNA nucleoprotein filaments as Gene Therapy vectors. *Biogenic Amines*, 14: 41-65.

Bertolotti, R. (1998b). Prospects of recombinase-mediated HPRT Gene Therapy for Lesch-Nyhan disease. In: *Neurochemical Markers of Degenerative Nervous Diseases and Drug Addiction*, G.A. Qureshi, H. Parvez, P. Caudy and S. Parvez (Eds), VSP Int. Science Press, Zeist, N.L., pp. 133-170.

Bertolotti, R. (1998c). Gene therapy 1998: transient or stable minigene expression and gene repair/inactivation. *Biogenic Amines*, 14: 389-406.

Bertolotti, R. (1999). Recombinase-DNA nucleoprotein filaments as vectors for gene repair /inactivation and targeted integration of minigenes. *Biogenic Amines*, 15: 169-195.

Bertolotti, R. and Lutfalla, G. (1983). Genes programming cell differentiation. *J. Cell. Biochem.*, 7A: 149.

Bertolotti, R., Armbruster-Hilbert, L. and Okayama, H. (1995). Liver fructose-1,6-bisphosphatase cDNA: trans-complementation of fission yeast and characterization of two human transcripts. *Differentiation*, 59: 51-60.

Bertolotti, R., Parvez, S.H. and Nagatsu, T. (Eds.) (1998). Gene Therapy 1998, Part 1. *Biogenic Amines*, 14: 387-572.

Bertolotti, R., Parvez, S.H. and Nagatsu, T. (Eds.) (1999). Gene Therapy 1998, Part 2. *Biogenic Amines*, 15: 1-195.

Bittner, R.E., Schofer, C., Weipoltshammer, K., Ivanova, S., Streubel, B., Hauser, E., Freilinger, M., Hoger, H., Elbe-Burger, A. and Wachtler, F. (1999). Recruitment of bone-marrow-derived cells by skeletal and cardiac muscle in adult dystrophic mdx mice. *Anat. Embryol. (Berl.)*, 199: 391-396.

Blaese, R.M., Culver, K., Miller, A.D., Carter, C., Fleisher, T., Clerici, M., Shearer, G., Chang, L., Chiang, Y., Tolstoshev, P., Greenblatt, J., Rosenberg, S., Klein, H., Berger, M., Mullen, C., Ramsey, W., Muul, L., Morgan, R. and Anderson, F.W. (1995). T lymphocyte-directed gene therapy for ADA- SCID: initial trial results after 4 years. *Science*, 270: 475-480.

Blomer, U., Naldini, L., Kafri, T., Trono, D., Verma, I.M. and Gage, F.H. (1997). Highly efficient and sustained gene transfer in adult neurons with a lentivirus vector. *J. Virol.*, 71: 6641-6649.

Bonadio, J., Smiley, E., Patil, P. and Goldstein, S. (1999). Localized, direct plasmid gene delivery *in vivo*: prolonged therapy results in reproducible tissue regeneration. *Nat. Med.*, 5: 753-759.

Bonini, C., Ferrari, G., Verzeletti, S., Servida, P., Zappone, E., Ruggieri, L., Ponzoni, M., Rossini, S., Mavilio, F., Traversari, C. and Bordignon, C. (1997). HSV-TK gene transfer into donor lymphocytes for control of allogeneic graft-versus-leukemia. *Science*, 276: 1719-1724.

Bordignon C, Notarangelo LD, Nobili N, Ferrari G, Casorati G, Panina P, Mazzolari E, Maggioni D, Rossi C, Servida P, Ugazio, A.G. and Mavilio, F. (1995). Gene therapy in peripheral blood lymphocytes and bone marrow for ADA- immunodeficient patients. *Science*, 270: 470-475.

Breathnach, R. and Chambon, P. (1981). Organization and expression of eucaryotic split genes coding for proteins. *Ann. Rev. Biochem.*, 50: 349-383.

Bulger, M., van Doorninck, J.H., Saitoh, N., Telling, A., Farrell, C., Bender, M.A., Felsenfeld, G., Axel, R. and Groudine, M. (1999). Conservation of sequence and structure flanking the mouse and human β-globin loci: the β-globin genes are embedded within an array of odorant receptor genes. *Proc. Natl. Acad. Sci. USA*, 96: 5129-5134.

Bukovsky, A., Song, J.P. and Naldini, L. (1999). Interaction of human immunodeficiency virus-derived vectors with wild-type virus in transduced cells. *J. Virol.*, 73: 7087-7092.

Capecchi, M. (1989). The new mouse genetics: altering the genome by gene targeting. *Trends Genet.*, 5: 70-76.

Case, S.S., Price, M.A., Jordan, C.T., Yu, X.J., Wang, L., Bauer, G., Haas, D.L., Xu, D., Stripecke, R., Naldini, L., Kohn, D.B. and Crooks, G.M. (1999). Stable transduction of quiescent CD34(+)CD38(-) human hematopoietic cells by HIV-1-based lentiviral vectors. *Proc. Natl. Acad. Sci. USA*, 96: 2988-2993.

Cech, T.R. (1988). Ribozymes and their medical implications. *J. Am. Med. Assoc.*, 260: 3030-3034.

Chen, C. and Okayama, H. (1987). High-efficiency transformation of mammalian cells by plasmid DNA. *Mol. Cell. Biol.*, 7: 2745-2752.

Cole-Strauss, A., Yoon, K., Xiang, Y., Byrne, B., Rice, M., Gryn, J., Holloman, W. and

Kmiec, E. (1996). Correction of the mutation responsible for sickle cell anemia by an RNA-DNA oligonucleotide. *Science*, 273: 1386-1389.

Critchlow, S.E. and Jackson, S.P. (1998). DNA end-joining: from yeast to man. *Trends Biochem. Sci.*, 23: 394-398.

Culver K.W., Ram, Z., Wallbridge, S., Ishii, H., Oldfield, E. and Blaese, R.M. (1992). *In vivo* gene transfer with retroviral vector-producer cells for treatment of experimental brain tumors. *Science*, 256: 1550-1552.

Culver, K.W., Hsieh, W-T., Huyen, Y., Chen, V., Liu, J., Khripine, Y. and Khorlin, A. (1999). Correction of chromosomal point mutations in human cells with bifunctional oligonucleotides. *Nat. Biotechnol.*, 17: 989993.

Davis, R.L., Weintraub, H. and Lassar, A.B. (1987). Expression of a single transfected cDNA converts fibroblasts to myoblasts. *Cell*, 51: 987-1000.

Deng, C. and Capecchi, M. (1992). Reexamination of gene targeting frequency as a function of the extent of homology between the targeting vector and the target locus. *Mol. Cell. Biol.*, 12: 3365-3371.

DiMaio, D., Treisman, R. and Maniatis, T. (1982). Bovine papilloma-virus vector that propagates as a plasmid in both mouse and bacterial cells. *Proc. Natl. Acad. Sci. USA.*, 79: 4030-4034.

Donnelly, J., Ulmer, J., Shiver, J. and Liu, M. (1997). DNA vaccines. *Annu. Rev. Immunol.*, 15: 617-648.

Doolan, D.L., Hedstrom, R.C., Gardner, M.J., Sedegah, M., Wang, H., Gramzinski, R.A., Margalith, M., Hobart, P. and Hoffman, S.L. (1998). DNA vaccination as an approach to malaria control: current status and strategies. *Curr. Top. Microbiol. Immunol.*, 226: 37-56.

Dull, T., Zufferey, R., Kelly, M., Mandel, R.J., Nguyen, M., Trono, D. and Naldini, L. (1998). A third-generation lentivirus vector with a conditional packaging system. *J. Virol.*, *72*: 8463-8471.

Edlund, H. (1999). Pancreas: how to get there from the gut? *Curr. Opin. Cell Biol.*, 11: 663-668.

Edwards, A., Voss, H., Rice, P., Civitello, A., Stegemann, J., Schwager, C., Zimmermann, J., Erfle, H., Caskey, C. and Ansorge, W. (1990). Automated DNA sequencing of the human *hprt* locus. *Genomics*, 6: 593-608.

Ely, J., Macpherson, J., Rigden, J., Gerlach, W., Sun, L.Q. and Symonds, G. (1999). Anti-HIV ribozymes in the inhibition of HIV and AIDS. *Biogenic Amines*, 15: 113-135.

Eming, S., Medalie, D., Tompkins, R., Yarmush, M. and Morgan, J. (1998). Genetically modified human keratinocytes overexpressing PDGF-A enhance the performance of a composite skin graft. *Hum. Gene Ther.*, 9: 529-539.

Evans, J.T., Kelly, P.F., O'Neill, E. and Garcia, J.V. (1999). Human cord blood CD34+CD38- cell transduction via lentivirus-based gene transfer vectors. *Hum. Gene Ther.*, 10: 1479-1489.

Felgner, P., Heller, M., Lehn, P., Behr, J-P. and Szoka, F. (Eds.) (1996). *Artificial self-assembling systems for gene delivery*. Am. Chem. Soc., Washington.

Ferrari, G., Cusella-De Angelis, G., Coletta, M., Paolucci, E., Stornaiuolo, A., Cossu,

G. and Mavilio, F. (1998). Muscle regeneration by bone marrow-derived myogenic progenitors. *Science*, 279: 1528-1530.

Festenstein, R., Tolaini, M., Corbella, P., Mamalaki, C., Parrington, J., Fox, M., Miliou, A., Jones, M. and Kioussis, D. (1996). Locus control region function and heterochromatin-induced position effect variegation. *Science*, 271: 1123-1125.

Folkman, J. (1998). Antiangiogenic gene therapy. *Proc. Natl. Acad. Sci. USA*, 95: 9064-9066.

Forrester, W.C., Takegawa, S., Papayannopoulou, T., Stamatoyannopoulos, G. and Groudine, M. (1987). Evidence for a locus activation region: the formation of developmentally stable hypersensitive sites in globin-expressing hybrids. *Nucleic Acids Res.*, 15: 10159-10177.

Forrester, W., Novak, U., Gelinas, R. and Groudine, M. (1989). Molecular analysis of the human β-globin locus activation region. *Proc. Natl. Acad. Sci. USA*, 86: 5439-5443.

Gerich, T.G., Fu, F.H., Robbins, P.D. and Evans, C.H. (1996). Prospects for gene therapy in sports medicine. *Knee Surg. Sports Traumatol. Arthrosc.*, 4: 180-187.

Ghivizzani, S., Lechman, E., Kang, R., Tio, C., Kolls, J., Evans, C. and Robbins, P. (1998). Direct adenovirus-mediated gene transfer of interleukin 1 and tumor necrosis factor α soluble receptors to rabbit knees with experimental arthritis has local and distal anti-arthritic effects. *Proc. Natl. Acad. Sci. USA*, 95: 4613-4618.

Goncz, K., Kunzelmann, K., Xu, Z. and Gruenert, D.C. (1998). Targeted replacement of normal and mutant CFTR sequences in human airway epithelial cells using DNA fragments. *Hum. Mol. Genet.*, 7: 1913-1919.

Gorman, L., Suter, D., Emerick, V., Schumperli, D. and Kole, R. (1998). Stable alteration of pre-mRNA splicing patterns by modified U7 small nuclear RNAs. *Proc. Natl. Acad. Sci. USA*, 95: 4929-4934.

Gorski, K., Carniero, M. and Schibler, U. (1986). Tissue-specific *in vitro* transcription from the mouse albumin promoter. *Cell*, 47: 767-776.

Gossen, M. and Bujard, H. (1992). Tight control of gene expression in mammalian cells by tetracycline-responsive promoters. *Proc. Natl. Acad. Sci. USA*, 89: 5547-5551.

Graham, F. and Van der Eb, J. (1973). A new technique for the assay of infectivity of human adenovirus 5 DNA. *Virology*, 52: 456-467.

Grosveld, F., van Assendelft, G.B., Greaves, D.R. and Kollias, G. (1987). Position-independent, high-level expression of the human β-globin gene in transgenic mice. *Cell*, 51: 975-985.

Gussoni, E., Soneoka, Y., Strickland, C.D., Buzney, E.A., Khan, M.K., Flint, A.F., Kunkel, L.M. and Mulligan, R.C. (1999). Dystrophin expression in the mdx mouse restored by stem cell transplantation. *Nature*, 401: 390-394.

Harrington, J.J., Van Bokkelen, G., Mays, R.W., Gustashaw, K. and Willard, H.F. (1997). Formation of de novo centromeres and construction of first-generation human artificial microchromosomes. *Nat. Genet.*, 15: 345-355.

Haseloff, J. and Gerlach, W.L. (1988). Simple RNA enzymes with new and highly specific endonuclease activities. *Nature*, 334: 585-591.

Henning, K.A., Novotny, E.A., Compton, S.T., Guan, X.Y., Liu, P.P. and Ashlock,

M.A. (1999). Human artificial chromosomes generated by modification of a yeast artificial chromosome containing both human alpha satellite and single-copy DNA sequences. *Proc. Natl. Acad. Sci.USA*, 96: 592-597.

Herzog, R., Yang, E., Couto, L., Hagstrom, J., Elwell, D., Fields, P., Burton, M., Read, M., Bellinger, D., Brinkhous, K., Podsakoff, G., Nichols, T., Kurtzman, G. and High, K. (1999). Long-term correction of canine hemophilia B by gene transfer of blood coagulation factor IX mediated by adeno-associated viral vector. *Nat. Med.*, 5: 56-63.

Hoogerbrugge, P.M., van Beusechem, V.W., Fischer, A., Debree, M., le Deist, F., Perignon, J.L., Morgan, G., Gaspar, B., Fairbanks, L.D., Skeoch, C.H., Moseley, A., Harvey, M., Levinsky, R.J. and Valerio, D. (1996). Bone marrow gene transfer in three patients with adenosine deaminase deficiency. *Gene Ther.*, 3: 179-183.

Huard, C., Moisset, P.A., Dicaire, A., Merly, F., Tardif, F., Asselin, I. and Tremblay, J.P. (1998). Transplantation of dermal fibroblasts expressing MyoD1 in mouse muscles. *Biochem. Biophys. Res. Commun.*, 248: 648-654.

Huygen, K., Content, J., Denis, O., Montgomery, D., Yawman, A.M, Deck, R., DeWitt, C., Orme, I., Baldwin, S., D'Souza, C., Drowart, A., Lozes, E., Vanden-bussche, P., Van Vooren, J-P., Liu, M.A. and Ulmer, J.B. (1996). Immunogenicity and protective efficacy of a tuberculosis DNA vaccine. *Nature Med.*, 2: 893-898.

Ikeno, M., Grimes, B., Okazaki, T., Nakano, M., Saitoh, K., Hoshino, H., McGill, N.I., Cooke, H. and Masumoto, H. (1998). Construction of YAC-based mammalian artificial chromosomes. *Nat. Biotechnol.*, 16: 431-439.

Isner, J., Pieczek, A., Schainfeld, R., Blair, R., Haley, L., Asahara, T., Rosenfield, K., Razvi, S., Walsh, K. and Symes, J. (1996). Clinical evidence of angiogenesis after arterial gene transfer of $phVEGF_{165}$ in patient with ischemic limb. *Lancet*, 348: 370-374

Itaya, T., Yamagiwa, S., Okada, F., Oikawa, T., Kuzumaki, N., Takeichi, N., Hosokawa, M. and Kobayashi, H. (1987). Xenogenization of a mouse lung carcinoma (3LL) by transfection with an allogenic class I major histocompatibility complex gene (H-2Ld). *Cancer Res.*, 47: 3136-3140.

Izant, J. and Weintraub, H. (1985). Constitutive and conditional suppression of exogenous and endogenous genes by anti-sens RNA. *Science*, 229: 345-352.

Jiang, A. and Dornburg, R. (1999). In vivo cell type-specific gene delivery with retroviral vectors that display single chain antibodies. *Gene Ther.*, 6: 1982-1987.

Jones J.T., Lee, S.W. and Sullenger, B.A. (1996). Tagging ribozyme reaction sites to follow *trans*-splicing in mammalian cells. *Nature Med.*, 2: 643-648.

Kessler, P.D., Podsakoff, G.M., Chen, X., McQuiston, S.A., Colosi, P.C., Matelis, L. A., Kurtzmann, G.J. and Byrne, B.J. (1996). Gene delivery to skeletal muscle results in sustained expression and systemic delivery of a therapeutic protein. *Proc. Natl. Acad.Sci., USA,* **93:** 14082-14087.

Klein, R.L., Meyer, E.M, Peel, A., Zolotukhin, S., Meyers, C., Muzyczka, N. and King, M. (1998). Neuron-specific transduction in the rat septohippocampal or nigrostriatal pathway by recombinant adeno-associated virus vectors. *Exp. Neurol.*, 150: 183-194.

Kircheis, R. and Wagner, E. (2000). Polycation/DNA complexes for *in vivo* gene delivery. *Gene Ther. Regul.*, 1: 95-114.

Kohn, D.B. and Parkman, R. (1997). Gene therapy for newborns. *FASEB J.*, 11: 635-639

Kohn, D.B., Weinberg, K.I., Nolta, J.A., Heiss, L.N., Lenarsky, C., Crooks, G.M., Hanley, M.E., Annett, G., Brooks, J.S., El-Khoureiy, A., Lawrence, K., Wells, S., Shaw, K., Moen, R.C., Bastian, J., Williams-Herman, D.E., Elder, M., Wara, D., Bowen, T., Hershfield, M.S., Mullen, C.A., Blaese, R.M. and Parkman, R. (1995). Engraftment of gene-modified cells from umbilical cord blood in neonates with adenosine deaminase deficiency. *Nature Med.*, 1: 1017-1026.

Kohn, D.B., Hershfield, M.S., Carbonaro, D., Shigeoka, A., Brooks, J., Smogorzewska, E.M., Barsky, L., Chan, R., Burotto, F., Annett, G., Nolta, J., Crooks, G., Kapoor, N., Elder, M., Wara, D., Bowen, T., Madsen, E., Snyder, F., Bastian, J., Muul, L., Blaese, R.M., Weinberg, K. and Parkman, R. (1998). T lymphocytes with a normal ADA gene accumulate after transplantation of transduced autologous umbilical cord blood CD34+ cells in ADA-deficient SCID neonates. *Nature Med.*, 4: 775-780.

Koivunen, E., Arap, W., Valtanen, H., Rainisalo, A., Medina, O.P., Heikkila, P., Kantor, C., Gahmberg, C.G., Salo, T., Konttinen, Y.T., Sorsa, T., Ruoslahti, E. and Pasqualini, R. (1999). Tumor targeting with a selective gelatinase inhibitor. *Nat. Biotechnol.*, 17: 768-774.

Kren, B., Bandyopadhyay, P. and Steer, C. (1998). *In vivo* site-directed mutagenesis of the *factor IX* gene by chimeric RNA/DNA oligonucleotides. *Nat. Med.*, 4: 285-290.

Lan, N., Howrey, R., Lee, S.W., Smith, C. and Sullenger, B. (1998). Ribozyme-mediated repair of sicle β globin mRNA in erythrocyte precursors. *Science*, 280: 1593-1596.

Lattanzi, L., Salvatori, G., Coletta, M., Sonnino, C., Cusella-De Angelis, G., Gioglio, L., Murry, C.E., Kelly, R., Ferrari, G., Molinaro, M., Crescenzi, M., Mavilio, F. and Cossu, G. (1998). High efficiency myogenic conversion of human fibroblasts by adenoviral vector-mediated MyoD gene transfer. An alternative strategy for *ex vivo* gene therapy of primary myopathies. *J. Clin. Invest.*, 101: 2119-2128.

Lee, M. and Nurse, P. (1987). Complementation used to clone a human homologue of the fission yeast cell cycle control gene *cdc2*. *Nature*, 327: 31-35.

Leff, S., Rosenfeld, M. and Evans, R. (1986). Complex transcriptional units: diversity in gene expression by alternative RNA processing. *Ann. Rev. Biochem.*, 55: 1091-1117.

Li, K., Zhang, J., Buvoli, M., Yan, X.D., Leinwand, L. and He, H. (1997). Ochre suppressor transfer RNA restored dystrophin expression in mdx mice. *Life Sci.*, 61: 15PL.

Li, L. and Hoffman, R.M. (1995). The feasability of targeted selective gene therapy of the hair follicule. *Nature Med.*, 1: 705-706.

Littlefied, J. (1964). Selection of hybrids from mating of fibroblasts *in vitro* and their presumed recombinants. *Science*, 145: 709-710.

Loser, P., Jennings, G., Strauss, M. and Sandig, V. (1998). Reactivation of the previously silenced cytomegalovirus major immediate-early promoter in the mouse liver: involvement of NFκB. *J. Virol.*, 72: 180-190.

Losordo, D.W, Vale, P.R., Symes, J.F., Dunnington, C.H., Esakof, D.D., Maysky, M., Ashare, A.B., Lathi, K. and Isner J.M. (1998). Gene therapy for myocardial angiogenesis: initial clinical results with direct myocardial injection of phVEGF165 as sole therapy for myocardial ischemia. *Circulation*, 98: 2800-2804.

Lutfalla, G., Blanc, H. and Bertolotti, R. (1985). Shuttling of integrated vectors from mammalian cells to *E. coli* is mediated by head-to-tail multimeric inserts. *Somat. Cell*

Mol. Genet., 11: 223-238.

Lutfalla, G., Armbruster, L., Dequin, S. and Bertolotti, R. (1989). Construction of an EBNA-producing line of well-differentiated human hepatoma cells and appropriate Epstein-Barr virus-based shuttle vectors. *Gene*, 76: 27-39.

Mann, R., Mulligan, R. and Baltimore, D. (1983). Construction of a retrovirus packaging mutant and its use to produce helper-free defective retrovirus. *Cell*, 33: 153-159.

McCown TJ, Xiao X, Li J, Breese GR, Samulski RJ. (1996).Differential and persistent expression patterns of CNS gene transfer by an adeno-associated virus (AAV) vector. *Brain Res.*, 713: 99-107.

Merrihew, R.V., Marburger, K., Pennington, S.L., Roth, D.B. and Wilson, J.H. (1996). High-frequency illegitimate integration of transfected DNA at preintegrated target sites in a mammalian genome. *Mol. Cell. Biol.*, 16: 10-18.

Miller, A.D., Miller, D., Garcia, J. and Lynch, C. (1993). Use of retroviral vectors for gene transfer and expression. *Methods Enzymol.*, 217: 581-599.

Mills, W., Critcher, R., Lee, C. and Farr, C.J. (1999). Generation of an approximately 2.4 Mb human X centromere-based minichromosome by targeted telomere-associated chromosome fragmentation in DT40. *Hum. Mol. Genet.*, 8: 751-761.

Miyoshi, H., Takahashi, M., Gage, F.H. and Verma, I.M. (1997). Stable and efficient gene transfer into the retina using an HIV-based lentiviral vector. *Proc. Natl. Acad. Sci. USA*, 94: 10319-10323.

Miyoshi, H., Blomer, U., Takahashi, M., Gage, F.H. and Verma, I.M. (1998). Development of a self-inactivating lentivirus vector. *J. Virol.*, 72: 8150-8157.

Miyoshi, H., Smith, K.A., Mosier, D.E., Verma, I.M. and Torbett, B.E. (1999). Transduction of human CD34+ cells that mediate long-term engraftment of NOD/SCID mice by HIV vectors. *Science*, 283: 682-686.

Monahan, P.E., Samulski, R.J., Tazelaar, J., Xiao, X., Nichols, T., Bellinger, D., Read, M. and Walsh, C. (1998). Direct intramuscular injection of recombinant AAV vectors results in sustained expression in a dog model of hemophilia. *Gene Ther.*, 5: 40-49.

Moss, B. (1996). Genetically engineered poxviruses for recombinant gene expression, vaccination and safety. *Proc. Natl. Acad. Sci. USA*, 93: 11341-11348.

Mulligan, R. and Berg, P. (1980). Expression of a bacterial gene in mammalian cells. *Science*, 209: 1422-1427.

Mulligan, R. and Berg, P. (1981). Selection for animal cells that express the Escherichia coli gene coding for xanthine-guanine phosphoribosyltransferase. *Proc. Natl. Acad. Sci. USA*, 78: 2072-2076.

Naldini, L., Blomer, U., Gage, F.H., Trono, D. and Verma, I.M. (1996a). Efficient transfer, integration, and sustained long-term expression of the transgene in adult rat brains injected with a lentiviral vector. *Proc. Natl. Acad. Sci. USA*, 93: 11382-11388.

Naldini, L., Blomer, U., Gallay, P., Ory, D., Mulligan, R., Gage, F.H., Verma, I.M. and Trono, D. (1996b). *In vivo* gene delivery and stable transduction of nondividing cells by a lentiviral vector [see comments]. *Science*, 272: 263-267.

Neumann, E., Schaefer-Ridder, M., Wang, Y. and Hofschneider, P. (1982). Gene transfer into mouse lyoma cells by electroporation in high electric fields.*EMBO J.*, 1: 841-845

Okayama, H. and Berg, P. (1983). A cDNA cloning vector that permits expression of cDNA inserts in mammalian cells. *Mol. Cell. Biol.*, 3: 280-289.

Palese, P. and Roizman, B. (Eds) (1996). Genetic engineering of viruses and of virus vectors. *Proc. Natl. Acad. Sci. USA*, 93: 11287-11425.

Pardoll, D.M. (1998). Cancer vaccines. *Nature Med.*, 4: 525-531.

Parks, R.J., Chen, L., Anton, M., Sankar, U., Rudnicki, M.A. and Graham, F.L. (1996). A helper-dependent adenovirus vector system: removal of helper virus by Cre-mediated excision of the viral packaging signal. *Proc. Natl. Acad. Sci. USA*, 93: 13565-13570.

Pinkert, C., Ornitz, D., Brinster, R. and Palmiter, R. (1987). An albumin enhancer located 10 kb upstream functions along with its promoter to direct efficient, liver-specific expression in transgenic mice. *Genes Dev.*, 1: 268-276.

Prioleau, M.N., Nony, P., Simpson, M. and Felsenfeld, G. (1999). An insulator element and condensed chromatin region separate the chicken beta-globin locus from an independently regulated erythroid-specific folate receptor gene. *EMBO J.*, 18: 4035-4048.

Puttaraju, M., Jamison, S.F., Mansfield, S.G., Garcia-Blanco, M.A. and Mitchell, L.G. (1999). Spliceosome-mediated RNA trans-splicing as a tool for gene therapy. *Nat. Biotechnol.*, 17: 246-252.

Ram, Z., Culver, K., Oshiro, E., Viola, J., DeVroom, H., Otto, E., Long, Z., Chiang, Y., McGarrity, G., Muul, L., Katz, D., Blaese, R.M. and Oldfied, E. (1997). Therapy of malignant brain tumors by intratumoral implantation of retroviral vector-producing cells. *Nat. Med.*, 3: 1354-1361.

Rosenberg, M., Friedmann, T., Robertson, R., Tuszynski, M., Wolff, J., Breakefield, X. and Gage, F. (1988). Grafting genetically modified cells to the damaged brain: restorative effects of NGF expression. *Science*, 242: 1575-1577.

Rosengart, T.K., Lee, L.Y., Patel, S.R., Sanborn, T.A., Parikh, M., Bergman, G.W., Hachamovitch, R., Szulc, M., Kligfield, P.D., Okin, P.M., Hahn, R., Devereux, R., Post, M., Hackett, N., Foster, T., Grasso, T., Lesser, M., Isom, O. and Crystal, R.G. (1999). Angiogenesis gene therapy: phase I assessment of direct intramyocardial administration of an adenovirus vector expressing VEGF121 cDNA to individuals with clinically significant severe coronary artery disease. *Circulation*, 100: 468-474.

Roth, J.A. and Cristiano, R. (1997). Gene therapy for cancer: what have we done and where are we going? *J. Natl. Cancer Inst.*, 89: 21-39.

Saeki, Y., Wataya-Kaneda, M., Tanaka, K. and Kaneda, Y. (1998). Sustained transgene expression *in vitro* and *in vivo* using an Epstein-Barr virus replicon vector system combined with HVJ-liposomes. *Gene Ther.*, 5: 1031-1037.

Saez, E., No, D., West, A. and Evans, R. (1997). Inducible gene expression in mammalian cells and transgenic mice. *Curr. Opin. Biotechnol.*, 8: 608-616.

Sarver, N., Cantin, E., Chang, P., Zaia, J., Ladne, P., Stephens, D. and Rossi, J. (1990). Ribozymes as potential anti-HIV-1 therapeutic agents. *Science*, 247: 1222-1225.

Schiedner, G., Morral, N., Parks, R., Wu, Y., Koopmans, S., Langston, C., Graham, F., Beaudet, A. and Kochanek, S. (1998). Genomic DNA transfer with a high-capacity adenovirus vector results in improved *in vivo* gene expression and decreased toxicity. *Nature Genet.*, 18: 180-183.

Shen, M.H., Mee, P.J., Nichols, J., Yang, J., Brook, F., Gardner, R.L., Smith, A.G. and Brown, W.R. (2000). A structurally defined mini-chromosome vector for the mouse germ line. *Curr. Biol.*, 10: 31-34.

Smerdou, C. and Liljeström, P. (2000). Alphavirus vectors: from protein production to gene therapy. *Gene Ther. Regul.*, 1: 33-63.

Snyder, R.O., Miao, C., Meuse, L., Tubb, J., Donahue, B.A., Lin, H.F., Stafford, D.W., Patel, S., Thompson, A.R., Nichols, T., Read, M.S., Bellinger, D.A., Brinkhous, K.M. and Kay, M.A. (1999). Correction of hemophilia B in canine and murine models using recombinant adeno-associated viral vectors. *Nat. Med.*, 5: 64-70.

Stief, A., Winter, D. , Strätling, W. and Sippel, A.E. (1989). A nuclear DNA attachment element mediates elevated and position-independent gene activity. *Nature*, 341: 343-345

Strauss, M. (1998). The site-specific correction of genetics defects. *Nat. Med.*, 4: 274-275

Sullenger, B.A. and Cech, T.R. (1994). Ribozyme-mediated repair of defective mRNA by targeted trans-splicing. *Nature*, 371: 619-622.

Sullenger, B.A., Gallardo, H.F., Ungers, G.E. and Gilboa, E. (1990). Overexpression of TAR sequences renders cells resistant to HIV replication. *Cell*, 63: 601-608.

Summerford, C., Bartlett, J.S. and Samulski, R.J. (2000). Adeno-associated viral vectors and successful gene therapy, the gap is closing. *Gene Ther. Regul.*, 1: 9-32.

Sun, L., Xu, L., Chang, H., Henry, F., Miller, R., Harmon, J. and Nielsen, T. (1997). Transfection with aFGF cDNA improves wound healing. *J. Invest. Dermatol.*, 108: 313-318.

Sung, P. (1994). Catalysis of ATP-dependent homologous DNA pairing and strand exchange by yeast RAD51 protein. *Science*, 265: 1241-1243.

Sung, P. and Robberson, D. (1995). DNA strand exchange mediated by a RAD51-ssDNA nucleoprotein filament with polarity opposite to that of RecA. *Cell*, 82: 453-461.

Suzuki, K., Nakashima, H., Sawa, Y., Morishita, R., Matsuda, H. and Kaneda, Y. (2000). Reconstitted fusion liposomes for gene transfer in vitro and in vivo. *Gene Ther. Regul.*, 1: 65-77.

Symes, J.F., Losordo, D.W., Vale, P.R., Lathi, K.G., Esakof, D.D., Mayskiy, M. and Isner, J.M. (1999). Gene therapy with vascular endothelial growth factor for inoperable coronary artery disease. *Ann. Thorac. Surg.*, 68: 830-836.

Szybalska, E. and Szybalski, W. (1962). Genetics of human cell lines: DNA-mediated heritable transformation of a biochemical trait. *Proc. Natl. Acad. Sci. USA*, 48: 2026-2034

Tascon, R.E., Colston, M.J., Ragno, S., Stavropoulos, E., Gregory, D. and Lowrie, D.B (1996). Vaccination against tuberculosis by DNA injection. *Nature Med.*, 2: 888-892.

Taubes, G. (1997). Salvation in a snippet of DNA? *Science*, 278: 1711-1714.

Temple, G.F., Dozy, A.M., Roy, K.L. and Kan, Y.W. (1982). Construction of a functional human suppressor tRNA gene: an approach to gene therapy for beta-thalassaemia. *Nature*, 296: 537-540.

Tepper, R., Pattengale, P. and Leder, P. (1989). Murine interleukine-4 displays potent anti-tumor activity *in vivo*. *Cell*, 57: 503-512.

Te Riele, H., Maandag, E. and Berns, A. (1992). Highly efficient gene targeting in embryonic stem cells through homologous recombination with isogenic DNA constructs. *Proc. Natl. Acad. Sci. USA*, 89: 5128-5132.

Uchida, N., Sutton, R.E., Friera, A.M., He, D., Reitsma, M.J., Chang, W.C., Veres, G., Scollay, R. and Weissman, I.L. (1998). HIV, but not murine leukemia virus, vectors mediate high efficiency gene transfer into freshly isolated G0/G1 human hematopoietic stem cells. *Proc. Natl. Acad. Sci. USA*, 95: 11939-11944.

Ueki, T., Kaneda, Y., Tsutsui, H., Nakanishi, K., Sawa, Y., Morishita, R., Matsumoto, K., Nakamura, T., Takahashi, H., Okamoto, E. and Fujimoto, J. (1999). Hepatocyte growth factor gene therapy of liver cirrhosis in rats. *Nat Med.*, 5: 226-230.

Ulmer, J., Donnelly, J., Parker, S., Rhodes, G., Felgner, P., Dwarki, V., Gromkowski, S., Deck, R., DeWitt, C., Friedman, A., Hawe, L., Leander, K., Martinez, D., Perry, H., Shiver, L., Montgomery, D. and Liu, M. (1993). Heterologous protection against influenza by injection of DNA encoding a viral protein. *Science*, 259: 1745-1749.

Verma, I.M. and Somia, N. (1997). Gene therapy - promises, problems and prospects. *Nature*, 389: 239-242.

Walker, M.D., Edlund, T., Boulet, A.M. and Rutter, W.J. (1983). Cell-specific expression controlled by the 5'-flanking region of insulin and chymotrypsin genes. *Nature*, 306: 557-561.

Wang, R., Doolan, D.L., Le, T.P., Hedstrom, R.C., Coonan, K.M., Charoenvit, Y., Jones, T.R., Hobart, P., Margalith, M., Ng, J., Weiss, W.R, Sedegah, M., de Taisne, C., Norman, J. and Hoffman, S.L. (1998). Induction of antigen-specific cytotoxic T lymphocytes in humans by a malaria DNA vaccine. *Science*, 282: 476-480.

Wang, Y., Xu, J., Pierson, T., O'Malley, B.W. and Tsai, S.Y. (1997). Positive and negative regulation of gene expression in eukaryotic cells with an inducible transcriptional regulator. *Gene Ther.*, 4: 432-441.

Waugh O'Neill, R., O'Neill, M and Marshall Graves, J. (1998). Undermethylation associated with retroele- ment activation and chromosome remodelling in an interspecific mammalian hybrid. *Nature*, 393: 68-72.

Wickelgren, I. (1998). A method in Ebola's madness. *Science*, 279: 983-984.

Wolff, J.A., Malone, R., Williams, P., Chong, W., Acsadi, G., Jani, A. and Felgner, P. (1990). Direct gene transfer into mouse muscle *in vivo*. *Science*, 247: 1465-1468.

Wu, G.Y. and Wu, C.H. (1988). Receptor-mediated gene delivery and expression in vivo. *J. Biol. Chem.*, 263: 14621-14624.

Xiao, X., Li, J. and Samulski, R. J. (1996). Efficient long term gene transfer into muscle tissue of immunocompetent mice by adeno-associated virus vector. *J. Virol.*, 70: 8098-8108.

Yamasaki, K., Edington, H., McClosky, C., Tzeng, E., Lizonova, A., Kovesdi, I., Steed, D. and Billiar, T. (1998). Reversal of impaired wound repair in iNOS-deficient mice by topical adenoviral-mediated iNOS gene transfer. *J. Clin. Invest.*, 101: 967-971.

Yanez, R.J. and Porter, A.C. (1998). Therapeutic gene targeting. *Gene Ther.*, 5: 149-159.

Yates, J., Warren, N. and Sugden, B. (1985). Stable replication of plasmids derived from Epstein-Barr virus in various mammalian cells. *Nature*, 313: 812-815.

Ye, X., Rivera, V.M., Zoltick, P., Cesaroli , F., Schnell, M.A., Gao, G-P., Hughes, J., Gilman, M. and Wilson, J.M. (1999). Regulated delivery of therapeutic proteins after *in vivo* somatic cell gene transfer. *Science*, 283: 88-91.

Yoder J., Walsh, C. and Bestor, T. (1997). Cytosine methylation and the ecology of intragenomic parasites. *Trends Genet.*, 13: 335-340.

Yoon, K. (1999). Single-base conversion of mammalian genes by an RNA-DNA oligonucleotide. *Biogenic Amines*, 15: 137-167.

Yoon, K., Cole-Strauss, A. and Kmiec, E. (1996). Targeted gene correction of episomal DNA in mammalian cells mediated by a chimeric RNA-DNA oligonucleotide. *Proc. Natl. Acad. Sci. USA*, ,93: 2071-2076.

Yuh, C-H., Bolouri, H. and Davidson, E.H. (1998). Genomic cis-regulatory logic: experimental and computational analysis of a sea urchin gene. *Science*, 279: 1896-1902.

Zaug, A.J., Been, M.D. and Cech, T.R. (1986). The tetrahymena ribozyme acts like an RNA restriction endonuclease. *Nature*, 324: 429-433.

Zufferey, R., Dull, T., Mandel, R.J., Bukovsky, A., Quiroz, D., Naldini, L. and Trono, D. (1998). Self-inactivating lentivirus vector for safe and efficient *in vivo* gene delivery. *J. Virol.*, 72: 9873-9880.

7. SUMMARY — From inherited diseases, gene therapy has now extended to acquired disorders; in addition, it has moved from therapeutic to prophylactic goals as illustrated by DNA vaccines. Indeed, gene therapy is expected to revolutionize the practice of Medicine. The breakthrough is that therapeutic benefits can result from transient expression of minigenes upon transfer into patients' somatic cells. Minigenes are devoided of most non-coding genomic sequences that interrupt and embed regular human genes, and are therefore small enough to be handled in bacterial plasmids or viral vectors. In addition, they provide an unlimited combination of expression cassettes and expressible sequences and are thus flexible enough to fit a broad field of applications. Such artificial genes, encoding therapeutic proteins and non-coding RNA, are experimented with viral and non-viral vectors on a variety of diseases either for transient expression (cancer, limb/heart ischemia, AIDS, wound healing, bone repair, DNA vaccines, ...) or for long-term expression (inherited diseases and different types of acquired disorders). However, although simple in principle, minigene expression therapy is still hampered by many problems, in particular to achieve 1) *in vivo* delivery to the correct cell/stem cell targets, 2) non-random integration for long-term expression and 3) physiological genetic regulation for inherited diseases. This is why, in 1991, we devised a new approach to gene therapy aimed at repairing mutant genes or inactivating deleterious ones. Based on the transfer of premade DNA-recombinase nucleoprotein complexes (presynaptic filaments), the idea is to master homologous recombination and substitute DNA segments from target cells or viruses by homologous genomic DNA from wild-type (gene repair) or mutant (gene inactivation) origin. Another complementary gene repair approach relies on RNA-DNA oligonucleotides for efficient correction of point mutations. Importantly enough, we have invented presynaptic filaments with a double-stranded DNA core that offer broad DNA exchange potentialities. Thus, in addition to broad gene repair for inherited diseases, these filaments may mediate efficient targeted (non-random) integration of therapeutic minigenes for acquired disorders.

Key Words: gene therapy; minigene expression therapy; transient or long-term transgene expression; gene repair/inactivation; inherited diseases, vaccines and acquired disorders

Progress in Gene Therapy: Basic and Clinical Frontiers, pp. 35-52
R. Bertolotti *et al.* (Eds)

High-titer retroviral pseudotype vectors for specific targeting of human CD4-positive cells

Jörn Stitz, Peter Muller, Heike Merget-Millitzer And Klaus Cichutek*

Department of Medical Biotechnology, Paul-Ehrlich-Institut, Paul-Ehrlich-Straße 51-59, D-63225 Langen, Germany

Table of Contents

* Corresponding author. E-mail: cickl@pei.de

1. INTRODUCTION

An ideal gene therapy protocol would comprise direct application of gene delivery vehicles (*e.g.* viral vector particles) into suitable sites of the patient's body, *e.g.* into the blood stream. Such gene delivery systems probably would have to allow 1) cell specific expression of the foreign gene in the cell type to be genetically modified, 2) long-term expression of the delivered gene, and 3) safe and easy clinical handling which should also be cost-effective.

The retroviral vectors which are until today most frequently employed in clinical trials are derived from the well characterised amphotropic murine leukaemia virus (MLV). These vector particles are able to transduce proliferating cells of rodent and human origin expressing the cellular receptor Ram-1. Besides serving as the receptor for amphotropic MLV, Ram-1 functions as a phosphate-channel (Kavanaugh *et al.*, 1994; Kozak *et al.*, 1995; Eiden *et al.*, 1996). As Ram-1 is expressed in all human cell species, vectors derived from amphotropic MLV are not suitable for *in vivo* gene delivery because unspecific transduction of cells not intended to be genetically modified is likely to occur. The unspecific transduction of cells from irrelevant tissues will presumably require higher vector doses to achieve transduction efficiencies in the target cell population that would allow the reconstitution of tissue specific functions previously disrupted by genetic defects. Higher vector doses will most likely result in a stronger immune response against the viral vector particles and thus, will make repeatable efficacious vector application less likely, as vector particles will probably be increasingly neutralised by antibodies directed against the vector proteins. Moreover, the risk of malignant cell transformation as a consequence of the unspecific integration of vector sequences into the host cell genome (insertional mutagenesis) may increase, if irrelevant tissues are transduced. To avoid these major drawbacks, efforts have been undertaken during the past years to develop MLV-derived vectors with a narrowed host cell range.

To alter vector tropism, several strategies have been invented, including the use of chimeric envelope proteins that antibody derived fragments or receptor ligands. These modifications are aimed at allowing the respective vector particles to specifically bind to the antigen that the antibody or ligand fused to Env is directed against and to specifically enter a certain cell type displaying the antigen. These approaches have been shown to be successful, but the gene transfer efficiencies reached so far have been relatively low (Chu *et al.*, 1994; Somia *et al.*, 1995; Ager *et al.*, 1996).

The cell tropism of enveloped viruses or viral vector particles is determined at the level of entry by the envelope proteins, which allow binding to as well as uptake into the cell. The tropism of a given virus can be conferred to a viral vector particle by incorporating heterologous envelope proteins into the virion, a procedure termed pseudotyping. This may result in the generation of virions with an altered host cell range. A number of MLV-derived pseudotype vector particles have been developed in recent years, among them several that were intended to further optimize gene delivery for the purpose of gene therapy. For example, capsid particles derived from MLV which are and pseudotyped using the envelope proteins of the gibbon ape leukaemia virus (GaLV) or with the G-protein of the vesicular stomatitis virus (VSV) are often used in today's *ex vivo* transduction of haematopoetic stem cells (Kalle *et al.*, 1994). These MLV(VSV-G) and MLV(GaLV) pseudotype vector particles seem to be superior with regard to the achievable gene delivery efficiency compared to amphotropic MLV-vectors, probably because the cellular receptors used by GaLV and VSV to enter the host cell are most likely expressed at higher levels than the amphotropic receptor Ram-1 (Richardson *et al.*, 1996). Pseudotype vector particles that display the envelope proteins of the feline endogenous virus RD114 (Cosset *et al.*, 1995) are resistent against inactivation by human complement and are therefore believed to be morc stable under *in vivo* conditions. Remarkably, none of the vector particles mentioned above show a narrowed host range to a limited cell species of human origin and are therefore only suitable for *ex vivo* gene delivery. Further pseudotype vectors derived from MLV have been developed, but are not necessarily suitable for gene therapy rather than for studying virological phenomena such as receptor choice or cell entry mechanisms of the parental viruses the envelope proteins were derived from (Vile *et al.*, 1990; Landau *et al.*, 1992).

We and others (Schnierle & Stitz *et al.*, 1997; Mammano *et al.*, 1997) decided to exploit the restricted tropism of lentiviruses by incorporating lentiviral envelope glycoproteins into MLV capsids. This was shown to result in the generation of retroviral pseudotype vector particles that selectively transduce CD4-positive human cells and accordingly promise to be valuable tools for gene therapy. Diseases that affect CD4+ cells could be possible targets for gene transfer using MLV(HIV-1) vector particles. These include severe combined immunodeficiency (Anderson *et al.*, 1984; Blaese *et al.*, 1995) and AIDS (reviews: Yu *et al.*, 1994; Pomerantz *et al.*, 1995).

2. MATERIALS AND METHODS

a) Bacteria and plasmids

All plasmids were purified from the *E. coli*-strain DH10B using plasmid purification kits (Diagen, Hilden, Germany). The plasmid pTr712 encoding a truncated Env-variant of HIV-1 was described elsewhere (Wilk *et al.*, 1992). The plasmid pCMV-*rev*/hyg. was derived from pCMV-*rev* (Lewis *et al.*, 1990) by insertion of a DNA fragment encompassing the hygromycin B gene via *Sph I*, which was amplified by PCR from DNA of the plasmid pRep4 (Invitrogen, Leek, Netherlands) by a standard protocol using the following oligonucleotides: Hygro 5'-Sph I (+), *i.e.* 5'-CATGC ATGCCTGCTTCATCCCCGTGGCCCG-3' and Hygro 3'-Sph I (-), *i.e.* 5'-ACATGCATGCCCAGACCCCAGGCAACG CCC-3'.

b) Cells and transfections

The env-negative MLV-derived packaging cell line TELCeB6 (Cosset *et al.*, 1995) was kindly provided by F.-L. Cosset (Centre National de la

Recherche Scientifique, Lyon, France) and Y. Takeuchi (Institute of Cancer Research, London, UK). TELCeB6 cells express the *gag/pol*-genes of MLV and the transfer vector MFGlnslacZ (Ferry *et al.*, 1991) containing the reporter gene *lacZ* that encodes the beta-galactosidase of *E.coli*. HeLaCD4+ cells (ADP047, ref. Chesebro *et al.*, 1991) and the T-cell lines Molt4.8 (175), Jurkat (177) and C8166 (404) were purchased from the NIH-AIDS Research and Reference Reagent-Program. All adherent cell lines were kept in Dulbecco's Modified Eagles Medium (DMEM) supplemented with 10% fetal calf serum (FCS, Biochrom KG, Berlin, Germany). The T-cell lines were expanded in RPMI 1648 containing 10% FCS. For the transfection of TELCeB6 and K52S cells, we employed the reagent Lipofectamine (Gibco/BRL, Eggenstein, Germany) following the manufacturer's instructions. Selection of transfected cells was performed using 800µg/ml G418 and 200µg/ml hygromycin B (Boehringer, Mannheim, Germany) respectively. Primary human peripheral blood mononuclear cells (PBMCs) were prepared from blood samples of healthy donors by Ficoll-gradient-centrifugation (Ficoll-Histopaque, Sigma, Deisenhofen, Germany). Cells were stimulated using 20 u/ml IL-2 (Eurocetus, Ratingen, Germany) and 1 µg/ml PHA (Murex Biotech, Dartford, UK). Monoclonal aSDF-1β-antibodies were purchased from R&D-System (Wiesbaden, Germany).

c) Determination of vector titers

The calculation of vector titers and the X-Gal staining procedure were described previously (Schnierle & Stitz *et al.*, 1997). Endpoint titrations of pseudotype vector stocks were performed using various dilutions of a total volume of 1 ml to infect 1 x 10^6 suspension cells that were washed and pelleted prior to transduction. Adherent target cells were seeded at a density of 2 x 10^5 cells per six-well (Nunc, Wiesbaden, Germany) 24 hours prior transduction. All target cells were exposed to vector particles for 2 hours followed by washing with PBS. Transduced cells were then further expanded for two days before *lacZ*-positive cells were detected by X-Gal staining.

d) Electron microscopy

Confluent cultures of the packaging cell line K52S were treated with fixation buffer (PBS / 2 % formaldehyde) for one hour at 4°C and then repeatedly washed with PBS and incubated with 1:500 dilutions of an anti-HIV-1 serum from a HIV-1-infected donor for one hour at 37°C. After further washing, a 1:50 dilution of gold particle-conjugated anti-human IgG (Biocell, Cardiff, UK) was added and left for one hour at 37°C on the samples. After extensive washing with PBS, cells were embedded into epoxyd according to standard procedures (Luft *et al.*, 1964).

e) Generation of vector stocks and ultrafiltration

Vector-containing supernatants (15-20 ml per 250 cm2 flask) were generated by incubation of confluent cultures of packaging cells for 6 to 12 hours with respective cell culture media. Supernatants were pooled and contaminating cells were removed by filtration (0.45 μm filter). These samples were then stored or further concentrated by ultrafiltration. We used Centriprep-devices (Amicon, Beverly, USA) with membranes that are pervious to molecules of a weight smaller than 30 kDa or 50 kDa according to the manufacturer's instructions. Usually, 17 ml vector-containing supernatants were reduced to a final volume of 1 ml by this procedure. Untreated and concentrated vector stocks were stored in liquid nitrogen.

3. RESULTS

a) Establishing high-titer packaging cell lines

The previously established MLV(HIV-1) vector producer cell lines TELCeB6/pTr712-K14 and TELCeB6/pTr712-9 (Schnierle & Stitz *et al.*, 1997) derived from the *env*-negative packaging cell line TELCeB6 (Cosset *et al.*, 1995) and transfected with HIV-1 *env*-gene variant pTr712 (Wilk *et al.*, 1992) enabled us to prepare vector stocks reaching titers of up to 10^5 i.u./ml. To test whether more efficient packaging cell lines could be established, we transfected plasmid pTr712 into TELCeB6 cells followed by G418-selection two days later. Further 10 days later, 200 cell clones were picked and screened for infectious vector particle release by titration of pseudotype vector containing supernatants in HeLaCD4+ cells (Chesebro *et al.*, 1990) that stably express human CD4. Cell clone K52S produced the highest vector titers and was expanded. Endpoint titrations repeatedly performed in HeLaCD4+ cells and various T-cell lines (Molt4.8, Jurkat and C8166) revealed pseudotype vector titers of up to 1.6 x 10^6 i.u./ml (Table 1).

Table 1. MLV(HIV-1) vector titers of supernatants from the packaging cells K52S and K52S/R20 in CD4-positive cell lines and HeLa cells*.

Target cells	Packaging cells	
	K52S	K52S/R20
HeLaCD4+	1.6 x 10^6 **	5 x 10^6 **
HeLa	n.d.	n.d.
Molt4.8	9 x 10^5	n.t.
C8166	9 x 10^5	n.t.
Jurkat	5 x 10^5	n.t.

* Gene transfer efficiencies were determined by X-Gal staining. Transduction of HeLa cells was not detected (n.d.). Titers shown resulted from three independent experiments (n.t. = not tested).
** i.u./ml

All following experiments described here were performed employing vector stocks prepared from the new packaging cell line K52S. The packaging cell line K52S showed no significant loss in its capacity to produce high-titer vector stocks over a period of several months. By immunostaining using HIV-1-specific anti-sera from HIV-1-infected human donors as previously described (Schnierle & Stitz *et al.*, 1997) only about 40% of the K52S cells were found to express the variant envelope proteins of HIV-1. Assuming that the titer of infectious vector particles should correlate with the percentage of *env*-positive packaging cells, we attempted to increase the titer by subcloning the K52S cells. To avoid the time-consuming procedure of biological subcloning and in attempt to enhance the expression levels of the HIV-1-derived envelope glyco-proteins, plasmid pCMV*rev*/hyg. encoding the HIV-1 *rev*-gene was transfected into K52S cells followed by hygromycin B-selection. Fifty resistant cell clones were screened for efficient vector particle release as described above. The most productive clone termed K52S/R20 was shown to produce three-times higher infectious vector titers. More than 95% of the K52S/R20 cells were found to express the HIV-1-derived envelope glycoproteins as shown by Env-specific immunostaining (data not shown).

b) Morphology of MLV(HIV-1) pseudotype vector particles

To directly demonstrate the presence of HIV-1-derived envelope proteins in MLV-capsid particles, we prepared confluent of K52S cells for electron microscopy. Specific labelling of the envelope proteins was performed using polyclonal antibodies from a HIV-1-infected human donor and α-human-IgG-gold conjugates as described in materials and methods. The electron microscopic images revealed C-type retroviral particles specifically labelled by anti-HIV-1-antibodies (Fig. 1). This showed the presence of HIV-1-derived envelope glycoproteins in the vector particles released from packaging cell line K52S.

c) MLV(HIV-1) pseudotype vectors produced by packaging cells (K52S) use coreceptor CXCR4 during entry in CD4+ cells

Especially for *in vivo* use, it is indispensable to exactly characterize the tropism of the vector particles to be used. As plasmid pTr712 employed to establish packaging cell line K52S encoding, in addition to the truncated transmembrane protein (TM), the gp120-SU of HIV-1 isolate BH10 known to be T-cell tropic (Wong-Staal *et al.*, 1985; Gene Bank accession numbers M15654, K02008, K02009, K02010), the respective MLV(HIV-1) pseudotype vector particles were thus expected to mediate CXCR4- dependent

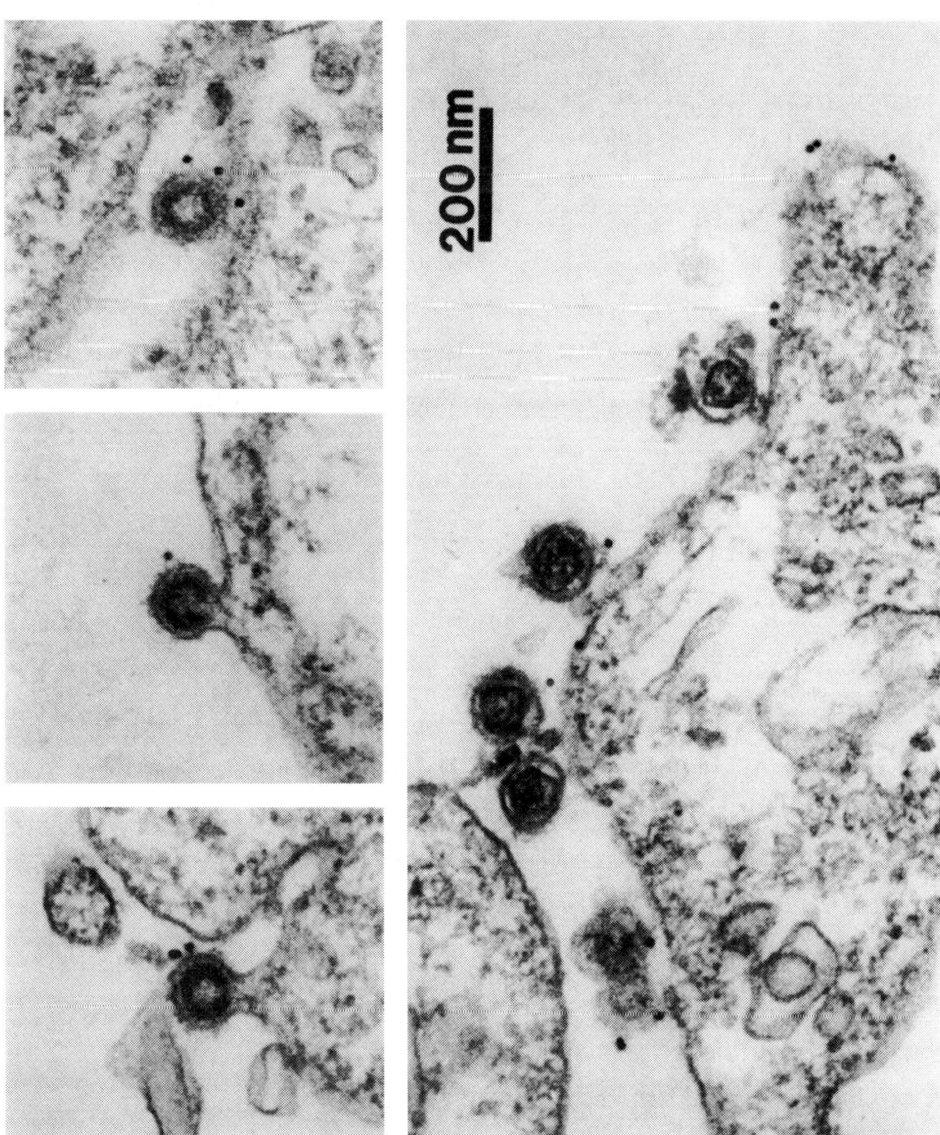

Figure 1. MLV(HIV-1) vector particles (electron micrographs). Packaging cells (K52S) were fixed and stained using an αHIV-1 serum and a human IgG-gold conjugate. Specifically labeled vector particles were found, demonstrating the presence of HIV-1-derived envelope proteins in the MLV-derived virions.

transduction of CD4-positive cells. To experimentally verify this hypothesis, we used a panel of cell lines (kindly provided by D.R. Littman, Skirball Institute, New York, through E.M. Fenyö, Karolinska Institute, Stockholm, Sweden) derived by Hill *et al.* (1997) from the megaglioma-asterocyte cell line U87 as target cells for transduction. In contrast to the parental U87 cells, the panel of U87.CD4 cell lines expresses either human CD4 alone or in conjunction with one of the chemokine receptors known to function as co-receptors for HIV-1 infection (reviewed by D'Souza *et al.*, 1996). 1×10^5 cells were used as target cells and transduced using MLV(HIV-1) vector containing supernatants of K52S cells. The X-Gal assay performed two days later revealed the exclusive transduction of U.87 cells expressing CD4 and the chemokine receptor CXCR4 (Fig. 2).

No transduction of the U87.CD4 cells expressing any other of the co-receptors was detected. It was thus demonstrated that the MLV(HIV-1) vector particles retained the tropism of the parental T-cell tropic HIV-1-isolate BH10, from which the surface protein gp120-SU was derived.

d) Optimising MLV(HIV-1) vector particle concentration

It is known that pseudotype vector particles containing the G-protein of VSV are very stable and can be efficiently concentrated by ultrafiltration or by ultracentrifugation. It was therefore also attempted to further increase the infectious titer of vector stocks by applying simple concentration procedures assuming sufficient stability. 17 ml of MLV(HIV-1) vector-containing supernatants obtained from confluent cultures of K52S packaging cells were usually concentrated to 1 ml using a single centrifugation step as described in materials and methods. The concentrated and, as controls, the untreated vector preparations were subsequently used to transduce C8166 T-cells. Transduction was detected by X-Gal staining two days post infection. In addition, differently conditioned cell culture media were compared with regard to their influence on vector production by the packaging cells. During the production of vector-containing supernatants, the packaging cells were

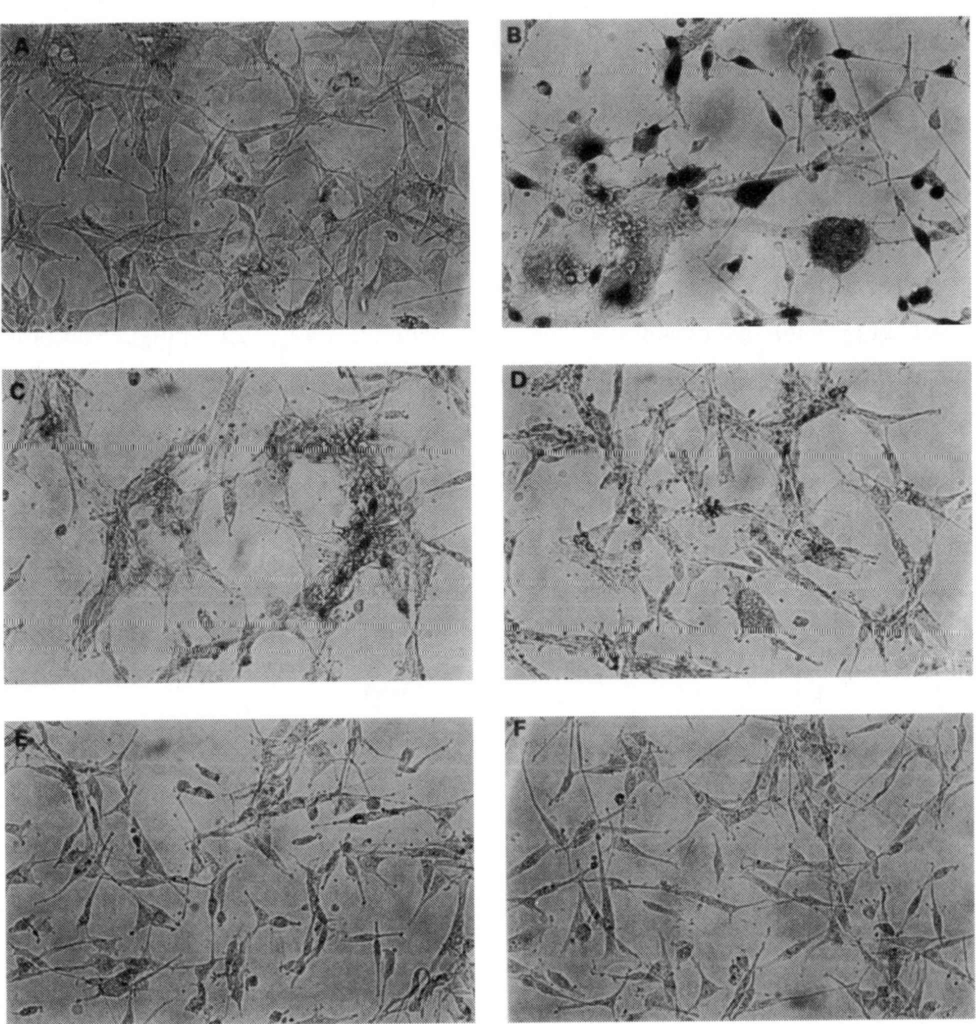

Figure 2. Co-receptor usage of MLV(HIV-1) vector particles. A panel of U87.CD4-derived cell lines was transduced by MLV(HIV-1) vector particles prepared from cell culture supernatants of K52S packaging cells. Successful gene delivery was revealed by X-Gal staining. Obviously, the respective vector particles use the co-receptor CXCR4 (B) to enter the target cells. No transduction of cells expressing any of the other co-receptors was detected.

A) U87.CD4 / —; B) U87.CD4 / CXCR4;
C) U87.CD4 / CCR1; D) U87.CD4 / CCR2b;
E) U87.CD4 / CCR3; F) U87.CD4 / CCR5.

cultured in DMEM with 10% FCS, glutamine and NSP (VZ), DMEM with glutamine and NSP (VZ-FCS) and DMEM without any supplements (pure), respectively. Unconcentrated and vector preparations ultrafiltrated using Centriprep30™- and Centriprep50™-devices were used in parallel to transduce C8166 T-cells. As shown in Fig. 3, the highest vector titers were reached using DMEM without any supplements during transduction of the unconcentrated vector particles. In contrast, the ultrafiltrated and thus concentrated packaging cell supernatants retained the highest vector titers when generated with complete cell culture media (VZ). This suggested that VZ allows the packaging cells to produce the largest amount of vector particles, but may somehow inhibit vector infectivity. The best vector preparations obtained using the described concentration procedures yielded vector stocks with titers of up to 2 x 10^8 i.u./ml (data not shown).

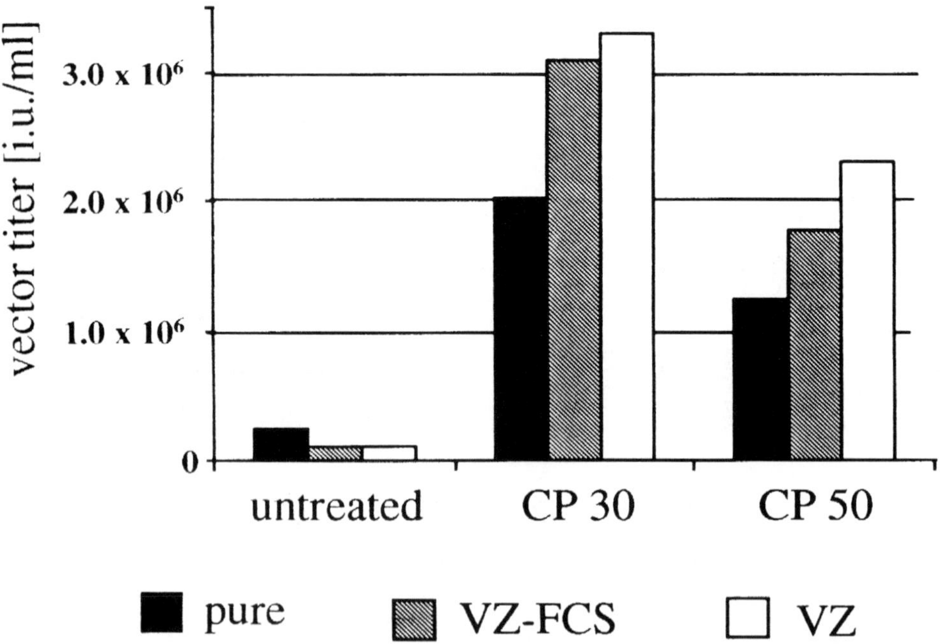

Figure 3. Concentrating vector particles by ultrafiltration. Supernatants of the packaging cell line K52S were prepared using different culture media: pure DMEM without any supplements (pure), DMEM supplemented with glutamine and NSP (VZ-FCS) and DMEM supplemented with glutamine, NSP and 10% FCS (VZ).

e) Transduction of primary T-cells

To test the potential of MLV(HIV-1) vector particles to mediate gene delivery into primary T-cells, experiments were performed using human peripheral blood mononuclear cells (PBMCs). PBMCs were obtained from healthy human donors by Ficoll gradient centrifugation. The cells were stimulated by IL-2 and PHA for 48 hours. Three separate transduction protocols were designed. A portion of the stimulated cells was cultured in the presence of 10 μg/ml monoclonal antibody directed against the CXCR4-ligand human stromal derived factor 1β (SDF-1β). The antibodies were added 4 or 24 hours prior to transduction. In each case, αSDF-1β-antibodies were also present during transduction (5 μg/ml). In parallel, stimulated PBMCs were transduced in the absence of additional αSDF-1β. Transductions were performed employing 1 x 10^6 PBMCs and 1 x 10^5 i.u. of MLV(HIV-1) vector particles titrated on HeLaCD4+ cells. The transduced target cells were expanded for 2 days before staining of lacZ-positive cells.

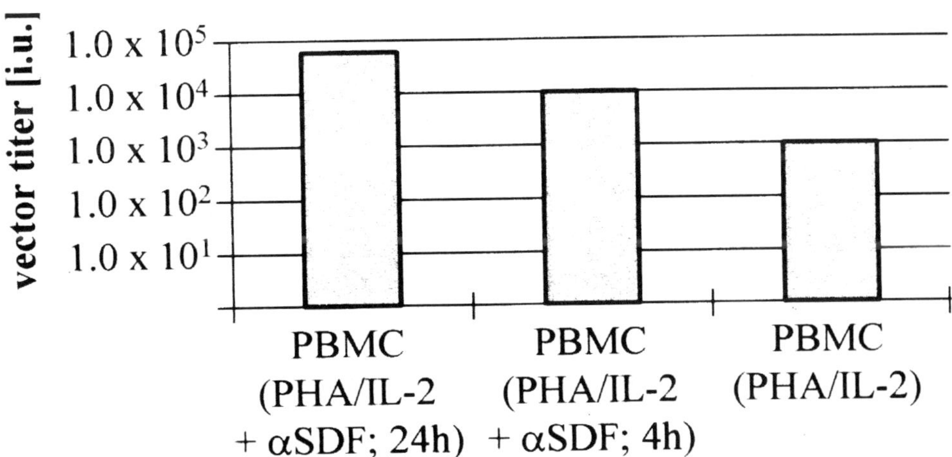

Figure 4. Transduction of peripheral blood mononuclear cells (PBMCs). Cells were stimulated by PHA and IL-2 for two days. A part of these cells was directly transduced by the respective MLV(HIV-1) vector particles, whereas some cells were preincubated in the presence of αSDF-1β-antibodies 4 or 24 hours prior transduction. Two days post transduction, successful transduction was detected by X-Gal staining. Vector titers from one representative experiment are shown.

Preincubation (24 h) of stimulated PBMCs with αSDF-1β -antibodies led to a 15-fold increase of transduction efficiency compared to stimulated PBMCs without αSDF-1β-preincubation (Fig. 4).

The evaluated vector titer was 6 x 10^4 i.u. and thus hardly as high as the vector amount employed in these experiments, indicating that almost all pseudotyped vector particles were able to infect the permissive T-cell subpopulation among the PBMCs. Using a higher multiplicities of infection (M.O.I. = infectious vector particle to target cell ratio), we were not able to achieve higher transduction efficiencies (data not shown), suggesting that only about 10% of the PBMC were susceptible to the respective MLV(HIV-1) pseudotype vectors. We further hypothezised that the observed enhanced titers on αSDF-1β-treated PBMCs reflected the up-regulation of the SDF-receptor CXCR4 induced by the decreased concentration of SDF in the culture medium. CXCR4 was shown to function as a coreceptor for the MLV(HIV-1) vector particles and is needed to allow efficient cell entry (Fig. 2). Therefore, the expression-level of CXCR4 may have an impact on the efficiency of gene transfer mediated by the respective particles.

4. DISCUSSION

We have previously shown that an *env* gene variant encoding a C-terminally truncated transmembrane protein and a full-length surface protein of HIV-1 was instrumental to generate MLV(HIV-1) pseudotype vector particles upon its expression in otherwise *env*-negative MLV-derived packaging cells (Schnierle & Stitz *et al.*, 1997). Previously described packaging cell lines produced only moderate vector titers of up to 9 x 10^4 i.u./ml. We report here the generation of new packaging cell lines that allow the harvest of supernatants containing more than 1 x 10^6 i.u./ml of respective vector particles. The efficiency of vector production is thus comparable with that of stable amphotropic packaging cell lines (Markowitz *et al.*, 1988; Cosset *et al.*, 1995) used in today's clinical gene therapy trials. The described MLV(HIV-1) vector particles were shown to allow simple concentration employing ultrafiltration devices. Using this method, vector stocks with titers of up to 2 x 10^8 i.u./ml could be generated that should be considered sufficient for possible future clinical trials. We also demonstrated the successful transduction of primary human T-cells. However, transduction seemed to be limited to a small subpopulation of the PBMC and could not be increased using higher vector doses. It is conceivable to conclude from these data that only a small portion of the primary T-cells used here expressed the CD4 receptor in conjunction with the CXCR4 co-receptor and was in a state of active proliferation. This is in agreement with reports showing that only a very small subpopulation of human T-cells can be infected with HIV (Bruunsgaard *et al.*, 1995; Zhang *et al.*, 1998). The MLV(HIV-1) vector particles described here were shown to specifically transduce human CD4/CXCR4-positive cells which are the natural host cells for T-cell tropic HIV-1 strains. Thus, MLV(HIV-1) vectors may be valuable tools to deliver anti-retroviral genes into the host cell compartment of HIV-1. Anti-retroviral therapeutic genes delivered by respective pseudotype vector particles could for example encode ribozymes, transdominant-negative mutant HIV proteins, RNA-decoys or antibody-fragments directed against HIV-1 proteins (reviewed in: Yu *et al.*, 1994; Pomerantz *et al.*, 1995). This would lead to intracellular immunisation of the natural host cells against infection by HIV. Unfortunately, the employment of the MLV(HIV-1) vector particles for *in vivo* applications would probably be restricted to HIV-1 seronegative individuals. The humoral anti-HIV-1-immune response of HIV-1-infected patients would presumably lead to the neutralisation of the MLV(HIV-1) vector particles *in vivo* before transduction could occur. In contrast, the presence of HIV-1 derived envelope proteins in the respective vectors could be of benefit when using them as vaccines to induce immunity in healthy recipients. These vector particles would induce a humoral immune response against HIV-1 Env and are thus potentially beneficial in uninfected individuals.

Acknowledgements

We would like to thank D. Bauer for excellent technical assistance, M. Selbert for expert automatic DNA sequencing and S. Norley, S. Ottmann and M. Grez for constructive discussions. We are grateful to S. Norley for the donation of αHIV-1-sera, F.-L. Cosset for the donation of TELCeB6 cells, B. Chesebro for HeLaCD4+ cells obtained through the NIH AIDS Research and Reference Reagent Program. This work was supported by grants # 01 Kl 9718 and # 01 KV 9550 of the Bundesministerium für Bildung, Wissenschaft, Forschung und Technologie to K. Cichutek.

5. REFERENCES

Ager, S., Nilson, B.H.K., Morling, F.J., Peng, K.W., Cosset, F.-L. and Russell, S.J. (1996). Retroviral display of antibody fragments; interdomain spacing strongly influences vector infectivity. *Hum. Gene Ther.*, 7: 2157-2164.

Anderson, F.W. (1984). Prospects for human gene therapy. *Science*, 226: 401-409.

Blaese, M.R., Culver, K.W., Miller, A.D., Carter, C.S., Fleisher, T., Clerici, M., Shearer, G., Chang, L., Chiang, Y., Tolstohev, P., Greenblatt, J.J., Rosenberg, S.A., Klein, H., Berger, M., Mullen, C.A., Ramsey, W.J., Muul, L., Morgan, R.A. and Anderson, W.F. (1995). T Lymphocyte-directed gene therapy for ADA- SCID: initial trial results after 4 years. *Science*, 270: 475-480.

Bruunsgaard, H., Pedersen, C., Scheibel, E. and Pedersen, B.K. (1995). Increase in percentage of CD45RO+ / CD8+ cells is associated with previous severe primary HIV infection. *JAIDS*, 10: 107-114.

Chesebro, B., Wehrly, K., Metcalf, J. and Griffin, D.E. (1991). Use of a new CD4-positive HeLa cell clone for direct quantivication of infectious human immunodeficiency virus from blood cells of AIDS patients. *J. Infect. Dis.*, 163: 64-70.

Chu, T., Martinez, I., Sheay, W. and Dornburg, R. (1994). Cell targeting with retroviral vector particles containing antibody-envelope fusion proteins. *Gene Ther.*, 1: 292-299.

Cosset, F.-L., Tackeuchi, Y., Battini, J.-L., Weiss, R.A. and Collins, M.K.I. (1995). High-titer packaging cells producing recombinant retroviruses resistent to human serum. *J. Virol.*, 69: 7430-7436.

D'Souza, M.P. and Harden V.A. (1996). Chemokines and HIV-1 second receptors. *Nature Med.*, 2: 1293-1300.

Eiden, M.V., Farrel, K.B. and Wilson, C.A. (1996). Substitution of a single amino acid residue is sufficient to allow the human amphotropic murine leukemia virus receptor to also function as gibbon ape leukemia virus receptor. *J. Virol.*, 70: 1080-1085.

Hill, C.M., Deng, H., Unutmaz, D., Kewalramani, V.N., Bastiani, L., Gorny, M.K., Zolla-Pazner, S. and Littman, D.R. (1997). Envelope glycoproteins from human immunodeficiency virus types 1 and 2 and simian immunodeficiency virus can use human CCR5 as a co-receptor for viral entry and make direct CD4-dependent interactions with this chemokine receptor. *J. Virol.*, 71: 6296-6304.

Kalle, C., Kiem, H.-P., Goehle, S., Darovsky, B., Heimfeld, S., Torok-Strob, B., Strob, R. and Schuening, F.G. (1994). Increased gene transfer into human hematopoietic progenitor cells by extended *in vitro* exposure to a pseudotyped retroviral vector. *Blood*, 84: 2890-2897.

Kavanaugh, M.P., Miller, D.G., Zhang, W., Law, W., Kozak, S.L., Kabat, D., Miller, D.A. (1994). Cell-surface receptors for gibbon ape leukemia virus and amphotropic murine retrovirus are inducible sodium-dependent phosphate symporters. *Proc. Natl. Acad. Sci. USA*, 91: 7071-7075.

Kozak, S.L., Siess, D.C., Kavanaugh, M.P., Miller, A.D. and Kabat, D. (1995). The envelope glycoproteins of an amphotropic murine retrovirus binds specifically to the cellular receptor/phosphate transporter of susceptible species. *J. Virol.*, 69: 3433-3440.

Landau, N.R. and Littamn, D.R. (1992). Packaging system for rapid production of murine leukemia virus vectors with variable tropism. *J. Virol.*, 66: 5110-5113.

Lewis, N., Williams, J., Rekosh, D. and Hammarskjöld, M.-L. (1990). Identification of a cis-acting element in human immunodeficiency virus type 2 (HIV-2) that is responsive to the HIV-1 rev and human T-cell leukemia virus types I and II rex protein. *J. Virol.*, 64: 1690-1697.

Luft, J.H. (1964). Improvements in epoxy resin embedding methods. *J. Biophys. Biochem. Cytol.*, 9: 109-113.

Mammano, F., Salvatori, F., Indraccolo, S., De Rossi, A., Chieco-Bianchi, L. and Göttlinger, H.G. (1997). Truncation of the immunodeficiency virus type 1 envelope allows efficient pseudotyping of murine leukemia virus particles and gene transfer into CD4+ cells. *J. Virol.*, 71: 3341-3345.

Markowitz, D., Goff, S. and Bank, A. (1988). Construction and use of a safe and efficient amphotropic packaging cell line. *Virology*, 167: 400-406.

Richardson, C. and Bank, A. (1996). Developmental-stage-specific expression and regulation of an amphotropic retroviral receptor in hematopoietic cells. *Mol. Cell. Biol.*, 16: 4240-4247.

Pomerantz, R.J. and Trono, D. (1995). Genetic therapies for HIV infections: promise for the future. *AIDS*, 9: 985-993.

Sattentau, Q.J., Dalgleish, A.G., Weiss, R.A. and Beverley P.C.L. (1986). Epitopes of the CD4 antigen and HIV infection. *Science*, 234: 1120-1123.

Schnierle, B.S., Stitz, J., Bosch, V., Nocken, F., Merget-Millitzer, H., Engelstädter, M., Kurth, R. and Cichutek, K. (1997). Pseudotyping of murine leukemia virus with the envelope glycoproteins of HIV generates a retroviral vector with specificity of infection for CD4-expressing cells. *Proc. Natl. Acad. Sci. USA*, 94: 8640-8645.

Somia, N.V., Zoppe, M. and Verma, I.M. (1995). Generation of targeted retroviral vectors by using single-chain variable fragment: an approach to *in vivo* gene delivery. *Proc. Natl. Acad. Sci. USA*, 92: 7570-7574.

Vile, R.G., Schulz, T.F., Danos, O.F., Collins, M.K.L. and Weiss, R.A. (1991). A murine cell line producing HTLV-I pseudotype virions carrying a selektable marker gene. *Virology*, 180: 420-424.

Wilk, T., Pfeifer, T. and Bosch, V. (1992). Retained *in vitro* infectivity and cytopathogenicity of HIV-1 despite truncation of the C-terminal tail of the env gene product. *Virology*, 189: 167-177.

Wong-Staal, F., Gallo, R.C., Chang, N.T., Ghrayeb, J., Papas, T.S., Lautenberger, J.A., Pearson, M.L., Petteway, S.R.Jr., Ivanoff, L., Baumeister, K., Whitehorn, E.A., Rafalski, J.A., Doran, E.R., Josephs, S.J., Starcich, B., Livak, K.J., Patarca, R., Haseltine, W.A. and Ratner, L. (1985). Complete nucleotide sequence of the AIDS virus, HTLV-III. *Nature*, 313: 277-284.

Yu, M., Poeschla, E. and Wong-Stall, F. (1994). Progress towards gene therapy for HIV infection. *Gene Ther.*, 1: 13-26.

Zhang, Z.-Q., Notermans, D.W., Sedgewick, G., Cavert, W., Wietgreffe, S., Zupancic, M., Gebhard, K., Henry, K., Boies, L., Chen, Z., Jenkins, M., Mills, R., McDade, H., Goodwin, C., Schuwirth C.M., Danner, S.A. and Haase, A.T. (1998). Kinetics of CD4+ T cell repopulation of lymphoid tissues after treatment of HIV-1 infection. *Proc. Natl. Acad. Sci. USA*, 95: 1154-1159.

6. SUMMARY — We (Schnierle & Stitz *et al.*, 1997) previously reported the generation of packaging cell lines that produced MLV(HIV-1) pseudotype vector particles at moderate titers of up to 9 x 10^4 infectious units per millilitre (i.u./ml). New packaging cell lines have now been established that enable us to create MLV(HIV-1) pseudotype vector preparations reaching a titer of more than 10^8 i.u./ml. Using these vectors, stimulated human primary CD4-positive T-cells were efficiently transduced.

Key Words: gene therapy; HIV-1; MLV-vectors; cell targeting; CD4-expressing human cells; pseudotyping; vector concentration procedures

Progress in Gene Therapy: Basic and Clinical Frontiers, pp. 53-65
R. Bertolotti *et al.* (Eds)

Lentiviral vectors for gene therapy

Antonia Follenzi and Luigi Naldini*

Laboratory for Gene Transfer & Therapy, IRCC, Institute for Cancer Research, University of Torino Medical School, Strada Provinciale 142, 10060 Candiolo (Torino), Italy

Table of Contents

1. INTRODUCTION

Vectors derived from retroviruses have been chosen for most *ex vivo* gene therapy clinical trials as they can efficiently integrate into the genome of target cells without expressing viral proteins in target cells. These characteristics are likely to be crucial for achieving_sustained expression of the transgene and reduce the risks of immume responses against transduced cells. Retroviral vectors can transfer up to 7.0 kilobases of foreign genetic material, a sizeable amount of DNA (Miller *et al.*, 1993). Furthermore, the biology of retroviruses is relatively simple and well understood, which has facilitated continuous improvement in vector design and the generation of stable producer systems amenable to characterization and scale-up.

---

* Corresponding author. E-mail: lnaldini@ircc.unito.it

To date, the retroviral vectors used in clinical trials are derived from onco-retroviruses such as the Moloney murine leukemia virus (MLV). A major disadvantage of these vectors is that they require cell division for transduction of target cells (Miller *et al.*, 1990) limiting their use for many important somatic cell targets. This block to onco-retroviral infection in quiescent cells has been characterized at the molecular level (Roe *et al.*, 1993; Lewis and Emerman, 1994). During infection, the virus delivers the viral core comprising the capsid proteins surrounding a nucleoprotein complex consisting of two identical strands of the viral genomic RNA and viral enzymes including reverse transcriptase and integrase, in the cytoplasm of the target cells. Here, uncoating and reverse transcription of the single-stranded RNA take place. The DNA-containing complex resulting from this process must enter the nucleus for integration to occur. However, most likely due to steric constraints, fragmentation of the nuclear membrane during prophase of mitosis is required to allow access of the preintegration complex to the chromatin.

While actively dividing cells are found in several tissues, they are short-lived and have a high turnover rate. The major target cells for gene transfer, *i.e.* long-lived specialized cells of the liver, heart, skeletal muscle, brain and the hematopoietic progenitor cells are mostly non-proliferating in mature animals, and thus severely limit the use of retroviral vectors for clinical applications.

One of the strategies being used to address this problem involves the use of vectors derived from lentiviruses, a family of complex retroviruses typically associated with infection of macrophages and lymphocytes. As most tissue macrophages are terminally differentiated quiescent cells, lentiviruses must circumvent the preintegration block, inherent to onco-retroviruses. Several researchers have shown that, at least in the case of the human lentivirus, HIV-1, the viral preintegration complex interacts with the nuclear import machinery of the infected cell, and it is actively transported to the nucleus through the nucleopores (Bukrinsky *et al.*, 1993; Heinzinger *et al.*, 1994; Gallay *et al.*, 1995a and 1995b; Popov *et al.*, 1998). The preintegration complex mimics other nuclear bound cellular proteins by displaying nuclear localization signals, and thus exploiting a cellular pathway to achieve progression of the viral life cycle.

Therefore vectors derived from the HIV-1 virus were developed, designed to retain the ability of the virus to infect quiescent cells while completely inactivating its replicative functions and pathogenesis.

2. GENE DELIVERY BY LENTIVIRAL VECTORS

The ability of lentiviral vectors to directly deliver genes *in vivo* was demonstrated by introducing marker genes expressing β-galactosidase and green fluorescent protein (GFP) into the central nervous system. After injection into the brain of adult rats, efficient transduction and stable expression of the transgene was observed in neurons, without any detectable pathology. Triple labeling confirmed that almost 90% of the cells transduced by the lentiviral vector were terminally differentiated neurons (Naldini *et al.*, 1996b and 1996a; Blomer *et al.*, 1997). Animals analyzed nine months after a single injection of the vector, the longest time tested so far, showed no apparent decrease in the average level of transgene expression and no sign of tissue pathology or immune reaction (Blomer *et al.*, 1997). It would appear that lifelong expression of an exogenous gene can be achieved in the brain of normal animals by a single injection of the lentiviral vector. Stable and efficient gene delivery was also shown in the retina. Lentiviral vectors carrying the GFP gene were injected into the subretinal space of rat eyes. The GFP gene under the control of the CMV promoter was efficiently expressed in both photoreceptor cells and retinal pigment epithelium. However, the use of the rhodopsin promoter resulted in expression predominantly in photoreceptor cells. Up to 80% of the area of whole retina expressed the GFP, and the expression persisted for at least 12 weeks with no apparent decrease (Miyoshi *et al.*, 1997).

Lentiviral vectors were reported to introduce genes directly into the liver and muscles of rats. Sustained expression of GFP could be observed for both tissues. Furthermore, no inflammation or recruitment of lymphocytes could be detected at the site of injection (Kafri *et al.*, 1997). More recently, the transduction of liver by intraportal delivery of vector was analyzed in greater detail (Park *et al.*, 2000). The transduction was limited to a small percentage of cells, and was significantly enhanced after partial hepatectomy. Moreover, the majority of transduced cells were shown to be cycling, suggesting that recruitment of hepatocytes into cell cycle, disruption of extracellular matrix and/or improved access to the liver are limiting factors after *in vivo* delivery of the vector. These results indicate that several factors affect the efficiency of gene delivery by lentiviral vectors in the target tissues.

Like the parental virus, HIV-derived vectors infected resting lymphocytes very poorly. However, treatment of lymphocytes with a variety of cytokines made them permissive to infection (Zhang *et al.*, 1999; Unutmaz *et al.*, 1999). These treatments induce activation of lymphocytes, without triggering proliferation. Here, it would appear that there is a threshold of cell activation that induces a critical level of cellular factors needed for transduction (Kinoshita *et al.*, 1998).

Other tissues in which successful gene transfer by HIV-derived vectors was reported are growth-arrested human primary fibroblasts and macrophages (Naldini *et al.*, 1996b; Reiser *et al.*, 1996; Poeschla *et al.*, 1996), pancreatic islets (Gallichan *et al.*, 1998; Giannoukakis *et al.*, 1999), airway-epithelia (Goldman *et al.*, 1997; Johnson *et al.*, 2000) and cardiomyocytes (Rebolledo *et al.*, 1998).

Studies examining the ability of lentiviral vectors to transduce primitive human hematopoietic progenitor cells have shown great promise. Whereas MLV-based vectors only transduce efficiently committed progenitor cells following prolonged stimulation with cytokines, lentiviral vectors have been shown to transduce the primitive progenitor cells which engraft in immune-deficient NOD-SCID mice and are capable of multi-lineage reconstitution (Miyoshi *et al.*, 1999). They also transduce quiescent CD34(+)CD38(-) progenitors, which initiate extended long-term culture (Uchida *et al.*, 1998; Case *et al.*, 1999; Evans *et al.*, 1999), and can do so following a single exposure of cells to vectors on the day of isolation.

In summary, gene transfer protocols must be optimized to the target by selecting the design of the expression cassette, the type of packaging system, and the possible requirements of external agents such as growth factors or cytokines.

3. DESIGN AND PRODUCTION OF LENTIVIRAL VECTORS

Lentiviral vectors are replication-defective, hybrid viral particles made by the core proteins and enzymes of a lentivirus, and the envelope of a different virus, most often the vesicular stomatitis virus (VSV) (Burns *et al.*, 1993). The general strategies employed in the design of lentiviral vectors involve segregation of *trans*-acting sequences that encode for viral proteins from *cis*-acting sequences involved in the transfer of the viral genome to target cells (see Fig. 1). The prototype vector particle is assembled by viral proteins expressed in the producer cell from constructs stripped of all viral *cis*-acting sequences. The viral *cis*-acting sequences are linked to the transgene, and are introduced into the same cell. As the vector particle can only transfer the latter construct, the infection process is limited to a single round without further spreading. The efficiency and biosafety of an actual vector system depends on the extent to which the ideal situation of complete segregation of *cis*- and *trans*-acting functions of the viral genome is achieved.

The packaging functions for the lentiviral vector are provided by at least two separate expression plasmids that use transcriptional signals different from those of the virus (Fig. 1). A "core" packaging construct, derived from the HIV-1 proviral DNA, expresses the viral proteins but not the *env* gene that has been deleted. A separate construct expresses a heterologous envelope that is incorporated into the vector particles (pseudotyping) and allows entry into the target cells. A third construct, the transfer vector, expresses RNA that contains the viral *cis*-acting sequences required for packaging by the vector particles, reverse transcription and integration in the target cells. It transfers into target cells the transgene that is under the control of an internal promoter, typically the promoter of the housekeeping phosphoglycerokinase gene (PGK) or the immediate early enhancer/promoter of the human cytomegalovirus (CMV). Alternatively, several others promoters can be used, including tissue-specific and regulatable ones.

4. IMPROVEMENTS IN THE BIOSAFETY OF LENTIVIRAL VECTORS

Prior to clinical evaluation of a new vector, its biosafety must be thoroughly demonstrated. Vectors derived from a lentivirus such as HIV-1, will undoubtedly require the most critical validation of their safety. Fortunately, the advances in the understanding of the molecular biology and pathogenesis of HIV have facilitated the design of a new generation of safer HIV-derived vectors.

Significant improvements in vector biosafety were achieved following the identification of the minimal genetic requirements for gene transduction by lentiviral vectors. Non-essential sequences were eliminated from the constructs used to package vectors, resulting in the production of so-called second-generation "minimal" lentiviral vectors. This was an important step to increasing biosafety and reducing the possible pathogenicity of replication competent recombinant viruses originating during vector production, as these sequences are important for the replication and pathogenesis of the parental virus. In addition to the structural *gag, pol* and *env* genes common to all retroviruses, lentiviruses contain two regulatory genes, *tat* and *rev*, essential for viral replication, and a variable set of accessory genes that are critical for pathogenesis but not required for viral growth. The four accessory genes of HIV-1, *vif, vpr, vpu* and *nef*, are

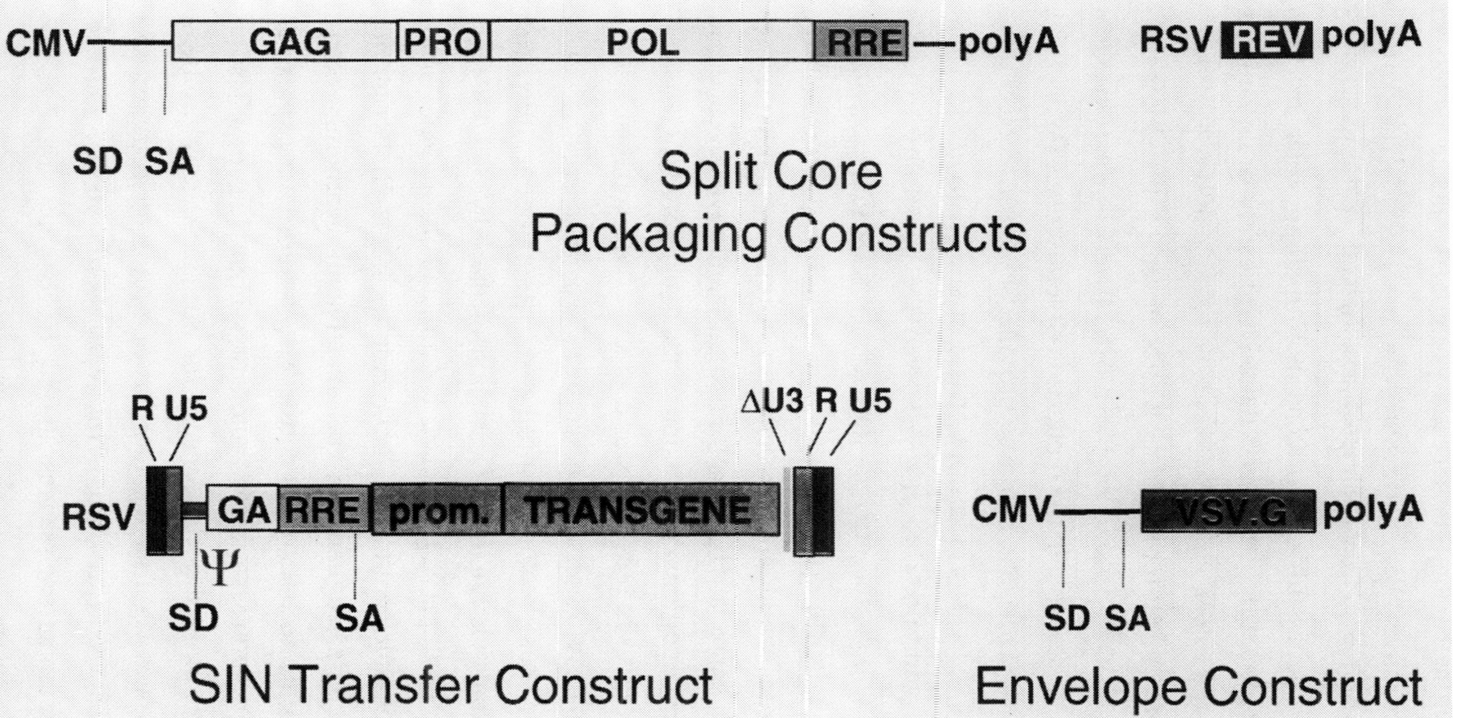

Figure 1. Schematic drawing of the four constructs used to make a lentiviral vector of the third generation. The splice donor and acceptor sites (SD and SA), the packaging sequence (Ψ), the Rev Response Element (RRE) and polyadenylation site (polyA) are indicated. The conditional packaging construct expresses the *gag* and *pol* genes from the CMV promoter. A non-overlapping construct, RSV-Rev, expresses the *rev* cDNA under the control of the RSV promoter. The SIN (self-inactivating) transfer construct contains HIV-1 *cis*-acting sequences and an expression cassette for the transgene. The 5' LTR is chimeric, with the enhancer-promoter of the Rous sarcoma virus replacing the U3 region. The 3' LTR has an almost complete deletion of the U3 region. A fourth construct encodes a heterologous envelope to pseudotype the vector, here shown coding for the protein G of the Vesicular Stomatitis Virus (VSV.G) under the control of the CMV promoter. Only the relevant part of the constructs is shown.

strictly conserved and their gene products are essential virulence factors *in vivo.* Their functions have been reviewed (Cullen, 1998). HIV-derived vectors made with inactive or deleted *vif, vpr, vpu* and *nef* genes in the packaging construct successfully transduced non-dividing cells *in vitro* (Reiser *et al.*, 1996; Kim *et al.*, 1998) and more significantly brain neurons *in vivo* (Zufferey *et al.*, 1997).

A further improvement for biosafety was achieved by the elimination of the *tat* gene. Its product is one of the most powerful transcriptional activators and plays a central role in the replication of lentiviruses (Wei *et al.*, 1998). The function of the *tat* gene could be provided by strong constitutive promoters upstream of the vector transcriptional start site (Kim *et al.*, 1998; Dull *et al.*, 1998). In addition, Dull *et al.* showed the possibility to use a packaging vector in which not only *tat* but also *rev* is deleted. *Rev* is provided in *trans* by a fourth construct driven by the Rous sarcoma virus promoter (Fig. 1). The resulting gene delivery system, which conserves only three of the nine genes of HIV-1, offers significant advantages for its predicted biosafetyand conditional expression of the packaging functions.

Significant progress in biosafety was also achieved by the successful generation of self-inactivating lentiviral vectors (Miyoshi *et al.*, 1998; Zufferey *et al.*, 1998). These vectors are produced by transfer constructs that carry an almost complete deletion in the U3 region of the HIV 3' LTR (Fig. 1). The U3 region contains the viral enhancer and promoter. Thus, transduction of vector deleted in the 3' U3 results in the transcriptional inactivation of both LTRs. Remarkably, self-inactivating lentiviral vectors could be produced with infectious titers and levels of transgene expression driven by an internal promoter similar to vectors carrying wild-type LTRs. A self-inactivating vector diminishes the concern for oncogenesis by promoter insertion, and it alleviates significantly the risk of vector mobilization and recombination with the wild-type virus (Bukovsky *et al.*, 1999). Furthermore, the lack of promoter sequences in the LTRs may improve the performance of an internal, tissue specific or regulatable promoter.

Combining a self-inactivating transfer vector with a packaging system of the third generation, it is now possible to achieve a high level of predicted biosafety, perhaps acceptable for future clinical experimentation. Previously developed constructs express toxic proteins such as Vpr that compromise cell viability in the long-term. Using the new minimal packaging constructs, stable producer cell lines can be generated allowing better characterization and scale-up of vector manufacturing. A number of steps are being taken towards this goal, including the use of inducible expression systems (Kaul *et al.*, 1998; Kafri *et al.*, 1999).

Vectors derived from a non-human lentivirus, the Feline Immunodeficiency Virus (FIV), were also shown to efficiently transduce non-dividing human cells *in vitro* (Poeschla *et al.*, 1998). Recently, a multiple attenuated FIV vector was produced, with a chimeric 5' LTR in the transfer vector and a packaging vector deleted of some accessory genes, and was shown to have an improved gene transfer efficiency (Johnston *et al.*, 1999) FIV-vectors were used to transduce the gene for the transmembrane conductance regulator in cystic fibrosis airway cells *in vitro* (Wang *et al.*, 1999). Another non-primate lentiviral vector, based on the equine infection anemia virus (EIAV), was reported to work in primary cells with efficiency comparable to the HIV vectors (Mitrophanous *et al.*, 1999). These positive results confirm the value of lentiviral vectors for gene transfer. However, the actual safety advantages provided by the non-primate lentiviral vectors remain to be evaluated.

In conclusion, several recent advances in the areas of vector design, safety and function may make lentiviral vectors the preferred tool for gene transfer in areas where other existing systems have demonstrated to be inadequate.

5. REFERENCES

Anderson, W.F. (1998). Human gene therapy. *Nature*, 392 Suppl.: 25-30.

Blomer, U., Naldini, L., Kafri, T., Trono, D., Verma, I.M. and Gage, F.H. (1997).

Highly efficient and sustained gene transfer in adult neurons with a lentivirus vector. *J. Virol.*, 71: 6641-6649.

Bukovsky, A.A., Song, J.P. and Naldini, L. (1999). Interaction of human immunodeficiency virus-derived vectors with wild- type virus in transduced cells. *J. Virol.*, 73: 7087-7092.

Bukrinsky, M.I., Haggerty, S., Dempsey, M.P., Sharova, N., Adzhubel, A., Spitz, L., Lewis, P., Goldfarb, D., Emerman, M. and Stevenson, M. (1993). A nuclear localization signal within HIV-1 matrix protein that governs infection of non-dividing cells [see comments]. *Nature*, 365: 666-669.

Burns, J.C., Friedmann, T., Driever, W., Burrascano, M., and Yee, J.K. (1993). Vesicular stomatitis virus G glycoprotein pseudotyped retroviral vectors: concentration to very high titer and efficient gene transfer into mammalian and nonmammalian cells [see comments]. *Proc. Natl. Acad. Sci. USA*, 90: 8033-8037.

Case, S.S., Price, M.A., Jordan, C.T., Yu, X.J., Wang, L., Bauer, G., Haas, D.L., Xu, D., Stripecke, R., Naldini, L., Kohn, D.B. and Crooks, G.M. (1999). Stable transduction of quiescent CD34(+)CD38(-) human hematopoietic cells by HIV-1-based lentiviral vectors. *Proc. Natl. Acad. Sci. USA*, 96: 2988-2993.

Cullen, B.R. (1998). HIV-1 auxiliary proteins: making connections in a dying cell. *Cell*, 93: 685-692.

Dull, T., Zufferey, R., Kelly, M., Mandel, R.J., Nguyen, M., Trono, D. and Naldini, L. (1998). A third-generation lentivirus vector with a conditional packaging system. *J. Virol.*, *72*: 8463-8471.

Evans, J.T., Kelly, P.F., O'Neill, E. and Garcia, J.V. (1999). Human cord blood CD34+CD38- cell transduction via lentivirus-based gene transfer vectors. *Hum. Gene Ther.*, 10: 1479-1489.

Gallay, P., Swingler, S., Aiken, C. and Trono, D. (1995a). HIV-1 infection of nondividing cells: C-terminal tyrosine phosphorylation of the viral matrix protein is a key regulator. *Cell*, 80: 379-388.

Gallay, P., Swingler, S., Song, J., Bushman, F. and Trono, D. (1995b). HIV nuclear import is governed by the phosphotyrosine-mediated binding of matrix to the core domain of integrase. *Cell*, 83: 569-576.

Gallichan, W.S., Kafri, T., Krahl, T., Verma, I.M. and Sarvetnick, N. (1998). Lentivirus-mediated transduction of islet grafts with interleukin 4 results in sustained gene expression and protection from insulitis. *Hum. Gene Ther.*, 9: 2717-2726.

Giannoukakis, N., Mi, Z., Gambotto, A., Eramo, A., Ricordi, C., Trucco, M. and Robbins, P. (1999). Infection of intact human islets by a lentiviral vector. *Gene Ther.*, 6: 1545-1551.

Goldman, M.J., Lee, P.S., Yang, J.S. and Wilson, J.M. (1997). Lentiviral vectors for gene therapy of cystic fibrosis. *Hum. Gene Ther.*, 8: 2261-2268.

Heinzinger, N.K., Bukinsky, M.I., Haggerty, S.A., Ragland, A.M., Kewalramani, V., Lee, M.A., Gendelman, H.E., Ratner, L., Stevenson, M. and Emerman, M. (1994). The Vpr protein of human immunodeficiency virus type 1 influences nuclear localization of viral nucleic acids in nondividing host cells. *Proc. Natl. Acad. Sci. USA*, 91: 7311-7315.

Johnson, L.G., Olsen, J.C., Naldini, L. and Boucher, R.C. (2000). Pseudotyped human lentiviral vector-mediated gene transfer to airway epithelia in vivo. *Gene Ther.*, in press.

Johnston, J.C., Gasmi, M., Lim, L.F., Elder, J.H., Yee, J.K., Jolly, D.J., Campbell, K.P., Davidson, B.L. and Sauter, S.L. (1999). Minimum requirements for efficient transduction of dividing and nondividing cells by feline immunodeficiency virus vectors. *J. Virol.*, 73: 4991-5000.

Kafri, T., Blomer, U., Peterson, D.A., Gage, F.H., and Verma, I.M. (1997). Sustained expression of genes delivered directly into liver and muscle by lentiviral vectors. *Nat. Genet.*, *17*: 314-317.

Kafri, T., van Praag, H., Ouyang, L., Gage, F.H. and Verma, I.M. (1999). A packaging cell line for lentivirus vectors. *J. Virol.*, 73: 576-584.

Kaul, M., Yu, H., Ron, Y. and Dougherty, J.P. (1998). Regulated lentiviral packaging cell line devoid of most viral cis-acting sequences. *Virology*, 249: 167-174.

Kim, V.N., Mitrophanous, K., Kingsman, S.M. and Kingsman, A.J. (1998). Minimal requirement for a lentivirus vector based on human immunodeficiency virus type 1. *J. Virol.*, 72: 811-816.

Kinoshita, S., Chen, B.K., Kaneshima, H. and Nolan, G.P. (1998). Host control of HIV-1 parasitism in T cells by the nuclear factor of activated T cells. *Cell*, 95: 595-604.

Lewis, P.F. and Emerman, M. (1994). Passage through mitosis is required for onco-retroviruses but not for the human immunodeficiency virus. *J.Virol.*, 68: 510-516.

Miller, A.D., Miller, D.G., Garcia, J.V. and Lynch, C.M. (1993). Use of retroviral vectors for gene transfer and expression. *Methods Enzymol.*, 217: 581-599.

Miller, D.G., Adam, M.A. and Miller, A.D. (1990). Gene transfer by retrovirus vectors occurs only in cells that are actively replicating at the time of infection [published erratum appears in Mol. Cell. Biol., 12: 433, 1992]. *Mol. Cell. Biol.*, 10: 4239-4242.

Mitrophanous, K., Yoon, S., Rohll, J., Patil, D., Wilkes, F., Kim, V., Kingsman, S., Kingsman, A. and Mazarakis, N. (1999). Stable gene transfer to the nervous system using a non-primate lentiviral vector. *Gene Ther.*, 6: 1808-1818.

Miyoshi, H., Takahashi, M., Gage, F.H. and Verma, I.M. (1997). Stable and efficient gene transfer into the retina using an HIV-based lentiviral vector. *Proc. Natl. Acad. Sci. USA*, 94: 10319-10323.

Miyoshi, H., Blomer, U., Takahashi, M., Gage, F.H. and Verma, I.M. (1998). Development of a self-inactivating lentivirus vector. *J. Virol.*, 72: 8150-8157.

Miyoshi, H., Smith, K.A., Mosier, D.E., Verma, I.M. and Torbett, B.E. (1999). Transduction of human CD34+ cells that mediate long-term engraftment of NOD/SCID mice by HIV vectors. *Science*, 283: 682-686.

Naldini, L., Blomer, U., Gage, F.H., Trono, D. and Verma, I.M. (1996a). Efficient transfer, integration, and sustained long-term expression of the transgene in adult rat brains injected with a lentiviral vector. *Proc. Natl. Acad. Sci. USA*, 93: 11382-11388.

Naldini, L., Blomer, U., Gallay, P., Ory, D., Mulligan, R., Gage, F.H., Verma, I.M. and Trono, D. (1996b). In vivo gene delivery and stable transduction of nondividing cells by a lentiviral vector [see comments]. *Science*, 272: 263-267.

Park, F., Ohashi, K., Chiu, W., Naldini, L. and Kay, M.A. (2000). Efficient lentiviral transduction of liver requires cell cycling in vivo. *Nat. Genet.*, 24: 49-52.

Poeschla, E., Corbeau, P. and Wong-Staal, F. (1996). Development of HIV vectors for anti-HIV gene therapy. *Proc. Natl. Acad. Sci. USA*, 93: 11395-11399.

Poeschla, E.M., Wong-Staal, F. and Looney, D.J. (1998). Efficient transduction of nondividing human cells by feline immunodeficiency virus lentiviral vectors. *Nat. Med.*, 4: 354-357.

Popov, S., Rexach, M., Zybarth, G., Reiling, N., Lee, M.A., Ratner, L., Lane, C.M., Moore, M.S., Blobel, G. and Bukrinsky, M. (1998). Viral protein R regulates nuclear import of the HIV-1 pre-integration complex. *EMBO J.*, 17: 909-917.

Rebolledo, M.A., Krogstad, P., Chen, F., Shannon, K.M., and Klitzner, T.S. (1998). Infection of human fetal cardiac myocytes by a human immunodeficiency virus-1-derived vector. *Circ. Res.*, 83: 738-742.

Reiser, J., Harmison, G., Kluepfel-Stahl, S., Brady, R.O., Karlsson, S. and Schubert, M. (1996). Transduction of nondividing cells using pseudotyped defective high- titer HIV type 1 particles. *Proc. Natl. Acad. Sci. USA*, 93: 15266-15271.

Roe, T., Reynolds, T.C., Yu, G. and Brown, P.O. (1993). Integration of murine leukemia virus DNA depends on mitosis. *EMBO J.*, 12: 2099-2108.

Uchida, N., Sutton, R.E., Friera, A.M., He, D., Reitsma, M.J., Chang, W.C., Veres, G., Scollay, R. and Weissman, I.L. (1998). HIV, but not murine leukemia virus, vectors mediate high efficiency gene transfer into freshly isolated G0/G1 human hematopoietic stem cells. *Proc. Natl. Acad. Sci. USA*, 95: 11939-11944.

Unutmaz, D., KewalRamani, V.N., Marmon, S. and Littman, D.R. (1999). Cytokine signals are sufficient for HIV-1 infection of resting human T lymphocytes. *J. Exp. Med.*, 189: 1735-1746.

Verma, I.M. and Somia, N. (1997). Gene therapy -- promises, problems and prospects [news]. *Nature*, 389: 239-242.

Wang, G., Slepushkin, V., Zabner, J., Keshavjee, S., Johnston, J.C., Sauter, S.L., Jolly, D.J., Dubensky, T.W.J., Davidson, B.L. and McCray, P.B.J. (1999). Feline immunodeficiency virus vectors persistently transduce nondividing airway epithelia and correct the cystic fibrosis defect [see comments]. *J. Clin. Invest.*, 104: R55-R62.

Wei, P., Garber, M.E., Fang, S.M., Fischer, W.H. and Jones, K.A. (1998). A novel CDK9-associated C-type cyclin interacts directly with HIV-1 Tat and mediates its high-affinity, loop-specific binding to TAR RNA. *Cell*, 92: 451-462.

Zhang, Z., Schuler, T., Zupancic, M., Wietgrefe, S., Staskus, K.A., Reimann, K.A., Reinhart, T.A., Rogan, M., Cavert, W., Miller, C.J., Veazey, R.S., Notermans, D., Little, S., Danner, S.A., Richman, D.D., Havlir, D., Wong, J., Jordan, H.L., Schacker, T.W., Racz, P., Tenner-Racz, K., Letvin, N.L., Wolinsky, S. and Haase, A.T. (1999). Sexual transmission and propagation of SIV and HIV in resting and activated CD4+ T cells. *Science*, 286: 1353-1357.

Zufferey, R., Nagy, D., Mandel, R.J., Naldini, L. and Trono, D. (1997). Multiply attenuated lentiviral vector achieves efficient gene delivery in vivo. *Nat. Biotechnol.*, 15: 871-875.

Zufferey, R., Dull, T., Mandel, R.J., Bukovsky, A., Quiroz, D., Naldini, L. and Trono,

D. (1998). Self-inactivating lentivirus vector for safe and efficient in vivo gene delivery. *J. Virol.*, 72: 9873-9880.

6. SUMMARY — Gene therapy is considered to be a promising approach to the treatment of inherited and acquired diseases. However, knowledge in gene delivery technology still remains a crucial limiting factor, and more effective gene delivery systems must be developed to overcome the inadequacies of existing gene transfer systems, particularly in the area of direct *in vivo* delivery (Verma and Somia, 1997; Anderson, 1998).

An ideal vector should have the ability to integrate and sustain expression of the transgene while eliciting no immunological and pathological consequences to the recipient. Up to now, vectors derived from lentiviruses such as Human Immunodeficiency Viruses (HIV) came close to these properties in gene marking experiments conducted in rodents. However, several issues have been raised pertaining to the actual biosafety of vectors derived from a human pathogenic virus. Here, we review recent progress in the design, biosafety and functional analysis of lentiviral vectors.

Key Words: lentiviral vectors; vector biosafety; *in vivo* gene transfer; transduction of hemopoietic stem cells

Progress in Gene Therapy: Basic and Clinical Frontiers, pp. 67-84
R. Bertolotti *et al.* (Eds)

Helper dependent adenoviral vectors — improved safety and expression

Manal A. Morsy* and C. Thomas Caskey

Merck Research Laboratories, Department of Human Genetics, WP26A-3000, Sumneytown Pike, West Point, PA 19486, USA

Table of Contents

1. INTRODUCTION

Gene therapy is a rapidly evolving technology for therapeutic intervention. Since the first human clinical trial has begun in 1989, there has been more than 100 gene transfer protocols approved by the FDA in the United States alone. This new approach involves delivery and expression of nucleic acid in a target cell to complement a genetic defect or deliver a new protein. The diversity in the nature of the expressed product is wide and

* Corresponding author. E-mail: manal_morsy@merck.com

ranges from expressing cellular enzymes, cellular or circulating proteins, secreted hormones, cytokines or growth factors, to immunogens, ribozymes, or antisense oligonucleotides. The main components of a gene therapy agent are the vector or delivery vehicle and the expression cassette, which is composed of the gene(s) of interest and the promoter elements controlling expression. Recently, intense research has evolved and is ongoing to identify the most suitable vector(s) for gene delivery. Viral and non-viral vectors are continuously being modified, examined and compared for safety, persistence and efficacy in delivery and mediation of gene expression. Among the most extensively studied viral vectors are the retroviral, adenoviral and adeno-associated, and among the non-viral vectors are the DNA-lipid complexes (cationic liposomes), DNA-polylysine conjugates or delivery of naked DNA (Morsy *et al.*, 1993b).

Over the past few years, adenoviruses (Ad) have taken a forefront position as gene delivery vehicles as a result of their numerous advantageous features. Their popularity as recombinant vectors is largely due to the successful and safe immunization of millions of US military recruits with enteric coated Ad4 and Ad7 to prevent against acute respiratory disease (ARD) outbreaks (Gaydos and Gaydos, 1995). In addition, the Ad genome is well characterized, and easily manipulated. Recombinant Ad vectors have been generated, with different deletions in their genome to render the vector replication deficient and to allow for insertion of foreign DNA sequences. Current generations of Ad vectors have insert capacity ranging from 7-9 Kb, and are deleted in one or more combinations of the early genes. Recombinant viruses are stable and stocks can be concentrated to titers as high as 10^{12} plaque forming units /ml. The virus has a broad cellular host range, its up take is not restricted to dividing cells, has natural tropism to liver, lung and intestine which is dictated by route of delivery (vascular, inhalation and oral, respectively) and persists as an episome in infected cells. It encodes for a cascade of polypeptides including the capsid structural proteins, the hexones, pentons (fiber and penton base). Recent studies have shown Ad vectors to be among the most efficient gene transfer vehicles for both *in vitro* and *in vivo* delivery. The general utilization, however, of current Ad vectors for many gene therapy applications is limited by the transient nature of transgene expression observed (Muzzin *et al.*, 1996; Stratford-Perricaudet *et al.*, 1990; Herz and Gerard, 1993; Morsy *et al.*, 1993b, 1993a and 1996). Several factors have been shown to contribute to and modulate the duration of Ad-mediated gene expression and the underlying immunogenicity of these vectors. These factors include "leaky" viral protein expression and/or the immunogenicity of the transgene that is delivered (Yang, 1995; Gahery-Segard *et al.*, 1997; Kaplan *et al.*, 1997; Tripathy *et al.*, 1996; Worgall *et al.*, 1997). The development of Ad vectors that are deleted in all viral protein-coding sequences offers the prospect of a potentially safer, less immunogenic vector with an insert capacity of up to approximately 37 kb (Mitani *et al.*, 1995; Kochanek *et al.*, 1996; Clemens *et al.*, 1996; Fisher *et al.*, 1996; Kumar-Singh and Chamberlain, 1996; Hardy *et al.*, 1997; Lieber *et al.*, 1996; Parks *et al.*, 1996; Schiedner *et al.*, 1998; Haecker *et al.*, 1996). This vector requires viral regulatory and structural proteins which, when supplied in *trans*, can support packaging and rescue, and is thus named helper-dependent (HD) (Parks *et al.*, 1996). It is noteworthy however, to emphasize that such modifications would modulate the toxicity and enhance the safety of the HD vehicle itself, yet may or may not have an effect on the impact or extent of transgene immunogenicity.

In the process of developing a safe gene delivery vehicle, it was important to select an animal model in which the developed vehicle can be assessed.

2. SELECTION OF AN ANIMAL MODEL FOR SCREENING AND COMPARING GENE DELIVERY VECTORS: ob/ob MOUSE MODEL

The appropriateness of animal model selection in research is pivotal to rapid, accurate and useful data collection.

The recent cloning of the ob gene identified the defect in the obese mouse model ob/ob to be a nonsense mutation at codon 105 (Halaas *et al.*, 1995). Leptin is undetectable in plasma of these animals and as a consequences the mice are extremely obese. They are hyperphagic, hyperglycemic, hypoactive and have elevated plasma insulin levels. Homozygotes increase rapidly in body weight and are recognized as early as 4 weeks after birth. Treatment of an ob/ob mouse with recombinant leptin results in a 10-20 % reduction in body weight that is a consequence of both decreased food intake and increased activity. In addition, leptin reduces the hyperglycemia and the hyperinsulinemia observed in the ob/ob animal. Significantly, the ability of leptin to reduce weight is also observed in wild type mice and in mice with diet induced obesity (Halaas *et al.*, 1995; Ingalls *et al.*, 1950; Campfield *et al.*, 1995; Charlton, 1984). We and others have previously shown that delivery of the leptin gene by first-generation Ad vectors may substitute for daily recombinant leptin protein treatment, although the effects were transient in both lean and ob/ob treated mice (Morsy *et al.*, 1998b; Muzzin *et al.*, 1996).

3. EVALUATION OF THE MODEL

The number of non-invasive parameters available for evaluating the ob/ob model rendered it very attractive. We have compared the efficiencies of 1st generation Ad vectors expressing mouse or human leptin, recombinant leptin protein and mouse or human leptin expression via plasmid DNA, in phenotypically correcting the ob/ob mouse defect as a first pass screening test for the validity of the model (Morsy *et al.*, 1998b).

The parameters used to evaluate the therapeutic effect of Ad-mediated leptin delivery, include daily monitoring of food intake and weight loss in addition to measurement of leptin, glucose and insulin levels in the sera of treated animals. Weight loss and appetite suppression, however, can be non-specific findings associated with Ad-mediated toxicity. At least 0.1-1 x 10^9 *pfu* of Ad vector were delivered by tail vein injection, to the different mouse treatment groups. Weight loss and reduction in food intake were not seen in animals receiving control buffer or a control Ad vector expressing β-galactosidase (Fig. 1). These findings were consistent and specifically associated with leptin delivery and were only observed in animals receiving

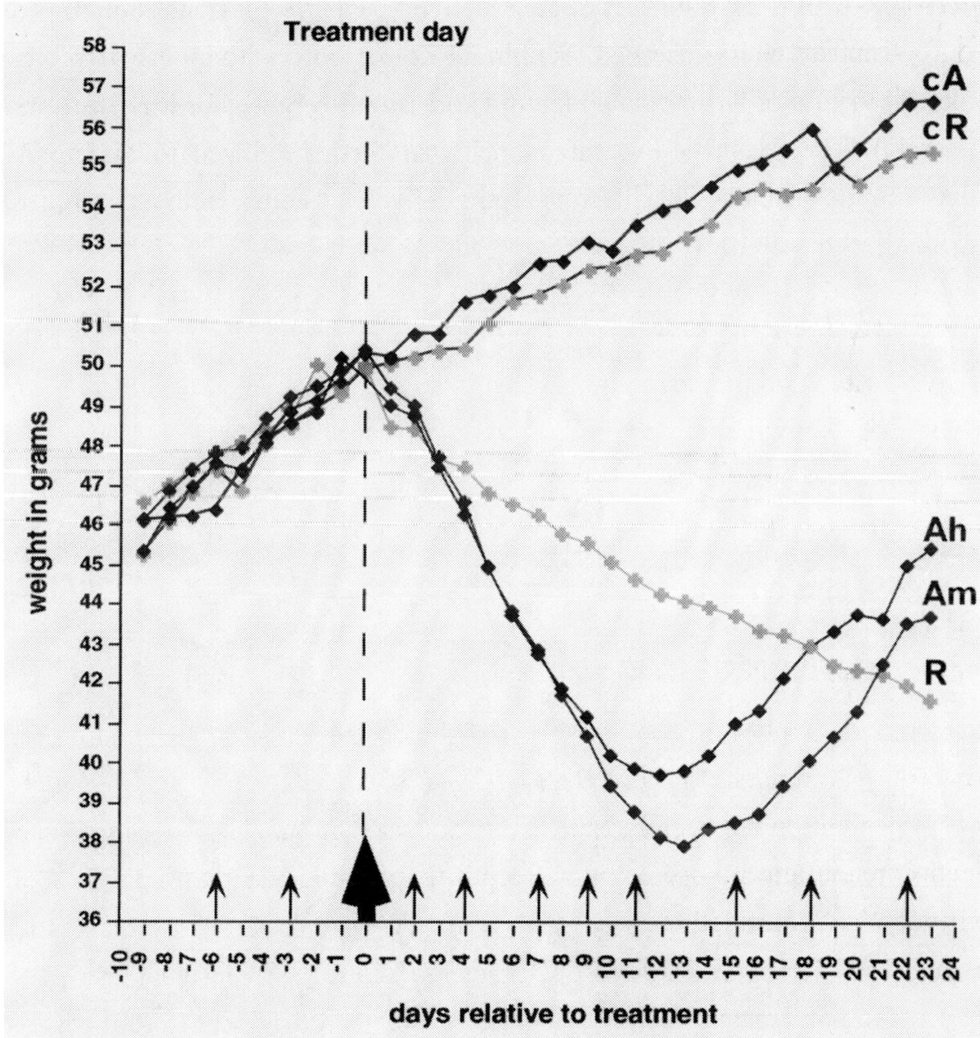

Figure 1. Effect of the different treatments on the weight of ob/ob mice. Data points represent mean body weight, calculated from daily measurements. The (thick) arrow marks day zero. Ad-human leptin (Ah) and Ad-murine leptin (Am) are recombinant E1 deleted, replication deficient, 1st generation Ad vectors expressing human or mouse leptin, respectively. R: recombinant leptin protein; c-A: Ad buffer contol; c-R: recombinant protein buffer control; thin arrows: serum collection time points.

leptin expressing Ad vectors or daily intra-peritoneal injections of recombinant leptin protein (Fig. 1). Furthermore, the results were reproducible, all animals treated with Ad-vectors expressing human or mouse leptin lost weight. Finally, the model was very sensitive to leptin delivery, within 24 hours from gene delivery, weight loss and suppression in food intake were observed. Within 24 hours from withdrawal of protein therapy or loss of Ad-mediated leptin expression, weight gain was measured (Fig. 2). The sensitivity of the model was further utilized to distinguish between the efficiency of different leptin delivery systems. Both recombinant

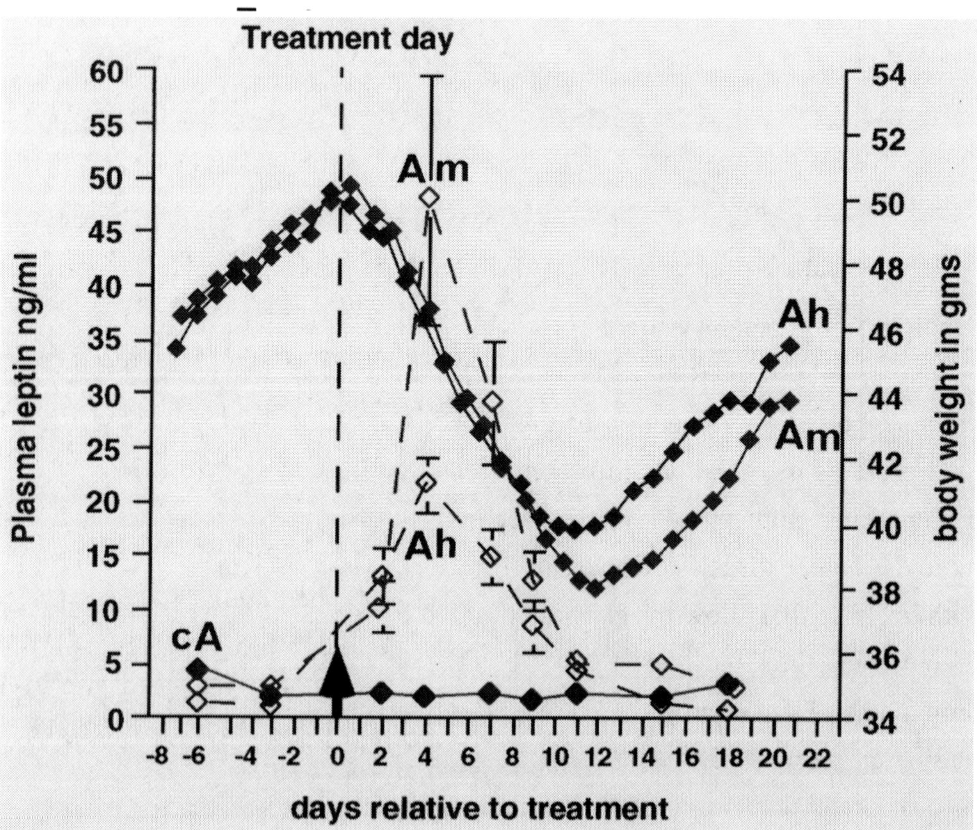

Figure 2. Plasma leptin levels in ob/ob Ad-leptin-treated or control-treated mice. Mean ± SEM values of leptin levels are represented by opened symbols (Am and Ah); control level is shown for mice treated with Ad buffer (cA) . The leptin data in this figure are overlapped with body weight data (filled symbols; Ah and Am) to show the time-dependent relationship between leptin expression and weight reduction. Ah: Ad-human leptin; Am: Ad-murine leptin; c-A: Ad buffer contol.

leptin protein and plasmid DNA coding for leptin were compared to Ad-mediated leptin delivery. In order of efficiency, Ad vectors (even though 1st generation were used in these studies) were the most efficient, correcting the phenotype at a significantly faster rate than daily recombinant leptin protein delivery followed by plasmid DNA delivery (data not shown). The out come of the study was characterized by 1) a specificity of effect, 2) a consistency of findings, 3) a reproducibility of data, and 4) a sensitivity of parameters. Factors, which not only supported the high efficiency of the Ad-mediated gene delivery system in this model, but also validated the model. This model was then used to evaluate the HD vector system as it compares to the 1st generation recombinant Ad virus (Morsy *et al.*, 1998a).

4. GENERATION OF THE HELPER DEPENDENT ADENO-VIRAL (HD) VECTOR

A modified Ad vector has been generated such that it is completely devoid of viral protein encoding sequences (Morsy *et al.*, 1998a). The new vector contains the ITR and packaging sequences, with an insert capacity of up to 37 kb. The propagation of this vector requires supplementation of the viral proteins in *trans*, which presently are supplied by co-propagation of a helper virus. Both viruses can be separated on a cesium gradient. Further more the helper virus is crippled by flanking the packaging signal sequence with lox sites that allow the excision of the intervening DNA, in the presence of cre, and thus the capacity of the helper virus to rescue itself as it propagates. This vector has been previously used to clone the full-length murine dystrophin cDNA (13.8 kb) under the control of the murine muscle specific creatinine kinease (6.5 kb) and a CMV promoter - *E.coli* LacZ gene cassette (4.6 kb) (Kochanek *et al.*, 1996). This completely debilitated, recombinant virus propagated efficiently in 293 cells (which complement E1 functions) in the presence of a helper mutant Ad virus (SV5). The yield after cesium chloride density gradient banding was ~ 5 x 10^9 *pfu* obtained from

1.4 x 10^8 293 cells with about 1% contamination of helper virus as determined by a plaque forming unit (*pfu*) assay on 293 cells and by southern blot analysis. The recombinant HD vector was efficient in co-expressing the dystrophin protein and β-gal in primary myoblasts derived from mdx mouse (a genetic and biochemical model for human DMD disease) and *in vivo* (Kochanek *et al.*, 1996; Clemens *et al.*, 1996).

In a more recent study, we delivered the leptin cDNA using the HD virus, testing the hypothesis that elimination of the viral protein coding sequences would diminish the vector's cellular immunogenicity and toxicity, and hence support its longevity *in vivo* (Morsy *et al.*, 1998a). Since both the viral proteins and the transgene were factors implicated in the cellular immunogenicity of recombinant Ad viruses, we designed experiments to compare the HD and Ad vectors in ob/ob mice that are naive to leptin (in which the protein is potentially immunogenic), as well as in lean mice that normally express leptin.

In this study, we showed that HD-leptin provided greater safety as reflected by absence of liver toxicity, cellular infiltrates, extended longevity of gene expression and stability of vector DNA in livers of treated mice over that observed with 1st generation Ad-mediated leptin treatment.

a) Safety of HD compared to first generation Ad vectors

Mice were treated with a single tail intravenous infusion of 1-2 x 10^{11} particles of either HD-leptin, Ad-leptin, control Ad-β-gal vector or an equal volume of control buffer. Toxicity was evaluated by measuring the levels of released liver enzymes in sera and by studying the histopathology of liver sections obtained from treated animals at successive intervals post treatment. Figure 3 shows the levels of aspartate aminotransferase (AST) and alanine aminotransferase (ALT) in the sera of lean mice at one, two and four weeks post-treatment (similar results were observed in treated ob/ob mice - data not shown). Liver toxicity, as reflected by the significant elevation in AST and ALT serum levels over basal control levels, was observed only in mice treated

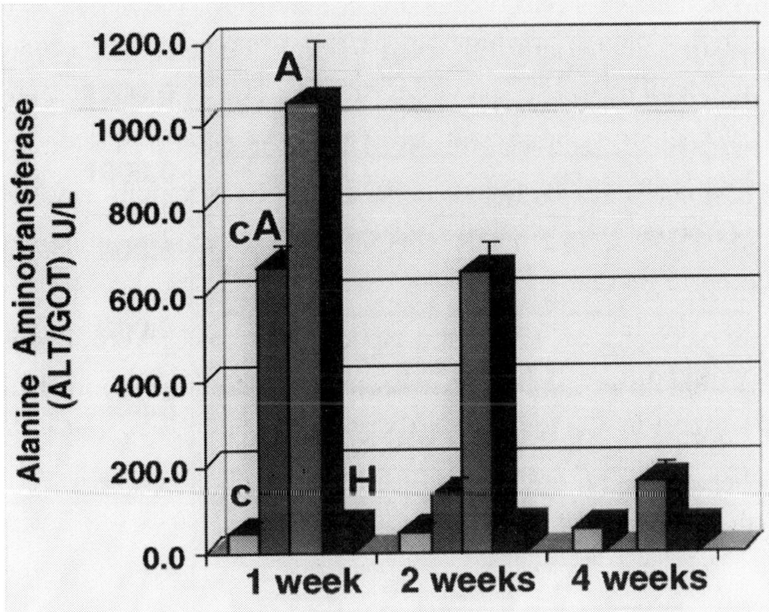

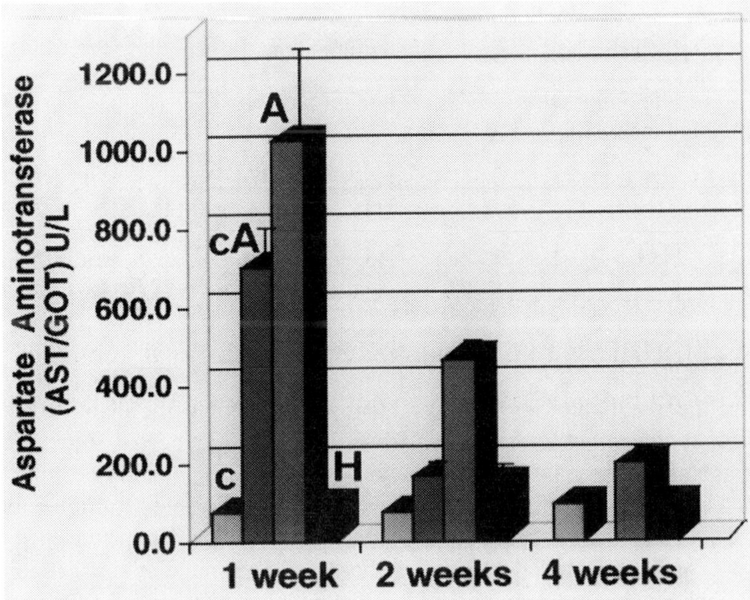

Figure 3. Aspartate aminotransferase (AST) and alanine aminotransferase (ALT) levels in the sera of lean control and treated mice. Mice were treated with Ad-β-gal (cA), Ad-leptin (A), HD-leptin (H) or dialysis buffer (c controls), and their serum AST and ALT levels were determined at one, two and four weeks post-treatment.

with Ad-β-gal and Ad-leptin (center lanes), but not HD-leptin (right lane). Ad-vector-associated toxicity observed in both the lean and ob/ob treated mice was most significant at one week, was present but to a less significant extent at two weeks, and was resolved by four weeks post-treatment. In contrast, HD-treatment was not associated with liver toxicity as reflected by the AST and ALT serum levels which were essentially indistinguishable from controls.

Liver sections of HD-leptin-treated lean mice were histologically indistinguishable from control liver sections at all time points tested post treatment (table 1). In contrast, Ad-leptin and Ad-13-gal treated mice displayed hepatic pathology (hepatopathy) throughout the first 1-2 weeks post-treatment, which resolved by week 4.

Table 1. Liver toxicity: histopathology*

Lean liver	1 week	2 weeks	4 weeks
⇒Untreated	Not remarkable	Not remarkable	Not remarkable
⇒Ad-βgal	Hepatopathy	Hepathopathy very slight	Not remarkable
⇒Ad-leptin	Hepathopathy	Hepathopathy very slight	Individual cell necrosis (trace)
⇒HD-leptin	Not remarkable	Not remarkable	Not remarkable

* Summary of the histopathology findings. Histopathy refers to the displayed degenerative hepatic pathology found in livers of Ad-leptin and Ad-β-gal treated mice. This hepatic pathology was characterized by foci of round cell infiltration composed almost entirely (>98%) of T-cells, individual liver cell necrosis, increased liver cell mitotic activity, and dissociation of hepatic cords. At two weeks post-treatment, Ad-leptin treated mice display a similar, but less pronounced hepatic pathology. The cellular infiltration observed resolved by 4 weeks post-treatment; there was almost an absence of lesions in the Ad-leptin treated mice, with only a trace of individual cell death present, which is within normal ranges. Examination of liver sections obtained from ob/ob mice reflected similar Ad-vector associated histopathology. Liver histology was indistinguish- able between HD-leptin-treated and untreated control mice.

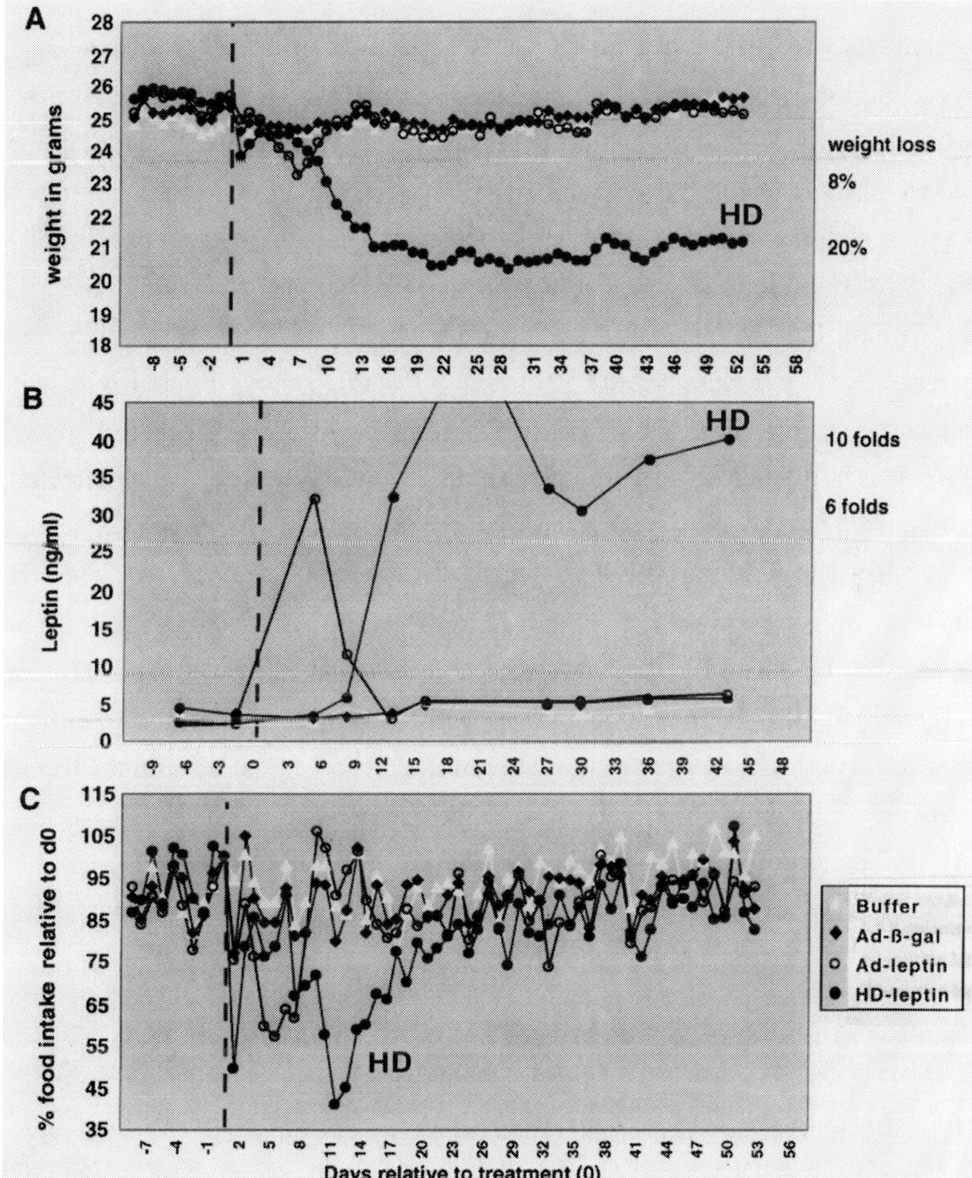

Figure 4. HD-leptin and Ad-leptin effects in lean mice. Animals were injected via the tail vein with a single dose of 1-2 x 10^{11} particles of HD-leptin (n=5), Ad-leptin (n=10), Ad-β-gal (n=10), or the equivalent volume of dialysis buffer (n=10). The time course shows: **A)** weight (grams) and the percent of maximum weight loss observed in Ad-leptin and HD-leptin treated mice (8% and 20%, respectively); **B)** serum leptin levels, collected 2-3 times weekly (ng/ml) and the maximum fold increase in serum leptin levels above basal observed in the Ad-leptin and HD-leptin treated mice (6 fold and 10 fold, respectively); **C)** the percentage of food intake relative to untreated control mice. The dashed line marks day 0 relative to day of injection. HD: mice injected with HD-leptin.

b) Efficacy of HD compared to first generation AD vectors

In the lean mice, treatment with Ad-leptin resulted in weight loss that lasted for only 7-10 days which was associated with a transient increase in serum leptin levels (Fig. 4 A and B). In contrast, treatment with HD-leptin resulted in approximately 20% weight loss that persisted at least two months and high serum leptin levels (6- to 10-fold over background) (Fig. 4 A and B). Weight loss in HD-leptin-treated mice was associated with satiety that persisted over a longer period (2-3 weeks) than in those treated with Ad-leptin (5-7 days) (Fig. 4 C). Vector DNA in the livers of Ad-leptin treated mice was rapidly lost and fewer than 0.2 copies per cell were detected, compared to 1-2 copies per cell following HD-leptin treatment at 8 weeks post-injection (data not shown). These effects can be correlated with the duration of gene expression obtained with these two vector types. Gene expression mediated by Ad-leptin was transient and almost undetectable as early as 1 week post treatment as seen by northern blot analysis of total liver RNA, whereas that mediated by HD-leptin persisted for at least eight weeks (data not shown).

The ob/ob mice are naive to leptin and thus transgene immunogenicity is not an unexpected finding. In these animals, similar to what was

Figure 5. HD-leptin and Ad-leptin effects in ob/ob mice. Essentially as described in Fig. 4, animals were injected in the tail vein with a single dose of 1-2 x 10^{11} particles of HD-leptin (n=5), Ad-leptin (n=10), Ad-β-gal (n=10), or the equivalent volume of dialysis buffer (n=10). Lean control values are plotted for comparison. The time course shows: **A)** weight (grams) and the percent of maximum weight loss observed in Ad-leptin and HD-leptin treated mice (20% and 60%, respectively); **B)** Phenotypic correction of HD-leptin-treated ob/ob mice. On the left is a representative ob/ob mouse treated with HD-leptin at day 54 post-treatment, next to a litter mate treated with Ad-leptin. The Ad-leptin-treated mouse initially lost weight during the first two weeks following the treatment, and subsequently gained weight. At 54 days post-Ad-leptin treatment, ob/ob mice are indistinguishable from untreated ob/ob control litter mates, whereas HD-leptin-treated mice remained indistinguishable from untreated lean control mice. Untreated ob/ob and lean control mice are shown for comparison as labeled. **C)** serum leptin levels, collected 2-3 times weekly (ng/ml); **D)** the percentage of food intake relative to untreated control mice. The dashed line marks day 0 relative to day of injection. HD: mice injected with HD-leptin.

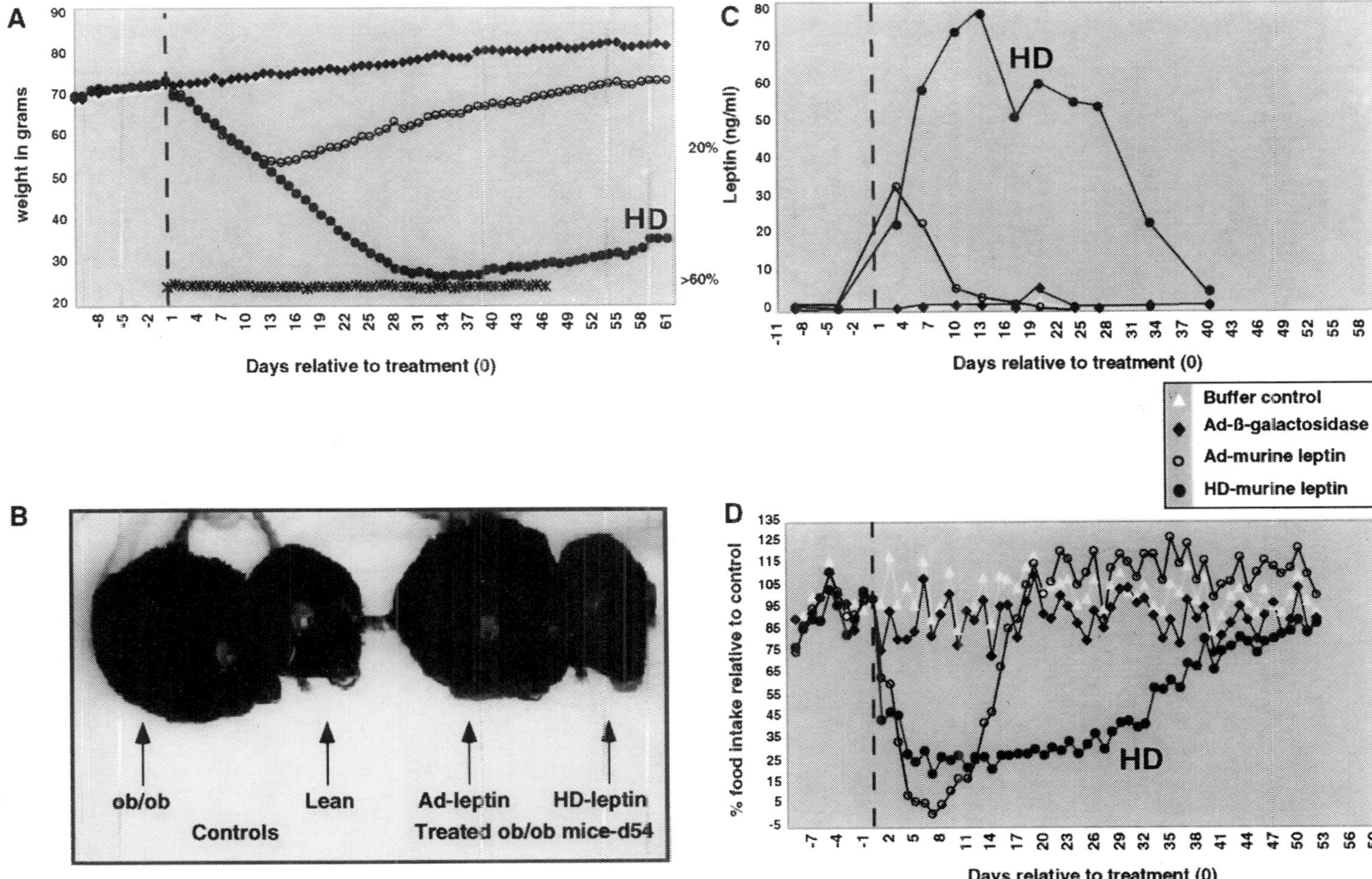
A
weight in grams
Days relative to treatment (0)
HD
20%
>60%
B
ob/ob
Lean
Ad-leptin
HD-leptin
Controls
Treated ob/ob mice-d54
C
Leptin (ng/ml)
HD
Days relative to treatment (0)
Buffer control
Ad-β-galactosidase
Ad-murine leptin
HD-murine leptin
D
% food intake relative to control
HD
Days relative to treatment (0)

observed in the lean mice, HD-leptin was found to be more effective than the first-generation Ad-leptin vector. In the ob/ob mice treated with Ad-leptin, transient body weight loss of ~25%, followed by weight gain, two weeks after treatment was observed (Fig. 5 A and B). Associated, serum levels of leptin increased only for a short period during the first 4 days of treatment, returning to baseline levels within ten days post-injection (Fig. 5 C). Similar to the results obtained in lean mice, the Ad-leptin vector DNA (data not shown) was also rapidly lost (< 0.2 copies per cell were detected by 2 weeks post treatment, and undetectable by 8). In contrast, the ob/ob HD-leptin-keated mice had increased serum leptin levels up to ~15 days post-treatment, after which the levels gradually dropped to baseline over the subsequent 25 days (Fig. 5 C). The initial rise in leptin levels correlated with rapid weight reduction resulting in >60% weight loss (reaching normal lean weight) by one month (Fig. 5B). Weight loss was maintained for a period of 6-7 weeks post-treatment. The overall HD-leptin-mediated prolonged effect was also reflected in the accompanying phenotypic correction, which lasted longer than that seen in litter mates treated with Ad-leptin (6-7 versus 2-3 weeks) (Fig. 5 B). As leptin levels dropped to baseline, a gradual increase in body weight was observed (Fig. 5 A). Satiety was observed in association with increased leptin levels, and appetite suppression was sustained for a longer period (~1 month) compared to the short transient effect induced by Ad-leptin (~10 days) (Fig. 5 D). Leptin-specific antibodies were detected in the sera of ob/ob Ad-leptin- and HD-leptin-treated mice (data not shown) suggesting immunogenicity of leptin in these naive animals. Results of southern blot analysis showed greater stability of HD-vector DNA over Ad-vector DNA in livers of ob/ob treated mice compared at similar time points, the analysis revealed eventual loss of the HD-vector DNA over the 8 week time interval (data not shown). Approximately 75% less vector DNA was detected in the livers of HD-leptin-treated ob/ob mice at 4 and 8 weeks post-treatment compared to the persistent levels found in the livers of HD-leptin-treated lean litter mates at similar time points (data not shown). Gene expression in ob/ob Ad-leptin-treated mice correlated with the DNA findings, RNA levels were below the sensitivity level of detection at one week post-treatment, whereas in HD-leptin-treated mice, gene expression was detected up to four weeks post-injection and was undetectable at eight weeks (data not shown).

5. CONCLUSION

The leptin model used in these studies provided a very instructive animal model to investigate the influence of both vector design and transgene product on the duration of expression after gene transfer. The HD-vector system is a significant advance over existing Ad vectors with regards to safety, vector-mediated immunogenicity and insert capacity (up to 37kb). In addition to the gain of these valuable properties, the HD-vectors have not lost the features that contributed to the general attractiveness of Ad vectors which include: 1) efficient *in vivo* gene delivery, and 2) high titer production. This system has come a long way in terms of development and ease of vector preparation and purification. Several studies involving the development of helper-dependent vectors were hindered by the complexity of the system (Mitani *et al.*, 1995; Kochanek *et al.*, 1996; Kumar-Singh and Chamberlain, 1996; Hardy *et al.*, 1997; Lieber *et al.*, 1996). The characterization of size requirements for efficient packaging and the generation of crippled helper viruses greatly enhanced the prospects of these new vectors in becoming a promising tool for gene delivery (Parks and Graham, 1997; Parks *et al.*, 1996). Further modifications and fine tuning are required to convert the HD vector system to an industrially scaleable system for clinical utility.

6. REFERENCES

Campfield, L.A., Smith, F.J., Guisez, Y., Devos, R. and Burn, P. (1995). Recombinant mouse OB protein: evidence for a peripheral signal linking adiposity and central neural networks. *Science*, 269: 546-549.

Charlton, H.M. (1984). Mouse mutants as models in endocrine research. *Q. J. Exp. Physiol.*, 69: 655-676.

Clemens, P.R., Kochanek, S., Sunada, Y., Chan, S., Chen, H.H., Campbell, K.P. and Caskey, C.T. (1996). *In vivo* muscle gene transfer of full-length dystrophin with an adenoviral vector that lacks all viral genes. *Gene Ther.*, 3: 965-972.

Fisher, K.J., Choi, H., Burda, J., Chen, S.J. and Wilson, J.M. (1996). Recombinant adenovirus deleted of all viral genes for gene therapy of cystic fibrosis. *Virology*, 217: 11-22.

Gahery-Segard, H., Juillard, V., Gaston J., Lengagne, R., Pavirani, A., Boulanger, P. and Guillet, J.G. (1997). Humoral immune response to the capsid components of recombinant adenoviruses: routes of immunization modulate virus-induced Ig subclass shifts. *Eur. J. Immunol.*, 27: 653-659.

Gaydos, C.A. and Gaydos, J.C. (1995). Adenovirus vaccines in the U.S. military. *Mil. Med.*, 160: 300-304.

Haecker, S.E., Stedman, H.H., Balice-Gordon, R.J., Smith, D.B., Greelish, J.P., Mitchell, M.A., Wells, A., Sweeney, H.L. and Wilson, J.M. (1996). *In vivo* expression of full-length human dystrophin from adenoviral vectors deleted of all viral genes. *Hum. Gen. Ther.*, 7: 1907-1914.

Halaas, J.L., Gajiwala, K.S., Maffei, M., Cohen, S.L., Chait, B.T., Rabinowitz, D., Lallone, R.L., Burley, S.K. and Friedman, J.M. (1995). Weight-reducing effects of the plasma protein encoded by the obese gene. *Science*, 269, 543-546.

Hardy, S., Kitamura, M., Harris-Stansil, T., Dai, Y. and Phipps, M.L. (1997). Construction of adenovirus vectors through Cre-lox recombination. *J. Virol.*, 71: 1842-1849.

Herz, J. and Gerard, R.D. (1993). Adenovirus-mediated transfer of low density lipoprotein receptor gene acutely accelerates cholesterol clearance in normal mice. *Proc. Natl. Acad. Sci. USA*, 90: 2812-2816.

Ingalls, A., Dickie, M. and Snell, G. (1950). *J Hered.*, 41: 317-318.

Kaplan, J.M., Armentano, D., Sparer, T.E., Wynn, S.G., Peterson, P.A., Wadsworth, S.C., Couture, K.K., Pennington, S.E., St George, J.A., Gooding, L.R. and Smith, A.E. (1997). Characterization of factors involved in modulating persistence of transgene expression from recombinant adenovirus in the mouse lung. *Hum. Gene Ther.*, 8: 45-56.

Kochanek, S., Clemens, P.R., Mitani, K., Chen, H.H., Chan, S. and Caskey, C.T. (1996). A new adenoviral vector: replacement of all viral coding sequences with 28 kb of DNA independently expressing both full-length dystrophin and beta-galactosidase. *Proc. Natl. Acad. Sci USA*, 93: 5731-5736.

Kumar-Singh, R. and Chamberlain, J.S. (1996). Encapsidated adenovirus minichromosomes allow delivery and expression of a 14 kb dystrophin cDNA to muscle cells. *Hum. Mol. Genet.*, 5: 913-921.

Lieber, A., He, C.Y., Kirillova, I. and Kay, M.A. (1996). Recombinant adenoviruses with large deletions generated by Cre-mediated excision exhibit different biological properties compared with first-generation vectors *in vitro* and *in vivo*. J. Virol., 70: 8944-8960.

Mitani, K., Graham, F.L., Caskey, C.T. and Kochanek, S. (1995). Rescue, propagation, and partial purification of a helper virus-dependent adenovirus vector. *Proc. Natl. Acad. Sci. USA*, 92: 3854-3858.

Morsy, M.A., Alford, E.L., Bett, A., Graham, F.L. and Caskey, C.T. (1993a). Efficient adenoviral-mediated ornithine transcarbamylase expression in deficient mouse and human hepatocytes. *J. Clin. Invest.*, 92: 1580- 1586.

Morsy, M.A., Mitani, K., Clemens, P. and Caskey, C.T. (1993b). Progress toward human gene therapy. *JAMA*, 270: 2338-2345.

Morsy, M.A., Zhao, J.Z., Warman, A.W., O'Brien, W.E., Graham, F.L. and Caskey, C.T. (1996). Patient selection may affect gene therapy success. Dominant negative effects observed for ornithine transcarbamylase in mouse and human hepatocytes. *J. Clin. Invest.*, 97: 826-832.

Morsy, M.A., Gu, M., Motzel, S., Zhao, J., Lin, J., Qin, S., Allen, H., Franlin, L., Parks, R., Graham, F., Kochanek, S., Bett, A. and Caskey, C.T. (1998a). An adenoviral vector deleted for all viral coding sequences results in enhanced safety and extended expression of a leptin transgene. *Proc. Natl. Acad. Sci. USA*, 95. 7866-7871.

Morsy, M.A., Gu, M., Zhao, J.Z., Holder, D.J., Rogers, I.T., Pouch, W., Motzel, S.L., Klein, H.J., Gupta, S.K., Liang, X., Tota, M.R., Rosenblum, C.I. and Caskey, C.T. (1998b). Leptin gene therapy and daily protein administration: a comparative study in the ob/ob mouse. *Gene Ther.*, 5: 8-18.

Muzzin, P., Eisensmith, R.C., Copeland, K.C. and Woo, S.L.C. (1996). Correction of obesity and diabetes in genetically obese mice by leptin gene therapy. *Proc. Nat. Acad. Sci. USA*, 93: 14804-14808.

Parks, R.J., Chen, L., Anton, M., Sankar, U., Rudnicki, M.A. and Graham, F.L. (1996). A helper-dependent adenovirus vector system: removal of helper virus by Cre-mediated excision of the viral packaging signal. *Proc. Natl. Acad. Sci. USA*, 93: 13565-13570.

Parks, R. and Graham, F. (1997). A helper-dependent system for adenovirus vector production helps define a lower limit for efficient DNA packaging. *J. Virol.*, 71: 3293-3298.

Schiedner, G., Morral, N., Parks, R.J., Wu, Y., Koopmans, S.C., Langston, C., Graham, F.L., Beaudet, A.L. and Kochanek, S. (1998). Genomic DNA transfer with a high-capacity adenovirus vector results in improved in vivo gene expression and decreased toxicity. *Nat. Genet.*, 18: 180-183.

Stratford-Perricaudet, L.D., Levrero, M., Chasse, J., Perricaudet, M. and Briand, P. (1990). Evaluation of the transfer and expression in mice of an enzyme-encoding gene using a human adenovirus vector. *Hum. Gene Ther.*, 1: 241-256.

Tripathy, S.K., Black, H.B., Goldwasser, E. and Leiden, J.M. (1996). Immune responses to transgene-encoded proteins limit the stability of gene expression after injection of replication-defective adenovirus vectors. *Nat. Med.*, 2: 545-550.

Worgall, S., Wolff, G., Falck-Pedersen, E. and Crystal, R.G. (1997). Innate immune mechanisms dominate elimination of adenoviral vectors following *in vivo* administration. Hum. Gene Ther., 8: 37-44.

Yang, Y., Li, Q., Ertl, H.C. and Wilson, J.M. (1995). Cellular and humoral immune responses to viral antigens create barriers to lung-directed gene therapy with recombinant adenoviruses. *J. Virol.*, 69: 2004-2015.

6. SUMMARY — The recent advances in vector development and gene delivery systems are significant. The knowledge acquired from the wealth of experiments performed over the past decade was instrumental in guiding the directions in which modifications were targeted. In case of adenoviral (Ad) vectors, which have been found to be especially useful as gene delivery vectors, the main draw back was vector-mediated immunogenicity. Recent modifications of the Ad backbone led to the development of helper-dependent Ad vectors (HD), which are completely devoid of all viral protein coding sequences. These modifications have significantly contributed to the safety and reduced immunogenicity of the new HD vector system. In this manuscript, the efficiency of Ad-mediated gene delivery compared to other therapeutic approaches, in the context of a leptin-deficient mouse model, will be presented. Furthermore, we will review our results from comparative studies which support the significant safety enhancement achieved by the use of HD vectors over that observed with 1st generation Ad vectors.

Key Words: HD vectors; leptin; gene therapy; immunogenicity; obesity

Progress in Gene Therapy: Basic and Clinical Frontiers, pp. 85-138
R. Bertolotti *et al.* (Eds)

Adeno-associated virus: basic biology and vector applications

Candace Summerford[1,2], Jeffrey S. Bartlett[1,3,4] and Richard Jude Samulski[1,2,*]

[1]*Gene Therapy Center,* [2]*Department of Pharmacology,* [3]*CF Pulmonary Research Center,* [4]*Department of Medicine, University of North Carolina at Chapel Hill, Chapel Hill, NC 27599, USA*

Table of Contents

---

* Corresponding author. E-mail: RJS@med.unc.edu

1. INTRODUCTION

The field of gene therapy is still in its infancy. Although significant progress has been made in development of gene transfer vectors, there remains a gap between the successful delivery in either *in vitro* systems or *in vivo* animal models and that of successful human gene therapy. Bridging this gap remains a challenge. It is clear that an ideal gene transfer vector must transduce both dividing and non-dividing cells, allow for long term gene expression without toxicity or a host immune response and be relatively easy to produce. In recent years, adeno-associated virus (AAV) has emerged as an attractive vector for gene therapy. The decision to pursue the development of AAV as a vector system was based on many characteristics of AAV biology that predicted potential success. First, AAV is a single stranded DNA human parvovirus that, to date, has not been associated with any human disease (Berns, 1996). The non-pathogenic nature of AAV substantially subdues safety concerns often encountered with other vector systems. AAV is also a defective virus that is dependent on the presence of a helper virus (i.e., adenovirus or herpes simplex virus) in order to undergo a productive viral infection. This is an attractive feature for, in the absence of helper virus, AAV is maintained as a latent provirus. AAV's ability to establish a latent infection is a quality clearly amenable to long term therapeutic gene expression. Other attractive characteristics lie in the ability of AAV to infect a variety of cell types, as well as to infect both mitotic and post-mitotic cells. Furthermore, AAV has a relatively simple genome that can be easily manipulated to generate recombinant AAV (rAAV) vectors that lack all wild type coding sequences. The absence of wild type proteins bypasses the cellular toxicity often observed with other viral vectors. Thus, when one considers the features that are attractive qualities for a viral gene transfer vector 1) efficient infection of target cells, 2) the ability to infect quiescent cells, 3) potential for long term therapeutic gene expression, 4) lack of viral induced cellular toxicity and 5) a relative ease in generation of recombinant virus, AAV meets all of these objectives.

In the following chapter, studies that have lead to AAV's emergence as a promising gene delivery vector will be discussed. The current research being carried out to better understand how to maximize the potential of this vector system will also be presented. This is not intended to be an exhaustive review of AAV literature; however, it will highlight many of the studies that support further testing of AAV for use as a human gene therapy vector. The following reviews may be consulted for more details on AAV genetics, AAV biology and AAV vector applications (Kotin, 1994; Muzyczka, 1992; Flotte and Carter, 1995; Samulski, 1997; Samulski, 1995; Berns, 1996; Samulski, 1999).

2. AAV BIOLOGY

a) Classification

Adeno-associated virus (AAV) is a single-stranded DNA (ssDNA) animal virus that is classified in the subfamily *Parvovirinae* (the parvoviruses) of the virus family *Parvoviridae* (Berns, 1996). Parvovirus AAV was first observed by electron microscopy in the late 1950's as an unknown contaminant in preparations of the double-stranded DNA virus, adenovirus (Berns and Bohenzky, 1987). Hence, the newly identified virus was named adeno-associated virus. By the mid 1960's, it was determined that AAV had characteristics of a parvovirus (Atchison *et al.*, 1965; Parks *et al.*, 1967a,b). All previously identified parvoviruses were known to replicate autonomously in their host. However, AAV was discovered to require the presence of another virus (a helper virus) in order to replicate and produce progeny virions (Atchison *et al.*, 1965; Buller *et al.*, 1981; Hoggan *et al.*,

1966; Melnick *et al.*, 1965). This unique property required placement of AAV in its own genus of the *Parvoviridae* family, the *Dependoviruses*. The *Parvovirinae* subfamily now represents 32 viruses that can be subdivided into three genera i) *Parvovirus*, ii) *Erythrovirus*, and iii) *Dependovirus*.

The AAVs represent the sole members of the *Parvoviridae* Genus *Dependovirus*. To date, there are 5 different human serotypes of AAV that have been identified (AAV types 1 to 3, 5 and 6). In addition, adeno-associated viruses have been isolated from other species, including avian, bovine, canine, equine and ovine and monkey (AAV-4) (Berns, 1996; Chiorini *et al.*, 1997; Parks *et al.*, 1970; Rutledge *et al.*, 1998). Of the AAVs, AAV serotype 2 (AAV-2) has been the most extensively characterized.

Adeno-associated viral infection is wide spread throughout the human population. It is estimated that up to 80% of children acquire antibodies to AAV within their first 10 years of life (Blacklow *et al.*, 1968; Parks *et al.*, 1970). Fortunately, AAV is not associated with any human disease and appears to be benign. However, the defective nature of AAV (dependence on other viruses) complicates separation of clinical symptoms, if any, caused by AAV replication.

b) AAV life-cycle

There are two aspects to AAV's life cycle i) a latent phase and, ii) a replicative lytic phase (Berns, 1996); these two distinct phases stem from the viruses dependence on a helper virus in order to undergo a productive infection (Fig. 1). In the absence of helper virus, AAV establishes a latent infection through integration into host chromosomal DNA (Berns, 1996; Cheung *et al.*, 1980; Handa *et al.*, 1977; Hoggan *et al.*, 1972); integration is site specific and targeted to human chromosome 19 (Kotin *et al.*, 1991; Kotin *et al.*, 1990; Samulski *et al.*, 1991). AAV remains as an integrated latent provirus until super-infection of cells with an AAV helper virus. Upon exposure of latently infected cells to helper virus, AAV proceeds to the lytic

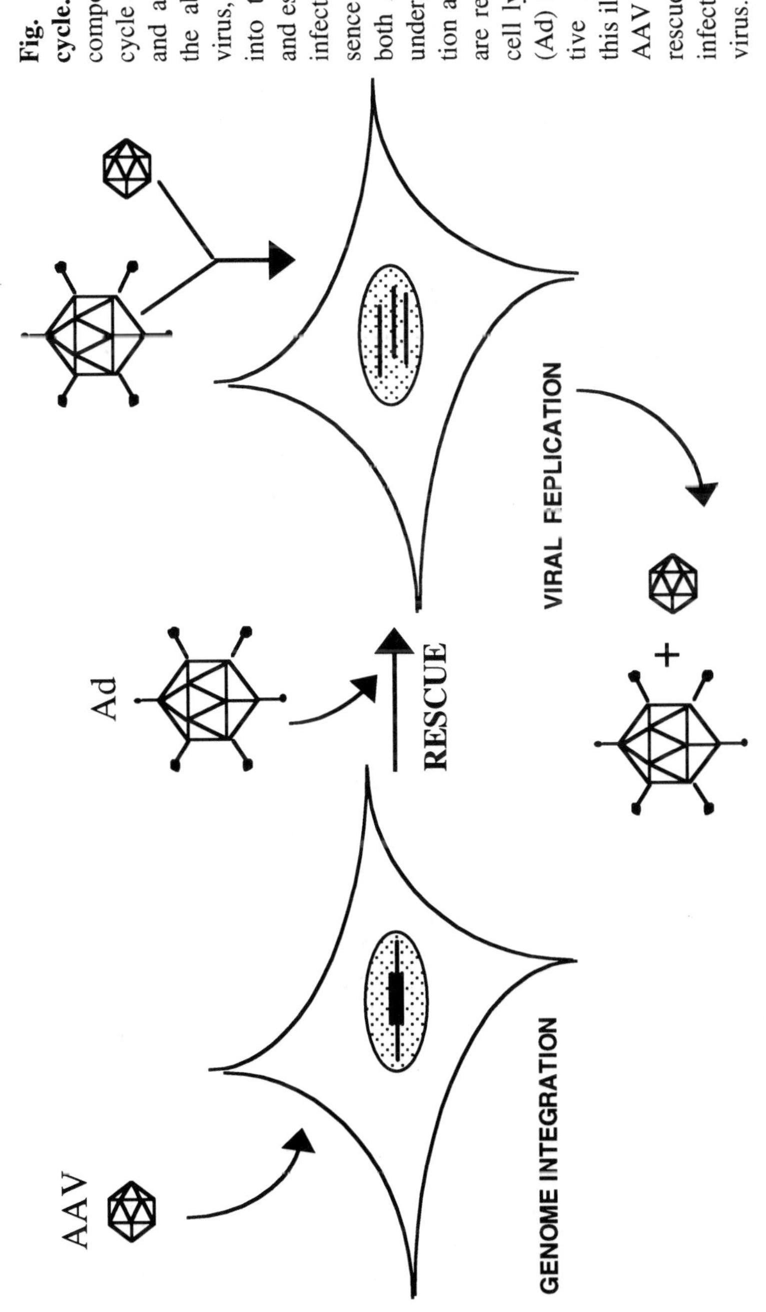

Fig. 1. AAV life cycle. There are two components to the life cycle of AAV, a latent and a lytic phase. In the absence of helper virus, AAV integrates into the host genome and establishes a latent infection. In the presence of helper virus, both AAV and helper undergo viral replication and progeny virus are released upon host cell lysis. Adenovirus (Ad) is the representative helper virus in this illustration. Latent AAV can later be rescued upon superinfection with helper virus.

(productive) phase of its life cycle, i) AAV genes are expressed, ii) the latent provirus is excised (rescued) from the host DNA, iii) there is replication of the AAV genome, iv) and progeny virions are produced (Berns, 1996). In nature, adenovirus (Ad) is the best known helper virus for AAV, however, *in vitro* studies have suggested that some members of the *Herpesviridae* family can also function as AAV helpers (Buller *et al.*, 1981; McPherson *et al.*, 1985; Schlehofer *et al.*, 1986). For a more detailed review of Ad helper functions readers are referred (Berns, 1996).

c) AAV genetics

The single-stranded AAV-2 genome has been completely sequenced and is 4680 nucleotides in length (Ruffing *et al.*, 1994; Srivastava *et al.*, 1983). Figure 2A shows a diagrammatic representation of the AAV genome, which is relatively simple in that it contains just three promoters at map positions 5 (p5), 19 (p19), and 40 (p40) (Green and Roeder, 1980; Green *et al.*, 1980; Laughlin *et al.*, 1979; Lusby and Berns, 1982). Furthermore, all mRNA transcripts are derived from one strand (the minus strand (antisense)) (Carter, 1976), terminate at a single polyadenylation site (Green *et al.*, 1980; Laughlin *et al.*, 1979; Srivastava *et al.*, 1983), and share a common intron sequence (p2107 to p2227) (Green *et al.*, 1980; Laughlin *et al.*, 1979). Transcripts from the p40 promoter encode the AAV capsid (Cap) structural genes; Vp1, Vp2 and Vp3. Each of the structural Cap genes are transcribed in the same open reading frame and differ only by their N-terminus. There are four non-structural proteins that are transcribed from the p5 and p19 promoters. These are referred to as Rep proteins and are necessary for viral replication.

The Rep and Cap genes represent the only AAV coding sequences and are flanked by terminal repeats (TRs) of 145 nucleotides (Fig. 2). The TRs are rich in GC bases (over 80%) and contain a 125 nucleotide (nt) palindromic sequence that consists of two small internal palindromes (BB' and CC') flanked by a larger palindromic sequence (AA') (Fig. 2B) (Lusby

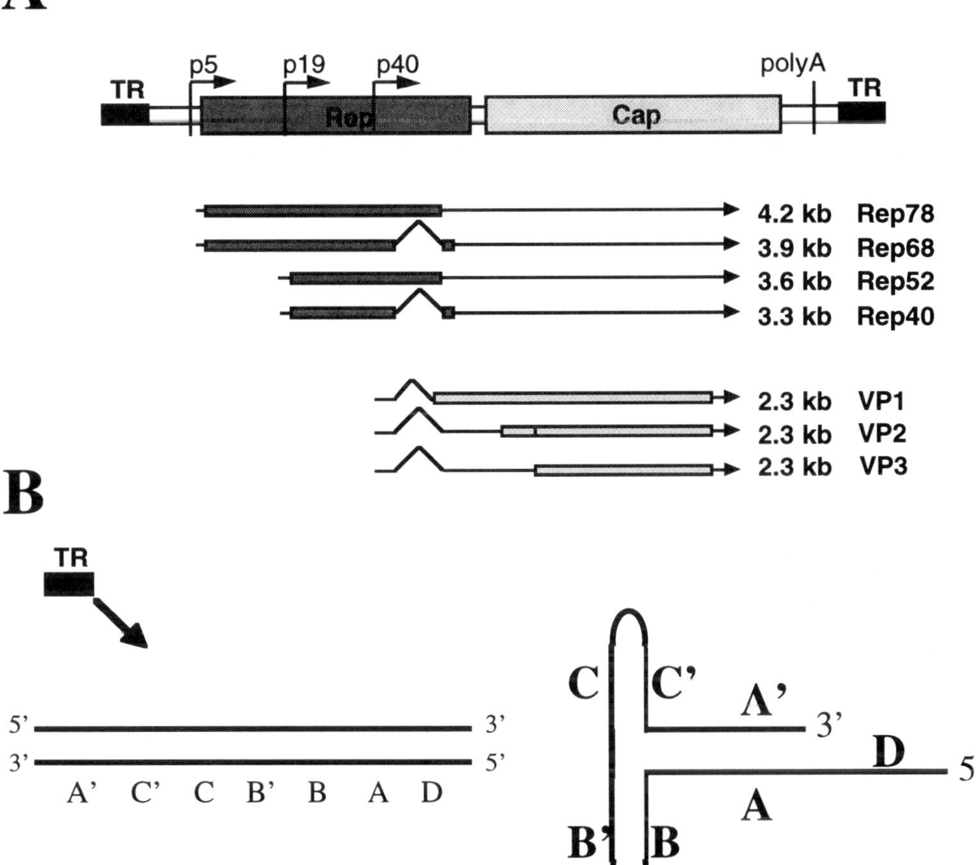

Fig. 2. The AAV genome. A) The genome consists of two large open reading frames (ORF). The non-structural Rep proteins are transcribed from the left hand ORF and the structural Cap proteins from the right hand ORF. The viral mRNA transcripts shown below are derived from three promoters (p5, p19 and p40) and share a common polyadenylation site. **B)** A diagrammatic representation of the AAV terminal repeat. In the single-stranded AAV genome, the palindromic sequences AA', BB', and CC', base-pair to form a T-shaped hairpin structure. The non-palindromic D sequence lies outside the putative T-shaped structure.

et al., 1980). The ssDNA palindromes are believed to base pair and adopt a stable secondary T-shaped hairpin structure at each end of the genome (Fig 2B) (Lusby *et al.*, 1980). This allows for the terminal 3' OH to provide the origin of self primed DNA replication. A non-palindromic sequence, the D sequence, lies outside the putative T-shape structure. This sequence is an important part of the terminal repeat and is required for replication of the AAV genome. The TRs are also known to represent the sole *cis*-acting sequences that are necessary for AAV integration, rescue of latent provirus, AAV DNA replication, and DNA encapsidation (McLaughlin *et al.*, 1988; Samulski *et al.*, 1989).

3. AAV VECTOR DEVELOPMENT

The simplicity of the AAV genome has led to the straight forward production of recombinant AAV (rAAV) vectors. After cloning of AAV into bacterial plasmids, it was determined that all viral coding sequences between the AAV TRs could be removed and replaced with foreign DNA (Samulski *et al.*, 1983, Samulski *et al.*, 1987). This represents removal of 96% of the AAV genome and allows for introduction of expression cassettes as large as 5 kb (Muzyczka, 1992). Since the foreign DNA is flanked by the parental AAV TR sequences, vector DNA can be 'rescued' from plasmid DNA upon exposure to adenovirus. The recombinant viral genome can then replicate and be encapsidated into AAV viral particles as long as the AAV Rep and Cap gene products are provided in *Trans*, usually by a second plasmid (Samulski *et al.*, 1987).

rAAV production and purification

There are several ways to generate rAAV vectors. A traditional method involves co-transfection of an AAV vector plasmid (containing an expression cassette flanked by the AAV TRs) and a Rep/Cap encoding AAV helper plasmid into adenovirus infected cells (Samulski *et al.*, 1989). The lack of sequence homology between AAV helper and vector plasmids precludes the generation of wild type AAV through homologous recombination. However, this procedure still results in the production of wild type adenovirus (Ad). To circumvent the co-production of wild type Ad along with rAAV, 'Ad mini plasmids' that carry all adenoviral genes required for efficient AAV production have been generated (Xiao *et al.*,

1998b). These mini plasmids do not encode Ad structural proteins. Therefore, by the use of a triple transfection procedure, it is now possible to generate rAAV preparations that are completely free of both infectious Ad virions and contaminating Ad proteins (Fig 4) (Xiao *et al.*, 1998b, Ferrari *et al.*, 1997). The ability to generate Ad free AAV stocks represents a significant advancement in AAV production technology, since several Ad proteins (fiber, hexon ect.) are known to produce a CTL immune response in humans (Yang *et al.*, 1994, Yang *et al.*, 1995, Yang and Wilson, 1995, Monahan *et al.*, 1998). Furthermore, the use of Ad mini plasmids can increase the production of rAAV by up to 40 fold (Xiao *et al.*, 1998b).

The co-transfection procedures, when optimized, can produce up to 10^{10} infectious units of rAAV per milliliter, which is equivalent to approximately 50-150 infectious virus particles per cell. However, the need to transfect plasmids is cumbersome and inevitably adds a degree of variability to viral production. To overcome this, several laboratories have developed producer cell lines that still require adenovirus infection but bypass the necessity to perform transfections (Clark *et al.*, 1995, Holscher *et al.*, 1994, Trempe and Yang, 1993, Liu *et al.*, 1999). The use of packaging cell lines represents a promising approach for vector production; however, it is still clear that there is room for optimization. It is expected that with the development of better AAV helper plasmids (Vincent *et al.*, 1997, Li *et al.*, 1997, Xiao *et al.*, 1998b, Flotte *et al.*, 1995, Wang *et al.*, 1998) stable cell lines that can produce even higher quantities of viral vector will be generated in the future. Novel systems have recently been established that use herpes simplex virus instead of adenovirus as a helper virus for AAV production. The AAV Rep and Cap genes have been cloned into a replication defective herpes simplex virus vector that can still serve as an AAV helper virus; this provides coordinated expression of AAV Rep and Cap (Conway and al, 1997; Zhang *et al.*, 1999) and helps to maximize viral production.

The manner in which rAAV is purified can also affect virus yield. A common procedure for purification of rAAV involves the use of ammonium sulfate to precipitate virus from crude cell lysate (Snyder *et al.*, 1996). The precipitated virus is then purified away from contaminating cellular proteins by several rounds of CSCl density centrifugation. Although this procedure isolates rAAV particles that are sufficiently pure for clinical studies, purification by CSCl density centrifugation i) requires at least 72 hours of centrifugation, ii) results in a significant loss of virus (Zolotukhin *et al.*, 1999), iii) involves the use of toxic reagents and, iv) is not practical for large scale preparations. Yet, we can now move away from traditional purification procedures. The recent identification of highly specific anti-AAV antibodies (Wistuba *et al.*, 1997) as well as the identification of heparan sulfate proteoglycan as a cellular receptor for AAV (Summerford and Samulski, 1998) have led to novel purification strategies that eliminate the need for CSCl density centrifugation.

These new purification procedures are based on the use of affinity column chromatography (Summerford and Samulski, 1999). In one approach heparin (an analog of heparan sulfate) is used as an affinity matrix for AAV purification. Two independent laboratories have used heparin columns with success (Clark *et al.*, 1999, Zolotukhin *et al.*, 1999). Importantly, when compared traditional purification procedures, virus purification by heparin matrix results in a higher yield of virus (greater than 10 fold) that is more infectious. The particle-to-infectivity ratios are less than 100:1, which is a significant improvement over the 1000:1 ratio often observed with CSCl density centrifugation (Zolotukhin *et al.*, 1999). The purification of a more infectious virus population is presumably due to the use of more benign conditions and/or the use of a molecule analogous to the AAV receptor.

The use of high affinity anti-AAV antibody as a purification matrix has also been successful (Grimm *et al.*, 1998). Grimm *et al.* (1998) has complexed an antibody that recognizes only assembled AAV virions (monoclonal antibody A20) to an NHS-activated HiTrap-Sepaharose matrix. Lysate of virus producing cells is passed over the column and the captured virus can be simply eluted with high salt. As with purification by heparin columns, this procedure can be performed within a working day. Thus, each of these new approaches for virus isolation have greatly simplified vector purification and provide an efficient, inexpensive procedure to isolate virus of high purity. Furthermore, the new rAAV purification strategies together with the advancements in production technology predict that the progression of rAAV to the clinic will not be slowed by the inability to obtain clinical grade high titer rAAV.

4. AAV *IN VIVO*

Concurrent with vector development has been the testing of AAV vectors *in vivo*. Through these studies an attractive feature of rAAV has emerged, and that is the ability to infect quiescent cells. Recombinant AAV vectors very efficiently transduce differentiated muscle (Xiao *et al.*, 1996, Kessler *et al.*, 1996, Snyder *et al.*, 1997b) and have recently shown success in hepatic gene transfer (Snyder *et al.*, 1997a, Snyder *et al.*, 1999, Xiao *et al.*, 1998a, Ponnazhagan *et al.*, 1997). In addition, efficient rAAV transduction has been reported in post mitotic neuronal cells of the brain (Kaplitt *et al.*, 1994, McCown *et al.*, 1996, During *et al.*, 1998, Mandel *et al.*, 1997) and photoreceptor cells of mouse and rat retina (Flannery *et al.*, 1997, Drenser *et al.*, 1998, Jomary *et al.*, 1997, Lewin *et al.*, 1998). Other cell types known to be efficiently transduced by rAAV vectors include heart muscle (Kaplitt *et al.*, 1996, Maeda *et al.*, 1998, Mohuczy *et al.*, 1999), vascular endothelial cells (Gnatenko *et al.*, 1997), and cochlear cells (Lalwani *et al.*, 1996, Lalwani *et al.*, 1998).

a) Long term transgene expression

In addition to efficient transduction of quiescent cells, rAAV vectors have shown the ability to mediate sustained transgene expression. Importantly, *in vivo* studies have documented that rAAV vectors mediate long-term gene expression in muscle (Xiao *et al.*, 1996, Kessler *et al.*,

1996). Since muscle tissue is able to serve as a reservoir for secreted proteins, rAAV vectors represent an attractive choice for the delivery of therapeutic gene products that can be secreted. *In vivo* studies have further demonstrated that rAAV vectors can mediate persistent transgene expression in the brain and retina. Transgene expression up to 1 year has been observed in the inferior colliculucs of neonatal and adult rat brain (Smith and McCown, 1997). In addition, sustained expression, up to 15 months, has been documented in photoreceptor cells of mouse, rat and guinea pig retina (Flannery *et al.*, 1997), W.W. Hauswirth, personal communication).

The initial observation that gene transfer to muscle with rAAV vectors could result in long term transgene expression was reported by two independent groups (Kessler *et al.*, 1996, Xiao *et al.*, 1996). After a single intramuscular injection of vector, Xiao *et al.* (1996) documented that a rAAV delivered transgene (beta-galactosidase) could be expressed for more than 1.5 years in immunocompetent mouse muscle. Importantly, although neutralizing antibody to the AAV virion shell could be detected, there was no significant immune response to vector transduced cells. A lack of an immune response to rAAV transduced muscle was also observed by Kessler *et al.* (1996). In this study, intramuscular injection of rAAV vector carrying the gene for human erythropoietin (h-EPO) resulted in stable expression of h-EPO for over 10 months, as measured by hematocrit production. These studies provided the first documentation that there is little immune response to vector transduced muscle and that rAAV vectors in muscle can give rise to sustained expression of a therapeutic protein. Many additional studies have further documented the success of rAAV vectors in muscle and are described under the following subheading of this review, *Success of rAAV vectors in animal models*.

The lack of an immune response to vector transduced cells was an unexpected, but welcomed, observation. There is a recent study that may help explain why muscle tissue transduced by rAAV vectors evade a cytotoxic T cell immune response (Jooss *et al.*, 1998). Immune responses induced by foreign transgene expression from either rAAV-lacZ or rAd-lacZ vectors were directly compared in immunocompetent mice after intramuscular injection. Recombinant Ad-lacZ transduction elicited a strong T-cell response that was responsible for extensive muscle fiber destruction, while rAAV-lacZ vector transduced muscle failed to elicit such a response. Since antigen presenting cells mediate cytotoxic T cell responses, Joss *et al.* (1998) performed adoptive transfer studies in which antigen presenting cells (dendritic cells) were infected ex vivo with either rAAV-lacZ or rAd-lacZ and the ability of these cells to induce an immune response was monitored upon transplantation. Surprisingly, adoptive-transfer of dendritic cells infected with rAd-lacZ was sufficient to mediate destruction of rAAV-lacZ transduced muscle fibers, while transfer of cells transduced with rAAV-lacZ failed to mediate cell death. Thus, the lack of the immune response to AAV transduced muscle is attributed to poor transduction of dendritic (antigen presenting) cells by rAAV-lacZ vector, an observation confirmed by enzymatic studies (Joos *et al.*, 1998). The ability of AAV vector delivered transgenes to evade destructive immune responses helps explain the observed success of rAAV vectors in mediating persistent transgene expression in muscle.

It should be noted that although rAAV vectors result in long term gene expression, persistence in some cell types can be due to the formation of rAAV episomes or concatameric structures and is not necessarily due to integration into host chromosomal DNA (Duan *et al.*, 1999a, Afione *et al.*, 1996, Duan *et al.*, 1998, Flotte *et al.*, 1994(Yang *et al.*, 1999)). Furthermore, there is often a lag phase (3 to 4 weeks) before transgene expression can be detected. This has been observed in the CNS, retina and muscle. Since it has been established that second strand synthesis of the rAAV genome can be a rate limiting step in vector transduction (Ferrari *et al.*, 1996, Fisher *et al.*, 1996), the slow onset of transgene expression may be due to a slow induction of AAV second strand synthesis. There have not been any formal studies that address the length of time that the rAAV genome remains single stranded *in vivo*. However, the use of agents known to increase AAV second strand synthesis *in vitro* can also enhance rAAV transduction *in vivo* (Fisher *et al.*, 1996), supporting that the observed delayed expression is due inefficient second strand synthesis.

b) Promoter contribution

There appears to be a role for specific promoters in governing persistence of transgene expression. McCowen *et al.* (1996) has documented long-term expression of beta-galactosidase in the inferior colliculus of rat brain. However, in the same study, there was a decline of gene expression over time in neurons of both the hippocampus and the cerebral cortex. It has been suggested that the differential expression is due to inactivation of the CMV promoter, since there was no indication of cellular toxicity or induced immune response to these transduced cells. Klein *et al.* (1998) recently addressed this possibility by directly comparing how different promoters affect the persistence of transgene expression in neuronal cells. Transgene expression driven by a CMV promoter declined over time, while expression from a neuronal specific promoter (neural specific endolase) was persistent for 3 months (Klein *et al.*, 1998). This result is consistent with observations of Peel *et al.* (1997) who has documented sustained expression (up to 15 weeks) in neuronal cells with neuronal specific promoters, while previous studies using the CMV promoter were less successful (Peel *et al.*, 1997). A similar phenomenon has been observed with rAAV delivery to photoreceptor cells of mouse retina (Flannery *et al.*, 1997). Flannery *et al.* (1997) has seen very efficient transduction of photoreceptor cells with rAAV vector carrying green fluorescent protein (GFP) driven by the photoreceptor specific promoter, opsin. However, earlier experiments by Zolotukhin *et al.* (1996) showed little rAAV transduction in these cell types when GFP expression was driven by the CMV promoter (Zolotukhin *et al.*, 1996). Thus, it appears that, in some cell types, there is inactivation of the CMV promoter.

c) Regulatable promoters

Clearly the choice of promoter can have a dramatic effect on rAAV transduction and when one considers *in vivo* gene therapy, regulatable and tissue specific promoters are of prime interest. There have been several reports of regulated gene expression using rAAV vectors (Ye *et al.*, 1999, Haberman *et al.*, 1998, Rendhal *et al.*, 1998, Bol *et al.*, 1998). Haberman *et al.* (1998) has controlled expression of green fluorescent protein (GFP) in the inferior colliculus of rat brain using a tetracycline (tet) regulated system. In this study, a duel cassette AAV vector that carries both a tet regulated GFP transgene and a modified tet promoter trans-activator gene was used. In the presence of deoxycycline (a tetracycline analog) GFP expression was successfully suppressed and expression resumed to normal levels upon removal of drug. The on/off transgene expression, controlled by the presence of deoxycycline in the rat's drinking water, was monitored for 6 weeks (the duration of the experiment).

Successful regulation of rAAV delivered transgenes has also been achieved in skeletal muscle (Ye *et al.*, 1999, Rendhal *et al.*, 1998, Bohl *et al.*, 1998). Bohl *et al.* (1998) has used a duel cassette AAV vector, similar to that described by Haberman *et al.* (1998), to regulate erythropoietin (EPO) expression by tetracycline (Bohl *et al.*, 1998). *In vivo* regulation of EPO expression was also documented by Rendhal *et al.* (1998) after co-injection of two separate rAAV vectors into skeletal muscle of immunocompetent mice. One rAAV vector encoded a tet inducible murine EPO transgene and the other vector encoded a transcriptional activator. Systemic administration and withdrawal of tetracycline was sufficient to regulate transcription of the EPO transgene demonstrating that the independent vectors were able to efficiently transduced the same cell. In a similar study Ye *et al.* (1999) has used a rapamycin-inducible system to regulate the expression of EPO delivered to muscle tissue of immunocompetent mice and rhesus monkey. Upon administration of rapamycin there was a 200 fold increase in plasma EPO levels and the regulatable promoter showed very little basal activity in the absence of drug. Thus, this is a very attractive system with great potential for regulation therapeutic transgenes. However, it must be remembered that the rapamycin system also requires the use of two AAV vectors, 1) a rAAV vector carrying two transcription factors that form an active complex in the presence of rapamycin and, 2) a rAAV vector carrying the therapeutic gene that is dependent on the complexed transcription factors for expression. Since it is essential that all elements be delivered to the same target cell, the muscle tissue, with large multinucleated syncytia, may be the most appropriate target tissue for systems that require the use of independent vectors.

5. SUCCESS OF rAAV VECTORS IN ANIMAL MODELS

With regards to the use of rAAV vectors for therapy, the most promising results have been obtained using animal disease models. Delivery of therapeutic transgenes with rAAV has been tested in animal models for Parkinson's disease, hypertension, recessive retinopathy, autosomal dominant retinitis pigmentosa, non-insulin dependent diabetes mellitus and hemophilia B.

a) rAAV in animal models for Parkinson's disease

Parkinson's disease is characterized by a degeneration of dopamine neurons found in the substantia nigra of the brain. The loss of dopamine neurotransmitter upon neuronal death severely affects voluntary movement, results in muscle stiffness, and induces tremors. There have been two gene therapeutic approaches for intervention of Parkinson's disease. The first involves introduction of rate limiting enzymes necessary for dopamine synthesis, in hopes to restore dopamine production, and the second involves delivery of neurotrophic factors known to protect dopaminergic neurons from degeneration. The use of rAAV for delivery of therapeutic genes for Parkinson's has meet with some success.

In early experiments, Kaplitt *et al.* (1994) observed behavioral improvements in a rat model for Parkinson's after rAAV mediated delivery of tyrosine hydroxylase (a rate limiting enzyme in dopamine synthesis) to rat brain striatum (Kaplitt *et al.*, 1994). The potential of AAV vectors for use in treatment of Parkinson's was further realized in a double blind study using a primate Parkinson's model in which African green monkeys were depleted of brain dopamine (During *et al.*, 1998). Monkeys that received rAAV-TH exhibited dramatic improvements in their disease related brain irregularities and behavior when compared to rAAV-*lacZ* controls. There has been similar success with rAAV vectors using the second approach for the treatment of Parkinson's (Choi-Lundberg *et al.*, 1997, Mandel *et al.*, 1997). Mandel *et al.* (1997) and Choi-Lundberg *et al.* (1997) observed protection of dopaminergic neurons in a degenerative rat model for Parkinson's after rAAV mediated delivery of GDNF, a neurotrophic survival factor, to the substantia nigra of rat brain. Yet, in both of these studies transgene expression was not sustained and the amount of GDNF decreased over time. In contrast, long term transgene expression was observed after delivery of rAAV vector carrying TH and GTP-cyclohydrolase to rat striatum (Mandel *et al.*, 1998). These enzymes are important in dopamine synthesis and susequent enzyme expression correlated with an increased production of L-DOPA (a dopamine precursor) for up to a year (Mandel *et al.*, 1998).

In addition to the potential of AAV vectors in the treatment of Parkinson's, there may also be a use for rAAV vectors in treatment of Alzheimer's disease and hypertension. Klein *et al.* (1998) has recently shown that rAAV vectors can efficiently deliver an Alzheimer's therapeutic, brain derived neurotrophic factor (BDNF), to hippocampal neurons. The effectiveness of gene delivery to the brain by rAAV vectors has also been seen in hypertensive rat model, where delivery of antisense to the angiotensin type 1 receptor (AT1-R) resulted in a reduction of rat hypertension (Phillips *et al.*, 1997). More recently, rAAV that carries antisense to AT1-R was delivered to vascular smooth muscle cells (Mohuczy *et al.*, 1999). This resulted in a significant reduction of AT1-R expression with transgene expression lasting up to 8 weeks (Mohuczy *et al.*, 1999).

b) rAAV in animal models for eye disorders

AAV vectors have shown recent success in mediating efficient transgene expression in both the photoreceptor and pigmented epithelial cell layers of retinal tissue (Zolotukhin *et al.*, 1996, Flannery *et al.*, 1997, Jomary *et al.*, 1997, Drenser *et al.*, 1998, Lewin *et al.*, 1998). The ability of AAV to mediate reporter gene transfer to the retina (Flannery *et al.*, 1997, Zolotukhin *et al.*, 1996) has provided a strategy for the treatment of retinal degeneration with rAAV vector. Retinal degeneration can be either associated with a disease (diabetes retinopathy) or inherited. Jomary *et al.* (1997) tested the therapeutic potential of rAAV vectors in treatment of recessive retinitis pigmentosa caused by a mutation in cGMP phosphodiesterase (PDE). A rAAV vector carrying the correct PDE gene was delivered by inraocular injection to a PDE recessive mouse model (rd/rd) for degenerative retinal disease. This vector was able to protect photoreceptor cells from degeneration and preserve retinal function (Jomary *et al.*, 1997). The potential of AAV vectors for the treatment of autosomal dominant retinitis pigmentosa (ADRP) has also been assessed. ADRP is a disorder that causes blindness and is due to a defective (dominant negative) rhodopsin gene product that causes photoreceptor cell death. Lewin *et al.* (1998) has shown that rAAV mediated delivery of a ribozyme designed to specifically cleave and destroy defective rhodopsin gene mRNA (Drenser *et al.*, 1998) was sufficient to slow photoreceptor degeneration in a transgenic rat model of ADRP (Lewin *et al.*, 1998). This study has provided the first example that ribozymes can be effective against a dominant negative allele *in vivo*.

c) rAAV in animal models for diabetes and hemophilia B

Some of the most impressive results come from studies in which rAAV vector was used to transfer secretable gene products for treatment of diabetes and hemophilia B. These studies took advantage of the observation that rAAV can efficiently and persistently transduce post-mitotic muscle (Xiao *et al.*, 1996; Fisher *et al.*, 1997; Clark *et al.*, 1997) and liver tissue (Koeberl *et al.*, 1997; Miao *et al.*, 1998; Snyder *et al.*, 1997a; Xiao *et al.*, 1998a).

Murphy *et al.* (1997) have looked at the potential of AAV vectors for treatment of non-insulin dependent diabetes melitus and obesity (Murphy *et al.*, 1997). Studies were performed using the *ob/ob* mouse that is genetically deficient in leptin (a 16 kDa secreted protein). The lack of serum leptin leads to increased food consumption and obesity. In addition, the leptin deficient mice present with lethargy, hyperglycemia, glucose intolerance, and a hyperinsulinemia that resembles human non-insulin dependent diabetes. Administration of recombinant leptin is known to reverse this phenotype, however the administration must be constant. Murphy *et al.* (1997) documented that a single intramuscular injection of rAAV encoding mouse leptin was sufficient to result in complete reversal of the *ob/ob* phenotype. All treated animals showed correction of hyperglycemia, insulin resistance, glucose intolerance and lethargy. Importantly, leptin expression was maintained at normal serum levels (2-5 ng/ml) and was sufficient to prevent obesity and non-insulin dependent diabetes for the duration of the experiment (6 months). This study clearly illustrates the potential of AAV vectors for delivery of secreted gene products to muscle.

Sustained delivery of secreted gene products is also attractive for the treatment of Hemophilia B. Hemophilia B is a bleeding disorder caused by an X-linked deficiency in functional coagulation factor IX. The disease is currently treated by administration of factor IX protein either derived from human blood products or is of recombinant origin. Numerous studies have confirmed that the plasma levels of factor IX protein need only to be above 1% (50 ng/ml) of normal to result in phenotypic correction and the reduction spontaneous bleeding into joints and other organs. Since treatment with factor IX protein is complicated by the proteins short half-life and the risk of acquiring a blood born disease, a gene therapeutic approach for treatment of hemophilia B is more attractive.

The first studies with rAAV carrying factor IX gene were performed in mice. Either direct injection of vector into skeletal muscle (Herzog *et al.*, 1997) or delivery of vector to hepatocytes by portal vein injection (Snyder *et al.*, 1997a, Nakai *et al.*, 1998) was sufficient to result in sustained therapeutic levels of factor IX. The first large animal study was performed by Monahan *et al.* (1998) who observed expression of human factor IX after direct intramuscular injection of rAAV into the hindlimbs of hemophilia B dogs. There was a transient reduction in whole blood clotting time (a measure of phenotypic correction) and expression of human factor IX was observed by immunohistochemistry 10 weeks post injection. Unfortunately, the development of circulating antibody to the human protein excluded assessment of therapeutic efficacy past the first week of the experiment. However, antibody to the human gene could be detected for the duration of the experiment (6 months) supporting sustained expression.

Recently, two independent groups have continued studies in the canine hemophilia B model (Snyder *et al.*, 1999, Herzog *et al.*, 1999). Herzog *et al.* (1999) has used rAAV to deliver canine factor IX by multiple intramuscular injections of vector at a single time point. Five dogs were treated with various vector-doses and a dose dependent partial correction of whole blood clotting time was observed. Furthermore, stable expression of factor IX protein was observed for greater than one year. The highest dosed dog (8.5×10^{12} particles/kg) presented with plasma factor IX levels in the therapeutic range (1.4% of normal) and showed partial correction of the activated partial thromboplastin time. In addition, as documented by others (Joos *et al.*, 1998, Fisher *et al.*, 1997, Xiao *et al.*, 1996) there was lack of a CTL immune response against AAV transduced muscle fibers. Furthermore, the immune responses against factor IX protein were minimal and transient.

Snyder *et al.* (1999) has delivered rAAV vectors carrying human factor IX (hFIX) and canine factor IX (cFIX) to the natural site for factor IX production, the liver. In addition to their extensive studies of rAAV-hFIX vector in a mouse model for hemophilia, rAAV-cFIX was delivered to hemophilic dogs by intraportal infusion. In both animal models, intraportal infusion of vector was not associated with any toxicity, as monitored by liver enzyme and bilirubin levels. Expression of hFIX in hemophilic mice was sustained and correction of their associated bleeding disorder was observed for over 17 months. In the same study, two hemophilic dogs were treated with 2×10^{12} particles (approximately 2×10^{11} particles/kg) of AAV-cFIX vector. A single intrahepatic administration of rAAV-cFIX was sufficient to result in FIX expression levels that reached 1% and 0.5% of normal. The whole blood clotting time in these dogs was reduced by 75% and 50%, respectively. In addition, the activated partial thromboplastin times were significantly reduced and corresponded to partial correction of the bleeding disorder. This partial correction of canine hemophilia B mediated by rAAV cFIX vector was observed for at least 8 months (Snyder *et al.*, 1999).

The recent success of rAAV in canine hemophilia B dogs (Herzog *et al.*, 1999, Snyder *et al.*, 1999) supports additional large animal studies and the eventual progression of this vector to the clinic for treatment of humans with hemophilia B.

6. CLINICAL TRIALS USING AAV VECTORS FOR THERAPEUTIC GENE TRANSFER

Ongoing clinical trials utilizing AAV vectors have been limited to those for the treatment of cystic fibrosis (CF). CF is an autosomally recessive genetic disease affecting between 1 and 2,000 and 1 in 4,500 children of Caucasian origin. The cause of cystic fibrosis has been shown to be faulty chloride secretion across the apical membrane of epithelial cells lining the airways and various exocrine organs. The gene responsible for CF, the cystic fibrosis transmembrane conductance regulator (CFTR) gene, was discovered in 1989 (Rommens *et al.*, 1989, Riordan *et al.*, 1989). The most common mutation of the CFTR gene is ΔF508, a deletion of the codon for Phe^{508}. This mutation accounts for about 70% of all CF mutations. Importantly, the CF defect can be corrected in CF epithelial cells when a normal CFTR cDNA is transferred into these cells *in vitro* (Drumm *et al.*, 1990, Rich *et al.*, 1990). Thus, gene therapy holds the promise of targeting the primary defect of CF by endowing epithelial cells with the ability to reconstitute proper chloride transport *in vivo*.

Several viral vectors have been proposed for the delivery of the normal CFTR gene to the respiratory epithelium. Initially, interest was focused on the use of adenovirus vectors and to date these vectors have been used in a number of clinical trials. However, Ad vectors have only been able to express CFTR cDNA in the CF respiratory epithelium for a short period of time and, at high doses these vectors cause severe inflammation of the lower respiratory tract (Crystal *et al.*, 1994). Thus, recent efforts have centered on the development of AAV as a vector for CF gene therapy.

There are several precedents for the use of AAV vectors for the treatment of CF lung disease. The construction of AAV vectors suitable for transduction of airway epithelium was first reported with an AAV vector containing the neomycin phosphotransferase reporter gene (Flotte *et al.*, 1992). This vector was used to transduce a CF bronchial epithelial cell line, IB3. Up to 70% of these cells expressed neo, as assessed by G418 resistance. Similar AAV vectors were then constructed containing the CFTR cDNA (Flotte *et al.*, 1993a). These vectors were shown to mediate CFTR protein expression and CFTR-mediated cAMP-dependent chloride conductance in CF bronchial epithelial cells in culture. Subsequently, the same AAV-CFTR vectors used for these *in vitro* studies were evaluated for biological efficacy *in vivo* after delivery to airways of New Zealand white rabbits (Flotte *et al.*, 1993b). Vector (10^{10} total particles) was administered to the lower bronchus of the rabbits via a fiberoptic bronchoscope. In these studies, AAV-mediated CFTR expression was assessed by RT-PCR and could be detected for up to 6 months following administration. CFTR protein expression was also detectable by immunohistochemistry. Importantly, there were no histopathologic changes seen in lung, liver, kidney, or heart samples from any of these animals. These findings demonstrated both biologic activity and safety of AAV-CFTR vector after administration to the airway surface. The vector was given in a manner analogous to that which would be used for clinical gene therapy in CF patients and thus paved the way for primate studies.

Rhesus monkeys were chosen for the final preclinical phase of AAV-CFTR vector testing because these primates provided a suitable host for AAV and adenovirus (Engelhardt *et al.*, 1993, Conrad *et al.*, 1996, Simon *et al.*, 1993). Thus, Rhesus are particularly relevant in the context of identifying potential problems related to recombinant virus shedding, immunologic reactions, or other potential adverse reactions from AAV-CFTR vectors delivered to the lung. For the monkey studies, animals received various amounts of vector, up to 10^{11} total particles, and were sacrificed at 10 days, 21 days, 90 days, or 180 days after vector administration to assess vector safety (Conrad *et al.*, 1996). There were no significant abnormalities noted in any of the safety assessments. However, spread of vector DNA to other organs was observed by PCR. This was the first time vector spread had been noted and the consequences of vector spread are unknown. Nonetheless, this phenomenon did not seem to be dose dependent and no vector could be detected in other organs at extended time points (i.e., past 6 weeks). Additionally, several experiments were performed to determine if vector could be rescued if the lungs were exposed to replication competent Ad. No rescue was seen in the presence of Ad. However, vector could be rescued when animals were infected with both wild-type AAV-2 and adenovirus (Conrad *et al.*, 1996).

Two clinical protocols are now underway to assess the safety of AAV-CFTR vectors for the treatment of human CF disease. The first study is a phase I trial of AAV-mediated CFTR gene transfer in adult CF patients with mild lung disease (Flotte, 1995). The results of the study have not yet been released. However, recruitment of patients for a phase II trial has already begun. The second trial is a phase I/II study of AAV-CFTR for the treatment of chronic sinusitis in patients with CF (Wagner *et al.*, 1998). As an alternative to pulmonary testing of CF gene therapy, these investigators have used the maxillary sinuses as a surrogate model of CF lung disease. The maxillary sinuses are attractive for evaluating new treatments for CF because their ion transport systems are similar to the lower respiratory tract. In addition, catheters can be used for both delivery of vectors as well as for withdrawal of samples for laboratory analysis. The clinical efficacy can also be measured by the recurrence of sinusitis. In the initial trial, ten patients received AAV-CFTR vector, in a randomized, non-blinded, dose escalation protocol. Five patients received one dose of vector and five were treated with one low dose followed by a higher dose in the contralateral sinus. All patients treated with doses of at least 10^4 infectious units of vector exhibited detectable gene transfer by PCR. However, no evidence of vector transcripts was obtained by RT-PCR after AAV-CFTR treatments. Since this may have been due to difficulties discriminating between vector DNA and vector RNA, and alternative measure of expression, transepithelial potential difference, was also used. The sinuses treated with the highest doses of vector all exhibited, by day 14, hyperpolarizations in response to superfusion with isoproterenol, amiloride, and low-chlorine-containing solutions, as would be expected if functional CFTR was being expressed. These effects were absent by day 28. No inflammation was observed in maxillary sinuses at day 14, and any observed adverse events did not seem to be related to AAV-CFTR. Thus, this study suggests that AAV-CFTR administration to the maxillary sinus results in safe, and possibly successful gene transfer. However, further characterization of AAV vector in additional human clinical trials is needed.

Failure to obtain direct evidence of AAV-mediated CFTR gene expression in the maxillary sinus could have been due to any number of reasons. Theoretical barriers to vector delivery might include the mucus layer of secretions, which protect the airway from the external environment, or the unique structure of the ciliated apical surface. The polarity of airway epithelia is maintained by tight junctions at the lateral surfaces separating apical from basolateral membranes (Cereijido *et al.*, 1989). These two membranes have distinct properties with regard to receptor composition. Previous studies evaluating gene transfer with recombinant adenovirus vectors have shown that the apical membrane of differentiated airways is highly refractory to infection (Grubb *et al.*, 1994, Pickles *et al.*, 1998, Zabner *et al.*, 1997). Recently it has been shown that the basolateral surface of cultured airway epithelial cells is more effectively infected with AAV than the apical surface, supporting the recurrent theme that apical membrane of polarized airway epithelia are infected poorly by a wide variety of different vectors (Duan *et al.*, 1999b). Heparan sulfate proteoglycan (HSPG) has recently been shown to be a cellular receptor for AAV-2 (Summerford and Samulski, 1998). Since, HSPG expression is limited to the basolateral surface of polarized airway epithelial cells, this may be the main reason for the observed disparity in infection between the two membranes (Duan *et al.*, 1999b). Recently, a number of co-receptors have been identified which facilitate either AAV attachment or entry in the presence of HSPG. These are integrin $\alpha v\beta 5$, which has been shown to promote AAV entry (Summerford *et al.*, 1999), and the fibroblast growth factor receptor-1 (FGFR-1), which has been suggested to form a dimer with HSPG to promote virus attachment (Qing *et al.*, 1999). Interestingly, each of these potential co-receptors is also absent from the apical surface of polarized airway epithelial cells (Pickles *et al.*, 1998)(J.S.B unpublished observations). Therefore, the absence of cellular receptors for AAV is a probable reason for the low efficiency of AAV CFTR gene transfer observed in the maxillary sinuses of the treated patients. However, there may be additional blocks to efficient transduction of these cells by AAV vectors as suggested by Duan *et al.* (Duan *et al.*, 1999a) and Teramoto *et al.* (Teramoto *et al.*, 1998). These reports suggest that vector mediated gene expression in polarized airway epithelial cells can also be blocked at the level of endosomal processing of internalized virus, nuclear entry, and/or subsequent molecular conversion of single stranded DNA to expressible forms.

7. LIMITING STEPS TO rAAV TRANSDUCTION

Although AAV vectors have shown great success *in vivo* and in the transduction of many cell types, it is clear that not all cells types can be transduced. It is through experimentation designed to decipher the barriers to rAAV transduction that new insights into AAV biology have emerged. In particular, it has become apparent that, in many cell types, a rate limiting step for rAAV transduction is the conversion of the single-stranded viral DNA to a double stranded template that is functional for transcription. Furthermore, it is evident that certain cell types lack the ability to support the primary step of AAV infection, adsorption and uptake of the virus. It is only through better understanding of the barriers to rAAV transduction that we will be able to broaden the potential of rAAV vectors for gene therapy.

a) AAV second strand synthesis

It is well established that wild-type AAV relies on the presence of adenovirus (Ad) or other helper viruses to undergo an efficient productive replicative life cycle. Yet, since rAAV vectors are replication incompetent and the transgenes of rAAV vectors are not driven by Ad or AAV promoters, it was unforeseen that Ad would also have an effect on recombinant viruses. The observation that the presence of Ad could increase rAAV transduction by up to 100-fold (Ferrari *et al.*, 1996) stimulated a series of studies designed to elucidate the mechanism by which Ad exerts this effect. Well characterized Ad mutants were used by independent laboratories to determine which Ad gene product(s) were responsible for the increase in rAAV transduction (Ferrari *et al.*, 1996, Fisher *et al.*, 1996). Through these studies, the enhancement of rAAV transduction was mapped. to the expression of the adenovirus E4 open reading frame 6 (E4 ORF 6). It was further demonstrated that the Ad E4 ORF 6 gene product induces second-strand synthesis of the rAAV single stranded DNA genome (Ferrari *et al.*, 1996, Fisher *et al.*, 1996). Since only the double stranded version of the AAV genome is transcriptionally active, the stimulated conversion of the single stranded virion DNA to duplex could account for enhanced rAAV transduction. The idea that adenovirus exerts its effect through the stimulation of second strand synthesis is further supported by the absence of an adenovirus effect when rAAV double-stranded plasmid DNA is transfected into tissue culture cells (Ferrari *et al.*, 1996).

The mechanism by which Ad E4 ORF6 induces synthesis of the rAAV second strand is unknown. Interestingly, alike Ad E4 ORF 6, heat shock treatment and several genotoxic agents including; hydroxyurea, topoisomerase inhibitors, ultra violet and X-ray irradiation have been found to also facilitate rAAV second strand synthesis (Russell *et al.*, 1995, Ferrari *et al.*, 1996, Fisher *et al.*, 1996, Alexander *et al.*, 1994). There is speculation that the genotoxic agents, physical stress, and the Ad E4 ORF 6 gene product are acting through a common mechanism which includes induction of the host cell DNA repair machinery. Since most of these treatments induce DNA damage, this represents a reasonable possibility. Although, a recent observation may shed some light on a more precise role of E4 ORF 6 in rAAV transduction. A host cellular protein that interacts with the single stranded TR of the AAV genome has been identified (Qing *et al.*, 1997). This protein has been designated the single stranded D-sequence binding protein (ssD-BP) and, when phosphorylated at tyrosine residues, appears to inhibit second strand synthesis of the rAAV genome. Furthermore, in the presence of Ad E4 ORF 6, hydroxyurea, and kinase inhibitors the ssD-BP converts to a dephosphorylated state and rAAV second strand synthesis incurs (Qing *et al.*, 1997). Whether the dephosphorylation process correlates with the induction of DNA damage by other agents remains to be determined but it appears that the phosphorylation state of the ssD-BP may play an important role in the life cycle of AAV.

What is the impact of these findings on the use of AAV vectors for gene therapy? The ability to enhance rAAV transduction *in vivo* remains to be fully explored. It is clear that coinfection of Ad with rAAV in-vivo can significantly enhance rAAV transduction. While the use of Ad coinfection is not plausible for a gene therapy protocol, one can envision the use of cytotoxic agents such as X-ray or chemotherapy to enhance rAAV transduction in cancer gene therapy (Alexander *et al.*, 1996). Furthermore, hydroxyurea is already currently being used in the treatment of sickle cell anemia and may represent a reagent that can safely enhance rAAV transduction *in vivo* (Charache *et al.*, 1995). Although the use of genotoxic agents *in vivo* would have inherent risks, their significant effect on rAAV transduction observed *in vitro* warrants further study for their possible use in human gene therapy. Identification of a host cellular factor involved in second strand synthesis is also a very important observation. Understanding the mechanism behind the induction of second strand synthesis may lead to the finding of less toxic manners in which enhancement of rAAV transduction can be achieved. Furthermore, it appears that the ratio of phosphorylated to unphosphorylated ssD-BP can differ between cell types. The amount of unphosphorylated ssD-BP in a panel of cell lines correlated with the efficiency by which AAV could transduce cells (Qing *et al.*, 1998). Therefore, monitoring the fate of ssD-BP in various cell types may help us better identify appropriate tissues for gene therapy by AAV vectors.

These observations illustrate that there are basic mechanisms of AAV virology that need to be understood in order to better exploit this virus as a vector for gene delivery. Regardless, rAAV vectors have been demonstrated to efficiently transduce several tissues *in vivo* and the results are very promising. Yet, if we want to maximize the potential of AAV vectors *in vivo*, it will be necessary to elucidate all aspects of AAV biology that can effect gene transfer, including the mechanism of AAV entry.

b) Cellular uptake of AAV virions

The primary event of a viral infection is attachment of the virus to the host cell. The initial attachment receptor often defines the tropism of a virus and may or may not be separated from a discrete entry (endocytosis/fusion) step. In the case of AAV, after endocytic uptake, the virus i) escapes the endosome, ii) uncoats and, iii) undergoes second strand synthesis before a productive viral infection can ensue. Thus the early stages of AAV infection; host cell attachment and endocytic uptake, serve as other potential barriers to rAAV transduction. Although AAV infects a very wide range of cell types, including human, nonhuman primate, canine, murine, and avian cells, there are cell lines that do not support the initial binding of AAV. These include the human megacaryocytic cell lines, MB-02 and MO7e (Ponnazhagan *et al.*, 1996, Mizukami *et al.*, 1996). In addition, primary cells that do not support AAV binding have been identified (differentiated polarized epithelium and some CD34+ cells) (Duan *et al.*, 1999b, Bals *et al.*, 1999, Ponnazhagan *et al.*, 1997).

Recent progress has been made in understanding factors that are important for cellular uptake of AAV virions (Summerford and Samulski, 1999, Summerford *et al.*, 1999, Qing *et al.*, 1999, Mizukami *et al.*, 1996). Heparan sulfate proteoglycan has been found to serve as a cellular receptor for AAV type 2 virions, a receptor that mediates attachment to- and subsequent infection of host cells (Summerford and Samulski, 1998). In addition, fibroblast growth factor receptor and/or $\alpha V\beta 5$ integrin receptor have been shown to be important factors in mediation of AAV 2 entry in the presence of proteoglycan (Summerford *et al.*, 1999, Qing *et al.*, 1999). Thus, it appears that AAV can be added to the list of viruses known to use multiple cell surface receptors to mediate the viral attachment and entry events required for infection; these include, adenovirus, HIV, and herpes simplex virus. For example, adenovirus initiates infection by attachment to CAR and then exhibits a subsequent interaction with secondary receptors, identified as α_V integrins, that facilitate virus internalization (Wicham *et al.*, 1993). Similarly, certain strains of HIV attach to host cells through the CD4 receptor yet require an additional cofactor molecule for infection. Several members of the chemokine receptor family have been identified as cofactor molecules required for mediation of envelope fusion of these HIV strains (for review see (Dimiter, 1997; Miller, 1996)). In addition, it is well established that the initial interaction of herpes simplex virus (HSV) with its host cell is mediated through heparan sulfate proteoglycans (Lycke *et al.*, 1991; Sheih *et al.*, 1992; WuDunn and Spear, 1989) and that another secondary event is responsible for promoting entry (Fuller *et al.*, 1989). A novel member of the tumor necrosis factor/nerve growth factor (TNF/NGF) receptor family, termed the herpesvirus entry mediator (HVEM), was shown to serve as a mediator of HSV entry (Montgomery *et al.*, 1996). Furthermore, two members of the immunoglobulin receptor superfamily, poliovirus receptor related protein 1 (HSV-1 & HSV-2) and poliovirus receptor related protein 2 (HSV-2), have recently been identified as alternative co-receptors capable of mediating HSV entry (Geraghty *et al.*, 1998; Warner, 1998).

Whether or not a co-receptor is absolutely required for AAV-2 infection is unknown. It is possible that AAV may use more than one mechanism of uptake to mediate infection: internalization mediated solely by HSPG and a more efficient uptake mediated by interaction of the AAV/HSPG complex with $\alpha_V\beta_5$ integrin and/or AAV/HSPG/FGFR-2 complex. HSPG can directly mediate entry of several heparan sulfate ligands, including basic fibroblast growth factor (bFGF) and lipoprotein lipase (Gleizes *et al.*, 1995; Roghani and Moscatelli, 1992; Saxena *et al.*, 1990; Williams and Fuki, 1997). Internalization of these HS ligands by HSPG raises the possibility that AAV internalization could also be directly mediated by HSPG. The experiments described by Summerford *et al* 1999, which monitored internalization of AAV into cell lines that either lack (CS1) or express cell surface $\alpha_V\beta_5$ integrin (CS1/β5), indicate that there are different mechanisms of virus uptake. It is clear that the presence of $\alpha_V\beta_5$ integrin promotes virus internalization, however, the presence of integrin is not absolutely required for virus infection. Instead, $\alpha_V\beta_5$ integrin significantly increases the rate of AAV uptake when compared to cells that lack integrin. The uptake rate of AAV into cells that lack integrin is consistent with the observed internalization rates for HSPGs. Thus, the slow uptake of AAV into CS1 cells supports the possibility that HSPG may be sufficient for virus entry.

Although endocytic uptake rates differ between cell types, receptor mediated clathrin coated pit endocytosis occurs generally at a $T_{1/2}$ less than 15 minutes (Mukherjee *et al.*, 1997; Williams and Fuki, 1997). The quick internalization of AAV into the CS1/β5 cell line is consistent with mediation of AAV uptake through clathrin coated pits. Duan *et al* . (1999) recently confirmed that the main pathway for AAV-2 entry into cells is through clathrin coated pit mediated endocytosis (Duan *et al.*, 1999). The cytoplasmic domain of $\alpha_V\beta_5$ integrin contains a NXPY motif (Ramaswamy and Hemler, 1990), which has been implicated as an endocytosis signal necessary for accumulation of similar receptors into clathrin coated pits (Mukherjee *et al.*, 1997). Thus, interaction of AAV with $\alpha_V\beta_5$ integrin may represent a mechanism by which AAV can be efficiently internalized by clathrin coated pit mediated endocytosis. When $\alpha_V\beta_5$ integrin is absent (in the CS-1 cell line) the uptake rate of AAV is much slower. The uptake of HSPGs have been documented to occur by a non-coated pit pathway and at endocytic rates from a $T_{1/2} \geq 1$ hour (oligomerized Syndecan) (Fuki *et al.*, 1996; Fuki *et al.*, 1997; Williams and Fuki, 1997), to a $T_{1/2} > 4$ hours (Perlecan) (Iozzo, 1987). The slow uptake rate of AAV in the CS-1 cell line is consistent with uptake through a non-coated pit pathway and/or bulk flow endocytosis that could potentially be mediated solely by HSPG. It will be interesting to determine if the different uptake rates of AAV into the CS-1 and CS1/β5 cell lines is a consequence of clathrin coated pit versus non-clathrin coated pit mediated endocytosis. The possibility that an, as yet, unidentified co-receptor is present on both the CS-1 and CS1/β5 cells has not been ruled out. However, the slow uptake rate of AAV-2 into the CS-1 cell line suggests that if such a co-receptor was present, then its expression level would have to be low.

The different mechanisms of AAV uptake may have significant consequences *in vivo* where the rate of virus entry is in direct competition with the rate of viral clearance. Interaction of AAV with HSPG may be very important in concentrating the virus on the cell surface. This may facilitate interaction with the co-receptors that play a critical role in mediating a rapid AAV-2 infection.

8. HEPARAN SULFATE PROTEOGLYCAN AS A VIRUS RECEPTOR

Proteoglycans are proteins modified by post-translational attachment of a polysaccharide glycosaminoglycan (GAG) moiety (for review, see references (Jackson *et al.*, 1991; Kjellen and Lindahl, 1991). There are 4 common classes of glycosaminoglycans (GAGs): 1) hyaluronan, 2) chondroitin sulfate, 3) heparan sulfate/heparin, and 4) keratan sulfate. Each are defined by the repeating disaccharide units that comprise their chains, by their specific sites of sulfation, and by their susceptibility to bacterial enzymes known to cleave distinct GAG linkages (Linhardt *et al.*, 1986).

Several members of the *Herpesviridae* family of viruses, including herpes simplex virus types 1 and 2 (Sheih *et al.*, 1992; WuDunn and Spear, 1989), cytomegalovirus (Compton *et al.*, 1993), pseudorabies virus (Mettenleiter *et al.*, 1990), bovine herpesvirus types 1 and 4 (Liang, 1991; Vanderplasschen *et al.*, 1993), and varcilla zoster virus (Zhu *et al.*, 1995), are known to attach to host cells through initial interaction with cell surface heparan sulfate proteoglycans. In addition, infectivity of two RNA viruses, foot-and-mouth disease type O virus (Jackson *et al.*, 1996), and dengue virus (Chen *et al.*, 1997), have recently been found to be dependent on binding to heparan sulfate proteoglycans. It is postulated that initial viral attachment to heparan sulfate moieties serves to concentrate virus on the cell surface, thereby increasing the probability that the virus will bind to a secondary receptor that mediates virus entry (Fuller *et al.*, 1989; Rostand and Esko, 1997; WuDunn and Spear, 1989). Recently, three co-receptors capable of mediating HSV strains have been identified (Geraghty *et al.*, 1998; Montgomery *et al.*, 1996; Warner, 1998).

Interestingly, the ability of foot-and-mouth disease type O virus to interact with heparan sulfate proteoglycan is a tissue culture adaptation of this virus (Neff *et al.*, 1998). Foot-and-mouth disease virus is a non-enveloped RNA virus that uses $\alpha_V\beta_3$ integrin as its natural receptor for cell attachment and entry (Berinstein *et al.*, 1995). However, upon passage in tissue culture the virus acquires adaptations (changes in capsid amino acids) that negate its dependence on $\alpha_V\beta_3$ integrin for replication in culture (Neff *et al.*, 1998). This adaptation results in a newly acquired ability of virus to bind cell surface heparan sulfate and ultimately enhances infection of tissue culture cells (Jackson *et al.*, 1996). Such an adaptation brings into question whether the RNA dengue virus has also adapted in tissue culture. The possibility that members of the *Herpesviridae* family of double-stranded DNA viruses or AAV-2 have acquired adaptations in culture is not as likely, since DNA replication fidelity is much greater than that observed with RNA replication mechanisms (Domingo *et al.*, 1996).

To date, all viral-proteoglycan interactions appear to be mediated by the GAG carbohydrate sequences and not the protein core of proteoglycans. Interestingly, analysis of HSV-1 and HSV-2 GAG binding properties indicate that these viruses prefer interactions with different sequences found in heparan sulfate. Competition experiments with modified heparin compounds indicate that both 6-O desulfated (6-O-DS) heparin and 2-,3-O-DS heparin can inhibit HSV-2 infection, but have little affect on HSV-1 infection (Herold *et al.*, 1996). This suggests that HSV-1 requires O-sulfation of HS for binding while this is not a requirement for HSV-2. Recently, affinity chromatography and virus binding assays indicate that the presence of a specific di-O-sulfated unit found in HS,{-Iduronic acid(2-OSO_3) & N-sulfated glucosamine(6- OSO_3)-}, is an important sequence for mediating HSV-1 binding (Feyzi *et al.*, 1997). This illustrates the importance of a specific carbohydrate backbone in mediating GAG/ligand interactions. Interestingly, mutational analysis of pseudorabies virus (PrV) glycoprotein C, which binds to heparin/heparan sulfate, indicate that separate regions within glycoprotein C may support binding to different structural features along the HS GAG chain (Trybala *et al.*, 1998; Trybala *et al.*, 1996). Virus interactions with distinct HS sequences may be a factor that helps explain, in part, the different tropisms of the *Herpesviridae* family of viruses and may be a factor with different AAV serotypes.

a) Identification of AAV Vector Targets

Identification of heparan sulfate as a factor important for AAV-2 entry should aid in the identification of cell types that are capable of supporting rAAV transduction. For example, the amount of heparan sulfate detected by fluorescent flow cytometric analysis (FACS) correlates with the ability of virus to bind various cell lines *in vitro* (unpublished observation). This correlation suggests that one could screen for the ability of virus to bind the cell surface by monitoring the presence or absence of heparan sulfate. However, it is important to keep in mind that there may be unidentified receptors that mediate AAV binding in the absence of heparan sulfate. Furthermore, heparan sulfate is a very heterogeneous molecule and it is possible that AAV exhibits an interaction with only a specific sequence of monosaccharides with specific sites of sulfation. Several ligands of HS including; HSV (Feyzi *et al.*, 1997; Herold *et al.*, 1996; Herold *et al.*, 1995), pseudorabies virus (Trybala *et al.*, 1998; Trybala *et al.*, 1996), bFGF (Maccarana *et al.*, 1993), and anti-thrombin (Atha *et al.*, 1984; Lindahl *et al.*, 1980) are known to bind specific sequences that are found in HS.

Further studies will be required to determine if AAV also interacts with HS in a 'sequence' specific manner.

In either case, preliminary data correlates the absence of HS with the inability of virus to bind the cell surface of CD34$^+$ human primary cells (unpublished observations). CD34$^+$ cells purified from multiple human bone marrow donors have been screened by FACS analysis for both the levels of cell surface HS and the ability to support binding of AAV-2 labeled with fluorescent Cy3 dye. To date, the CD34$^+$ cells from all donors that lack cell surface expression of HS do not support virus binding, whereas CD34$^+$ cells positive for HS expression are positive for virus binding. Furthermore, significant AAV internalization is observed in donor cells that are positive for HS (unpublished observations).

The observation that AAV binds only to CD34$^+$ cells that have cell surface HS is an interesting finding since, throughout the literature, there is discrepancy as to whether or not CD34$^+$ cells can be infected by rAAV vectors. The reports of rAAV transduction in CD34$^+$ cells vary from 0-80% transduction (Fisher-Adams *et al.*, 1996; Goodman *et al.*, 1994; Ponnazhagan *et al.*, 1997). It was suggested that this discrepancy is due to donor variability and that it is the ability of virus to bind the cell surface is responsible for this variation (Ponnazhagan *et al.*, 1997). This raises the possibility that the ability of CD34$^+$ cells to be transduced by rAAV is dependent on whether or not HS is present. Although a more in depth analysis will be required to confirm this notion, if true, one can envision screening donors for their responsiveness to rAAV gene therapy of hematopoietic stem/progenitor cells by monitoring the presence or absence of HS.

Interestingly, AAV binds to cells from populations that represent the monocyte (M) and granulocyte/neutrophil (G/N) bone marrow cells and these cell populations do not express detectable levels of heparan sulfate (unpublished data). This suggests that AAV binds to a distinctly different cell surface receptor on these cell types. It is not uncommon for viruses to use alternative receptors on different cell types. For example, the attachment receptor for adenovirus (Ad) is primarily CAR, yet $\alpha_M\beta_2$ integrin serves as an attachment receptor on monocytes (Huang *et al.*, 1996). Further studies will be required to assess the presence of alternative receptors that mediate a productive AAV infection.

b) Mediating AAV infection of cells which lack HS

Heparan sulfate as a primary attachment molecule for AAV raises the possibility that one can mediate binding to cells that lack HS proteoglycan by use of soluble HS as an artificial receptor. Cells are known to express specific cellular receptors for HS (Jackson, *et al.*, 1991). Furthermore, membrane associated heparan sulfate proteoglycans have been implicated in associations with other cell surface receptors through their HS moieties. Such an interaction has been proposed for diphtheria toxin receptor (Shishido *et al.*, 1995) and vascular endothelial derived growth factor receptor (FLK-1) (Chiang and Flanagan, 1995). Since FLK-1 is present on hematopoietic progenitor cells (Kaburn *et al.*, 1997), it may be possible to use soluble HS as a bridge to mediate AAV binding to the cell surface. This could be done either by coincubation of AAV with low concentrations of heparin/HS prior to infection or by pre-incubation of the cells with heparin/HS. A similar mechanism has been used to mediate HSV-1 binding and infection of a poorly permissive murine fibroblast cell line (sog9) that lacks the ability to synthesize glycosaminoglycans (Dyer *et al.*, 1997). By this manner, HSV-1 infection efficiency could be increased by 35 fold. If AAV interacts with a specific sequence of HS than it may be necessary to fractionate HS preparations and use only those fractions capable of binding AAV before efficient binding can be mediated. However, in the event that binding of AAV to the cell surface is sufficient to allow for productive viral entry, this could prove to be a powerful technique to transduce cells previously unable to efficiently bind and uptake virus.

9. MODIFIED VECTORS TO MEDIATE VIRAL UPTAKE

Another manner in which cells that lack viral receptor(s) could be infected involves the development of specific targeting vectors. The ability of AAV to infect a non-permissive megakaryocyte cell line (Mo7e) that does not support virus binding was enhanced by more than 70 fold through the use of bispecific $F(ab'\gamma)_2$ antibodies that recognizes both the AAV virion and the megakaryocyte specific $\alpha_{IIb}\beta_3$ integrin receptor (Bartlett *et al.*, submitted). This suggests that modulation of virus binding can be sufficient to mediate a productive viral infection. The approach also has an added advantage in that decorating the virus with the antibody appears to mask the natural receptor binding site and results in a decreased ability of AAV to infect cells that are normally permissive to the virus. Unfortunately, use of the bispecific antibody is still only partially restrictive in regards to AAV infection of permissive cells. However, the observation that heparan sulfate proteoglycan is responsible for mediating AAV attachment should facilitate the identification of the amino acids responsible for adsorbing virus to the cell surface. Typically ligand-proteoglycan interactions are mediated by electrostatic interactions between the high density charge of sulfated GAG moieties and a cluster of basic amino acids displayed by the ligand (Jackson *et al.*, 1991; Kjellen and Lindahl, 1991). Systematic mutation of basic amino acid residues suspected to be exposed on the virion surface should lead to the isolation of mutant viruses that are not capable of binding HS. Only after restricting the naturally broad tropism of AAV, will we be able to develop vectors that possess truly specific targeting abilities.

Recently, it was shown that the capsid sequence of AAV-2 can be genetically modified to change the viral tropism (Girod *et al.*, 1999). Girod *et al.* successfully inserted the L14 peptide into capsid AAV-2 so that it is expressed on the surface of the virion. L14 is a 14 amino acid peptide that contains the RGD motif of laminin fragment P1, and can thus target several integrin receptors. In this case, L14 was able to mediated AAV-2 infection of cells that normally are resistant to AAV-2 infection, B16F10 and RN22 cells (Girod *et al.*, 1999). In addition, the requirement for heparan sulfate proteoglycan in mediating efficient viral infection was circumvented.

10. IMPACT OF AAV SEROTYPES ON AAV VECTORS

The use of different AAV serotypes as vectors may also represent a way in which differential targeting can be achieved. There are at least 6 serotypes of AAV that have been identified (AAV types 1 to 6) (Atchison *et al.*, 1965; Bantel-Schaal and Hausen, 1984; Hoggan *et al.*, 1966; Parks *et al.*, 1967b; Rutledge *et al.*, 1998). To date, infectious clones have been generated for the widely used AAV-2 and recently the AAV-3, AAV-4, and AAV-6 serotypes (Chiorini *et al.*, 1997; Muramatsu *et al.*, 1996; Rutledge *et al.*, 1998). AAV-2, AAV-3, and AAV-6 serotypes have 82% nucleotide sequence identity while AAV-4 has only 75-78% sequence identity with the others (Chiorini *et al.*, 1997; Muramatsu *et al.*, 1996; Rutledge *et al.*, 1998). As observed with AAV-2, each of these serotypes can infect a broad host range of cells. However, the serotypes exhibit differences in the efficiency of vector transduction of various cell types (Chiorini *et al.*, 1997; Rutledge *et al.*, 1998). Furthermore, preliminary experiments illustrate that the efficiency by which recombinant AAV type 3 (rAAV3-LacZ) and recombinant AAV type 2 (rAAV2-LacZ) infects CHO cells differs by more than 5 fold (unpublished observation). This suggests that the different serotypes may use different cellular receptors. Further supporting that the viruses have evolved to use distinct receptors, AAV-3 and AAV-1 serotypes exhibit limited competition with AAV-2 in binding experiments (Mizukami *et al.*, 1996).

Interestingly, preliminary data suggests that AAV serotype 3 also requires HS for efficient infection (unpublished observations). Although a direct association of AAV-3 with HSPG remains to be determined, this raises the possibility that, as observed with AAV-2, AAV-3 may also interact with HSPG. The inability of AAV-2 to efficiently compete AAV-3 binding may be reminiscent of HSV-1 and HSV-2 binding to specific sequences found in HS (Herold *et al.*, 1996). Preferential binding to distinct sequences found in HS may ultimately effect the efficiency by which AAV can infect cells. For example, the HS moieties displayed by syndecan proteoglycan are known to differ in their fine structure on different cell types (Kato *et al.*, 1994). The differences in infection efficiency could also be explained if AAV-2 and AAV-3 use different co-receptors capable of promoting viral entry. Alternatively, AAV-3 may not interact with HS and may bind to a distinctly different receptor that requires association with cell surface HS proteoglycan in order to mediate binding and a productive infection. Although further studies will be required to address these issues, it will be important to characterize what is responsible for the different serotype "tropisms". The use of different serotypes may eventually expand AAV vector targets for gene therapy. Furthermore, each serotype elicits a distinct humoral immune response and their use should circumvent the effect of neutralizing antibodies upon repeat administration of vector (Beck *et al.*, 1999).

AAV serotypes may exhibit overlap in the use of viral receptors, as observed with different HSV serotypes. For example, HSV-1 and HSV-2 both bind to HSPG and use HVEM (herpes virus entry mediator) as a principal entry mediator in lymphoid cells (Montgomery *et al.*, 1996). HSV-1 and HSV-2 entry also can be mediated by poliovirus receptor related protein 1, a protein that is expressed in cells of epithelial and neuronal origin (Geraghty *et al.*, 1998). In contrast, poliovirus receptor related protein 2 only mediates HSV-2 entry and fails to mediate entry of HSV-1 strains (Warner, 1998). The high amino acid sequence homology between the different AAV serotypes suggests that some AAV serotypes may exhibit a certain amount of overlap in their mechanisms of entry. It is possible that different AAV serotypes may have acquired additional mechanisms of entry without the loss of entry mechanisms used by their predecessors. It will be important to elucidate the entry mechanisms for the different AAV serotypes in order to maximize their potential in gene therapy applications.

11. CONCLUSION

In closing, *in vivo* animal studies with rAAV have supported the prediction that AAV has the properties necessary for effective therapeutic gene delivery, i) rAAV can efficiently infect a wide variety of target cells including quiescent cells ii) delivery by rAAV vectors results in long-term expression of therapeutic genes, iii) there is a lack of immune response to vector transduced cells and, iv) there is an absence of vector induced cellular toxicity. However, the true therapeutic potential of AAV vectors for gene transfer in humans has yet to be firmly established. Although the ongoing human clinical trials have demonstrated that AAV vectors are safe and well tolerated (a key drawback to other vector systems), efficient AAV-mediated gene delivery in humans awaits further experiments. However, with the recent success of AAV vectors in large animal models, it is predicted that rAAV vectors will soon be tested in the clinical arena for treatment of other disease states. A trial for treating FIX deficiency has initiated and reliminary results appear to have exceeded preclinical data established in the dog model. Although this trial is still on going, it points to the predicted results expected from AAV gene delivery in humans. Furthermore, it is anticipated that more effective AAV delivery systems will be generated as the mechanisms of AAV cellular uptake, second strand synthesis are better understood, all of these contributions should only improve clinical testing of this novel vector system.

Acknowledgements

This work was supported by NIH grants DK518809 and DK54419.

12. REFERENCES

Afione, S.A., Conrad, C.K., Kearns, W.G., Chunduru, S., Adams, R., Reynolds, T.C., Guggino, W.B., Cutting, G.R., Carter, B.J. and Flotte, T.R. (1996). *In vivo* model of adeno-associated virus vector persitence and rescue. *J. Virol.*, 70: 3235-3241.

Alexander, I.E., Russell, D.W. and Miller, A.D. (1994). DNA-damaging agents greatly increase the transduction of nondividing cells by adeno-associated virus vectors. *J. Virol.*, 68: 8282-8287.

Alexander, I.E., Russle, D.W., Spence, A.M. and Miller, A.D. (1996). Effects of gama irradiation on the transduction of dividing and nondividing cells in brain and muscle of rats by adeno-associated virus vectors. *Hum. Gene Ther.*, 7: 841-850.

Atchison, R.W., Castro, B.C. and Hammond, W.M. (1965). Adeno-associated defective virus particles. *Science,* 149: 754-759.

Atha, D.H., Stephens, A.W., Rimon, A. and Rosenberg, R.D. (1984). Sequence variation in heparin octasaccharides with high affinity for antithrombin III. *Biochemistry,* 23: 5801-5812.

Bals, R., Xiao, X., Sang, N., wiener, D., Meegalla, R.L. and Wilson, J. (1999) Transduction of well-differentiadted airway epithelium by recombinat adeno-associated virus is limited by vector entry. *J. Virol.*, 73: 6085-6088.

Bartlett, J.S., Boucher, R.S. and Samulski, R.J. (2000). Cell-targeted gene therapy vectors; adeno-associated virus vector transduction of non-permissive cells mediated by bispecific $F(ab'\_)_2$ antibody. Submitted.

Beck, S.E., Jones, L.A., Chesnut, K., Walsh, S.M., Reynolds, T.C., Carter, B.J., Askin, F.B., Flotte, T.R. and Guggino, W.B., (1999) Repeated delivery of Adeno-associated virus vectors to the rabbit airway. *J. Virol.*, 73: 9446-9455.

Bergelson, J.M., Cunningham, J.A., Droguett, G., Kurt-Jones, E.A., Krithivas, A., Hong, J.S., Horwitz, M.S., Crowell, R.L. and Finberg, R.W. (1997). Isolation of a Common Receptor for Coxsakie B viruses and Adenovirus 2 and 5. *Science,* 275: 1320-1323.

Berns, K.I. (1996). Parvoviridae: the viruses and their replication. In: *Fields Virology, Vol. 3*, Fields, B.N., Knipe, D.M. and Howley, P.M. (Eds.), Lippincott-Raven, Philadelphia, Pa., pp. 2173-2197.

Berns, K.I. and Bohenzky, R.A. (1987) Adeno-associated viruses: an update. *Adv. Virus. Res.*, 32: 243-306.

Blacklow, N.R., Hoggan, M.D. and Rowe, W.P. (1968) Serologic evidence for human with adenovirus-associated viruses. *J. Natl. Cancer Inst.*, 40: 319-327.

Blackow, N.R., Cukor, G., Kibrick, S. and Quinman, G. (1978). Interactions of adeno-associated viruses with cells transformed by herpes simplex virus. In: *Replication of mammalian parvoviruses*, D.C. Ward and P. Tattersall (Eds.), Cold Spring Harbor Laboratory Press, NY, pp. 87.

Bohl, D., Salvetti, A., Moullier, P. and Heard, J.M. (1998). Control of erythropoietin delivery by deoxycycline in mice after intramuscular injection of adeno-associated vector. *Blood,* 62: 1512-1517.

Buller, R.M., Janik, J.E., Sebring, E.D. and Rose, J.A. (1981). Herpes simplex virus types 1 and 2 completely help adenovirus-associated virus replication. *J. Virol.,* 40: 241-247.

Cereijido, M., Ponce, A. and Gonzales-Mariscal, L. (1989). Tight junctions and apical/ basolateral polarity. *J. Memb. Biol.,* 100: 1-9.

Charache, S., Terrin, M.L., Moore, R.D., Dover, G.J., Barton, F.B., Eckert, S.V., McMahon, R.P. and Bonds, D.R. (1995). Effect of hydroxyurea on the frequency of painful sickle cell anemia. *New Engl. J. Med.,* 332: 1317-1322.

Chen, Y., Maguire, T., Hileman, R.E., Fromm, J.R., Esko, J.D., Linhardt, R.J. and Marks, R.M. (1997). Dengue virus infectivity depends on envelope protein binding to target cell heparan sulfate. *Nature Med.,* 3: 866-871.

Cheung, A.K., Hoggan, M.D., Hauswirth, W.W. and Berns, K.I. (1980) Integration of the adeno-associated virus genome into cellular DNA in latently infected human Detroit 6 cells. *J. Virol.,* 33: 739-748.

Chiang, M.K. and Flanagan, J.G. (1995). Interactions between FLK-1 receptor, vascular endothelial derived growth factor, and cell surface proteoglycan identified with a souble receptor reagent. *Growth Factors,* 12: 1-10.

Chiorini, J.A., Yang, L., Liu, Y., Safer, B. and Kotin, R.M. (1997). Cloning of adeno-associated virus type 4 (AAV4) and generation of recombinant AAV4 particles. *J. Virol.,* 71: 6823-6833.

Choi-Lundberg, D.L., Lin, Q., Chang, Y.N., Chaing, Y.L., Hax, C.M., Mohajen, H., Davidson, B.L. and Bohn, M.C. (1997). Dopaminergic neurons are protected from degeneration by GDNF gene therapy. *Science,* 275: 838-841.

Clark, K.R., Voulgaropoulou, F., Fraley, D.M. and Johnson, P.R. (1995). Cell lines for the production of recombinant adeno-associated virus. *Hum. Gene Ther.,* 6: 1329-1341.

Clark, K.R., Sferra, T. and Johnson, P.R. (1997). Recombinant adeno-associated viral vectors mediate long-term transgene expression in muscle. *Hum. Gene Ther.,* 8: 659-669.

Clark, K.R., Liu, X., McGrath, J.P. and Johnson, P.R. (1999). Highly purified recombinant adeno-associated virus vectors are biologically active and free of detectable helper and wild type viruses. *Hum. Gene Ther.,* 10: 1031-1039.

Compton, T., Nowlin, D.M. and Cooper, N.R. (1993). Initiation of human cytomegalovirus infection requires initial interaction with cell surface heparan sulfate. *Virology,* 193: 834-841.

Conrad, C.K., Allen, S.S., Afione, S.A., Reynolds, T.C., Beck, S.E., Fee-Maki, M., Barrazza-Ortiz, X., Adams, R., Askin, F.B., Carter, B.J., Guggino, W.B. and Flotte, T.R. (1996). Safety of single-dose administration of an adeno-associated virus (AAV)-CFTR vector in the primate lung. *Gene Ther.,* 3: 658-668.

Conway, J.E., Zolotukhin, S., Muzyczka, N., Hayward, G.S. and Byrne, B.J. (1997) Recombinant adeno-associated virus type 2 replication and packaging is entirely supported by a herpes simplex virus type 1 amplicon expressing Rep and Cap. *J. Virol.*, 71, 8780-8789.

Crystal, R.G., McElvaney, N.G., Rosenfeld, M.A., Chu, C.S., Mastrangeli, A., Hay, J.G., Brody, S.L., Jaffe, H.A., Eissa, N.T. and Danel, C. (1994). Administration of an adenovirus containing the human CFTR cDNA to the respiratory tract of individuals with cystic fibrosis. *Nature Genet.,* 8: 42-51.

Dimiter, S.D. (1997). How do viruses enter cells? The HIV co-receptors teach us a lesson in complexity. *Cell,* 91: 721-730.

Domingo, E., Escarmis, C., Sevilla, N., Moya, A., Elena, S.F., Quer, J., Novella, I. and Holland, J.J. (1996). Basic concepts in RNA virus evolution. *FASEB J.,* 10: 859-864.

Drenser, K., Hauswirth, W. and Lewin, A. (1998). Ribozyme-targeted destruction of RNA's associated with autosomal-dominant retinitis pigmentosa. *Investg. Opthamol. Vis. Science,* 39: 681-689.

Drumm, M.L., Frizzell, R.A. and Wilson, J.M. (1990). Correction of the cystic fibrosis defect *in vitro* by retrovirus-mediated gene transfer. *Cell,* 62: 1227-1233.

Duan, D., Sharma, P., Yang, J., Yue, Y., Dudus, L., Zhang, Y., Fisher, K.J. and Engelhardt, J.F. (1998). Circular intermediates of recombinant adeno-associated virus have defined structural characteristics responsible for long-term episomal persistence in muscle. *J. Virol.,* 72: 8568-8577.

Duan, D., Sharma, P., Dudus, L., Zhang, Y., Sanlioglu, S., Ziying, Y., Yue, Y., Lester, R., Yang, J., Fisher, K.J. and Engelhardt, J.F. (1999a). Formation of adeno-associated virus circular genomes is differentially regulated by adenovirus E4 ORF6 and E2a gene expression. *J. Virol.,* 73: 161-169.

Duan, D., Yue, Y., Yan, Z.Y., McCray Jr, P.B. and Engelhardt, J.F. (1999b). Polarity influences the efficiency of recombinant adeno-associated virus infection in differentiated airway epithelia. *Hum. Gene Ther.,* 9: 2761 2776.

Duan, D., Li, Q., Kao, A.W., Yue, Y., Pessin, J.E. and Engelhardt, J.F. (1999c). Dynamin is required for recombinant adeno-associated virus type 2 infection. *J. Virol.,* 73: 10371-10376.

During, M., Samulski, R.J., Elsworth, J.D., Kaplitt, M.G., Leone, P., Xiao, X., Li, J., Freese, A., Taylor, J.R., Roth, R.H., Sladek, J.R., O'Malley, K.L. and Redmond, D.E. Jr. (1998). *In vivo* expression of therapuetic genes for dopamine production in caudates of MPTP-treated monkeys using an AAV vector. *Gene Ther.,* 5: 820-827.

Dyer, A.P., Banfield, B.W., Martindale, D., Spannier, D. and Tufaro, F. (1997). Dextran sulfate can act as an artificial receptor to mediate a type-specific herpes simplex virus infection via glycoprotein B. *J. Virol,* 71: 191-198.

Engelhardt, J.F., Simon, R.H., Yang, Y., Zepeda, M., Weber-Pendleton, S., Doranz, B., Grossman, M. and Wilson, J.M. (1993). Adenovirus-mediated transfer of the CFTR gene to lung of nonhuman primates: biological efficacy study. *Hum. Gene Ther.,* 4: 759-769.

Ferrari, F.K., Samulski, T., Shenk, T. and Samulski, R.J. (1996). Second-strand synthesis is a rate-limiting step for efficient transduction by recombinant adeno-associated virus vectors. *J. Virol.,* 7: 3277-3234.

Ferrari, F.K., Xiao, X., McCarty, D. and Samulski, R.J. (1997). New developments in the generation of Ad-free high titer rAAV gene therapy vectors. *Nature Med.,* 3: 1295-1297.

Feyzi, E., Trybla, E., Bergstrom, T., Lindahl, U. and Spillman, D. (1997). Structural requirement of heparan sulfate for interaction with herpes simplex virus type 1 virions and isolated glycoprotein c. *J. Biol. Chem.,* 272: 24850-24857.

Fisher, K.J., Gao, G.P., Weitzman, M.D., DeMatteo, R., Burda, J.F. and Wilson, J.M. (1996). Transduction with recombinant adeno-associated virus for gene therapy is limited by leading-strand synthesis. *J. Virol.,* 70: 520-532.

Fisher, K.J., Jooss, K., Alston, J., Yang, Y., Haecker, S.E., High, K., Pathak, R., Raper, S.E. and Wilson, J.M. (1997). Recombinant adeno-associated virus for muscle directed gene therapy. *Nature Med.,* 3: 306-312.

Fisher-Adams, G., Wong, K.K., Podsakoff, G., Forman, S.J. and Chatterjee, S. (1996). Integration of adeno-associated virus vectors in CD34+ human hematopoetic progenitor cells after transduction. *Blood,* 88: 492-504.

Flannery, J.G., Zolotukhin, M., Vaquero, M.I., Lavail, M.M., Muzyczka, N. and Hauswirth, W.W. (1997). Efficient photoreceptor-targeted gene expression *in vivo* by recombinant adeno-associated virus. *Proc. Natl. Acad. Sci. USA,* 94: 6916-6921.

Flotte, T.R. (1995). A phase I study of an adeno-associated virus CFTR gene vector in adult CF patients with mild lung disease. *Hum. Gene Ther.,* 6: 225.

Flotte, T.R. and Carter, B.J. (1995). Adeno-associated virus vectors for gene therapy. *Gene Ther.,* 2: 357-362.

Flotte, T.R., Solow, R., Owens, R.A., Afione, S., Zeitlin, P.L. and Carter, B.J. (1992). Gene expression from adeno-associated virus vectors in airway epithelial cells. *Am J. Respir. Cell Mol .Biol.,* 7: 349-356.

Flotte, T.R., Afione, S.A., Conrad, C., McGrath, S.A., Solow, R., Oka, H., Zeitlin, P.L., Guggino, W.B. and Carter, B.J. (1993a). Expression of the cystic fibrosis transmembrane conductance regulator from a novel adeno-associated virus promoter. *J. Biol. Chem.,* 268: 3781-3790.

Flotte, T.R., Afione, S.A., Conrad, C., McGrath, S.A., Solow, R., Oka, H., Zeitlin, P.L., Guggino, W.B. and Carter, B.J. (1993b). Stable *in vivo* expression of the cystic fibrosis transmembrane conductance regulator with an adeno-associated virus vector. *Proc. Natl. Acad. Sci. USA,* 90: 10613-10617.

Flotte, T.R., Afione, S.A. and Zeitlin, P.L. (1994). Adeno-associated virus vector gene expression occurs in non-dividing cells in the absence of vector DNA integration. *Am. J. Respir. Cell Mol. Biol.,* 11: 517-521.

Flotte, T.R., Barraza-Oritz, X., Solow, R., Afione, S.A., Carter, B. and Guggino, W.B. (1995). An improved system for packaging recombinant adeno-associated virus vectors capable of *in vivo* transduction. *Gene Ther.,* 2: 29-37.

Fuller, A.O., Santos, R.E. and Spear, P.G. (1989). Neutralizing antibodies specific for glycoprotein H of herpes simplex virus permit viral attachment to cells but prevent penetration. *J. Virol.,* 63: 3435-3443.

Fuki, I.V., Iozzo, R.V. and Williams, K.J. (1996) Perlecan heparan sulfate proteoglycan (HSPG) a novel, distinct receptor for atherogenic lipoproteins. *Circulation*, 94: 698-699.

Fuki, I.V., Kuhn, K.M., Lomazov, I.R., Rothman, V.L., Tuszynski, G.P., Iozzo, R.V., Swenson, T.L., Fisher, E.A. and Williams, K.J. (1997). The syndecan family of proteoglycans: novel receptors mediating internalization of atherogenic lipoproteins. *J. Clin. Invest.*, 100: 1611-1622.

Geraghty, R.J., Krummenacher, C., Cohen, G.H., Eisenberg, R.J. and Spear, P.G. (1998)

Entry of alphaherpesviruses mediated by poliovirus-receptor related protein and polivirus receptor. *Science*, 280: 1618-1620.

Girod, A., Reid, M., Wobus, C., Lahm, H., Leike, K., Kleinschmidt, J., Deleage, G. and Hallek, M., (1999) Genetic capsid modifications allow efficient re-targeting of adeno-associated virus type 2. *Nature Med.*, 5: 1052-1056.

Gleizes, P.E., Noailac-Depeyre, J., Amalric, F. and Gas, N. (1995). Basic fibroblast growth factor (FGF-2) internalization through heparan sulfate proteoglycans-mediated pathway: an ultrastructural approach. *Eur. J. Cell Biol.*, 66: 47-59.

Gnatenko, D., Arnold, T.E., Zolotukhin, S., Nuovo, G.J., Muzyczka, N. and Bahou, W.F. (1997). Characterization of recombinant adeno-associated virus-2 as a vehicle for gene delivery and expression into vascular cells. *J. Inves. Med.*, 45: 87-98.

Goodman, S., Xiao, X., Donahue, R.E., Moulton, A., Miller, J., Walsh, C., Young, N. S., Samulski, R.J. and Nienhuis, A.W. (1994). Recombinant adeno-associated virus-mediated gene transfer into hematopoietic progenitor cells. *Blood,* 84: 1492-1500.

Green, M.R. and Roeder, R.G. (1980). Definition of a novel promoter for the major adenovirus-associated virus mRNA. *Cell,* 1: 231-242.

Green, M.R., Straus, S.E. and Roeder, R.G. (1980). Transcripts of the adenovirus-associated virus genome: multiple polyadenylated RNAs including a potential primary transcript. *J. Virol.*, 35: 560-565.

Grimm, D., Kern, A., Rittner, K. and Kleinschmidt, J.A. (1998). Novel tools for production and purification of recombinant adeno-associated virus vectors. *Hum. Gene Ther.*, 9: 2745-2760.

Grubb, B.R., Pickles, R.J., Ye, H., Yankaskas, J.R., Vick, R.N., Engelhardt, J.F., Wilson, J.M., Johnson, L.G. and Boucher, R.C. (1994). Inefficient gene transfer by adenovirus vector to cystic fibrosis airway epithelia of mice and humans. *Nature,* 371: 802-806.

Haberman, R.P., McCown, T.J. and Samulski, R.J. (1998). Inducible long-term gene expression in brain with adeno-associated virus gene transfer. *Gene Ther.*, 5: 1604-1611.

Handa, H., Shiroki, K. and Shimojo, H. (1977) Establishment and characterization of KB cell lines latently infected with adeno-associated virus type 1. *Virology*, 82: 84-92.

Herold, B.C., Gerber, S.I., Polonsky, T., Belval, B.J., Shaklee, P.N. and Holme, K. (1995). Identification of structural features of heparin required for inhibition of herpes simplex virus type 1 binding.*Virology,* 206: 1108-1116.

Herold, B.C., Gerber, S.I., Belval, B.J., Siston, A.M. and Shulman, N. (1996). Differences in the susceptibility of herpes simplex virus types 1 and 2 to modified heparin compounds suggest serotype differences in viral entry. *J. Virol.,* 70: 3461-3469.

Herzog, R.W., Hagstrom, J.N., Kung, S.H., Tai, S.J., Wilson, J.M., Fisher, K.J. and High, K.A. (1997). Stable gene transfer and expression of human blood coagulation factor IX after intramuscular injection of recombinant adeno-associated virus. *Proc. Natl. Acad.Sci., USA,* 94: 5804-5809.

Herzog, R.W., Yang, E.Y., Couto, L.B., Hangstrom, J.N., Elwell, D., Fields, P.A., Burton, M., Bellinger, D.A., Read, M.S., Brinkhous, K.M., Podsakoff, G.M., Nichols, T.C., Kurtzman, G.J. and High, K. (1999). Long-term correction of canine hemophilia B by gene transfer of blood coagulation factor IX mediated by adeno-associated viral vector. *Nature Med.,* 5: 56-63.

Hoggan, M.D., Blacklow, N.R. and Rowe, W.P. (1966). Studies of small DNA viruses found in various adenovirus preparations: physical, biological, and immunological characteristics. *Proc. Natl. Acad. Sci. USA,* 55: 1457-1471.

Hoggan, M.D., Thomas, G.F., Thomas, F.B. and Johnson, F.B. (1972). Continuous of adenovirus associated virus genome in cell culture in the absence of helper adenovirus. In: *Proceedings of the Fourth Lepetite Colloquium*, Cocoyac, Mexico, pp. 243-249.

Holscher, C., Horer, M., Kleinschmidt, J.A., Zentgraf, H., Burkle, A. and Heilbronn, R. (1994). Cell lines inducibly expressing the adeno-associated virus (AAV) rep gene: requirements for productive replication of rep-negative AAV mutants. *J. Virol.,* 68: 7169-7177.

Iozzo, R.V. (1987) Turnover of heparan sulfate proteoglycan in human colon carcinoma cells. *J. Biol. Chem.*, 262: 1888-1900.

Jackson, R.L., Busch, S.J. and Cardin, A.D. (1991). Glycosaminoglycans: molecular propeties, protein interactions, and role in physiological processes. *Physiol. Reviews,* 71: 481-539.

Jackson, T., Ellard, F.M., Ghazaleh, R.A., Brookes, S.M., Blakemore, W.E., Corteyn, A.H., Stuart, D.I., Neewman, J.W.I. and King, A.M. (1996). Efficient infection of cells in culture by type O foot-and-mouth disease virus requires binding to cell surface heparan sulfate. *J. Virol.,* 70: 5282-5287.

Jomary, C., Vincent, K.A., Grist, J., Neal, M.J. and Jones, S.E. (1997). Rescue of photoreceptor function by AAV-mediated gene transfer in a mouse model of inherited retinal degeneration. *Gene Ther.,* 4: 683-690.

Jooss, K., Yang, Y., Fisher, K. and Wilson, J. (1998). Transduction of dendritic cells by DNA viral vectors directs the uimmune response to transgene products in muscle fibers. *J. Virol.,* 72: 4212-4223.

Kaburn, N., Buhring, H.J., Choi, K., Ulrich, A., Risau, W. and Kerler, G. (1997). FLK-1 expression defines a population of early embryonic hematopoetic precursors. *Development,* 124: 2039-2048.

Kaplitt, M.G., Leone, P., Samulski, R.J., Xiao, X., Pfaff, D.W., O'Malley, K.L. and During, M.J. (1994). Long-term gene expression and phenotypic correction using adeno-associated virus vectors in the mammalian brain. *Nature Genet.,* 8: 148-154.

Kaplitt, M.G., Xiao, X., Samulski, R.J., Li, J., Ojamaa, K., Klein, I., Makimura, H., Kaplitt, M.J., Strumpf, R.K. and Diethrich, E.B. (1996). Long-term gene transfer in porcine myocardium after coronary infusion of adeno-associated virus vector. *Ann. Thrac. Surg.,* 62: 1669-1676.

Kessler, P.D., Podsakoff, G.M., Chen, X., McQuiston, S.A., Colosi, P.C., Matelis, L.A., Kurtzmann, G.J. and Byrne, B.J. (1996). Gene delivery to skeletal muscle results in sustained expression and systemic delivery of a therapeutic protein. *Proc. Natl. Acad.Sci., USA,* 93: 14082-14087.

Kjellen, L. and Lindahl, U. (1991). Proteoglycans: structures and interactions. *Annu. Rev. Biochem.,* 60: 443-475.

Klein, R., Myer, E., Peel, A., Zolotukhin, S., Muzyczka, N., Myers, C. and King, M.

(1998). Neuron-specific transduction in the rat hippocampal or nigrosriatal pathway by recombinant adeno-associated virus vectors. *Exp. Neurol.,* 150: 183-194.

Koeberl, D.D., Alexander, I.E., Halbert, C.L., Russell, D.W. and Miller, A.D. (1997). Persistent expression of human clotting factor IX from mouse liver after intravenous injection of adeno-associated virus vectors. *Proc. Natl. Acad.Sci. USA,* 94: 1426-1431.

Kotin, R.M. (1994). Prospects for the use of adeno-associated virus as a vector for human gene therapy. *Hum. Gene Ther.,* 5: 793-801.

Kotin, R.M., Siniscalco, M., Samulski, R.J., Zhu, X., Hunter, L., Laughlin, C.A., McLaughlin, S., Muzyczka, N., Rocchi, M. and Berns, K.I. (1990) Site-specific integration by adeno-associated virus. *Proc. Natl. Acad. Sci. USA*, 87: 2211-2215.

Kotin, R.M., Menninger, J.C., Ward, D.C. and Berns, K.I. (1991) Mapping and direct visualization of a region-specific viral DNA integration site on chromosome 19q13-qter. *Genomics*, 10: 831-834.

Lalwani, A.K., Walsh, B.J., Reilly, P.G., Muzyczka, N. and Mhatre, A.N. (1996). Development of *in vivo* gene therapy for hearing disorders: introduction of adeno-associated virus into the cochlea of the ginea pig. *Gene Ther.,* 3: 588-592.

Lalwani, A., Han, J., Walsh, B., Zolotukhin, S., Muzyczka, N. and Mhatre, A. (1998). Green fluorescent protein as a reporter for gene transfer studies in the cochlea. *Gene Ther.,* 5: 272-276.

Laughlin, C.A., Westphal, H. and Carter, B.J. (1979) Spliced adenovirus-associated virus RNA. *Proc. Natl. Acad. Sci. USA*, 76: 5567-5571.

Laughlin, C.A., Tratschin, J.-D., Coon, H. and Carter, B.J. (1983). Cloning of infectious adeno-associated virus genomes in bacterial plasmids. *Gene,* 23: 65-73.

Lewin, A., Drenser, K., Hauswirth, W., Nishikawa, S., Yaumura, D., Flannery, J. and LaVail, M. (1998). Ribozyme rescue of photoreceptor cells in a transgenic rat model of autosomal dominant retinitis pigmentosa. *Nature Med.,* 4: 967-971.

Li, J., Samulski, R.J. and Xiao, X. (1997). Role of highly regulated *rep* gene expression in adeno-associated virus vector production. *J. Virol.,* 71: 5236-5243.

Liang, X. (1991). Pseudorabies virus gIII and bovine herpesvirus 1 gIII share complementary functions. *J. Virol.*, 65: 5553-5555.

Lindahl, U., Backstrom, G., Thunberg, L. and Leder, I.G. (1980). Evidence for a 3-O-sulfated D-glucosamine residue in the anti-thrombin-binding sequence of heparin. *Proc. Natl. Acad. Sci. USA,* 77: 6551-6555.

Linhardt, R.J., Cooney, C.L. and Galliher, P.M. (1986). Polysaccharide Lyases. *Appl. Biochem. Biotechnol.,* 12: 135-177.

Liu, X., Clark, K.R. and Johnson, P.R. (1999). Production of recombinant adeno-associated virus vectors using a packaging cell line and a hybrid recombinant adenovirus. *Gene Ther.,* 6: 293-299.

Lusby, E., Fife, K.H. and Berns, K.I. (1980). Nucleotide sequence of the inverted terminal repetition in adeno-associated virus DNA. *J. Virol.*, 34: 402-409.

Lycke, E., Johansson, M., Svennerholm, B. and Lindah, U. (1991). Binding of herpes simplex virus to cellular heparan sulfate, an initial step in the adsorption process. *J. Gen. Virol.,* 72: 1131-1137.

Maccarana, M., Casu, B. and Lindahl, U. (1993). Minimal sequence in heparin/heparan sulfate required for binding of basic fibroblast growth factor. *J. Biol. Chem.,* 268: 23898-23905.

Maeda, Y., Ikeda, U., Shimpo, M., Ueno, S., Ogasawara, Y., Urabe, M., Kume, A., Takizawa, T., Saito, T., Colosi, P., Kurtzman, G., Shimada, K. and Ozawa, K. (1998). Efficient gene transfer into cardiac myocytes using adeno-associated virus (AAV) vectors. *J. Mol. Cell Cardiol.,* 30: 1341-1348.

Mandel, R.J., Spratt, S.K., Snyder, R.O. and Leff, S.E. (1997). Midbrain injection of recombinant adeno-associated virus encoding rat glial cell line-derived neurotrophc factor protects nigral neurons in progressive 6-hydroxydopamine-induced degeneration model of Parkinson's disease in rats. *Proc. Natl. Acad. Sci. USA,* 94: 14083-14088.

Mandel, R.J., Rendahl, R.J., Spratt, S.K., Snyder, R.O., Choen, L.K. and Leff, S.E. (1998). Characterization of intrastriatal recombinant adeno-associated virus mediated gene transfer of human tyrosine hydroxylase and human GTP-cyclohydrolase I in a rat model of Parkinsons's disease. *J. Neuroscience,* 18: 4271-4284.

McCown, T.J., Xiao, X., Li, J., Breese, G.R. and Samulski, R.J. (1996). Differential and persistent expression patterns of CNS gene transfer by an adeno-associated virus (AAV) vector. *Brain Res.,* 713: 99-107.

McLaughlin, S.K., Collis, P., Hermonat, P.L. and Muzyczka, N. (1988) Adeno-associated virus general transduction vectors: analysis of proviral structures. *J. Virol.,* 62: 1963-1973.

McPherson, R.A., Rosenthal, L.J. and Rose, J.A. (1985). Human cytomegalovirus completely helps adeno-associated virus replication. *Virology,* 147: 217-222.

Melnick, J.L., Mayor, H.D., Smith, K.O. and Rapp, F. (1965). Association of 20 millimicron particles with adenoviruses. *J. Bacteriol.,* 90: 271-274.

Miao, C.H., Snyder, R.O., Schowalter, D.B., Patijin, G.A., Donahue, B., Winther, B., and Kay, M.A. (1998). The kinetics of rAAV integration in the liver. *Nature Genet.,* 19: 13-15.

Miller, A.D. (1996). Cell-surface receptors for retroviruses and implications for gene transfer. *Proc. Natl. Acad. Sci. USA,* 93: 11407-11413.

Mizukami, H., Brown, K. E. and Young, N. (1996). Adeno-Associated Virus type 2 binds to a 150-kilodalton cell membrane glycoprotein. *Virology,* 217: 124-130.

Mohuczy, D., Gelband, C.H. and Phillips, M.L. (1999). Antisense inhibition of AT1 receptor in vascular smooth muscle cells using adeno-associated virus-based vector. *Hypertension,* 33: 354-359.

Monahan, P.E., Samulski, R.J., Tazelaar, J., Xiao, X., Nichols, T.C., Bellinger, D.A., Read, M.S. and Walsh, C.E. (1998). Direct intramuscular injection with recombinant AAV vectors results in sustained expression in a dog model of hemophilia. *Gene Ther.,* 5: 40-49.

Montgomery, R.I., Warner, M.S., Lum, B.J. and Spear, P.G. (1996). Herpes Simplex Virus-1 entry into cells mediated by a novel member of the TNF/NGF receptor family. *Cell,* 87: 427-436.

Mukherjee, S., Gosh, R.N. and Maxfield, F.R. (1997) Endocytosis. *Physiol. Rev.,* 77: 759-803.

Muramatsu, S., Mizukami, H., Young, N.S. and Brown, K.E. (1996). Nucleotide sequencing and generation of an infectious clone of adeno-associated virus 3. *Virology,* 221: 208-217.

Murphy, J.E., Zhou, S., Giese, K., Williams, L.T., Escobedo, J.A. and Dwarki, V.J. (1997). Long-term correction of obesity and diabetes in genetically obese mice by single intramuscular injection of recombinant adeno-associated virus encoding mouse leptin. *Proc. Natl. Acad. Sci. USA,* 94: 13921-13926.

Muzyczka, N. (1992). Use of adeno-associated virus as a general transduction vector for mammalian cells. *Curr. Top. Microbiol. Immunol.,* 158: 97-129.

Nakai, H., Herzog, R.W., Hagstrom, J.N., Walter, J., Kung, S.H., Yang, E.Y., Tai, S.J., Iwaki, Y., Kurtzman, G.J., Fisher, K.J., Colosi, P., Couto, L.B. and High, K.A. (1998). Adeno-associated viral vector mediated gene transfer of human blood coagulation factor IX into mouse liver. *Blood,* 91: 4600-4607.

Neff, S., Sa-Carvalho, D., Rieder, E., Mason, P.W., Blystone, S.D., Brown, E.J. and Baxt, B. (1998). Foot-and mouth disease virulent for cattle utilizes integrin $\alpha v\beta 3$ as its receptor. *J. Virol.*, 72: 3587-3594.

Parks, W.P., Green, M., Pina, M. and Melnick, J.L. (1967a). Physicochemical characterization of adeno-associated satellite virus type 4 and its nucleic acid. *J. Virol.*, 1: 980-987.

Parks, W.P., Melnick, J.L., Rongey, R. and Mayor, H.D. (1967b). Physical assay and growth cycle studies of a defective adeno-satellite virus. *J. Virol.*, 1: 171-180.

Parks, W.P., Boucher, D.W., Melnick, J.L., Taber, L.H. and Yow, M.D. (1970). Seroepidemiological and ecological studies of the adenovirus–associated satellite viruses. *Infect. Immun.*, 2: 716–722.

Peel, A.L., Zolotukhin, S., Schrimscher, G.W. and Muzycka, N. (1997). Efficient transduction of green fluorescent protein in spinal chord neurons using adeno-associated virus vectors containing cell type specific promoters. *Gene Ther.,* 4: 16-24.

Phillips, M.I., Mohuczy Dominiak, D., Coffey, M., Galli, S.M., Kimura, B., Wu, P. and Zelles, T. (1997). Prolonged reduction of high blood pressure with an *in vivo*, nonpathogenic, adeno-associated viral vector delivery of AT1-R mRNA antisense. *Hypertension,* 29: 374-380.

Pickles, R.J., McCarty, D., Matsui, H., Hart, P.J., Randell, S.H. and Boucher, R.C. (1998). Limited entry of adenovirus vectors into well-differentiated airway epithelium is responsible for inefficient gene transfer. *J. Virol.,* 72: 6014-6023.

Ponnazhagan, S., Wang, X., Woody, M.J., Luo, F., Kang, L.Y., Madahavi, L.N., Munshi, C.M., Shang, Z.Z. and Srivasta, A. (1996). Differential expression in human cells from the p6 promoter of human parvovirus B19 following plasmid transfection and recombinant adeno-associated virus 2 (AAV) infection: human megacaryocytic leukaemia cells are non-permissive for AAV infection. *J. Gen. Virol.,* 77: 1111-1122.

Ponnazhagan, S., Mukherjee, P., Yoder, M. C., Wang, X. S., Zhou, S. Z., Kaplan, J., Wadsworth, S. and Srivastava, A. (1997a). Adeno-associated virus 2 mediated gene transfer *in vivo*: organ-tropism and expression of transduced sequences in mice. *Gene,* 190: 203-210.

Ponnazhagan, S., Mukherjee, P., Wang, X., Qing, K., Kube, D., Mah, C., Kurpad, C.,

Yoder, M., Srour, E. F. and Srivastava, A. (1997b). Adeno-associated virus type 2-mediated transduction in primary human bone marrow derived CD34+ hematopoetic progenitor cells: donor variation and correlation of transgene expression with cellular differentiation. *J. Virol.*, 71: 8262-8267.

Qing, K., Wang, X., Kube, D., Ponnazhagan, S., Bajpai, A. and Srivastava, A. (1997). Role of tyrosine phosphorylation of a cellular protein in adeno-associated virus 2-mediated transgene expression. *Proc. Natl. Acad. Sci. USA*, 94: 10879-10884.

Qing, K., Khuntirait, B., Mah, C., Kube, D.M., Wang, X., Ponnazhagan, S., Zhou, S., Dwarki, V., Yoder, M.C. and Srivasava, A. (1998). Adeno-associated type 2-mediated gene transfer: correlation of tyrosine phophorylation of the cellular single-stranded D sequence-binding protein with transgene expression in human cells *in vitro* and murine tissues *in vivo*. *J. Virol*, 72: 1593-1599.

Qing, K., Mah, C., Hansen, J., Zhou, S., Dwarki, V. and Srivastava, A. (1999). Human fibroblast growth factor receptor 1 is a co-receptor for infection by adeno-associated virus 2. *Nature Med.*, 5: 71-77.

Rendhal, K.G., Leff, S.E., Spratt, G.R., Bohl, D., Roey, M.V., Donahue, B.A., Choen, L.K., Mandel, R.J., Danos, O. and Snyder, R.O. (1998). Regulation of gene expression *in vivo* following transduction by two sepate rAAV vectors. *Nature Biotech.*, 16: 757-761.

Roghani, M. and Moscatelli, D. (1992). Basic Fibroblast Growth Factor is internalized through both receptor- mediated and heparan sulfate mediated mechanisms. *J. Biol. Chem.*, 267: 22156-22162.

Rich, D.P., Anderson, M.P., Gregory, R.J., Cheng, S.H., Paul, S., Jefferson, D.M., McCann, J.D., Klinger, K.W., Smith, A.E. and Welsh, M.J. (1990). Expression of cystic fibrosis transmembrane conductance regulator corrects defective chloride channel regulation in cystic fibrosis airway epithelial cells. *Nature*, 347: 358-363.

Riordan, J.R., Rommens, J.R., Kerem, B.-S., Alon, N., Rozmahel, R., Grzelczak, Z., Zielenski, J., Plavsic, N., Chou, J.-L., Drumm, M.L., Iannuzzi, M.C., Collins, F.S. and Tsui, L.-C. (1989). Identification of the cystic fibrosis gene: cloning and characterization of complementary DNA. *Science*, 245: 1066-1073.

Rommens, J.M., Iannuzzi, M.C., Kerem, B., Drumm, M.L., Melmer, G., Dean, M., Rozmahel, R., Cole, J.L., Kennedy, D., Hidaka, N. *et al.* (1989). Identification of the cystic fibrosis gene: chromosome walking and jumping. *Science*, 245: 1059-1065.

Rostand, K.S. and Esko, J.D. (1997). Microbial adherence and invasion through proteoglycans. *Infect. Immun.*, 65: 1-8.

Ruffing, M., Heid, H. and Kleinschmidt, J.A. (1994). Mutations in the carboxy terminus of adeno-associated virus 2 capsid proteins affect viral infectivity: lack of an RGD binding motif. *J. Gen. Virol.*, 75: 3385-3392.

Russell, D.W., Alexander, I.A. and Miller, A.D. (1995). DNA synthesis and topoisomerase inhibitors increase transduction by adeno-associated virus vectors. *Proc. Natl. Acad. Sci. USA*, 92: 5719-5723.

Rutledge, E.A., Halbert, C.L. and Russle, D.W. (1998). Infectious clones and vectors derived from adeno-associated virus (AAV) serotypes other than AAV type 2. *J. Virol*, 72: 309-319.

Samulski, R.J. (1995). Adeno-associated virus-based vectors for human gene therapy. In: *Gene therapy: from laboraatory to the clinic*, Hui, K.M. (Ed.), World Scientific Publishing Co., Singapore, Singapore, pp. 232-271.

Samulski, R.J. (1997). Development of Adeno-associated virus as a vector for *in vivo* gene therapy. In: *Transgenic animals: generation and use*, Houdebine, L.M. (Ed.), Harwood Academic Publishers, Chur, Switzerland, pp. 197-203.

Samulski, R.J., Berns, K.I., Tan, M. and Muzyczka, N. (1982). Cloning of adeno-associated virus into pBR322: rescue of intact virus from the recombinant plasmid in human cells. *Proc. Natl. Acad. Sci. USA,* 79: 2077-2081.

Samulski, R.J., Srivastava, A., Berns, K.I. and Muzyczka, N. (1983). Rescue of adeno-associated virus from recombinant plasmids: gene correction within the terminal repeats of AAV. *Cell,* 33: 135-143.

Samulski, R.J., Chang, L.-S. and Shenk, T. (1987). A recombinant plasmid from which an infectious adeno-associated virus genome can be excised *in vitro* and its use to study viral replication. *J. Virol.,* 61: 3096-3101.

Samulski, R.J., Chang, L.-S. and Shenk, T. (1989). Helper-free stocks of recombinant adeno-associated viruses: normal integration does not require viral gene expression. *J. Virol.,* 63: 3822-3828.

Samulski, R.J., Sally, M. and Muzycka, N. (1999). Adeno-associated viral vectors. In: *The Development of Human Gene Therapy*, Friedmann, T. (Ed.), Cold Spring Harbor laboratory press, Cold Spring Harbor, pp. 131 172.

Saxena, U., Klein, M.G. and Goldberg, I.J. (1990). Metabolism of endothelial cell-bound lipoprotein lipase. *J. Biol. Chem.,* 265: 12880-12886.

Schlehofer, J.R., Ehrbar, M. and zur Hausen, H. (1986). Vaccinia virus, herpes simplex virus, and carcinogens induce DNA amplification in a human cell line and support replication of a helpervirus dependent parvovirus. *Virology,* 152: 110-117.

Sheih, M., Montgomery, R.I., Esko, J.D. and Spear, P. (1992). Cell surface receptors for herpes simplex virus are heparan sulfate proteoglycans. *J. Cell Biol.,* 116: 1273-1281.

Shishido, Y., Sharma, K., Higashiyama, S., Klagsbrun, M. and Mekada, E. (1995). Heparin-like molecules on the cell surface potentiate binding of diptheria toxin receptor/membrane-anchored heparin binding epidermal growth factor-like growth factor. *J. Biol. Chem.,* 270: 29578-29585.

Simon, R.H., Engelhardt, J.F., Yang, Y., Zepeda, M., Pendelton, S.W., Grossman, M. and Wilson, J.M. (1993). Adenovirus-mediated transfer of the CFTR gene to lung of nonhuman primates: toxicity study. *Hum. Gene Ther.,* 4: 821-836.

Smith, F.E. and McCown, T.J. (1997). AAV vectors: general characteristics and potential use in the central nervous system. In: *Gene transfer and therapy for neurological disorders* , Chiocca, E.A. and Breakfield, X. O. (Eds.), Humana Press, pp. 79-88.

Snyder, R.O., Xiao, X. and Samulski, R.J. (1996). Production of recombinant adeno-associated virus vectors. In: *Current protocols in human genetics*, Dracopoli, N., Haines, J., Krof, B., Moir, D., Seidman, C. and Seidman, J. S. (Eds.), John Wiley & Sons Ltd., New York, pp. 12.11.11-12.12.23.

Snyder, R.O., Miao, C., Patijn, G., Spratt, S., Danos, O., Nagy, D., Gown, A., Winter, B., Meuse, L., Cohen, L., Thompson, A. and Kay, M. (1997a). Persistent and therapeutic concentrations of human factor IX in mice after hepatic gene transfer of recombinant AAV vectors. *Nature Genet.,* 16: 270-276.

Snyder, R.O., Spratt, S.K., Lagarde, C., Bohl, D., Kaspar, B., Sloan, B., Cohen, L.K. and Danos, O. (1997b). Efficient and stable adeno-associated virus mediated transduction in skeletal muscle of adult immunocompetent mice. *Hum. Gen. Ther.,* 8: 1891-1900.

Snyder, R.O., Miao, C., Meuse, L., Tubb, J., Donahue, B.A., Lin, H., Stafford, D., Patel, S., Thompson, A.R., Nichols, T., Read, M.S., Bellinger, D.A., Brinkhous, K.M. and Kay, M.A. (1999). Correction of hemophilia B in canine and murine models using recombinant adeno-associated viral vectors. *Nature Med.,* 5: 64-70.

Srivastava, A., Lusby, E.W. and Berns, K.I. (1983) Nucleotide sequence and organization of the adeno-associated virus 2 genome. *J. Virol.*, 45: 555-564.

Summerford, C. and Samulski, R.J. (1998). Membrane-associated heparan sulfate proteoglycan is a receptor for adeno-associated virus type 2 virions. *J. Virol.,* 72: 1438-1445.

Summerford, C. and Samulski, R.J. (1999). Viral receptors and vector purification: new approaches for generating clinical grade reagents. *Nature Med.,* 5: 587-588.

Summerford, C., Bartlett, J.S. and Samulski, R.J. (1999). $\alpha V\beta 5$ integrin: a co-receptor for adeno-associated virus type 2 infection. *Nature Med.,* 5: 78-81.

Teramoto, S., Bartlett, J.S., McCarty, D., Xiao, X., Samulski, R.J. and Boucher, R.C. (1998). Factors influencing adeno-associated virus-mediated gene transfer to human cystic fibrosis airway epithelial cells: comparison with adenovirus vectors. *J. Virol.,* 72: 8904-8912.

Trempe, J.P. and Yang, Q. (1993). In *Fifth Parvovirus Workshop*, Crystal River, FL.

Trybala, E., Bergestrom, T., Spillman, D., Svennerholm, B., Olofsson, S., Flynn, S.J. and Ryan, P. (1996). Mode of interaction between pseudorabies virus and heparan sulfate/heparin. *Virology,* 218: 35-42.

Trybala, E., Bergestrom, T., Spillman, D., Svennerholm, B., Flynn, S. J. and Ryan, P. (1998). Interaction between psuedorabies virus and heparin/heparan sulfate. *J. Biol. Chem.,* 273: 5047-5052.

Vanderplasschen, A., Bublot, M., Dubuisson, J., Pastoret, P. and Thiry, E. (1993). Attachment of the gammaherpesvirus bovine herpesvirus 4 is mediated by the interaction of gp8 glycoprotein with heparinlike moieties on the cell surface. *Virology*, 196: 232-240.

Vincent, K.A., Piraino, S.T. and Wadsworth, S.C. (1997). Analysis of recombinant adeno-associated virus packaging and requirements for rep and cap gene products. *J. Virol.,* 71: 1897-1905.

Wagner, J.A., Moran, M.L., Messner, A.H., Daifuku, R., Conrad, C., Reynolds, T., Guggino, W.B., Moss, R.B., Carter, B.J., Wine, J.J., Flotte, T.R. and Gardner, P. (1998). A Phase I/II study of tgAAV-CF for the treatment of chronic sinusitis in patients with cystic fibrosis. *Hum. Gene Ther.,* 9: 889-909.

Wang, X.S., Khunitraty, B., Qing, K., Ponnazhagan, S., Kube, D.M., Zhou, S., Dwarki, V.J. and Srivastava, A. (1998). Characterization of wild-type adeno-associated virus type 2-like particles generated during recombinant viral vector production and stratagies for their elimination. *J. Virol.,* 72: 5472-5480.

Warner, M.W., Geraghty, R.J., Martinez, W.M., Montgomary, R.I., Whitbaeck, C., Xu, R., Eisenberg, R.J., Choen, G.H. and Spear, P. (1998) A cell surface protein with herpesvirus entry activity (HveB) confers susceptibility to infection by mutants of herpes simplex virus type 1, herpes simplex virus type 2, and pseudorabies virus. *Virology* , 246: 179-189.

Wickham, T.J., Mathias, P., Cheresh, D.A. and Nemerow, G.R. (1993). Integrins αVβ3 and aVβ5 promote adenovirus internalization but not virus attachmemt. *Cell,* 73: 309-319.

Williams, K.J. and Fuki, V. (1997). Cell surface heparan sulfate proteoglycans: dynamic molecules for mediating catabolism. *Curr. Opin. Lipidol.*, 8: 253-262.

Wistuba, A., Kern, A., Weger, S., Grimm, D. and Kleinschmidt, J.A. (1997). Subcellular compartmentalization of adeno-associated virus type 2 assembly. *J. Virol.,* 71: 1341-1352.

WuDunn, D. and Spear, P. (1989). Initial interaction of Herpes Simplex Virus with cells is binding to heparin sulfate. *J. Virol.,* 63: 52-58.

Xiao, W., Berta, S.C., Lu, M.M., Moscioni, D., Tazelarr, J. and Wilson, J. (1998a). Adeno-associated virus as a vector for liver-directed gene therapy. *J. Virol.,* 72: 10222-10226.

Xiao, X., Li, J. and Samulski, R.J. (1996). Efficient long term gene transfer into muscle tissue of immunocompetent mice by adeno associated virus vector. *J. Virol.,* 70: 8098-8108.

Xiao, X., Li, J. and Samulski, R. J. (1998b). Production of high titer recombinant adeno-associated virus vectors in the absence of helper adenovirus. *J. Virol,* 72: 2224-2232.

Yang, J., Zhou, W., Zhang, Y., Zidon, T., Ritchie, T. and Englehardt, J.F., (1999). Concatamerization of adeno-associated virus circular genomes occurs through intermolecular recombination. *J. Virol,* 73: 9468-9477.

Yang, Y. and Wilson, J.M. (1995). Clearance of adenovirus-infected hepatocytes by MHC class I-restricted CD4+ CTLs in vivo. *J. Immunol.,* 155: 2564-2570.

Yang, Y., Nunes, F.A., Berencsi, K., Furth, E.E., Gonczol, E. and Wilson, J.M. (1994). Cellular immunity to viral antigens limits E1-deleted adenoviruses for gene therapy. *Proc. Natl. Acad.Sci. USA,* 91: 4407-4411.

Yang, Y., Li, Q., Ertl, H.C. and Wilson, J.M. (1995). Cellular and humoral immune responses to viral antigens create barriers to lung-directed gene therapy with recombinant adenoviruses. *J. Virol.,* 69: 2004-2015.

Ye, X., Rivera, V.M., Zoltick, P., Cerasoli, F.C., Schnell, M.A., Gao, G., Hughes, J.V., Gilman, M. and Wilson, J.M. (1999). Regulated delivery of therapuetic proteins after *in vivo* gene transfer. *Science,* 283: 88-91.

Zabner, J., Freimuth, P., Puga, A., Fabrega, A. and Welsh, M.J. (1997). Lack of high affinity fiber receptor activity explains the resistance of ciliated airway epithelia to adenovirus infection. *J. Clin. Invest.,* 100: 1144-1149.

Zhang, H., DeAlwis, M., Hart, S.L., Fitzke, F.W., Inglis, S.C., Boursnell, M.E., Levinsky, R.J., Kinnon, C., Ali, R.R. and Thrasher, A.J. (1999). High-Titer recombinant adeno-associated virus production from replicating amplicons and herpes vectors deleted for glycoprotein H. *Hum Gene Ther*, 10: 2527-2537.

Zhu, Z., Gershon, M.D., Ambron, R., Gabel, C. and Gershon, A.A. (1995). Infection of cells by varcilla zoster virus: inhibition of viral entry by mannose 6-phosphate and heparin. *Proc. Natl. Acad. Sci. USA*, 92: 3546-3550.

Zolotukhin, S., Potter, M., Hauswirth, W.W., Guy, J. and Muzyczka, N. (1996). A "humanized" green fluorescent protein cDNA adapted for high level expression in mammalian cells. *J. Virol.*, 70: 4646-4654.

Zolotukhin, S., Byrne, B.J., Mason, E., Zolotukhin, I., Summerford, C., Samulski, R.J. and Muzyczka, N. (1999). Recombinant adeno-associated virus purification using novel methods improves infectious titer and yield. *Gene Ther.*, 6: 973-985.

13. SUMMARY — Many *in vivo* studies illustrate that rAAV vectors have the necessary properties for effective gene transfer. This review will focus on studies which have fueled the emergence of AAV as an attractive vector for human gene delivery. The ability of AAV to mediate long term transgene expression in dividing and non-dividing cells, the consistent success of AAV vectors in animal models, and the initiation of clinical trials designed to test rAAV vectors will be discussed. Limitations of AAV vectors and future studies to better understand how to maximize the potential of this vector system will be presented.

Key Words: rAAV; viral vector; long term gene expression; clinical trials

Progress in Gene Therapy: Basic and Clinical Frontiers, pp. 139-159
R. Bertolotti *et al.* (Eds)

Herpes simplex virus vectors for gene therapy

David S. Latchman*

Department of Molecular Pathology, Windeyer Institute of Medical Sciences, University College London Medical School, Windeyer Bldg., Cleveland Street, London W1P 6DB, UK

Table of Contents

1. INTRODUCTION

The ability to deliver a specific gene to an intact organ *in vivo* evidently has numerous therapeutic applications. In particular, it is evident that where an individual is suffering from the effects of a genetic disease caused by a mutation in a specific gene, delivery of functional copy of that gene would, in many cases, provide effective relief of the disease. It should

---

* Present address: Institute of Child Health, University College London, UK. E-mail: d.latchman@ucl.ac.uk

also be noted however, that gene therapy approaches are not confined to situations where the primary genetic defect is understood. Thus, for example, gene therapy for cancer can involve the delivery of genes which stimulate an immune response to the tumour cells or which allow a non toxic pro-drug to be converted to a toxic form by the cancer cell, thereby killing the cell. Hence gene therapy can involve the delivery of genes which are of therapeutic benefit in the disease regardless of whether the disease has a genetic component or whether such a genetic component is fully understood.

The second type of gene therapy is particularly applicable in diseases of the nervous system where many of the complex diseases either do not have a simple genetic basis or that basis is not understood. Thus, for example, it has been shown that the symptoms of Parkinson's disease can be relieved both in animal models and human individuals by the transplantation of foetal neurons, because such neurons provide a source of dopamine to replace the deficit of this substance caused by the loss of dopaminergic neurons during the course of the disease (Backlund *et al.*, 1985). Hence, gene therapy for Parkinson's disease could be attempted by delivering the gene encoding tyrosine hydroxylase which is the rate limiting enzyme for dopamine production. Similarly, transgenic mice over expressing the gene for GAP-43 have been shown to show enhanced neurite outgrowth following nerve damage (Aigner *et al.*, 1995) suggesting that delivery of the gene encoding this factor could play a therapeutic role in promoting nerve outgrowth following spinal injury in humans.

These possibilities, have led numerous laboratories to attempt to develop means of effectively delivering genes to the nervous system and several of these are discussed in other articles in this book. We believe that vectors based on herpes simplex virus (HSV) have unique advantages compared to other systems and the reasons for this are discussed in the next section.

2. ADVANTAGES OF HSV-BASED VECTORS

Although the simplest method of delivering genes to any organ, would be the direct injection of DNA in a standard plasmid vector, such methods have proved to be of relatively low efficiency for organs other than skeletal muscle. Indeed, even when the DNA is coated with cationic liposomes to enhance gene delivery, only relatively low levels of gene expression are obtained following injection into the brain (Tsuda and Imaokea, 1996).

Thus, many laboratories have turned their attention to virus vectors since viruses naturally deliver their genetic material into the cells of the organisms which they infect. Unfortunately, the retroviral vectors based on Moloney murine leukaemia virus which are used in many gene delivery experiments will not infect non dividing cells and hence cannot be used for gene delivery to non dividing neurons (Miller *et al.*, 1990). As discussed in other articles however, lentiviral vectors based on the human immunodeficiency virus (HIV) type of retrovirus, adenovirus, adeno-associated virus and herpes simplex virus have all been used as gene delivery vectors capable of delivering genes to non dividing cells in the brain.

Table 1. Genome size of potential virus vectors for the nervous system

Virus	Genome size
Adeno-associated virus	8,500 base pairs
Adenovirus	35,000 base pairs
Herpes simplex virus	150,000 base pairs
Lentivirus	10,000 base pairs

In comparing such vectors, in terms of their genome size (table 1), it is immediately apparent that HSV has a much larger genome than any of the other viruses. This is of particular importance since a virus will only package the appropriate size of nucleic acid into a virus particle. Hence, viruses with small genome sizes cannot accept very large DNA inserts even when the majority of their genes are eliminated. This is of particular significance when it is necessary to introduce a large gene into cells or when effective therapy may require the delivery of more than one gene. Thus, for example, when adenovirus vectors were used in an animal model of motor neuron disease, effective therapy required both neurotrophin-3 and ciliary neurotrophic factors which had to be delivered in two separate adenovirus vectors (Haase *et al.*, 1997). This would not have been necessary with HSV due to its much larger genome size. In addition, whilst all the viruses can infect neuronal cells, only HSV naturally infects neuronal cells as part of its normal life cycle. Indeed, following initial infection at the periphery, the virus migrates up the sensory nerve processes innervating the site of initial infection and establishes life long latent infections of the neuronal cell bodies without in any way damaging the neuron or compromising neurological function (for review see Roizman and Sears, 1987; Latchman, 1990). Such latent infections serve as a reservoir for further infections at the periphery following reactivation of the infection but such reactivation once again does not apparently damage the neuronal cell.

Hence, this ability to deliver its DNA to neuronal cells without causing any damage renders the virus ideal for similarly delivering a foreign gene which it is necessary to express within the neuronal cell. Of course, in order to achieve this, it will be necessary to disable the virus so that it can no longer conduct the lytic replication which produces peripheral facial and genital sores. This is particularly true since direct injection of virus into the brain results in lytic replication leading to encephalitis and rapid death. The means of disabling the virus in order to achieve this are discussed in a subsequent section.

Despite this potential difficulty however, the ability of the virus to establish latent infections and its very large genome size mean that it has considerable potential as a gene delivery vector and the means of utilizing the virus as such a vector are discussed in the next section.

3. METHODS FOR USING HSV AS A GENE THERAPY VECTOR

Two basic methods have been devised for using HSV as a gene delivery vehicle and these will be discussed in turn (for other reviews see: Latchman, 1994; Coffin and Latchman, 1996; Leib and Olivo, 1993; Fink *et al.*, 1996; Kennedy, 1997).

a) Amplicon vectors

The simplest means of using HSV as a vector involves the construction of a plasmid vector which contains an HSV origin of replication and a packaging signal together with the gene of interest driven by an appropriate promoter (Spaete and Frenkel, 1982). If such an amplicon plasmid is introduced into cells infected with HSV, the amplicon will replicate because of the presence of a virus origin of replication and will produce a concatameric molecule containing numerous linked copies of the amplicon each having a copy of the foreign gene (Fig. 1). In an HSV infected cell, this DNA concatamer will be cleaved into lengths of approximately 150 kilobases (the genome size of the virus) and packaged into a virus particle. In this way, an infectious virus particle will be produced which does not contain the normal virus genome but instead contains numerous tandemly linked copies of the amplicon plasmid each containing the gene of interest. Upon infection into cells either *in vitro* or *in vivo*, the gene of interest will be delivered and will be expressed, thereby fulfilling the aim of a gene delivery vector. This system was originally shown to result in the high level expression of a chicken ovalbumin gene (Kwong and Frenkel, 1985) and has subsequently been extensively developed by the group of Geller. Thus, initially, they were able to show that it was able to deliver a β–galactosidase gene to neuronal cells from both the central and peripheral nervous systems (Geller and Breakefield, 1990; Kaplitt *et al.*, 1991).

More importantly, it has subsequently been possible to use this system to deliver genes of potential therapeutic benefit and show specific effects in animal model systems. Thus, for example, by introducing the gene encoding tyrosine hydroxylase into such an amplicon vector, it was possible to produce behavioural recovery for up to one year in a rat model of Parkinson's disease (During *et al.*, 1994). Similarly, a vector of this type expressing the anti-apoptotic protein Bc1-2 was able to protect neurons *in vivo* against neurological insults (Lawrence *et al.*, 1996) and most significantly protected them against ischaemia even when delivered after the ischaemic insult (Lawrence *et al.*, 1997). Other examples, include the protection of neuronal cells from the effects of axotomy and the promotion of neurite outgrowth by delivery of neurotrophic factors such as NGF (Fedderof *et al.*, 1992) or BDNF (Geschwind *et al.*, 1996).

Despite such successes, vectors of this type suffer from difficulties in terms of their ultimate human use. Thus, in order for the amplicon vector to replicate and be packaged, virus replication and packaging proteins must be supplied since these are not encoded in the amplicon vector which contains only the packaging signal and origin of replication. The only high efficiency means of doing this is to utilise a helper HSV virus to provide these functions. The final amplicon vector preparation therefore contains, in addition to the packaged amplicon, packaged helper virus (Fig. 1). It is evident therefore that this helper virus must be disabled if it is not to cause damaging effects when the preparation is injected into the brain. Although

Figure 1. Schematic representation of two methods for generating HSV vectors: Recombinant HSV vectors (A) and Amplicon system (B). **A).** HSV DNA is cotransfected into a permissive cell line together with a plasmid containing the therapeutic gene. The viral lytic cycle is initiated and recombination occurs between viral genome and the virus DNA in the plasmid. Recombined and replicated HSV is packaged into new viral particles and those containing recombinant HSV can be plaque purified. **B).** The therapeutic gene is cloned into an amplicon plasmid which is then transfected into a permissive cell line along with a helper virus. The helper virus provides the replication requirements for both its own DNA and that of the amplicon plasmid. DNA is then packaged separately into new viral particles that arising from the amplicon can contain multiple copies of the therapeutic gene. Initially the helper virus is in greater abundance but after passage an equal proportion of helper and defective virus is produced. It is important to note that the two virus types cannot be separated.

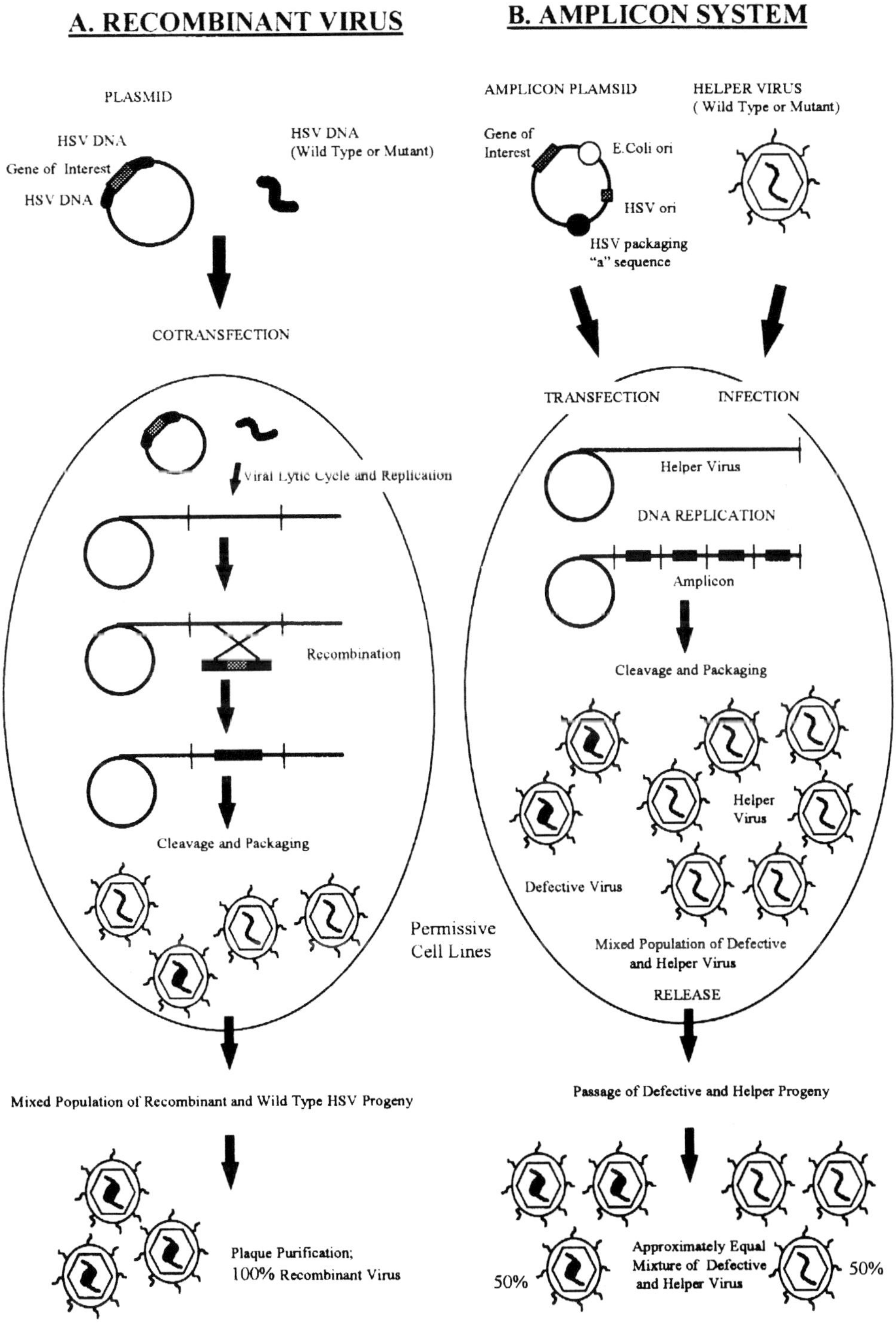

A. RECOMBINANT VIRUS

PLASMID

HSV DNA
Gene of Interest
HSV DNA

HSV DNA
(Wild Type or Mutant)

COTRANSFECTION

Viral Lytic Cycle and Replication

Recombination

Cleavage and Packaging

Permissive
Cell Lines

Mixed Population of Recombinant and Wild Type HSV Progeny

Plaque Purification;
100% Recombinant Virus

B. AMPLICON SYSTEM

AMPLICON PLAMSID

Gene of Interest
E.Coli ori
HSV ori
HSV packaging
"a" sequence

HELPER VIRUS
(Wild Type or Mutant)

TRANSFECTION

INFECTION

Helper Virus

DNA REPLICATION

Amplicon

Cleavage and Packaging

Helper
Virus

Defective Virus

Mixed Population of Defective
and Helper Virus

RELEASE

Passage of Defective and Helper Progeny

50%

Approximately Equal
Mixture of Defective
and Helper Virus

50%

this can be achieved by the means discussed in the next section, it can lead to serious problems.

Thus, in the experiments of During *et al.* (1994) discussed above, although therapeutic benefit was achieved, 10% of the animals died presumably due to reversion of the mutant helper virus to wild type which is capable of causing encephalitis.

Although this can be avoided by more extensively disabling the helper virus, an additional problem remains. Thus, amplicon virus is only able to infect cells and deliver the amplicon DNA but cannot replicate in these cells or form an infectious plaque. It is therefore not possible to purify the packaged amplicon away from the helper virus. This results, in all such preparations consisting of a mixture of the packaged amplicon and the packaged helper whose proportion will vary during growth of virus stocks in culture. Hence each batch of this reagent will vary in the proportion between the desirable amplicon particles and the helper virus which is unnecessary and indeed undesirable for the final gene delivery stage. This means that it will be difficult to obtain approval from regulatory authorities for the use of this reagent in human individuals due to batch variation between different samples of the amplicon.

b) Defective recombinant vectors

The problem of having two different types of particle in amplicon preparations can be overcome by use of the alternative method of constructing HSV vectors. In this method, the gene of interest is introduced directly into the virus genome by recombination. In this method (Fig. 1), the foreign gene of interest together with an appropriate promoter is flanked by two adjacent regions of the virus genome in a plasmid vector. If this vector is then introduced into cells by transfection together with full length viral DNA, recombination will occur between the viral DNA in the plasmid and the equivalent sequences in the virus, resulting in the insertion of the gene into the virus. This method was first used to introduce marker genes such as the β-galactosidase gene (Ho and Mocarski, 1988) or the hypoxanthine phosphoryibosyltransferase gene (Palella *et al.*, 1988) into the virus genome and viruses constructed in this manner have been shown to effectively deliver such genes to the central and peripheral nervous systems *in vivo* (Bloom *et al.*, 1995; Dobson *et al.*, 1990).

Although these recombinant viruses evidently need to be disabled prior to use *in vivo* (see next section), they do have one significant advantage over the amplicon system. Thus, although following recombination, some virus particles will contain recombinant virus having the gene of interest whilst others will contain the original non recombinant virus, it is relatively easy to separate these two types of viruses. Thus, both the recombinant and non recombinant viruses will form plaques on suitable cell lines (see below) and the recombinant virus can therefore be purified provided it can be distinguished from the non recombinant virus. This can be achieved, for example, by introducing the gene of interest into the thymidine kinase locus of the virus thereby inactivating this locus and rendering the resulting virus resistant to the effects of the antiherpes drug acyclovir (Ho and Mocarski, 1988; Palella *et al.*, 1988). Alternatively, the recombinant virus can be distinguished on the basis that it contains a readily assayable marker gene such as β–galactosidase either alone or in association with the gene of interest. If this cannot be achieved it is also possible in all cases to grow virus from individual plaques and identify those carrying the gene of interest by southern blot hybridization.

In whatever way the purification is achieved, it is thus possible with this technique to obtain a single purified preparation of the virus of interest rather than the mixed population obtained in the amplicon case. Although, the recombinant virus method has been much less used since it requires more specialist skills than the amplicon method, it is likely therefore that this method will be of increasing importance as gene therapy trials in human individuals are contemplated.

4. DISABLING OF HSV VECTORS

It is evident from the previous section that in both the amplicon and recombinant virus situations it will of critical importance to develop means of effectively disabling HSV. Thus, in the case of the recombinant virus, the actual vector virus must be disabled whilst in the case of the amplicon it will be equally essential to disable the helper virus. This is especially true since, as described above, wild type HSV virus will replicate lytically upon direct injection into the brain resulting in a lethal encephalitis. Therefore if advantage is to be taken of the ability of the virus to enter a latent state in neuronal cells, the potentially damaging effects of lytic replication must first be eliminated.

Of course however, it is essential to do this in a manner which still allows the recombinant or helper virus to be grown up so that stocks can be prepared for *in vivo* injection. It is this requirement to disable the virus for use *in vivo* whilst still allowing stocks to be prepared *in vitro* which constitutes the essential problem in effectively disabling the virus. Two basic methods of achieving this have been proposed and these will be discussed in turn.

The first method involves the inactivation of a virus gene which is essential for lytic replication in all cell types. The most frequently inactivated gene in vectors of this type is that encoding the essential immediate early protein ICP4 (Chiocca *et al.*, 1990; Dobson *et al.*, 1990). It is also possible to inactivate the other essential immediate early gene which encodes the viral ICP27 protein (Sacks *et al.*, 1985). Removal of either of these genes effectively renders the virus incapable of lytic replication in any cell type and it will not therefore produce encephalitis when injected into the brain. The problem evidently remains however, of growing these viruses in order to produce sufficient stocks. This is achieved by growing the virus on cell lines which have been artificially engineered by introduction of the gene encoding ICP4 or ICP27 and therefore allow lytic growth of the virus. Although this procedure does allow the virus stocks to be grown up, it suffers the disadvantage that recombination can occur between the defective virus genome lacking functional ICP4 or ICP27 and the functional ICP4 or ICP27 gene in the cell line. This will evidently result in the regeneration of dangerous wild type virus. Indeed, it is likely that such recombination was responsible for the death of 10% of the animals in the experiments of During *et al.* (1994) which was described above.

To overcome this problem, an alternative approach is to inactivate a virus gene which is necessary for replication in non dividing cells such as neurons but is not required for replication in dividing cell types. Hence, the virus will not produce encephalitis since it will not replicate lytically in the brain but stocks can readily be grown on dividing cells *in vitro*. Because of the advantages of this method, our laboratory has utilized a virus strain lacking the gene encoding ICP34-5. The removal of this gene renders the virus non neurovirulent upon injection into the brain but continues to allow it to replicate on dividing cells (Maclean *et al.*, 1991). We have introduced a marker β–galactosidase into this vector and shown that it is able to deliver genes to the central and peripheral nervous system following direct injection into the brain or the footpad of mice respectively (Coffin *et al.*, 1996b).

This system thus combines the ability of the virus to be grown up in culture without a complementing virus gene expressed in a cell line, with the lack of damaging encephalitis upon injection into the brain allowing gene transfer without damaging side effects. Unfortunately however, in these experiments only a relatively small number of cells in the brain successfully expressed the marker β–galactosidase following injection (Coffin *et al.*, 1996b). We therefore developed this vector further by inactivating also the gene encoding the viral ICP27 protein thereby introducing two defects into the virus genome. Very surprisingly, the resulting virus not only produced considerably less damage when injected into the brain *in vivo* but it was also capable of a much higher efficiency of gene delivery with approximately one thousand more neurons expressing the marker gene than with the single mutant virus lacking only functional ICP34-5 (Howard *et al.*, 1998). This double mutant virus therefore has both added safety since it contains two disabling mutations but is also a much higher efficiency gene delivery vector.

This virus does however, have to be grown on a cell line which artificially expresses ICP27 since this protein is essential for lytic growth of the virus on all cell types (Sacks *et al.*, 1985). The possibility of recombination resulting in a functional ICP27 gene being regenerated within the virus is minimised however by completely deleting the ICP27 gene from the virus and ensuring that there is no overlap between the DNA sequences remaining in the virus genome and those contained within the cell line. Moreover, even if a rare recombination event were to occur, this would still not regenerate a wild type virus since the resulting virus would continue to lack ICP34-5 and would still not cause a damaging encephalitis.

Hence this doubly disabled virus represents a relatively safe and efficient gene delivery vector which can be used as a platform for further disablement by deleting, for example, the gene encoding the other essential immediate early protein ICP4 and may therefore result ultimately in a virus which is safe enough for therapeutic use in humans.

5. LONG TERM EXPRESSION WITH HSV VECTORS

Many of the applications of gene therapy in the nervous system such as, for example, the treatment of Parkinson's disease will require long term expression of the therapeutic gene. Unfortunately however, during the onset of latent infection, the virus shuts down the expression of virtually all of its own genes in order to enter a virtually silent latent state (for review see: Roizman and Sears, 1987; Latchman, 1990). This shut down of gene expression also occurs for any exogenous genes which are introduced into the virus genome either under the control of the viruses own promoters, other virus promoters such as the immediate early promoter of cytomegalovirus or a variety of cellular promoters (Ho and Mocarski, 1988; Palella *et al.*, 1988; Fink *et al.*, 1992; Lokensgard *et al.*, 1994). This results in expression of the foreign gene being observed for only a few days at the most.

Clearly, some means must be found of overcoming this problem if HSV vectors are to be used successfully in a variety of diseases where long term expression of the therapeutic gene will be required. One approach to this problem involves taking advantage of the fact that a small region of the HSV genome remains active during latency producing the so called latency associated transcripts (LATs) (Croen *et al.*,1987; Stevens *et al.*, 1987). The expression of these LAT transcripts is driven by two adjacent promoters known as LATP1 and LATP2 (Fig. 2) and a number of laboratories have attempted to use these to drive long term expression of a foreign gene. This has been achieved either by inserting the foreign gene into the LAT region hence placing it under the control of the LAT promoters or by introducing the foreign gene linked to LATP1 or P2 into another region of the genome.

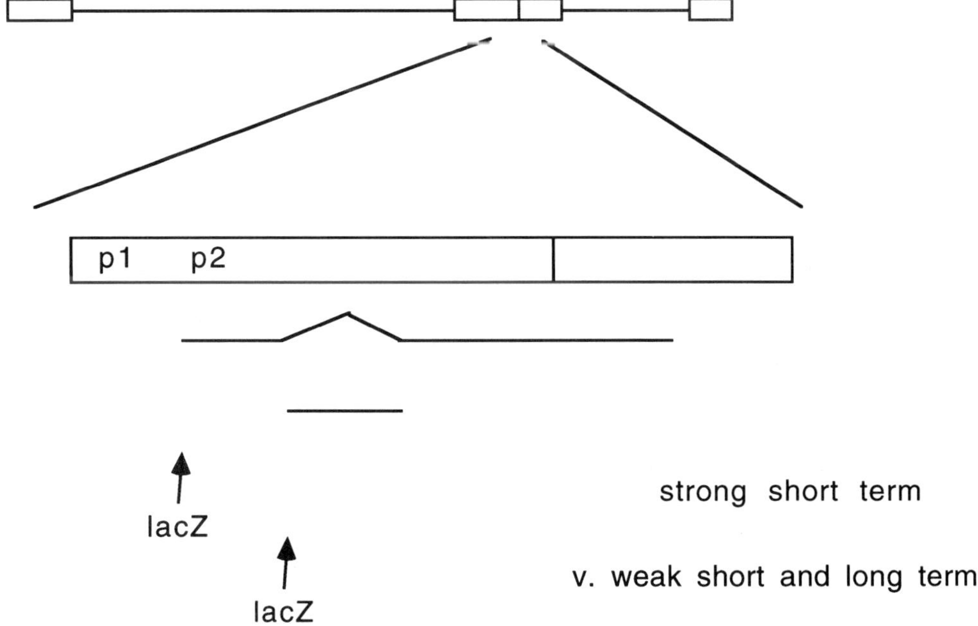

Figure 2. Schematic diagram showing the position of the Lat region and its two promoters P1 and P2 in the HSV genome.

In general, experiments with LATP1 have indicated that it is not sufficient for long term expression either when a gene is inserted directly downstream of it into the LAT locus or when a LATP1 promoter linked to the β–galactosidase gene is inserted into another site in the HSV genome (Margolis *et al.*, 1992; Margolis *et al.*,1993; Dobson *et al.*, 1995). In contrast, the LATP2 promoter does appear to give much longer term expression either within the LAT region or elsewhere, although such expression is relatively weak (Ho *et al.*, 1989; Goins *et al.*, 1994).

To overcome these difficulties, our laboratory has explored the possibility of whether LATP2 could act as an enhancer element conferring long term expression upon a strongly active promoter. Thus, we have linked LATP2 to the very strong cytomegalovirus immediate early promoter which normally only gives only very short term expression within an HSV vector and shown that this combination can direct strong long term expression when introduced into both the central and peripheral nervous systems. This is likely to be achieved by the LATP2 promoter acting as a signal to maintain the viral chromatin in its immediate vicinity in an open configuration consistent with transcription, thereby allowing the CMV IE promoter to produce high levels of transcription in the long term.

We are currently exploring the possibility of also linking LATP2 to cellular promoters with cell-type specific patterns of activity in order to determine whether such cell type specific activity could be produced in the long term within a virus vector by use of the P2 promoter. The use of the P2 promoter has another advantage however, in that like classical enhancer elements it can act bi-directionally and can therefore produce long term expression of two genes each under the control of their own promoters in opposite orientations with the P2 element contained at the centre of the construct (Fig. 3). This is of particular importance where it is desired to produce long term expression of two genes and could be used, for example, where maximal therapeutic efficiency will require two genes to be successfully expressed.

In addition however, it could be used to express both a marker gene and a gene of therapeutic interest. This would, for example, facilitate purification of the virus away from non recombinant virus (see previous section) on the basis of its expression of the marker gene with its subsequent use to express the therapeutic gene. Another example of the use of such a virus containing both a marker gene and a therapeutic gene would be in situations where it is necessary *in vivo* to deliver a gene to a population of cells in culture with subsequent re-insertion of these cells into a patient. Thus, many cells for which it is desired to achieve this, such as bone marrow stem cells or dentritic cells, are extremely difficult to transduce with exogenous genes. In the system proposed, the virus expressing two genes would be used to infect such cells and the successfully infected cells could be sorted on the basis of their expression of a reporter gene producing a 100% pure population expressing the therapeutic gene for reintroduction into the patient. We have recently shown the feasibility of this procedure by utilizing a virus which expresses both green fluorescent protein (GFP) and β–galactosidase and showing that sorting of the infected cells on the basis of their expression of GFP results in a 100% pure population of cells expressing β–galactosidase (Coffin *et al.*, 1998).

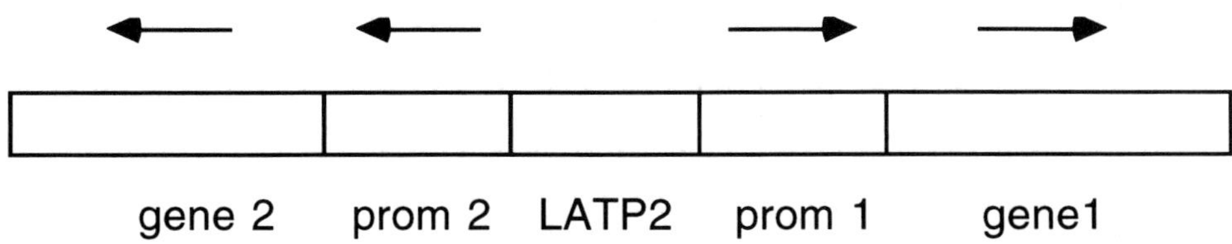

Figure 3. Use of the LAT P2 element as an enhancer allowing long term expression from two promoters in opposite orientations.

6. APPLICATIONS OF HSV VECTORS

Although, thus far, the easy to manipulate amplicon vectors have been most widely used to express genes of potentially therapeutic benefit in appropriate animal models, the progress made with disabled recombinant vectors indicates that they also represent an effective system for gene delivery. This is particularly so with the recent finding that such viruses do not recombine or reactive endogenous latent HSV upon introduction into the brain (Wang *et al*, 1997). This indicates that such a reactivation of wild type virus already contained within the patient is unlikely to be a problem with such gene therapy vectors. Hence, the progress made with the disabling of the virus whilst retaining its efficiency in gene delivery and in obtaining long term expression offers hope that such viruses may be of ultimate use therapeutically.

Indeed, in our laboratory we have already constructed viruses expressing both tyrosine hydroxylase and the neurotrophic factor GDNF which has a protective effect on dopaminergic neurons (Opacka-Juffry *et al.*, 1995). These viruses have been successfully used to express the appropriate genes and produce a reduction in rotational behaviour in the 6-hydroxydopamine model of Parkinson's disease. Similarly, we have utilised these viruses to express the genes encoding individual heat shock proteins. As with delivery of hsp genes by standard transfection of plasmid constructs (Amin *et al.*, 1996; Wyatt *et al.*, 1996) such over expression, for example, of the 70kDa heat shock protein (hsp70) can protect cultured neuronal cells from subsequent exposure to thermal stress or ischaemia. Moreover, over-expression of the 27 kDa heat shock protein (hsp27) protected the cells not only against thermal or ischaemic stress but also against stimuli which induce apoptosis whereas this anti-apoptotic effect was not observed with the hsp70 encoding virus (Wagstaff *et al.*, 1999). A similar anti-apoptotic effect in neuronal cells has also been observed when our vector is used to express the neuronal Brn-3a transcription factor (Smith *et al.*, 1998). Most importantly, the protective effect observed in these experiments was significant not only compared to a virus expressing a marker gene alone but also to samples of cells which were not treated prior to exposure to the stressful stimuli. Hence, the virus does not produce significant damaging effects itself and can deliver a foreign gene with an appropriate protective effect.

In similar experiments, we also observed a protective effect when these viruses expressing heat shock proteins were used to infect cells of cardiac origin (Brar *et al.*, 1999). This illustrates that HSV can also be used to deliver genes to other cell types and organs. Thus, we have previously shown that the virus can effectively deliver a marker gene to the heart *in vivo* (Coffin *et al.*, 1996a) and similar results have also been observed in skeletal muscle using a disabled virus vectors (Huard and Glorioso, 1995) and in liver cells using an amplicon vector (Lu *et al.*, 1995). Moreover, the ability to deliver genes to dendritic cells (see above and Coffin *et al.*, 1998) is of particular importance in view of the high efficiency of antigen presentation by these cells which could be use in cancer therapy.

Hence, whilst this review has concentrated on gene delivery to neuronal cells because of the particular applicability of HSV to this relatively difficult problem, the development of HSV vectors may also be of use for gene delivery to a variety of other tissues and cell types.

7. CONCLUSIONS

Gene delivery to the nervous system represents perhaps the ultimate challenge of gene therapy in view of the complexity of this system, the wide variety of intractable neurological diseases and the need to deliver the gene to non dividing cells. Although a variety of systems for such gene delivery are under development and are discussed in other chapters, we believe that HSV has unique advantages in terms of its large genome size and its ability to enter a latent state in neuronal cells. Considerable progress has been made in the effective disablement of this virus whilst retaining its ability to deliver genes and in producing long term expression of the foreign gene. Although much remains to be achieved in the further disablement of the virus and its testing in rodent and primate models of human diseases, it is likely that these viruses may ultimately be of use in human gene therapy procedures for otherwise intractable neurological diseases.

8. REFERENCES

Aigner, L., Arber, S., Kapfhammer, J.P., Laux, T., Schneider, G., Botteri, F., Brenner, H-R. and Caroni, P. (1995). Over expression of the neural growth associated protein GAP-43 induces nerve sprouting in the adult nervous system of transgenic mice. *Cell*, 83: 269-278.

Amin, V., Cumming, D.V.E. and Latchman, D.S. (1996). Over-expression of heat shock protein 70 protects neuronal cells against both thermal and ischaemic stress but with different efficiencies. *Neurosci. Lett.*, 206: 45-48.

Backlund, E-O., Grandberg, P-O., Hamberger, B., Knutsson, E., Martensson, A., Sedirall, G., Seigar, A. and Olson, L. (1985). Transplantation of adrenal medulla tissue to striatum in Parkinsonism. *J. Neurosurgery*, 62:169-173.

Bloom, D.C., Maidment, N.tT, Tan, A., Dissette, V.B., Feldman, L.T. and Stevens, J.G. (1995). Long term expression of a reporter gene from latent herpes simplex virus in the rat hippocampus. *Mol. Brain Res.*, 31: 48-60.

Brar, B.K., Stephanou, A., Wagstaff, M.J., Coffin, R.S., Marber, M.S., Engelmann, G. and Latchman, D.S. (1999). Heat shock proteins delivered with a virus vector can protect cardiac cells against apoptosis as well as against thermal or hypoxic stress. *J. Mol. Cell. Cardiol.*, 31: 135-146.

Chiocca, A.E., Choi, B.B., Cai, W., DeLuca, N., Schaffer, P.A., DiFiglia, M., Breakefield, X.O. and Martuzaj, R.L. (1990). Transfer and expression of the lac Z gene in rat brain neurons by herpes simplex virus mutants. *New Biol.*, 266: 739-746.

Coffin, R.S. and Latchman, D.S. (1996). Herpes Simplex Virus based vectors. In: *Genetic manipulation of the nervous system*, D.S. Latchman (Ed.), Academic Press, London, pp. 235-248.

Coffin, R.S., Howard, M.K., Cumming, D.V.E., Dollery, C.M., McEwan, J., Yellon, D.M., Marber, M.S., Maclean, A.R., Brown, S.M. and Latchman, D.S. (1996a). Gene delivery to cardiac cells *in vitro* and *in vivo* using herpes simplex virus vectors. *Gene Ther.*, 3: 560-566.

Coffin, R.S., Maclean, A.R., Latchman, D.S. and Brown, S.M. (1996b) Gene delivery to the central and peripheral nervous systems of mice using HSV1 ICP34-5 deletion mutant vectors. *Gene Ther.*, 3: 886-891.

Coffin, R.S., Thomas, S.K., Thomas, N.S.B., Lilley, C.E., Pizzey, A.R., Griffiths, C.H., Gibb, B.J., Wagstaff, M.J.D., Inges, S.J., Binks, M.H., Chain, B.M., Thrasher, A.J., Rutault, K. and Latchman, D.S. (1998). Pure populations of transduced primary human cells can be produced using GFP expressing herpes virus vectors and flow cytometry. *Gene Ther.*, 5: 718-722.

Croen, K.D., Ostrove, J.M., Dragovic, L.J., Smialek, J.E and Straus, S.E. (1987). Latent herpes simplex virus in human trigeminal ganglia. Detection of an immediate early gene antisense transcript by *in situ* hybridization. *New Eng. J. Med.*, 317:1427-1432.

Dobson, A.T., Margolis, T.P., Sedarati, F., Stevens, J.G. and Feldman, L.T. (1990). A latent nonpathogenic HSV-1 derived vector stably expresses β galactosidase in mouse neurons. *Neuron*, 5: 353-360.

Dobson, A.T., Margolis, T.P., Gomes, W.A. and Feldman, L.T. (1995). *In vivo* deletion analysis of the herpes simplex virus type 1 latency associated transcript promoter. *J.Virol.*, 69: 2264-2270.

During, M.J., Naegele, J.R., O'Malley, K.L. and Geller, A.I. (1994). Long-term behavioural recovery in parkinsonian rats by an HSV vector expressing tyrosine hydoxylase. *Science*, 266:1399-1403.

Fedderoff, H.J., Geschwind, M.D., Geller, A.I. and Kessler, J.A. (1992). Expression of nerve growth factor *in vivo* from a defective herpes simplex virus 1 vector prevents effects of axotomy on sympathetic ganglion. *Proc.Natl.Acad.Sci.USA*, 89:1636-1640.

Fink, D.J., Sternberg, L.R., Weber, P.C., Mata, M., Goins, W.F. and Glorioso, J.C. (1992). *In vivo* expression of β–galactosidase in hippocampal neurons by HSV-mediated gene transfer. *Hum. Gene Ther.*, 3: 11-19.

Fink, D.J., DeLuca, N.A., Goins, W.F. and Glorioso, J.C. (1996). Gene transfer to neurons using herpes simplex virus-based vectors. *Annu. rev. Neurosci.*, 19: 265-287.

Geller, A.I. and Breakefield, X.O. (1990). A defective HSV-1 vector expresses *E. coli* β–galactosidase in cultured CNS neurons. *Proc.Natl.Acad.Sci.USA*, 87: 1149-1153.

Geschwind, M.D., Hartnick, C.J., Liu, W., Amat, J., Van de Water, T.R. and Federoff, H.J. (1996). Defective HSV 1 vector expressing BDNF in auditory ganglia elicits neurite outgrowth: Model for treatment of neuron loss following cochlear degeneration. *Hum. Gene Ther.*, 7: 173-182.

Goins, W.F., Stenberg, L.R., Croen, K.D., Krause, P.R., Hendricks, R.L. Fink, D.J., Straus, S.E., Levine, M. and Gloriso, J.C. (1994). A novel latency-associated promoter is continued within the herpes simplex virus type 1 UL flanking repeats. *J.Virol.*, 68: 2239-2252.

Haase, G., Kennel, P., Pettmann, B., Vigne, E., Akli, A., Revah, F., Schmalbruch, H. and Kahn, A. (1997). Gene therapy of murine motor neuron disease using adenoviral vectors for neurotrophic factors. *Nature Med.*, 3, 429-436.

Ho, D.Y. and Mocarski, E.S. (1988). B-galactosidase as a marker in the peripheral and neural tissues of the herpes simplex virus infected mouse. *Virology*, 167: 279-283.

Ho, D.Y. and Mocarski, E.S. (1989). Herpes Simplex virus latent RNA (Lat) is not required for latent infection in the mouse. *Proc. Natl. Acad. Sci. USA*, 86: 7596-7600.

Howard, M.K., Kershaw, T., Gibb, B., Storey, N., MacLean, A.R., Zeng, B.Y., Tel, B.C., Jenner, P., Brown, S.M., Woolf, C.J., Anderson, P.N., Coffin, R.S. and Latchman, D.S. (1998). High efficiency gene transfer to the central nervous system of rodents and primates using herpes virus vectors lacking functional ICP27 and ICP34.5. *Gene Ther.*, 5: 1137-1147.

Huard, J. and Glorioso, J.C. (1995). Herpes simplex virus type 1 vector mediated gene transfer to muscle. *Gene Ther.*, 2: 385-392.

Kaplitt, M.G., Pfaus, J.G. , Kleopoulous, S.P., Hanlon, B.A., Rabkin, S.D. and Pfaff, D.W. (1991). Expression of a functional foreign gene in adult mammalaian brain following *in vivo* transfer *via* a herpes simplex virus type 1 defective viral vector. *Mol. Cell. Neurosci.*, 2: 320-330.

Kennedy, G.E. (1997). Potential use of herpes simplex virus (HSV) vectors for gene therapy of neurological disorders. *Brain*, 120: 1245-1259.

Kwong, A.D. and Frenkel, N. (1985). The herpes simplex virus amplicon IV. Efficient expression of a chimeric chicken ovalbumin gene amplified within defective virus genomes. *Virology*, 142: 421-425.

Latchman, D.S. (1990). Molecular biology of herpes simplex virus latency. *Int. J. Exp. pathol.*, 71: 133-141.

Latchman, D.S. (1994). Herpes Simplex virus vectors for gene therapy. *Mol. Biotechnol.*, 2: 179-195.

Lawrence, M.S., Ho, D.Y., Sun, G.H., Steinberg, G.K. and Sapolsky, R.M. (1996). Over expression of Bc1-2 with herpes simplex virus vectors protects CNS neurons against neurological insults *in vitro* and *in vivo*. *J. Neuroscience*, 16: 486-496.

Lawrence, M., McLaughlin, J., Sun, G., Ho, D., McIntosh, L., Kunis, D., Sapolsky, R. and Steinberg, G. (1997). Herpes simplex viral vectors expressing Bc1-2 are neuro-protective when delivered after a stroke. *J. Cereb. Blood Flow Metab.*, 17: 740-744.

Leib, D.A. and Olivo, P.D. (1993). Gene delivery to Neurons: is herpes simplex virus the right tool for the job? *BioEssays*, 15: 547-554.

Lokensgard, J.R., Bloom, D.C., Dobson, A.T. and Feldman, L.T. (1994). Long term promoter activity during herpes simplex virus latency. *J.Virol.*, 68: 7148-7158.

Lu, B., Gupta, S. and Federoff, H. (1995). *Ex vivo* hepatic gene transfer in mouse using a defective herpes simplex virus-1 vector. *Hepatology*, 21: 752-759.

Maclean, A.R., Fareed, M.U., Robertson, L., Harland, J. and Brown, S.M. (1991). Herpes Simplex Virus type 1 deletion variants 1714 and 1716 pinpoint neurovirulence related sequences in Glasgow strain 17+ between immediate-early gene 1 and the 'a' sequence. *J. Gen.Virol.*, 72: 631-639.

Margolis, T.P., Sederati, F., Dobson, A.T., Feldman, L.T. and Stevens, J.G. (1992). Pathways of viral gene expression during acute neuronal infection with HSV-1. *Virology*, 189: 150-160.

Margolis, T.P., Bloom, D.C., Dobson, A.T., Feldman, L.T. and Stevens, J.G. (1993). Decreased reporter gene expression during latent infection with HSV LAT promoter constructs. *Virology*, 197: 585-592.

Miller, A.G., Adam, M.A. and Miller, A.D. (1990). Gene transfer by retrovirus vectors occurs only in cells that are actively replicating at the time of infection. *Mol. Cell. Biol.*, 10: 4239-4242.

Opacka-Juffry, J., Ashworth, S., Hume, S., Martin, D., Brooks, D. and Blunt, S. (1995). rhGDNF protects nigrostriatal DA neurons against the neurotoxin 6-OHDA: an *in vivo* study using microdialysis and positron emission tomography. *Neuro. Report*, 7: 348-352.

Palella, T.D., Silverman, L.J., Schroll, C.T., Homa, F.L., Levine, M. and Kelley, W.N.

(1988). Herpes simplex virus mediated human hypoxanthine-guanine phosphoribosyl transferase gene transfer into neuronal cells. *Mol. Cell. Biol.*, 8: 457-460.

Roizman, B. and Sears, A.E. (1987). An inquiry into the mechanisms of herpes simplex virus latency. *Ann. Rev. Microbiol.*, 41: 543-571.

Sacks, W.R., Greene, C.C., Aschmann, D.P. and Schaffer, P.A. (1985). Herpes simplex virus type 1 ICP27 is an essential regulatory protein. *J.Virol.*, 55: 796-805.

Smith, M.D., Ensor, E.A., Coffin, R.S., Boxer, L.M. and Latchman, D.S. (1998). Bcl-2 transcription from the proximal p2 promoter is activated in neuronal cells by the Brn-3a POU family transcription factor. *J. Biol. Chem.*, 273: 16715-16722.

Spaete, R.R. and Frenkel, N. (1982). The herpes simplex virus amplicon: a new eukaryotic defective virus cloning amplifying vector. *Cell*, 30: 295-304.

Stevens, J.G., Wagner, E.K., Devi-Rao, G.B., Cook, M.L. and Feldman, L.T. (1987). RNA complementary to a herpes virus alpha gene mRNA is prominent in latently infected neurons. *Science*, 235: 1056-1059.

Tsuda, M. and Imaokea, T. (1996). Direct Injection of Plasmid DNA into the Brain. In: *Genetic Manipulation of the Nervous System*, D.S. Latchman (Ed.), Academic Press, London, pp. 235-248.

Wagstaff, M.J., Collaco-Moraes, Y., Smith, J., de Belleroche, J.S., Coffin, R.S. and Latchman, D.S. (1999). Protection of neuronal cells from apoptosis by Hsp27 delivered with a herpes simplex virus-based vector. *J. Biol. Chem.*, 274: 5061-5069.

Wang, Q., Guo, J. and Jia, W. (1997). Intracerebral recombinant HSV-1 vector does not reactivate latent HSV-1. *Gene Ther.*, 4:1300-1304, 1997.

Wyatt, S., Mailhos, C. and Latchman, D.S. (1996). Trigeminal ganglion neurons are protected by the heat shock proteins hsp70 and hsp90 from thermal stress but not from programmed cell death following NGF withdrawal. *Brain Res.*, 39: 52-56.

9. SUMMARY — A number of different viruses have been proposed as potential vectors for gene delivery to the nervous system. Herpes simplexvirus (HSV) has particular advantages in this regard since it establishes life long asymptomatic infections of the nervous systems and has a large genome size capable of accepting large amounts of foreign DNA. Two methods of using HSV-based vectors in gene delivery have been proposed. These involve either the insertion of the foreign gene directly into the virus to create a recombinant virus or its insertion into an amplicon plasmid vector which contains an HSV origin of replication and packaging signal and which requires a helper HSV virus in order to be propagated. In both cases, it is necessary to disable either the recombinant virus itself or the helper virus to prevent damaging effects due to lytic replication by the virus whilst maintaining the efficiency of gene delivery. This disabling process is one of the major challenges still to be overcome before HSV-based vectors can be used in clinical therapy, the other being the development of means to ensure that the foreign gene is expressed in the long term after the virus has entered latency. The potential means of overcoming these problems and the potential use of HSV-based vectors in a number of different situations are discussed.

Key Words: virus vectors; herpes simplex virus; gene therapy; neurological disease

Progress in Gene Therapy: Basic and Clinical Frontiers, pp. 161-181
R. Bertolotti *et al.* (Eds)

Liver-selective nucleic acid delivery using the asialoglycoprotein receptor

Toshiko Fukuma, George Y. Wu and Catherine H. Wu*

Division of Gastroenterology-Hepatology, Department of Medicine, University of Connecticut Health Center, 263 Farmington Avenue, Farmington, CT 06030-1845, USA

Table of Contents

* Corresponding author. E-mail: cwu@nso1.uchc.edu

1. INTRODUCTION: TARGETED GENE DELIVERY OF GENES AND OLIGONUCLEOTIDES

Gene therapy, the idea that gene transfer for effective management and possible cure of human disease, has attracted considerable interest and effort in the basic science and clinical medicine. However, substantial problems have to be solved before it can be routinely applied in the clinical practice. Currently, no single perfect vector is available, either viral or non-viral, with all the properties of an ideal gene transfer vehicle for gene therapy: efficiency, safety, lack of immunogenicity, unlimited size capability, and targetability.

An attractive approach for targeted delivery of genes and oligonucleotides is to take advantage of naturally existing cell surface receptors that are capable of internalization of specific ligands by receptor-mediated endocytosis. Receptor-mediated endocytosis is the process by which the cell promotes the selective uptake and processing of macro-molecules. It involves the interaction of ligand with the receptor on the plasma membrane, clustering of the ligand-receptor complexes in coated pits and internalization into membrane-limited cytoplasmic compartments. One of the extensively investigated endocytotic systems is that of the asialoglyco-protein receptor, which recognizes and internalizes glycoproteins that have exposed terminal galactose or N-acetylgalactosamine residues.

This review will cover general aspects related to targeting to the liver via the asialoglycoprotein receptor, and the relationship between gene expression, persistence, the structure and size of DNA complexes, as well as the current strategies aimed to improve the overall efficiency of receptor-mediated gene and oligonucleotide delivery.

2. STRUCTURE OF THE ASIALOGLYCOPROTEIN ASGP RECEPTOR

The hepatocellular membrane contains approximately 150,000 to 250,000 asialoglycoprotein (ASGP) receptors. The surface membrane population represents approximately 35% of the entire cellular content of ASGP receptors. The ASGP receptor, a C-type hepatic lectin receptor, is a transmembrane glycoprotein hetero-oligomer, composed of two structurally different subunits, H1 and H2. Both subunits are similar in the molecular weight of 46 and 50 kD respectively, and have an amino-terminal cytoplasmic tails, a transmembrane domains, that functions as an internal signal sequence, and a carboxyl-terminal extracellular domain. In addition, they contain two (HL1) or three (HL2) N-linked oligosaccharides in the extracellular domain and a phosphate on a serine residue of the cytoplasmic domain (Stockert, 1995). H1, the major species of the receptor, is seven times more abundant than H2, but the two subunits are necessary for the ligand-binding function. Mice lacking the minor subunit (MHL-2) have shown a significant reduction in major subunit (MHL-1) expression and do not clear asialoorosomucoid, indicating that H2 may promote stability of H1 (Ishibashi *et al.*, 1994).

Analysis of structural requirements for binding to the ASGP receptor by studying oligosaccharide structures found in desialylated serum glycoproteins have shown that ASGP receptor strongly binds desialylated triantennary oligosaccharides by interaction with the three galactose residues (Lee *et al.*, 1983; Rice *et al.*, 1990). Recently it was shown that lactoferrin (Lf) bound the rat major subunit of ASGP receptor by a mechanism that is independent of galactose-, N-acetylgalactosamine-terminated glycoproteins (McAbee *et al.*, 1998).

3. ENDOCYTOTIC PATHWAY OF THE ASGP RECEPTOR

Upon binding to ligands, the receptors concentrate in coated pits prior to internalization in clathrin-coated vesicles, which rapidly lose their clathrin coat and become endosomes. Microtubule function is essential for receptor-mediated endocytosis by translocation of endosomes to lysosomes and recycling of the receptor to the plasma membrane. Endosomes fuse with each other and some smooth-membrane vesicles forming a structure that has been termed a compartment of uncoupling receptor and ligand (CURL). Acidification of this compartment leads to segregation of ligand from receptor, with receptor molecules recycling back to the plasma membrane. The time calculated for an ASGP receptor to complete the constitutive cycle is 7-8 minutes (Schwartz *et al.*, 1982). It was subsequently reported that the endocytotic system behaves as through there are two distinct subpopulations of ASGP receptors, each of which mediates ligand uptake and processing in its distinct pathways. One of them, termed State 2 pathway, is considered more biologically active because of the shorter recycling time (Weigel *et al.*, 1998). Recently, it has also been shown that the two receptor subunits were endocytosed at different average rates, and lower than that of ligand. In the absence of ligand, subunit H1 was internalized twice as fast as H2, and ligand binding increased the turnover rates of both subunits approximately twofold. This suggests heterogeneity of oligomer formation of the receptor molecule, and potentially of ligand specificity (Bider *et al.*, 1998).

4. GENE TRANSFER VIA ASGP RECEPTORS

In spite of the considerable knowledge of how the receptor works, the physiological function of ASGP receptor has still not been clearly established. It is thought to be have a regulatory role in removal of glycoproteins from plasma after loss of sialic acid residues. A recent report presented the possibility of circulating cellular fibronectin as a naturally occurring ligand for the receptor in rat liver (Rotundo *et al.*, 1998). On the other hand, the special characteristics of the ASGP receptor, its large number on the surface of hepatocytes and the selectivity of its location, represent a useful system for targeting of chemotherapeutic agents and biologically active molecules including foreign transgenes to the liver.

Targeting of hepatocytes by the asialoglycoprotein receptor was originally shown by chemically coupling antiviral agents (trifluorothymidine or adenine 9-β-D-arabinofuranoside; Fiume *et al.*, 1979 and 1980) or anti-malarial drugs (primaquine; Trouet *et al.*, 1981) to asialoglycoproteins.

Radiolabeled galactosyl-albumin was also used to identify space-occupying lesions in the liver due to the fact that tumors do not contain significant numbers of asialoglycoprotein receptors (Vera *et al.*, 1985). The same phenomenon was the basis of a technique called "targeted rescue" consisting of treatment of liver tumors by the induction of liver damage with a cyto-toxin (methotrexate) and the protection of normal hepatocytes (expressing the asialoglyocoprotein receptor) by targeting a specific antagonist (folinic acid) coupled to asialoglycoprotein (Wu *et al.*, 1983 and 1985).

While targeted delivery of small molecules was achieved by chemical linkage to an asialoglycoprotein carrier, similar techniques could not be used for reactive macromolecules such as DNA because of potential damage resulting from covalent binding to the asialoglycoprotein. To solve this problem, a DNA carrier system was developed consisting of 2 components: the ligand possessing exposed terminal galactose residues linked covalently to a molecule that can bind DNA in an electrostatic, non-damaging manner. These conjugates can form soluble complexes with DNA that could be recognized and internalized by hepatocytes via ASGP receptor-mediated endocytosis.

The first successful demonstration that the DNA carrier system could deliver and express DNA in ASGP receptor-bearing cells was conducted in late 1980s. Wu and Wu prepared a DNA-protein complex containing asialo-orosomucoid (ASOR) covalently linked to polylysine and mixed with plasmid DNA coding for the chloramphenicol acetyltransferase (CAT) gene driven by an SV-40 viral promoter. Incubation of this complex with two hepatoma cell lines, HepG2 (ASGP receptor positive) and SK-Hep 1 (ASGP receptor negative), showed that only HepG2 cells produced detectable CAT activity, and that this activity could be blocked by an excess of ASOR (Wu and Wu, 1987). Targeted gene delivery and expression *in vivo* were also demonstrated by intravenous injection of the complex and subsequent detection of CAT DNA and activity exclusively in liver extracts (Wu and Wu, 1988b).

5. COMPONENTS OF THE DNA CARRIER SYSTEM FOR TARGETED GENE DELIVERY

a) Ligands

Natural ligands. Glycoproteins purified from serum and desialylated in the laboratory have been used as natural ligands for the ASGP receptor. Asialoorosomucoid (ASOR) has been commonly used (Wu and Wu, 1987, 1988a and 1988b; Wu *et al.*, 1989, 1991 and 1992; Wilson *et al.*, 1992; Chowdhury *et al.*, 1993 and 1996; Stankovics *et al.*, 1994). Another common natural ligand is asialofetuin (AF), with three galactose terminal motifs (Townsend *et al.*, 1986; Koike *et al.*, 1994; Hara *et al.*, 1995a, 1995b and 1996). Other asialoglycoproteins have not been tested extensively or have lower affinity for the receptor, such as asialotransferrin (AsTf) that has only two galactose residues (Rice and Lee, 1990).

Artificial ligands. Proteins or peptides that do not naturally possess exposed terminal galactose residues can be transformed into a ligand for the ASGP receptor by the addition of galactose residues. A minimum of three branched galactose residues is necessary to provide the ligand with high affinity for the receptor. A potential advantage of synthetic compounds over natural ligands is the possibility to use the same molecule as ligand and carrier, thus reducing the number of components of the complex. Substances such as galactosylated polylysine, polyethylenimine, histones, peptides or lipopolyamines have been used.

Viral fusion proteins. Sendai virus F protein is a glycoprotein that has terminal biantennary galactose moieties which can bind specifically to asialoglycoprotein receptor on the membrane of HepG2 cells *in vitro* (Markwell, 1985). Recent studies using reconstituted Sendai virus envelopes containing only the fusion protein (F-virosome) as biological carriers for the delivery of foreign gene have shown the possibility of application of this vehicle for gene delivery *in vitro* and *in vivo* (Ramani *et al.*, 1997 and 1998).

b) DNA binding molecules

Polylysine. This polycation molecule is a commonly used carrier in DNA-protein complexes, either covalently linked to asialoglycoproteins or modified itself to present galactose residues (Ferkol *et al.*, 1993; Perales *et al.*, 1994 and 1997; Martinez-Fong *et al.*, 1994). Polylysine not only binds DNA in a strong non-damaging manner, but also can protect DNA from degradation by serum nucleases. Experiments reproducing conditions encountered during transfection *in vitro* and *in vivo* (incubations in culture medium containing fetal bovine serum, fresh whole rat serum or crude cell lysates) have shown efficient protection of double stranded DNA and oligonucleotides bound to complexes while non-bound nucleic acids were degraded by nucleases (Chiou *et al.*, 1994).

Polyethylenimine (PEI). This highly branched polycationic polymer has the capability of binding DNA more efficiently than linear polycations such as polylysine. In addition, PEI has a high acid buffering capacity that protects DNA from lysosomal nucleases and induces osmotic swelling and lysosomal disruption. PEI itself has been shown to be an efficient transfection agent both *in vitro* and *in vivo* (Boussif *et al.*, 1995). Galactosylated PEI has also been used *in vitro* (Zanta *et al.*, 1997) and recently *in vivo* to deliver chimeric RNA/DNA oligonucleotides for site directed mutagenesis (Kren *et al.*, 1998).

Histones. Galactosylated histones offer some advantages over natural asialoglycoprotein ligands including simplicity, natural DNA-binding capacity and nuclear localization that enhances foreign DNA entry into the cell nucleus. They have been successfully used to deliver a CAT expressing plasmid into HepG2 cells (Chen *et al.*, 1994).

Liposomes. Cationic liposomes can be used to substitute for polycations and associate DNA by adsorbing DNA molecules on the surface and/or encapsulating them into the liposome interior. Asialofetuin-liposomes have been effective delivering reporter genes (β-galactosidase or CAT) *in vitro* to established cell lines and primary hepatocytes and *in vivo* into the liver after intraportal injection (Koike *et al.*, 1994; Hara *et al.*, 1995a, 1995b and 1996). Cationic liposomes have also been mixed with regular asialoglycoprotein-polylysine-DNA complexes, enhancing *in vitro* transfection without altering receptor specificity (Mack *et al.*, 1994). Galactosylated lipopolyamines (lipids with endosome buffering capacity and cell membrane destabilization properties) provided an efficient transfection system *in vitro* (Remy *et al.*, 1995). Recently, liposome/DNA complexes prepared using novel galactosylated cholesterol derivatives were shown to be efficiently recognized by ASGP receptors and internalized and led to gene expression (Kawakami *et al.*, 1998).

Other carrier molecules. Intercalating agents like bisacridine have been galactosylated and used *in vitro* (Haensler and Szoka, 1993) or glycopeptides containing three or four galactose terminal residues have shown to be effective delivering the luciferase gene *in vitro* (HepG2) and *in vivo* (Plank *et al.*, 1992; Merwin *et al.*, 1994; Wadhwa *et al.*, 1995). A synthetic polypeptide, a galactosyl-D-lysine/D-serine copolymer has been reported as an effective reagent facilitating formation of DNA-ligand complex for subsequent receptor-mediated gene transfer *in vivo* (Hisayasu *et al.*, 1998).

6. DNA CONDENSATION, COMPLEX STRUCTURE AND ASGP RECEPTOR TARGETING

Few studies have tried to correlate the size and/or structure of the DNA-protein complexes, internalization into cells via receptors and transgene expression efficiency. In fact, the size of the DNA-ligand complex is a fundamental factor in efficient internalization. It has been shown that the ASGP receptor internalizes ligands 12 nm in diameter or smaller (Schlepper-Schafer *et al.*, 1986; Bijsterbosch and Van Berkel, 1992), while macrophages and Kupffer cells bind larger asialoglycoproteins and other galactose-terminal glycoproteins by particle galactose receptors on their surface.

Perales *et al.* have demonstrated that the behavior of DNA-polycation complexes in solution depends on the method of condensing DNA with the polycation (Perales *et al.*, 1994a). DNA condensed with ligand-polylysine in a step-down dialysis method from high (2-3 M) to low (0.015 M) NaCl concentrations which resulted in the formation of multimolecular aggregates with a toroid shape under electron microscopy (Y DNA) and a median diameter of 100-200 nm. Most complexes used to date belong to this class. However, unimolecular complexes of small, defined size (12 nm) can be prepared by the slow addition of the polycation to a solution of DNA at high concentration of NaCl (0.7 M) using a vigorous mixing to achieve a nucleus of condensation with individual molecules of DNA. Aggregated particles were dissolved to small complexes by slowly adjusting NaCl concentration to approximately 1 M. Systemic injection, into caudal vena cava of adult rats, of these small complexes (12 nm) carrying a plasmid coding for the human Factor IX led to persistence of DNA in the liver, and detection of mRNA and functional Factor IX in serum up to 140 days, the longest expression reported by receptor-mediated endocytosis gene delivery (Perales *et al.*, 1994b). Subsequent studies have confirmed the greater efficacy of small DNA-ligand complexes over toroid complexes. Using the luciferase gene as a reporter, only small complexes were found to produce significant levels of luciferase expression in the liver (Perales *et al.*, 1997).

7. THE FATE OF TRANSGENE AFTER ASGP RECEPTOR MEDIATED ENDOCYTOSIS

Time course experiments performed with asialoorosomucoid-polylysine complexes carrying DNA injected by jugular vein showed that twenty minutes after systemic injection more than 85% of complexed DNA was taken up by the liver, 80% of which was found within hepatocytes. However, the transgene concentration decreased to 10% and 5% of the initial levels in 4 and 24 hours, respectively, and became undetectable at 7 days

(Chowdhury *et al.*, 1993). The short duration of gene expression and the poor overall efficiency of gene delivery based on receptor-mediated endocytosis was attributed to rapid degradation. However, according to other studies, despite a high degree of degradation, some DNA molecules might persist in plasma/endosome-enriched fractions where they are not degraded and could release DNA molecules in a controlled fashion, leading to prolonged gene expression (Edwards *et al.*, 1996).

8. STRATEGIES TO PROLONG THE GENE EXPRESSION OF ASGP RECEPTOR MEDIATED GENE TRANSFER

a) Partial hepatectomy

In vitro observations that exogenous DNA was more frequently integrated in the host genome in dividing cells led to the idea that stimulation of liver cell replication in the presence of targeted transgene might result in persistent the gene expression. The liver represented an interesting organ to test this hypothesis because of its capability of rapid regeneration after injury. The regeneration occurs as a result of proliferation of residual hepatocytes to completely replace those that are lost. The classic model of partial hepatectomy inducing liver regeneration was employed. Wu *et al.* showed persistent CAT enzyme activity and presence of injected plasmid DNA in rat liver up to 11 weeks when the DNA complexes were injected intravenously followed 30 minutes later by a two-thirds hepatectomy (Wu *et al.*, 1989). Other studies have confirmed this observation and demonstrated that the majority of transgene DNA are accumulated in cytoplasmic vesicles and retained as stabilized episomal DNA (Wilson *et al.*, 1992). This increase in DNA persistence has been attributed to disruption of the microtubule network by hepatocyte replication induced by hepatectomy (Bommineni *et al.*, 1994). Although partial hepatectomy was effective, it is not a practical method to apply routinely and other strategies have been devised.

b) Pharmacological strategies

Interference with lysosomal fusion or acidification can increase the proportion of DNA that escapes degradation in this compartment.. The discovery of the important role played by the microtubule network in the progression of endosomes to lysosomes has led to the use of compounds known to inhibit microtubule assembly. One such compound, colchicine, is known to cause depolymerization of microtubules and blockade of the endocytotic pathway. Using bilirubin-UDP-glucuronosyltransferase-deficient rats (Gunn rats), Chowdhury *et al.* have shown that pretreatment of animals with colchicine 30 minutes before DNA injection resulted in persistence of DNA, and reduction of bilirubin levels for more than 8 weeks (Chowdhury *et al.*, 1996). Chloroquine is another interesting pharmacological agent which is able to block lysosomal acidification. DNA complexes containing a galactosylated glycopeptide carrier (Wadhwa *et al.*, 1995) as well as a transferrin model of receptor-mediated endocytosis (Cotten *et al.*, 1990) resulted in enhancement of marker gene expression *in vitro.*

c) Structural modifications of the complex

Several compounds can be added to the basic components of the DNA carrier system to provide additional virus-like properties that improve the efficiency of the system. Most of these compounds have endosomolytic activity, disrupting endosomes-lysosomes and decreasing the amount of DNA that is degraded. These agents include fusogenic peptides, endosomolytic proteins or whole adenoviral particles.

Fusogenic peptides. Hemagglutinin peptides of various pathogens (such as the HA-2 N-terminal peptide from the Influenza virus) undergo a rearrangement in structure in acidic environments resulting in membrane agglutination activity. This property has been exploited by including these peptides in complexes containing transferrin-polylysine-DNA, significantly enhancing gene delivery *in vitro* (Wagner *et al.*, 1992). When the fusogenic peptide, HA-2 N-terminal peptide from the Influenza virus, was incorporated into the DNA complex with the galactose-polylysine conjugate, the reporter gene expression in HepG2 cells was more than 500-fold greater than the control without the peptide (Miyauchi *et al.*, 1997). Peptides used in combination with complexes targeting the ASGP receptor include gramicidin S, an amphipathic cyclic decapeptide that destabilizes biological membranes (Hara *et al.*, 1996) and a 25 amino acid peptide from vesicular stomatitis virus G protein (VSVG) (Wu *et al.*, 1997; Schuster *et al.*, 1999). Schuster *et al.* showed that the expression of the reporter gene by a conjugate containing ASOR L-lysine-methylester-VSVG, was 10^3-fold higher in mouse liver compared to that of control without VSVG component. The amount of the intact transgene within isolated liver cell nuclei was increased by up to 10^2-fold over that of control using semiquantitative competitive PCR analysis. These results are consistent with the putative improved endosomal release of delivered DNA by documented pH-dependent hemolytic properties toward the endosome of the VSVG peptide.

Proteins. Listeriolysin O (LLO). A protein from *Listeria monocytogenes* has a critical role in the escape of bacteria from the phagolysosome though lysing endosomes in a pH-dependent manner. When incorporated into an ASGP-polylysine-DNA complex, a 100-fold increase in reporter gene expression was observed *in vitro* (Zhang *et al.*, 1994). It was also demonstrated an LLO-containing ASOR-polylysine-DNA complex enhanced reporter gene expression only in ASGP receptor positive cells (Walton *et al.*, 1999).

Adenoviral particles. Adenoviruses are capable of inducing endosomal lysis during the process of adenoviral infection through the interaction of the endosomal membrane and capsid proteins on the surface of the viral particle. Co-internalization of compounds together with adenoviruses was found to release that compound from the endosome. Several groups have shown a significant increase in the efficiency of gene delivery when complexes for ligand-mediated transfection have been co-administered with adenoviruses (Curiel *et al.*, 1991; Cotten *et al.*, 1992). Conjugated adenovirus-ASOR-polylysine-DNA complexes have also been tested, allowing a significant decrease in the viral titer needed to improve transgene expression (Cristiano *et al.*, 1993a and 1993b). A problem related with these studies is that adenoviral particles retained their own binding capacity to their cognate receptors on the cell, making it difficult to evaluate their contribution to internalization. Wu *et al.* approached this problem by inactivating the viral fiber that governs the infection specificity of adenovirus through chemical coupling of asialoglycoproteins to these structures. Using this system, a 200-fold increase in gene expression was obtained while retaining binding specificity by the asialoglycoprotein ligand (Wu *et al.*, 1994).

9. APPLICATIONS OF ASGP RECEPTOR TARGETING IN CLINICAL DISORDERS *IN VIVO* AND *IN VITRO*

a) Gene transfer

Analbuminemia. The gene for human serum albumin has been transferred to the livers of Nagase analbuminemic rats using an asialoglycoprotein-polylysine conjugate followed by partial hepatectomy (Wu *et al.*, 1991). Human albumin was detected in serum 48 hours after injection (0.05 μg/ml), reached a maximum by two weeks post-injection (34 μg/ml) and remained stable four weeks.

Hypercholesterolemia. The Watanabe heritable hyperlipidemic rabbit is an animal model for homozygous familial hypercholesterolemia, that is due to a deficiency in the low-density lipoprotein (LDL) receptors. Watanabe rabbits injected with DNA-ASGP-polylysine complexes carrying a normal LDL receptor gene, showed a 25% decrease in total serum cholesterol levels for 5 days, followed by a return to pretreatment levels.

Hyperbilirubinemia. The bilirubin-UDP-glucuronosyltransferase-deficient Gunn rat constitutes an animal model of Crigler-Najjar syndrome type I. As in the clinical course of human disease, this animal is characterized by hyperbilirubinemia and spontaneous development of bilirubin encephalopathy. Injection of a plasmid encoding the human bilirubin-UDP-glucuronosyl-transferase-1 gene as a complex to ASGP-polylysine carrier, with pre-administration of colchicine, led to a 25-35% decline in serum bilirubin levels, excretion of bilirubin glucuronides in bile, whereas DNA persisted in the liver for 8-10 weeks (Chowdhury *et al.*, 1996).

HCC (Hepatocellular Carcinoma). Although it has been shown the ASGP receptor expression and binding to ligand tended to decrease during the malignant transformation, Lauer suggested the possibility of developing a hepatoma-restricted gene transfer system using the interaction of Sendai virus F protein and ASGP receptor in combination with hepatoma specific expression cassettes (Lauer *et al.*, 1998).

b) Oligonucleotide delivery

Hemophilia. A recent study has used receptor-mediated endocytosis through the asialoglycoprotein receptor to target rat liver and induce site-directed mutagenesis of the factor IX gene (Kren *et al.*, 1998). The approach is based on the observation that RNA/DNA hybrids are highly active in homologous pairing *in vitro* and can introduce single-nucleotide mutations when a mismatch exists with the homologous genomic DNA sequence. A chimeric RNA/DNA oligonucleotide homologous to the rat factor IX gene has been complexed with lactosylated-polyethylenimine and used *in vitro* and *in vivo* to transfer the oligonucleotide to liver cells. DNA/RNA analysis by polymerase chain reaction followed by cloning and sequencing has shown site-directed mutagenesis both *in vitro* and *in vivo*. In addition, a significant reduction in factor IX activity, and a marked prolongation of the activated thromboplastin time in the animals have demonstrated the biological consequences of the induced mutation.

Hepatitis B. Woodchucks chronically infected with woodchuck hepatitis virus (WHV) are an excellent model of human hepatitis B virus infection. Antisense DNA oligonucleotides against the polyadenylation region of WHV gene have been complexed with ASOR-polylysine, and injected intravenously into chronically infected animals (0.1 mg/kg/day antisense for 5 consecutive days) (Bartholomew *et al.*, 1995). A 5- to 10-fold decrease in circulating WHV-DNA was seen in treated animals 25 days after treatment. The decline lasted two weeks, after which viremia returned to pretreatment levels. This report confirms the studies *in vitro* which showed an inhibition of hepatitis B virus gene expression by complexed antisense oligonucleotides (Wu and Wu, 1992; Nakazono *et al.*, 1995).

Hepatitis C. To determine whether antisense oligonucleotides against the 5'-non-translated region (NTR) of HCV genome could be targeted to inhibit HCV gene expression, antisense oligonucleotides directed against a sequence in the internal ribosomal binding site of the NTR, and against a portion of the NTR overlapping the core protein translational start site of HCV were prepared. In transient transfections of a plasmid containing a luciferase gene immediately downstream from an HCV NTR insert, oligonucleotides in the form of asialoglycoprotein-polylysine complexes were administered to Huh7 cells, and luciferase activity generated by CMV HCVluc measured. The complexed antisense inhibited luciferase activity by 75%, and 99% at 0.01 μM, and 0.1 μM, respectively. In cell lines stably transfected with CMV HCVluc reporter plasmid, complexed antisense DNA inhibited luciferase activity in Huh7 cells by 20% at 10 μM and 85% at 60 μM, and was competable by an excess asialoglycoprotein. Controls had no detectable inhibitory effects (Wu and Wu, 1998).

Hepatic fibrosis. Recently, targeted inhibition of type I procollagen synthesis by antisense DNA oligonucleotides was shown (Wu *et al.*, 2000). Antisense DNA oligonucleotides directed against specific sequences within α1(I) and α2(I) mRNA of type I procollagen were complexed to a cell-type specific carrier, human ASOR coupled to poly L-lysine, and the effectiveness of the complexes was demonstrated in decreasing α1(I) and α2(I) mRNA levels and total collagen accumulation in NIH 3T3 mouse fibroblast cells.

10. CONCLUSIONS

Receptor-mediated gene and oligonucleotide transfer via asialoglycoprotein receptor represents a promising delivery approach which has potential for the gene therapy of clinically important diseases. Current limitations include variability, low efficiency and transient expression of the transgene. However, advances in our knowledge of the mechanisms involved in complexes preparation, and their relationships with transgene expression will help design more effective targeting vehicles. In addition, use of replication competent episomal vectors, and development of new strategies aimed at decreasing DNA degradation after internalization will increase the effectiveness of the systems.

Acknowledgments

Some of the results described were supported in part by grants from the NIH DK-42182 (GYW), the Immune Response Corp. (CHW) and the Herman Lopata Chair in Hepatitis Research (GYW).

11. REFERENCES

Bartholomew, R.M., Carmichael, E.P., Findeis, M.A., Wu, C.H. and Wu, G.Y. (1995). Targeted delivery of antisense DNA in woodchuck hepatitis virus-infected woodchucks. *J. Viral. Hepatitis*, 2: 273-278.

Bider, M.D. and Spiess, M. (1998). Ligand-induced endocytosis of the asialoglycoprotein receptor: evidence for heterogeneity in subunit oligomerization. *FEBS lett.*, 434: 37-41.

Bijsterbosch, M.K. and Van Berkel, T.J. (1992). Lactosylated high density lipoprotein: a potential carrier for the site-specific delivery of drugs to parenchymal liver cells. *Mol. Pharmacol.*, 41: 404-411.

Bommineni, V.R., Chowdhury, N.R., Wu, G.Y., Wu, C.H., Franki, N., Hays, R.M. and Chowdhury JR. (1994). Depolymerization of hepatocellular microtubules after partial hepatectomy. *J. Biol. Chem.*, 269: 25200-25205.

Boussif, O., Lezoualc̀h, F., Zanta, M.A., Mergny, M.D., Scherman, D., Demeneix, B. and Behr, J.P. (1995). A versatile vector for gene and oligonucleotide transfer into cells in culture and in vivo: polyethylenimine. *Proc. Natl. Acad. Sci. USA*, 92: 7297-7301.

Chen, J., Stickles, R.J. and Daichendt, K.A. (1994). Galactosylated histone-mediated gene transfer and expression. *Hum. Gene Ther.*, 5: 429-435.

Chiou, H.C., Tangco, M.V., Levine, S.M., Robertson, D., Kormis, K., Wu, C.H. and Wu, G.Y. (1994). Enhanced resistance to nuclease degradation of nucleic acids complexed to asialoglycoprotein-polylysine carriers. *Nucleic Acid Res.*, 22: 5439-5449.

Chowdhury, N.R., Wu, C.H., Wu, G.Y., Yerneni, P.C., Bommineni, V.R. and Chowdhury, J.R. (1993). Fate of DNA targeted to the liver by asialoglycoprotein receptor-mediated endocytosis in vivo . Prolonged persistence in cytoplasmic vesicles after partial hepatectomy. *J. Biol. Chem.*, 268: 11265-11271

Chowdhury, N.R., Hays, R.M., Bommineni, V.R., Franki, N., Chowdhury, J.R., Wu, C.H. and Wu, G.Y. (1996). Microtubular disruption prolongs the expression of human bilirubin-uridinediphosphoglucuronate-glucuronosyltransferase-1 gene transferred into Gunn rat livers. *J. Biol. Chem.*, 271: 2341-2346.

Cotten, M., Langle Rouault, F., Kirlappos, H., Wagner, E., Mechtler, K., Zenke, M., Beugh, H. and Birnstiel, M.L. (1990). Transferrin-polycation-mediated introduction of DNA into human leukemic cells: stimulation by agents that affect the survival of transfected DNA or modulate transferrin receptor levels. *Proc. Natl. Acad. Sci. USA*, 87: 4033-4037.

Cotten, M., Wagner, E., Zatloukal, K., Phillips, S., Curiel, D.T. and Birnstiel, M.L. (1992). High-efficiency receptor-mediated delivery of small and large (48 kilobase) gene constructs using the endosome-disruption activity of defective or chemically inactivated adenovirus particles. *Proc. Natl. Acad. Sci. USA*, 89: 6094-6098.

Cristiano, R.J., Smith, L.C. and Woo, S.L.C. (1993a). Hepatic gene therapy: adenovirus enhancement of receptor-mediated gene delivery and expression in primary hepatocytes. *Proc. Natl. Acad. Sci. USA*, 90: 2122-2126.

Cristiano, R.J., Smith, L.C., Kay, M.A., Brinkley, B.R. and Woo, S.L.C. (1993b). Hepatic gene therapy: efficient gene delivery and expression in primary hepatocytes utilizing a conjugated adenovirus-DNA complex. *Proc. Natl. Acad. Sci. USA*, 90: 11548-11552.

Curiel, D.T., Agarwal, S., Wagner, E. and Cotten, M. (1991). Adenovirus enhancement of transferrin-polylysine-mediated gene delivery. *Proc. Natl. Acad. Sci. USA*, 88: 8850-8854.

Edwards, R.J., Carpenter, D.S. and Minchin, R.F. (1996). Uptake and intracellular trafficking of asialoglycoprotein-polylysine-DNA complexes in isolated rat hepatocytes. Gene Ther., 3: 937-940.

Ferkol, T., Lindberg, G.L., Chen, J., Perales, J.C., Crawford, D.R., Ratnoff, O.D. and Hanson RW. (1993). Regulation of the phosphoenolpyruvate carboxykinase/human factor IX gene introduced into the livers of adult rats by receptor-mediated gene transfer. *FASEB J.*, 7: 1081-1091.

Fiume, L., Mattioli, A., Balboni, P.G., Tognon, M., Barbanti-Brodano, G., De Vries, J. and Wieland T. (1979). Enhanced inhibition of virus DNA synthesis in hepatocytes by trifluorothymidine coupled to asialofetuin. *FEBS Lett.*, 103: 47-51.

Fiume, L., Mattioli, A., Busi, C., Balboni, P.G., Barbanti-Brodano, G., De Vries, J., Altmann, R. and Wieland T. (1980). Selective inhibition of ectromelia virus DNA synthesis in hepatocytes by adenine-9-β-D-arabinofuranoside (ARA-A) and adenine-9-β-D-arabinofuranoside 5'-monophosphate (ARA-AMP) conjugated to asialofetuin. *FEBS*

Lett., 116: 185-188.

Haensler, J. and Szoka, F.C. (1993). Synthesis and characterization of a trigalactosylated bisacridine compound to target DNA to hepatocytes. *Bioconjug. Chem.*, 4: 85-93.

Hara, T., Aramaki, Y., Takada, S., Koike, K. and Tsuchiya, S. (1995a). Receptor-mediated transfer of pSV2CAT DNA to a human hepatoblastoma cell line HepG2 using asialofetuin-labeled cationic liposomes. *Gene*, 159: 167-174.

Hara, T., Aramaki, Y., Takada, S., Koike, K. and Tsuchiya S. (1995b). Receptor-mediated transfer of pSV2CAT DNA to mouse liver cells using asialofetuin-labeled liposomes. *Gene Ther.*, 2: 784-788.

Hara, T., Kuwasawa, H., Aramaki, Y., Takada, S., Koike, K., Ishidate, K., Kato, H. and Tsuchiya, S. (1996). Effects of fusogenic and DNA-binding amphiphilic compounds on the receptor-mediated gene transfer into hepatic cells by asialofetuin-labeled liposomes. *Biochim. Biophys. Acta*, 1278: 51-58.

Hisayasu, S., Miyauchi, M., Akiyama, K., Gotoh, T., Satoh, S., Shimada, T. (1999) In vivo targeted gene transfer into liver cells mediated by novel galactosyl-D-lysine/D-serine copolymer. *Gene Ther.*, 6: 689-693.

Ishibashi, S., Hammer, R.E., Herz, J. (1994). Asialoglycoprotein receptor deficiency in mice lacking the minor receptor subunit. *J. Biol. Chem.*, 269: 27803-27806

Kawakami, S., Yamashita, F., Nishikawa, M., Takakura, Y., Hashida, M. (1998) Asialoglycoprotein receptor-mediated gene transfer using novel galactosylated cationic liposomes. *Biochem. Biophys. Res. Commun.*, 252: 78-83.

Koike, K., Hara, T., Aramaki, Y., Takada, S. and Tsuchiya, S. (1994). Receptor-mediated gene transfer into hepatic cells using asialoglycoprotein-labeled liposomes. *Ann. N. Y. Acad. Sci.*, 716: 331-333.

Kren, B.T., Bandyopadhyay, P. and Steer C.J. (1998). In vivo site-directed mutagenesis of the factor IX gene by chimeric RNA/DNA oligonucleotides. *Nature Med.*, 4: 285-290.

Lauer, U., Spiegel, M., Bitzer, M., Wybranietz, W.A., Gross, Ch. D., Prinz, F., Graepler, F., Neubert, W.J., Gregor, M. (1998). New strategies for the genetic therapy of primary liver carcinoma. *Min. Invas. Ther. and Allied Technol.*, 7: 567-571.

Lee, Y.C., Townsend, R.R., Hardy, M.R., Lonngren, J., Arnarp, J. and Lohnn, J. (1983). Binding of synthetic oligosaccharides to the hepatic gal/galNAc lectin. *J. Biol. Chem.*, 262: 199-202.

Mack, K.D., Walzem, R. and Zeldis, J.B. (1994). Cationic lipid enhances in vitro receptor-mediated transfection. *Am. J. Med. Sci.*, 307: 138-143

Markwell, M.A.K., Portner, A. Schwartz, A.L. (1985). An alternative route of infection for viruses: entry by means of the asialoglycoprotein receptor of a Sendai virus mutant lacking its attachment protein. *Proc. Natl. Acad. Sci. USA*, 82: 978-982.

McAbee, D.D., Bennatt, D.J., Ling, Y.Y. (1998). Identification and analysis of a Ca(2+) -dependent lactoferrin receptor in rat liver. *Adv. Exp. Med. Biol.*, 443: 113-121.

Miyauchi, M., Hisayasu, S., Shimada, T. (1997). Influenza fusogenic peptide in DNA complex enhances asialoglycoprotein receptor-mediated gene transfer to hepatoma cells: strategy for liver targeting gene therapy. *J. Clin. Biochem. Nutr.*, 23: 85-93.

Nakazono, K., Ito, Y., Wu, C.H. and Wu, G.Y. (1996). Inhibition of hepatitis B virus replication by targeted pretreatment of complexed antisense DNA in vitro. *Hepatology*, 2: 1297-1303.

Perales, J.C., Ferkol, T., Molas, M. and Hanson, R.W. (1994a). An evaluation of receptor-mediated gene transfer using synthetic DNA-ligand complexes. *Eur. J. Biochem.*, 226: 255-266.

Perales, J.C., Ferkol, T., Beegen, H., Ratnoff, O.D. and Hanson, R.W. (1994b). Gene transfer in vivo: sustained expression and regulation of genes introduced into the liver by receptor-targeted uptake. *Proc. Natl. Acad. Sci. USA*, 91: 4086-4090.

Perales, J.C., Grossmann, G.A., Molas, M., Liu, G., Ferkol, T., Harpst, J., Oda, H. and Hanson, R.W. (1997). Biochemical and functional characterization of DNA complexes capable of targeting genes to hepatocytes via the asialoglycoprotein receptor. *J. Biol. Chem.*, 272: 7398-7407.

Plank, C., Zatloukal, K., Cotton, M., Mechtler, K. and Wagner, E. (1992). Gene transfer into hepatocytes using asialoglycoprotein receptor mediated endocytosis of DNA complexed with an artificial tetra-antennary galactose ligand. *Bioconjug. Chem.*, 3: 533-539.

Ramani, K., Bora, R.S., Kumar, M., Tyagi, S.K. and Sarkar, D.P. (1997). Novel gene delivery to liver cells using engineered virosomes. *FEBS Lett.*, 404: 164-168.

Ramani, K., Hassan, Q., Venkaiah, B., Hasnain, S.E., Sarkar, D.P. (1998). Site-specific gene delivery in vivo through engineered Sendai viral envelopes. *Proc. Natl. Acad. Sci. USA*, 95: 11886-11890.

Remy, J.S., Kichler, A., Mordvinov, V., Schuber, F. and Behr, J.P. (1995). Targeted gene transfer into hepatoma cells with lipopolyamine-condensed DNA particles presenting galactose ligands: a stage toward artificial viruses. *Proc. Natl. Acad. Sci. USA*, 92: 1744-1748.

Rice, K.G. and Lee, Y.C. (1990). Modification of triantennary glycopeptides into probes for the asialoglycoprotein receptor of hepatocytes. *J. Biol. Chem.*, 265: 18423-18428.

Rice, K.G., Weisz, O.A., Barthel, T., Lee, R.T. and Lee, Y.C. (1990). Defined geometry of binding between triantennary glycopeptides and the asialoglycoprotein receptor of rat hepatocytes. *J. Biol. Chem.*, 265: 18429-18434.

Rotundo, R.F., Rebres, R.A., Mckeown-Longo P.J. Blumenstock, F.A., Saba, T.M. (1998). Circulating fibronectin may be a natural ligand for the hepatic asialoglycoprotein receptor: possible pathway for fibronectin deposition and turnover in the rat liver. *Hepatology*, 28: 475-485.

Schlepper-Schafer, J., Hulsmann, D., Djovkar, A., Meyer, H.E., Herbertz, L., Kolb, H. and Kolb-Bachofen, V. (1986). Endocytosis via galactose receptors in vivo: ligand size directs uptake by hepatocytes and/or liver macrophages. *Exp. Cell Res.*, 165: 494-506.

Schuster, M.J., Walton, C.M., Wu, G.Y. and Wu, C.H. (1999). Multicomponent DNA carrier with a Vasicular Stomatitis Virus G-peptide greatly enhances liver-targeted gene expression in mice. *Bioconjug. Chem.*, 10: 1075-1083.

Schwartz, A.L., Fridovich, S.E. and Lodish, H.F. (1982). Kinetics of internalization and recycling of the asialoglycoprotein receptor in a hepatoma cell line. *J. Biol. Chem.*, 257: 4230-4237.

Stankovics, J., Crane, A.M., Andrews, E., Wu, C.H., Wu, G.Y. and Ledley, F.D. (1994).

Overexpression of human methylmalonyl CoA mutase in mice after in vivo gene transfer with asialoglycoprotein/polylysine /DNA complexes. *Hum. Gene Ther.*, 5: 1095-1104.

Stockert, R.J. (1996). The asialoglycoprotein receptor: relationships between structure, function, and expression. *Physiol. Rev.*, 75: 591-609.

Townsend, R.R., Hardy, M.R., Wong, T.C. and Lee, Y.C. (1986). Binding of N-linked bovine fetuin glycopeptides to isolated rabbit hepatocytes: Gal/GalNAc hepatic lectin discrimination between Gal b (1,4) GlcNAc and Gal b (1,3) GalcNAc in a triantennary structure. Biochemistry, 25: 5716-5725.

Trouet, A., Baurain, R., Deprez de Campeneere, D., Masquelier, M. and Pirson, P. (1981). Targeting of antitumor and antiprotozoal drugs by covalent linkage to protein carriers. In: *Targeting of Drugs*, G. Gregoriadis, J. Senior and A. Trouet (Eds.), Plenum, New York, pp. 28.

Vera, D.R., Stadalnik, R.C. and Krohn, K.A. (1985). Technetium 99m galactosyl-neoglycoalbumin: preparation and preclinical studies. *J. Nucl. Med.*, 26: 1157-1167.

Wadhwa, M.S., Knoell, D.L., Young, A.P. and Rice KG. (1995). Targeted gene delivery with a low molecular weight glycopeptide carrier. *Bioconjug. Chem.*, 6: 283-291.

Wagner, E., Plank, C., Zatloukal, K., Cotten, M. and Birnstiel, M.L. (1992). Influenza virus hemagglutinin HA-2 N-terminal fusogenic peptides augment gene transfer by transferrin-polylysine -DNA-complexes: toward a synthetic virus-like gene-transfer vehicle. *Proc. Natl. Acad. Sci. USA*, 89: 7934-7938.

Walton, C.M., Wu, C.H. and Wu, G.Y. (1999). A DNA delivery system containing listeriolysin O results in enhanced hepatocyte-directed gene expression. *World J. Gastroenterol.*, 5: 465-469.

Weigel, P.H. and Oka, J.A. (1998). The dual coated pit pathway hypothesis: vertebrate cells have both ancient and modern coated pit pathways for receptor mediated endocytosis. *Biochem. Biophys. Res. Commun.*, 246: 563-569.

Wilson, J.M., Grossman, M., Cabrera, J.A., Wu, C.H. and Wu, G.Y. (1992). A novel mechanism for achieving transgene persistence in vivo following somatic gene transfer into hepatocytes. *J. Biol. Chem.*, 267: 11483-11489.

Wu, G.Y., Wu, C.H. and Stockert, R.J. (1983). A model for the specific rescue of normal hepatocytes during methotrexate treatment of hepatic malignancy. *Proc. Natl. Acad. Sci. USA*, 80: 3878-3880.

Wu, G.Y., Wu, C.H. and Rubin, M.I. (1985). Acetaminophen hepatotoxicity and targeted rescue: a model for specific chemotherapy for hepatocellular carcinoma. *Hepatology*, 5: 709-713.

Wu, G.Y. and Wu, C.H. (1987). Receptor-mediated in vitro gene transformation by a soluble DNA carrier system. *J. Biol. Chem.*, 262: 4429-4432.

Wu, G.Y. and Wu C.H. (1988a). Evidence for targeted gene delivery to Hep G2 hepatoma cells in vitro. *Biochemistry*, 27: 887-892.

Wu, G.Y. and Wu, C.H. (1988b). Receptor-mediated gene delivery and expression in vivo. *J. Biol. Chem.*, 263: 14621-14624.

Wu, G.Y., Wilson, J.M. and Wu, C.H. (1989). Targeting genes: delivery and persistent expression of a foreign gene driven by mammalian regulatory elements in vivo. *J. Biol. Chem.*, 264: 16985-16987.

Wu, G.Y., Wilson, J.M., Shalaby, F., Grossman, M., Shafritz, D.A. and Wu, C.H. (1991). Receptor-mediated gene delivery in vivo. Partial correction of genetic analbuminemia in Nagase rats. *J. Biol. Chem.*, 266: 14338-14342.

Wu, G.Y. and Wu, C.H. (1992). Specific inhibition of hepatitis B viral gene expression in vitro by targeted antisense oligonucleotides. *J. Biol. Chem.*, 267: 12436-12439.

Wu, G.Y., Zhan, P., Sze, L.L., Rosenberg, A.R. and Wu, C.H. (1994). Incorporation of adenovirus into a ligand-based DNA carrier system results in retention of original receptor specificity and enhances targeted gene expression. *J. Biol. Chem.*, 269: 11542-11546.

Wu, C.H., Zhan, P., Carmichael, E., Chowdhury, J.R., Chowdhury, N.R. and Wu, G.Y. (1997). A tri-component DNA carrier: progress toward a synthetic virus. *Hepatology*, 26 (suppl.): 129A.

Wu, C.H. and Wu, G.Y. (1998). Targeted inhibition of hepatitis C virus-directed gene expression in human hepatoma cell lines. *Gastroenterology*, 114: 1304-1312.

Wu, C.H., Walton, C.M. and Wu, G.Y. (2000). Targeted inhibition of type I procollagen synthesis by antisense DNA oligonucleotides. *Gene Ther. Regulation* (In Press).

Zanta, M.A., Boussif, O., Adib, A. and Behr, J.P. (1997). In vitro gene delivery to hepatocytes with galactosylated polyethylenimine. *Bioconjug. Chem.*, 8: 839-844.

Zhang, Y., Wu, C.H. and Wu. G.Y. (1994). Enhancement of targeted gene expression by incorporation of a listerial lytic peptide into protein-DNA complexes. *Hepatology*, 20 (suppl.): 271A.

12. SUMMARY — Gene transfer to the liver by receptor-mediated endocytosis represents a promising method for therapeutic intervention of genetic disorders and acquired diseases affecting the liver. Based on the observation that asialoglycoprotein receptors are abundantly expressed on the surface of human hepatocytes *in vivo*, gene and oligonucleotide delivery and expression by DNA-ligand complexes targeted to liver have been demonstrated in several different experimental models. These include genetic metabolic deficiencies where gene delivery provided a normal functional gene, as well as delivery of oligonucleotides that corrected a mutation in non-functioning genes. The inhibition of endogenous gene expression has potential in the treatment of acquired conditions such as infectious diseases or neoplastic diseases. While much progress has been made, further improvements in efficiency and duration of targeted gene expression are needed.

Key Words: gene therapy; receptor-mediated endocytosis; asialoglycoprotein recepror; asialoorosomucoid; endosomolytic agents; fusogenic peptide; hepatocyte; DNA-protein complexes

Progress in Gene Therapy: Basic and Clinical Frontiers, pp. 183-203
R. Bertolotti *et al.* (Eds)

Receptor-mediated gene transfer based on plasmid DNA complexes

Ralf Kircheis[1], Lionel Wightman[1], Manfred Ogris[2], Sylvia Brunner[2], Margaretha Kursa[1], Susanne Schüller[2] and Ernst Wagner[1,2,*]

[1]*Boehringer Ingelheim Austria, Dr. Boehringer-Gasse 5-11, A-1121 Vienna, Austria, and* [2]*Institute of Biochemistry, Vienna University Biocenter, Dr. Bohr-Gasse 9/3, A-1030 Vienna, Austria*

Table of Contents

1. INTRODUCTION

At present, the deciding factor for progress in somatic gene therapy is the development of efficient and safe delivery systems. Nonviral gene transfer formulations (Felgner *et al.*, 1997; Ledley, 1994; Perales *et al.*, 1994b; Cotten and Wagner, 1993 and 1998) are attractive for the following

* Corresponding author. E-mail: Ernst.Wagner@vie.boehringer-ingelheim.com

reasons: 1) the chemical/physical diversity in synthesis and assembly of transfection complexes, 2) they can be generated by assembly of few defined components and can be designed, if desired, protein-free/non-immunogenic, 3) they can be very flexible regarding the size of DNA to be transported, and 4) plasmid DNA and transfection reagents can be produced at large scale with low costs. The major disadvantage of nonviral systems is a still unsatisfactory efficacy for most *in vivo* applications. Furthermore, input and expertise from biophysics, formulation, cell biology and virology will be required to understand and improve the charateristics of the current generation of nonviral delivery particles.

Many transfection strategies involve active cellular uptake of suitably complexed DNA. Transfection agents like cationic liposomes (Felgner *et al.*, 1994; Loeffler and Behr, 1993), polycationic polypeptides or macromolecules like polyethyleneimine (Boussif *et al.*, 1995) and dendrimers (Haensler and Szoka, 1993b; Tang and Szoka, 1997) exploit the anionic character of nucleic acids to form a transfection complex. These reagents share several functions that are essential for transfection: they condense the nucleic acid into particles that can be taken up by cells, and they protect the nucleic acid against degradation by cellular enzymes. Often, the cationic carrier mediates binding of the transfection complex to the cell, and also the transfer of the nucleic acid to the cytoplasm, either by fusion with the cell surface or vesicular membranes, or by disruption of vesicular membranes.

Viral vectors also utilize these types of functions, although in much more refined methods. Most importantly, they utilize specific cellular receptors for uptake into specific target cells. Current efforts consider the incorporation of virus-like entry functions into synthetic vectors, such as ligands binding to specific receptors or domains responsible for the delivery across cellular membranes and into the nucleus.

In systemic and other *in vivo* applications on the extracellular pathways, additional delivery barriers have to be considered: unspecific interaction and aggregation with blood components, and physical restrictions (limited extravasation and distribution of the transfection agent within the tissue). Formulations of DNA particles with a defined, preferentially small size, high solubility and reduced unspecific interactions have to be developed.

2. RECEPTOR-MEDIATED UPTAKE OF DNA COMPLEXES

Cells use receptor-mediated endocytosis for the uptake of proteins such as asialoglycoproteins, transferrin, epidermal growth factor, insulin, and small vitamins such as folic acid. The first steps in this process include binding of the ligand to specific cell-surface receptors and, in most cases, internalization into endosomal acidic compartments. The subsequent pathways are dependent on the type of ligand/receptor pair; in many cases, however, sorting processes lead to degradative lysosomal compartments.

In attempting to target DNA to a given cell population, a suitable receptor-ligand has to be attached covalently or noncovalently by conjugation to a carrier. The delivery of DNA to cells by direct linkage to alpha2-macroglobulin was suggested by Pastan and collaborators (Cheng *et al.*, 1983). Most receptor-mediated gene transfer systems apply the polycationic carrier polylysine conjugated to the cell-binding ligand. Wu and colleagues employed complexes of asialoorosomucoid-poly(L)lysine conjugates with DNA plasmids encoding marker genes or therapeutically relevant genes which were shown to result in gene expression both *in vitro* in cultured HepG2 hepatoma cells (Wu and Wu, 1987) and *in vivo* in the liver of rats or rabbits (Wu and Wu, 1988; Wu *et al.*, 1991; Wilson *et al.*, 1992; Chowdhury *et al.*, 1993; Grove and Wu, 1998). Other ligands already tested are listed in Table 1.

Receptor-specific uptake was tested by binding studies of the DNA complex to the target receptor, enhanced gene expression after upregulation of the cellular receptor, reduced transfection efficiency in cells without receptor, reduced expression with DNA complexes lacking the ligand, or in

Table 1. Ligands used in receptor-mediated gene transfer

Ligand	Selected references
Asialoglycoproteins	Wu and Wu, 1987, 1988; Wu *et al.,* 1991 Wilson *et al.,* 1992; Chowdhury *et al.,* 1993
Transferrin	Wagner *et al.,* 1990, 1994; Kircheis *et al.*, 1997, 1999
Insulin	Huckett *et al.,* 1990; Rosenkranz *et al.,* 1992
Folate	Mislick *et al.,* 1995; Gottschalk *et al.,* 1994
Lectins	Cotten *et al.*, 1993b; Batra *et al.,* Yin and Cheng, 1994
Alpha2 macroglobulin	Schneider *et al.,* 1996
Malarial circumsporozoite protein	Ding *et al.,* 1995
RGD-motif (integrin binding)	Hart *et al.,* 1995; Harbottle, 1998; Erbacher *et al.*, 1999
Surfactant proteins A and B	Baatz *et al.,* 1994; Ross *et al.,* 1995
Synthetic glycosylated ligands	Diebold *et al.*, 1999; Erbacher *et al.,* 1995 Ferkol *et al.,* 1996; Haensel and Szoka, 1993a Kollen *et al.,* 1996; Merwin *et al.,* 1994 Midoux *et al.,* 1993; Perales *et al.,* 1994a Plank *et al.,* 1992; Wadhwa *et al.,* 1995
Anti-secretory component Fab	Ferkol *et al.,* 1993, 1995
Anti-Tn	Thurnher *et al.,* 1994
Anti-CD3	Buschle *et al.,* 1995; Kircheis *et al.*, 1997
Anti-CD5	Merwin *et al.,* 1995
Anti-CD117	Zauner *et al.,* 1998
Steel factor (CD117 ligand)	Schwarzenberger *et al.,* 1996
Anti-EGF	Chen *et al.,* 1994
EGF	Cristiano and Roth, 1996
Anti-HER2	Foster and Kern, 1997
Anti-thrombomodulin	Trubetskoy *et al.,* 1992a, 1992b
Antibody ChCE7	Coll *et al.,* 1997
Anti-IgG	Curiel *et al.,* 1994
IgG (FcR ligand)	Curiel *et al.,* 1994; Rojanasakul *et al.,* 1994

competition experiments with free ligand. Endocytosis has been demonstrated by the time-course and temperature-dependence of internalization into endocytic vesicles as well as the effect of lysosomotropic agents. For example, intensive uptake of transferrin-coated DNA complexes into the transferrin receptor-rich cell line K562 correlates with high expression levels, but is strongly dependent on the presence of chloroquine (Cotten *et al.*, 1990).

3. NUCLEIC ACID BINDING AND CONDENSING POLYMERS

Poly(L)lysines of different sizes (20 to more than 600 L-lysines per polymer) have been successfully used for binding various ligands to the DNA. Neutralization of the charges of DNA with concomitant condensation (Wagner *et al.,* 1991a) and stabilization of the DNA against nucleases facilitates the endocytic uptake and survival of DNA on the way to the nucleus of the cell. Whether DNA-bound polylysine is replaced by cellular basic proteins, before or after nuclear entry, is unclear. The use of the metabolically more stable poly(D)lysine did not alter transfection efficiency (Wagner *et al.,* 1991a; Plank *et al.*, 1994). Obviously, release of polylysine by enzymatic degradation does not play a major role in the assembly of transfected DNA into functional chromatin. Polylysine however cannot be considered a biologically inert material. Polylysine is able to enhance pinocytosis (Shen and Ryser, 1978), as well as phagocytosis (Pruzanski and Saito, 1978). It can activate membrane phospholipase and hence may affect membrane permeability (Arnold *et al.*, 1979), it activates the complement system (Plank *et al.,* 1996) and may induce an immune response upon repeated injection (Vermeersch and Remon, 1994; Findeis *et al.,* 1993). Especially when polylysine is incorporated into DNA complexes in excessive amounts, interactions with proteins (cellular or in biological fluids) and phospholipid bilayers (of internal compartments or at the cell surface) must be considered.

Beside polylysines, a series of other DNA-binding compounds have been used in ligand/DNA complexes: polycations such as polyarginine, protamine, histones (Wagner *et al.,* 1991a) or polyethylenimine (Kircheis *et al.,* 1997; Zanta *et al.,* 1997; Erbacher *et al.,* 1999), and intercalative compouds like bisacridine (Haensler and Szoka, 1993a) or ethidium dimers (Wagner *et al.,* 1991b). Use of the intercalators in transfections in the absence of polycations was largely unsuccessful; essential requirements like condensation of the DNA and interactions with cellular lipids may be missing. In contrast, polyethylenimine (PEI; Boussif *et al.,* 1995) and dendrimers (Haensler and Szoka, 1993b; Kukowska-Latallo *et al.,* 1996) have a particularly high transfection potential for reasons described below.

4. ASSEMBLY OF TRANSFECTION COMPLEXES

The assembly of one or several plasmid molecules and a large number of ligand-polycation conjugates into one supramolecular complex has to be considered a most critical step in receptor-mediated gene transfer. The ratio of positively charged polycation to negatively charged DNA, the size and modification of the polycation, and also the procedure of complex formation may strongly influence the *in vitro* and *in vivo* gene transfer efficiency (Ferkol *et al.,* 1995; Kabanov and Kabanov, 1995; McKee *et al.,* 1994; Ogris *et al.,* 1998; Perales *et al.,* 1994a and 1994b; Tang and Szoka, 1997; Wagner *et al.,* 1991a; Wolfert and Seymour, 1996).

DNA polycation complexes have been characterized by: their electrophoretic mobility (reflects charge and size of complexes), electron microscopy and atomic force microscopy (shape, size), laser light scattering (size), zeta potential measurements (charge), circular dichroism (conformation of DNA), ethidium bromide exclusion assay (condensation) or centrifugation techniques (molecular weight and condensation). Measurements are often complicated by the difficulty of generating homogenous complexes at high concentrations.

The net charge of the DNA/polycation complex affects its solubility. Complexes with either excess of DNA or polycation are stabilized in solution by the negative or positive charges. At charge ratios close to 1, hydrophobic domains of polycations like polylysine are considered responsible for low solubility in water. This may lead to aggregation and precipitation of complexes. Also disproportionation into nonstoichometric soluble complexes and stoichometric insoluble complexes has been observed (Kabanov and Kabanov, 1995).

DNA/polylysine-conjugate complexes have been prepared in several ways. Wu and Wu (1987 and 1988) mixed the compounds at high salt concentration where electrostatic binding is strongly reduced. Slow reduction of the salt concentration by dialysis into physiological buffer results in a thermodynamically controlled complex formation. Charge ratios of polylysine/DNA <<1 and enhanced hydrophilicity due to the conjugated asialoglycoprotein presumably are essential for the solubility of the complex. We described a different approach (Wagner *et al.*, 1991a) using transferrin-polylysine DNA complexes. Flash mixing of rather dilute compounds in physiological phosphate-free buffer results in the formation of kinetically controlled complexes. The rationale of the protocol is to prevent microheterogenicity of compound solutions and subsequently complexes. Otherwise, initial generation of unevenly (*e.g.* positively and negatively) charged complexes inevitably would result in the formation of aggregates. Charge ratios of polylysine/DNA of <1/2 to >2/1 have been applied. At ratios of electroneutrality or higher, rod-like and donut-like particles of 70-120 nm diameter are formed. Complexes containing transferrin-conjugated polylysine have increased solubility compared to the use of unmodified polylysine (Zauner *et al.*, 1998).

Interestingly, donuts of similar sizes are formed, irrespective of the size of the DNA molecules used in the complex formation, *e.g.* 700 bp DNA, ~5 kbp plasmid or 48 kbp cosmid. Using a standard expression plasmid of 5-6 kbp, obviously several DNA molecules are incorporated into one particle. In an attempt to generate unimolecular DNA complexes, Perales and colleagues (Perales *et al.*, 1994a and 1994b; Ferkol *et al.*, 1995) added polylysine conjugates slowly, in several small portions, to a vortexing solution of DNA in approx. 0.5-0.9M sodium chloride until a charge ratio of polylysine/DNA of approx. 0.7 is reached. The slow addition of polylysine has been reported to generate monomeric DNA complexes, with sizes of approx. 15-30 nm. These findings look promising and have been reported to be applicable for *in vivo* gene transfer protocols (Perales *et al.*, 1994a; Ferkol *et al.*, 1995).

As described in recent reports, the size of DNA complexes with polyethylenimine or transferrin-polyethylenimine is strongly influenced by the previously mentioned parameters such as DNA concentration and charge ratio, and also further parameters such as ionic strength of solution, or serum content of culture medium. Mixing DNA/PEI complexes at N/P charge ratios below 6 in HEPES buffered 150 mM saline results in rapid aggregation; aggregation can be avoided by complex formation at low ionic strength (25 mM aqueous HEPES buffer), generating particles with an average diameter of approx. 40-50 nm (Ogris *et al.*, 1998).

Recently, the monomolecular collapse of DNA has been described using sulfhydryl-containing oligocations or detergents (Blessing *et al.*, 1998a and 1998b). Using the detergent C10CG+, DNA is condensed to approx. 25 nm small complexes without aggregate formation. These complexes, though small, would not be stable under physiological conditions. However, the detergent used contains a mercapto group which, upon exposure to air, results in oxidation and dimerization to a cationic lipid. By these means, small and stable particles can be generated.

5. DELIVERY ACROSS CELLULAR MEMBRANES

Although complexes of DNA and polylysine-conjugated ligands can be efficiently delivered into cells, the accumulation in internal vesicles may strongly reduce the efficiency of gene transfer. For instance, transferrin or asialoglycoprotein receptor-mediated delivery into cultured hepatocytes results in uptake of DNA into practically all cells, but only few cells express the delivered gene (Plank *et al.*, 1992). The addition of the lysosomotropic agent chloroquine or glycerol to the transfection medium may increase transfection efficiency, probably by interfering with lysosomal degradation and enhancing the release of the DNA into the cytoplasm. In K562 cells, the positive effect of chloroquine is especially strong (Cotten *et al.*, 1990). This effect can be explained by the unusually low pH in K562 early endosomes (Sipe *et al.*, 1991). Consequently, high amounts of the weak base accumulate in the endocytic vesicles, leading to the swelling and finally to the destabilization of the endosomes. However, the use of chloroquine is limited, because 1) only a limited number of cell lines respond to chloroquine, and 2) because of its toxicity towards cells. Incubation of cells with transferrin-polylysine/DNA or polylysine/DNA complexes in the presence of 1-1.5 M glycerol results in a strongly enhanced transfection efficiency in several cell lines and primary fibroblasts (Zauner *et al.*, 1996 and 1997). However, several cell lines, including suspension cells such as K562, were refractory to this technique.

The inclusion of replication-defective adenoviruses has been shown to augment the levels of transferrin-mediated gene transfer up to more than a thousand-fold in cell lines that express high levels of both adenovirus and transferrin receptor (Curiel *et al.*, 1991; Cotten *et al.*, 1992). Based on these observations, ligand-decorated DNA complexes were provided with an endosome-destabilizing domain as further important function of viral entry. For this purpose, inactivated adenovirus (Cotten *et al.*, 1993b; Baatz *et al.*, 1994; Curiel *et al.*, 1992; Wagner *et al.*, 1992a; Cristiano *et al.*, 1993; Wu *et al.*, 1994) and rhinovirus (Zauner *et al.*, 1995) particles, or synthetic peptides derived from sequences of the influenza virus hemagglutinin N-terminus (Midoux *et al.*, 1993; Plank *et al.*, 1992 and 1994; Oberhauser *et al.*, 1995; Wagner *et al.*, 1992b) and the rhinovirus VP-1 N-terminus

(Zauner *et al.*, 1995) have been incorporated into the DNA complex. Incorporation of these agents has been achieved by chemical or enzymatic linkage to polylysine (Cristiano *et al.*, 1993; Wu *et al.*, 1994; Fisher and Wilson, 1994), by biotinylation (Wagner *et al.*, 1992a; Zauner *et al.*, 1995) which allows subsequent binding to streptavidin-polylysine, by an antibody bridge (Curiel *et al.*, 1992), or simply by ionic interaction with polylysine (Plank *et al.*, 1994). As a consequence, gene transfer to cell lines or cultured primary cells was strongly enhanced. In Jurkat cells (a human acute T cell leukemia line), adenovirus-enhanced and peptide-enhanced gene transfer have similar efficiencies. In most cases, the adenovirus-linked transfection complex was found more efficient.

The high transfection potential of the cationic polymer polyethylenimine PEI (Boussif *et al.*, 1995) is based on another mechanism for endosomal escape. PEI within the DNA complex is only partially protonated at physiological pH (only one of 2-3 amino nitrogens). Upon delivery of the DNA particle into the endosome/lysosome, PEI presumably acts as proton sponge, with the protonation triggering osmotic swelling and destabilization of the endosomal/lysosomal vesicle. Notably, high level gene expressions were found in the absence of chloroquine or DNA complex-linked adenovirus (Boussif *et al.*, 1995; Kircheis *et al.*, 1997).

6. TARGETED GENE TRANSFER *IN VITRO* AND *IN VIVO*

The concept of utilizing the specificity of a ligand-receptor interaction for targeted gene delivery is very attractive, most challenging, but contains numerous pitfalls. For instance, special care has to be taken when polycations like polylysine are part of the DNA complexes. We observed in some cell cultures that minor changes in the DNA/polylysine-conjugate ratio of the complex (resulting in positively charged complexes) may convert a ligand-specific gene transfer into a completely unspecific process (Zauner *et al.*, 1998). Furthermore, even only slightly positively charged DNA/

polylysine complexes strongly interact with serum proteins such as complement resulting in inactivation of the complexes (Plank *et al.*, 1996). Another important parameter for *in vivo* targeted gene transfer is the size of the complex. Phagocytotic clearance of large particles by macrophages hampers the targeting specificity and efficiency of gene transfer. Large DNA complexes might be applicable in administrations where the target cells are directly accessible (*e.g.* local injection into tumor tissue). However, upon systemic administration, only very small complexes might have a chance to leave the circulation and migrate into a specific target tissue.

In vivo gene transfer using polylysine condensed DNA was reported for the first time by Wu and colleagues (Wu and Wu, 1988; Wu *et al.*, 1991; Wilson *et al.*, 1992; Chowdhury *et al.*, 1993). They showed that reporter gene expression after i.v. injection of asialoorosomucoid-polylysine/DNA complexes in rats was highest after 24 hours. Activity then declined and after 96 hours none was detectable. However, when partial hepatectomy (two-thirds of the liver) was performed 30 minutes after injection of the complexes, transgene expression was shown to persist for at least 11 weeks post-injection. There was no evidence of integration of the foreign gene and plasmid DNA appeared to be present extra-chromosomally. Using a different protocol for complex formation, Perales and colleagues (Perales *et al.*, 1994a) have reported prolonged expression of a reporter gene in hepatocytes without liver surgery. Systemic delivery of anti-pIgR Fc-polylysine coated DNA resulted in *in vivo* expression in the lung (Ferkol *et al.,* 1995).

Local injection of naked DNA or DNA complexes directly into subcutaneously growing tumors produced significant reporter gene expression, with DNA/transferrin-PEI complexes or adenovirus-linked DNA/transferrin-polylysine complexes being 10 to 100-fold more efficient than naked DNA (Kircheis *et al.*, 1999; Wightman *et al.*, submitted). Intravenous application of Tf-PEI/DNA complexes through the tail vein into the tumor-bearing mice resulted in gene expression in the tail and lung, but also serious toxicity. Occlusion of pulmonary vessels by aggregates is most probably responsible for frequent lethal events. By first coating the DNA complexes with polyethylene glycol (PEG) through covalent coupling to PEI, plasma protein binding is strongly reduced and complexes are sterically stabilized, resultingt in prolonged circulation following *in vivo* application (Ogris *et al.*, 1999). The complexes, when injected intravenously into mice, were far less toxic and gene expression was almost exclusively found at the application site (in the tail) and the tumor (Ogris *et al.*, 1999; Kircheis *et al.*, 1999). This kind of stealth coating of particles may also have the advantage to shield from interaction with the host immune system. Despite these first encouraging results, further optimization on *in vivo* gene delivery systems is required for the development of therapies in humans.

7. FUTURE PROSPECTS

The first prototypes of receptor-mediated transfection complexes have come into existence. Several functional domains involved in viral entry have already been incorporated into DNA complexes, and their influence on gene transfer has been tested. Nuclear entry and maintenance of the transfected DNA (in integrated or episomal form) in the cell are further steps where nonviral systems may benefit from including viral elements (see Table 2). Specific protein domains have been identified that are responsible for active transport across the nuclear membrane, such as the SV40 large T antigen nuclear localization signal (NLS) peptide (Kalderon *et al.,* 1984). These signal peptides have been shown to carry artificial substrates (conjugates of proteins or gold particles) from the cytoplasm to the nucleus. Previously, no convincing evidence had been reported that nuclear localization domains enhance transfection efficiency. Most recent work has demonstrated a 10 to 1000-fold increase in transfection efficiency by a single NLS peptide covalently linked to a terminal DNA hairpin (Zanta *et al.*, 1999). In replicating cells, the DNA may enter the nucleus during mitosis, after break-down of the nuclear envelope; nuclear import mechanisms are considered to be more critical in nondividing cells. Retroviruses such as MLV

Table 2. Viral entry functions

Entry function	Virus	Viral protein domain, mechanism
Packaging of genome	Adenovirus Retroviruses	Mu peptide core particle, gag proteins
Binding to cell surface	Influenza virus Rhinoviruses Retroviruses	HA-1, binding to sialic acid major group: ICAM; minor group: LDL-R HIV: gp120 - CD4 of T-cells MLV: gp70 - phosphate transporter
Internalization into the cell	Influenza virus, SFV, rhinovirus, adenovirus	endocytosis into endosomes
	HVJ, Herpes viruses	fusion at cell surface
Release into cytoplasm	Adenovirus Rhinovirus	endosome disruption formation of endosomal pore; VP-1
	Influenza virus, HIV Sendai virus, SFV	fusion: influenza HA-2, HIV gp41 fusion: Sendai F1, SFV E1 protein
Transfer into nucleus	Adenovirus Influenza virus HIV	injection of DNA through nuclear pore transport of 8 small RNP into nucleus nuclear localization of HIV core particle
Maintenance of expression	retroviruses AAV Herpes virus, EBV transposons	integration (integrase, LTR elements) integration (rep proteins, ITR elements) episomal persistence (*e.g.* oriP, EBNA-1) sleepy beauty

lack specific nuclear transport domains and can infect only replicating cells, whereas the HIV core particle is actively transported into the nucleus and HIV can infect nondividing cells. The size of the transported nucleic acid particle may be critical. In the case of influenza virus, the viral genome is distributed over eight small RNPs that can be translocated into nucleus. In the case of adenovirus, the capsid is too large to be translocated and accumulates at the nuclear pore; the DNA is subsequently released from the capsid and enters through the nuclear pore. Other proteins, like the Herpes VP22 may also be able to transport the DNA into the nucleus.

The transgene itself must be carefully designed, in terms of its control and expression. Several options exist to achieve long-term gene maintenance. For example, retroviruses and adenovirus-associated virus stably insert their genome into the host genome. The integration mechanisms have been characterized (retrovirus: LTR sequences, integrase protein; AAV: ITR sequences, rep proteins) and may be exploited by incorporation of the corresponding nucleic acid and protein elements into a synthetic transfection complex. Herpes viruses or Epstein Barr Virus (EBV) can persist in infected cells without integrating their genome into the host. The copy number of the viral genome is controlled and maintained by a viral origin of replication and a viral nuclear retention sequence. This mechanism is utilized in episomal vector plasmids containing the appropriate elements in their DNA sequence (*e.g.* EBV oriP, EBNA-1). Another option is the generation of human autonomous replicons (Haase *et al.,* 1994; Sclimenti and Calos, 1998) and artificial chromosomes (Huxley, 1994; Harrington *et al.,* 1997; Ikeno *et al.*, 1998) for maintenance of large genomic sequences.

A major challenge is the assembly of the gene transfer particles in a way that the individual entry functions synergize but do not interfere with each other. Also, the appropriate timing of the fonctions during cell entry is important. For instance, membrane-disruption domains should be active only after internalization into endosomes and not at the cell surface; incorporation of this function should not result in large complexes that would limit specificity and *in vivo* applications. Some of the receptor-mediated transfection systems have already been found highly efficient in *ex vivo* gene transfer, some have been successfully applied for *in vivo* gene transfer. It can be expected that optimized versions of synthetic target-specific complexes will play a major role in gene therapy.

Acknowledgements

This work was supported by grant S07405 from the Austrian Science Foundation and by the EC Biotech programm.

8. REFERENCES

Arnold, L., Dagan, A., Gutheil, J. and Kaplan, N. (1979). Antineoplastic activity of poly (L-lysine) with some ascites tumor cells. *Proc. Natl. Acad. Sci. USA*, 76: 3246-3250.

Baatz, J.E., Bruno, M.D., Ciraolo, P., Glasser, S., Stripp, B., Smyth, K.L., Korfhagen, T. (1994) Utilization of modified surfactant-associated protein B for delivery of DNA to airway cells in culture. *Proc. Natl. Acad. Sci. USA*, 91: 2547-2551.

Batra, R., Wang-Johanning, F., Wagner, E., Garver, R. and Curiel, D. (1994). Receptor-mediated gene delivery employing lectin-binding specificity. *Gene Ther.*, 1: 255-260.

Blessing, T., Remy, J.S. and J.P. Behr (1998a). Monomolecular collapse of plasmid DNA into stable virus-like particles. *Proc. Natl. Acad. Sci. USA*, 95: 1427-1431.

Blessing, T., Remy, J.S and Behr, J.P. (1998b). Template oligomerization of DNA-bound cations produces calibrated nanometric particles. *J. Am. Chem. Soc.*, 120: 8519-8520.

Boussif, O., Lezoualc'h, F., Zanta, M.A., Mergny, M., Scherrnan, D., Demeneix, B. and Behr, J.P. (1995). A novel, versatile vector for gene and oligonucleotide transfer into cells in culture and *in vivo*: polyethylenimine. *Proc. Natl. Acad. Sci. USA*, 92: 7297-7301.

Buschle, M., Cotten, M., Kirlappos, H., Mechtler, K., Birnstiel, M.L. and Wagner, E. (1995). Receptor-mediated gene transfer into T-lymphocytes via binding of DNA/CD3 antibody particles to the CD3 protein complex. *Hum. Gene Ther.*, 6, 753-761.

Chen, J., Gamou, S., Takayanagi, A. and Shimizu, N. (1994). A novel gene delivery system using EGF receptor-mediated endocytosis. *FEBS Lett.*, 338: 167-169.

Cheng, S.Y., Merlino, G.T. and Pastan, I.H. (1983). A method for coupling of proteins to DNA: synthesis of alpha2-macroglobulin-DNA conjugates. *Nucleic Acids Res.*, 11: 659-669.

Chowdhury, N.R., Wu, C.H., Wu, G.Y., Yerneni, P.C., Bommineni, V.R. and Chowdhury, J.R. (1993). Fate of DNA targeted to the liver by asialoglycoprotein receptor-mediated endocytosis *in vivo*: prolonged persistence in cytoplasmic vesicles after partial hepatectomy. *J. Biol. Chem.*, 268: 11265-11271.

Coll, J.L, Wagner, E., Combaret, V., Mechtler, K., Amstutz, H., Iacono-DiCacito, I., Simon, N. and Favrot, M.C. (1997). *In vitro* targeting and specific transfection of human neuroblastoma cells by chCE7 antibody-mediated gene transfer. *Gene Ther.*, 4: 156-161

Cotten, M. and Wagner, E. (1993). Non-viral approaches to gene therapy. *Current Op. Biotech.*, 4: 705-710.

Cotten, M. and Wagner, E. (1998). Receptor-mediated gene delivery strategies. In: *The Development of Human Gene Therapy*, Friedmann, T. (Ed.), Cold Spring Harbor Laboratory Press, Cold Spring Harbor, pp. 261-277.

Cotten, M., Langle-Rouault, F., Kirlappos, H., Wagner, E., Mechtler, K., Zenke, M., Beug, H. and Birnstiel, M.L. (1990). Transferrin-polycation-mediated introduction of DNA into human leukemic cells: stimulation by agents that affect the survival of transfected DNA or modulate transferrin receptor levels. *Proc. Natl. Acad. Sci. USA*, 87: 4033-4037.

Cotten, M., Wagner, E., Zatloukal, K., Phillips, S., Curiel, D.T. and Birnstiel, M.L.

(1992). High-efficiency receptor-mediated delivery of small and large (48kb) gene constructs using the endosome disruption activity of defective or chemically inactivated adenovirus particles. *Proc. Natl. Acad. Sci. USA*, 89: 6094-6098.

Cotten, M., Wagner, E. and Birnstiel, M.L. (1993a). Receptor mediated transport of DNA into eukaryotic cells. *Methods Enzymol.*, 217: 618-644.

Cotten, M., Wagner, E., Zatloukal, K. and Birnstiel, M.L. (1993b). Chicken adenovirus (CELO virus) particles augment receptor-mediated DNA delivery to mammalian cells and yield exceptional levels of stable transformants. *J. Virol.*, 67: 3777-3785.

Cristiano, R.J. and Roth, J.A. (1996). Epidermal growth factor mediated DNA delivery into lung cancer via the epidermal growth factor receptor. *Cancer Gene Ther.*, 3: 4-10.

Cristiano, R.J., Smith, L.J., Kay, M.A., Brinkley, B.R. and Woo, S.L. (1993). Hepatic gene therapy: Efficient gene delivery and expression in primary hepatocytes utilizing a conjugated adenovirus-DNA complex. *Proc. Natl. Acad. Sci. USA*, 90: 11548-11552.

Curiel, D.T., Agarwal, S., Wagner, E. and Cotten, M. (1991). Adenovirus enhancement of transferrin-polylysine-mediated gene delivery. *Proc. Natl. Acad. Sci. USA*, 88: 8850-8854.

Curiel, D.T., Wagner, E., Cotten, M., Birnstiel, M.L., Agarwal, S., Li, C.M., Loechel, S. and Hu, P.C. (1992). High-efficiency gene transfer by adenovirus coupled to DNA-polylysine complexes. *Hum. Gene Ther.*, 3: 147-154.

Curiel, T.J., Cook, D.R., Bogedain, C., Jilg, W., Harrison, G.S., Cotten, M., Curiel, D.T. and Wagner, E. (1994). Foreign gene expression in Epstein-Barr virus transformed human B cells. *Virology*, 198: 577-585.

Diebold, S.S., Lehrmann, H., Kursa, M., Wagner, E., Cotten, M., and Zenke, M. (1999). Efficient Gene Delivery into Human Dendritic Cells by Adenovirus Polyethylenimine and Mannose Polyethylenimine transfection. *Hum. Gene Ther.*, 10: 775-786.

Ding, Z.M., Cristiano, R.J., Roth, J.A., Takacs, B. and Kuo, M.T. (1995). Malarial circumsporozoite protein is a novel gene delivery vehicle to primary hepatocyte cultures and cultured cells. *J. Biol. Chem.*, 270: 3667-3676.

Erbacher, P., Roche, A.C., Monsigny, M. and Midoux, P. (1995). Glycosylated polylysine/DNA complexes: gene transfer efficiency in relation with the size and the sugar substitution level of glycosylated polylysines and with the plasmid size. *Bioconjug. Chem.*, 6: 401-410.

Erbacher, P., Remy, J.S. and Behr, J.P. (1999) Gene transfer with synthetic virus-like particles via the integrin-mediated endocytosis pathway. *Gene Ther.*, 6: 138-145.

Felgner, J.H., Kumar, R., Sridhar, C., Wheeler, C., Tsai, Y.J., Border, R., Ramsey, P., Martin, M. and Felgner, P.L. (1994). Enhanced gene delivery and mechanism studies with a novel series of cationic lipid formulations. *J. Biol. Chem.*, 269: 2550-2561.

Felgner, P.L., Barenholz, Y., Behr, J.P., Cheng, S.H., Cullis, P., Huang, L., Jessee, J.A., Seymour, L., Szoka, F., Thierry, A.R., Wagner, E. and Wu, G. (1997). Nomenclature for synthetic gene delivery systems. *Hum. Gene Ther.*, 8: 511-512.

Ferkol, T., Kaetzel, C. and Davis, P. (1993). Gene transfer into respiratory epithelial cells by targeting the polymeric immunoglobulin receptor. *J. Clin. Invest.*, 92: 2394-2400.

Ferkol, T., Perales, J.C., Eckman, E., Kaetzel, C.S., Hanson, R.W. and Davis, P.B.

(1995). Gene transfer into the airway epithelium of animals by targeting the polymeric immunoglobulin receptor. *J. Clin. Invest.*, 95: 493-502.

Ferkol, T., Perales, J.C., Mularo, F. and Hanson, R.W. (1996). Receptor-mediated gene transfer into macrophages. *Proc. Natl. Acad. Sci. USA*, 93: 101-105.

Findeis, M.A., Merwin, J.R., Spitalny, G.L. and Chiou, H.C. (1993). Targeted delivery of DNA for gene therapy via receptors. *Trends Biotechnol.*, 11: 202-205.

Fisher, K.J. and Wilson, J.M. (1994). Biochemical and functional analysis of an adenovirus-based ligand complex for gene transfer. *Biochem. J.*, 299: 49-58.

Foster, B.J. and Kern, J.A. (1997). HER2-targeted gene transfer. *Hum. Gene Ther.*, 8: 719-727.

Gottschalk, S., Cristiano, R.J., Smith, L.C. and Woo, S.L.C. (1994). Folate receptor mediated DNA delivery into tumor cells: potosomal disruption results in enhanced gene expression. *Gene Ther.*, 1: 185-191.

Grove, R.I. and Wu, G.Y. (1998) Pre-clinical trials using hepatic gene delivery. *Adv. Drug. Del. Rev.*, 30: 199-204.

Haase, S.B., Heinzel, S.S., Calos, M.P. (1994). Transcription inhibits the replication of autonomously replicating plasmids in human cells. *Mol. Cell. Biol.*, 14: 2516-2524.

Haensler, J. and Szoka, F.C. (1993a). Synthesis and characterization of a trigalactosylated bisacridine compound to target DNA to hepatocytes. *Bioconjug. Chem.*, 4, 85-93.

Haensler, J. and Szoka, F.C. (1993b). Polyamidoamine cascade polymers mediate efficient transfection of cells in culture. *Bioconjug. Chem.*, 4: 372-379.

Harbottle, R.P., Cooper, R.G., Hart, S.L., Ladhoff, A., McKay, T., Knight, A., Wagner, E., Miller, A.D. and Coutelle, C. (1998). An RGD-oligolysine peptide: a prototype construct for integrin-mediated gene delivery. *Hum. Gene Ther.*, 9: 1037-1047.

Harrington, J.J., Bokkelen, G.V., Mays, R.W., Gustashaw, K. and Willard, H.F. (1997). Formation of *de novo* centromeres and construction of first-generation human artificial microchromosomes. *Nature Genet.*, 15: 345-355.

Hart, S.L., Harbottle, R.P., Cooper, R., Miller, A., Williamson, R. and Coutelle, C. (1995). Gene delivery and expression mediated by an integrin-binding peptide. *Gene Ther.*, 2: 552-554.

Huckett, B., Ariatti, M. and Hawtrey, A.O .(1990). Evidence for targeted gene transfer by receptor-mediated endocytosis: Stable expression following insulin-directed entry of *neo* into HepG2 cells. *Biochem. Pharmacol.*, 40: 253-263.

Huxley, C. (1994). Mammalian artificial chromosomes: a new tool for gene therapy. *Gene Ther.*, 1: 7-12.

Ikeno, M., Grimes, B., Okazaki, T., Nakano, M., Saitoh, K., Hoshino, H., McGill, N.I., Cooke, H., Masumoto, H. (1998). Construction of YAC-based mammalian artificial chromosomes. *Nature Biotech.*, 16: 431-439

Kabanov, A.V. and Kabanov, V.A. (1995). DNA Complexes with polycations for the delivery of genetic material into cells. *Bioconjug. Chem.*, 6: 7-20.

Kalderon, D., Richardson, W., Markham, A. and Smith, A.E. (1984). Sequence requirements for nuclear location of simian virus 40 large-T antigen. *Nature*, 311: 33-38.

Kircheis, R., Kichler, A., Wallner, G., Kursa, M., Ogris, M., Felzmann, T., Buchberger,

M., and Wagner, E. (1997). Coupling of cell-binding ligands to polyethylenimine for targeted delivery. *Gene Ther.*, 4: 409-418.

Kircheis, R., Schueller, S., Brunner, S., Ogris, M., Heider, K.-H., Zauner, W. and Wagner, E. (1999). Polycation-based DNA complexes for tumor-targeted gene delivery in vivo. *J. Gene Medicine*, 1: 111-120.

Kollen, W., Midoux, P., Erbacher, P., Yip, A., Roche, A., Monsigny, M., Glick, M. and Scanlin, T. (1996). Gluconoylated and glycosylated polylysines as vectors for gene transfer into cystic fibrosis airway epithelial cells. *Hum. Gene Ther.*, 7: 1577-1586.

Kukowska-Latallo, J.F., Bielinska, A.U., Johnson, J., Spindler, R., Tomalia, D.A. and Baker, J.R. (1996). Efficient transfer of genetic material into mammalian cells using starburst polyamidoamine dendrimers. *Proc. Natl. Acad. Sci. USA*, 93: 4897-4902.

Ledley, F. (1994). Non-viral gene therapy. *Current Op. Biotech.*, 5: 626-636.

Loeffler, J.P. and Behr, J.P. (1993). Gene transfer into primary and established mammalian cell lines with lipopolyamine-coated DNA. *Methods Enzymol.*, 217: 599-618.

McKee, T.D., DeRome, M.E., Wu, G.Y. and Findeis, M.A. (1994). Preparation of asialoorosomucoid-polylysine conjugates. *Bioconjug. Chem.*, 5: 306-311.

Merwin, J.R., Noell, G.S., Thomas, W.L., Chiou, H.C., DeRome, M.E., McKee, T.D., Spitalny, G.L. and Findeis, M.A. (1994). Targeted delivery of DNA using YEE(GalNAcAH)3, a synthetic glycopeptide ligand for the asialoglycoprotein receptor. *Bioconjug. Chem.*, 5: 612-620.

Merwin, J., Carmichael, E., Noell, G., DeRome, M., Thomas, W., Robert, N., Spitalny, G. and Chiou, H. (1995). CD5-mediated specific delivery of DNA to T lymphocytes: compartmentalization augmented by adenovirus. *J. Immunol. Methods*, 186: 257-266.

Midoux, P., Mendes, C., Legrand, A., Raimond, J., Mayer, R., Monsigny, M. and Roche, A.C. (1993). Specific gene transfer mediated by lactosylated poly-L-lysine into hepatoma cells. *Nucleic Acids Res.*, 21: 871-878.

Mislick, K.A., Baldeschwieler, J.D., Kayyem, J.F. and Meade, T.J. (1995). Transfection of folate-polylysine DNA complexes: evidence for lysosomal delivery. *Bioconjug. Chem.*, 6: 512-515.

Oberhauser, B., Plank, C. and Wagner, E (1995). Enhancing endosomal exit of nucleic acids using pH-sensitive viral fusion peptides. In:: *Delivery Strategies for Antisense Oligonucleotide Therapeutics,* S. Akhtar (Ed.), CRC Press, Florida, pp. 247-268.

Ogris, M., Steinlein, P., Kursa, M., Mechtler, K., Kircheis, R. and Wagner, E. (1998). The size of DNA/transferrin-PEI complexes is an important factor for gene expression in cultured cells. *Gene Ther.*, 5: 1425-1433.

Ogris, M., Brunner, S., Schueller, S., Kircheis, R. and Wagner, E. (1999). PEGylated DNA/Transferrin-PEI complexes: Reduced interaction with blood components, extended circulation in blood and potential for systemic gene delivery. *Gene Ther.*, 6: 595-605.

Perales, J.C., Ferkol, T., Beegen, H., Ratnoff, O.D. and Hanson, R.W. (1994a). Gene transfer in vivo: sustained expression and regulation of genes introduced into the liver by receptor-targeted uptake. *Proc. Natl. Acad. Sci. USA*, 91: 4086-4090.

Perales, J.C., Ferkol, T., Molas, M. and Hanson, R.W. (1994b). An evaluation of receptor-mediated gene transfer using synthetic DNA-ligand complexes. *Eur. J. Biochem.*, 226: 255-266.

Plank, C., Zatloukal, K., Cotten, M., Mechtler, K., and Wagner, E. (1992). Gene transfer into hepatocytes using asialoglycoprotein receptor mediated endocytosis of DNA complexed with an artificial tetra-antennary galactose ligand. *Bioconjug. Chem.*, 3: 533-539.

Plank, C., Oberhauser, B., Mechtler, K., Koch, C. and Wagner, E. (1994). The influence of endosome-disruptive peptides on gene transfer using synthetic virus-like gene transfer systems. *J. Biol. Chem.*, 269: 12918-12924.

Plank, C., Mechtler, K., Szoka, F. and Wagner, E. (1996), Activation of the complement system by synthetic DNA complexes: A potential barrier for intravenous gene delivery. *Hum. Gene Ther.*, 7: 1437-1446.

Pruzanski, W. and Saito, S. (1978). The influence of natural and synthetic cationic substances on phagocytic activity of human polymorphonuclear cells. *Exp. Cell Res.*, 117: 1-13.

Rojanasakul, Y., Wang, L.Y., Malanga, C.J., Ma, J.K.H. and Liaw, J. (1994). Targeted gene delivery to alveolar macrophages via Fc receptor-mediated endocytosis. *Pharmaceutical Res.*, 11: 1731 -1736.

Rosenkranz, A.A., Yachmenev, S.V., Jans, D.A., Serebryakova, N.V., Murav'ev, V.I., Peters, R. and Sobolev, A.S. (1992). Receptor-mediated endocytosis and nuclear transport of a transfecting DNA construct. *Exp. Cell Res.*, 199: 323-329.

Ross, G.F., Morris, R.E., Ciraolo, G., Huelsman, K., Bruno,M., Whitsett, J., Baatz, J. and Korfhagen, T. (1995). Surfactant protein A-polylysine conjugates for delivery of DNA to airway cells in culture. *Hum. Gene Ther.*, 6, 31-40.

Schneider, H., Huse, K, Birkenmeier, G., Otto, A., and Scholz, G.H. (1996). Gene transfer mediated by alpha2-macroglobulin. *Nucleic Acids Res.*, 24: 3873-3874.

Schwarzenberger, P., Spence, S.E., Gooya, J.M., Michiel, D., Curiel, D.T., Ruscetti, F.W. and Keller, J.R. (1996). Targeted gene transfer to human hematopoietic progenitor cell lines through the c-kit receptor. *Blood*, 87: 472-478.

Sclimenti, C.R. and Calos, M.P. (1998). Assaying extrachromosomal gene therapy vectors that carry replication/persistence elements. *Adv. Drug. Del. Res.*, 30: 13-22.

Shen, W.C. and Ryser, H.J.P. (1978). Conjugation of poly-L-lysine to albumin and horseradish peroxidase: a novel method of enhancing the cellular uptake of proteins. *Proc. Natl. Acad. Sci. USA*, 75: 1872-1876.

Sipe, D.M., Jesurum, A. and Murphy, R.F. (1991). Absence of Na^{+},K^{+}-ATPase regulation of endosomal acidification in K562 erythroleukemia cells. *J. Biol. Chem.*, 266: 3469-3474.

Tang, M.X. and Szoka, F.C. (1997). The influence of polymer structure on the interactions of cationic polymers with DNA and morphology of the resulting complexes. *Gene Ther.*, 4: 823-832.

Thurnher, M., Wagner, E., Clausen, H., Mechtler, K., Rusconi, S., Dinter, A., Berger, *E.G.*, Birnstiel, M.L. and Cotten, M. (1994). Carbohydrate receptor-mediated gene transfer to human T-leukemic cells. *Glycobiology*, 4: 429-435.

Trubetskoy, V.S., Torchilin, V.P., Kennel, S.J. and Huang, L. (1992a). Use of N-terminal modified poly(L-lysine)-antibody conjugates as a carrier for targeted gene delivery in mouse lung endothelial cells. *Bioconjug. Chem.*, 3: 323-327.

Trubetskoy, V.S., Torchilin, V.P., Kennel, S. and Huang, L. (1992b). Cationic liposomes enhance targeted delivery and expression of exogenous DNA mediated by N-terminal modified poly(L-lysine)-antibody conjugate in mouse lung endothelial cells. *Biochim. Biophys. Acta*, 1131: 311-313.

Vermeersch, H. and Remon, J.P. (1994). Immunogenicity of poly-D-lysine, a potential polymeric drug carrier. *J. Controlled Release*, 32: 225-229.

Wadhwa, M.S., Knoell, D.L., Young, A.P., Rice, K.G. (1995). Targeted gene delivery with a low molecular weight glycopeptide carrier. *Bioconjug. Chem.*, 6: 283-291.

Wagner, E., Zenke, M., Cotten, M., Beug, H. and Birnstiel, M.L. (1990). Transferrin-polycation conjugates as carriers for DNA uptake into cells. *Proc. Natl. Acad. Sci. USA*, 87: 3410-3414.

Wagner, E., Cotten, M., Foisner, R., and Birnstiel, M.L. (1991a). Transferrin-polycation-DNA complexes: The effect of polycations on the structure of the complex and DNA delivery to cells. *Proc. Natl. Acad. Sci. USA*, 88: 4255-4259.

Wagner, E., Cotten, M., Mechtler, K., Kirlappos, H. and Birnstiel, M.L. (1991b). DNA-binding transferrin conjugates as functional gene-delivery agents: synthesis by linkage of polylysine or ethidium homodimer to the transferrin carbohydrate moiety. *Bioconjug. Chem.*, 2: 226-231.

Wagner, E., Zatloukal, K., Cotten, M., Kirlappos, H., Mechtler, K., Curiel, D.T. and Birnstiel, M.L. (1992a). Coupling of adenovirus to transferrin-polylysine/DNA complexes greatly enhances receptor-mediated gene delivery and expression of transfected genes. *Proc. Natl. Acad. Sci. USA*, 89: 6099-6103.

Wagner, E., Plank, C., Zatloukal, K., Cotten, M. and Birnstiel, M.L. (1992b). Influenza virus hemagglutinin HA-2 N-terminal fusogenic peptides augment gene transfer by transferrin-polylysine/DNA complexes: Towards a synthetic virus-like gene transfer vehicle. *Proc. Natl. Acad. Sci. USA*, 89: 7934-7938.

Wagner, E., Curiel, D. and Cotten, M. (1994). Delivery of drugs, proteins and genes into cells using transferrin as a ligand for receptor-mediated endocytosis. *Adv. Drug Del. Rev.*, 14: 113-136.

Wightman, L., Patzelt, E., Wagner, E. and Kircheis, R. (1999). Development of transferrin-polycation/DNA based vectors for gene delivery to melanoma cells. *J. Drug Targeting* (submitted).

Wilson, J.M., Grossman, M., Wu, C.H., Chowdhury, N.R., Wu, G.Y. and Chowdhury, J.R. (1992). Hepatocyte-directed gene transfer *in vivo* leads to transient improvement of hypercholesterolemia in low density lipoprotein receptor-deficient rabbits. *J. Biol. Chem.*, 267: 963-967.

Wolfert, M.A. and Seymour, L.W. (1996). Atomic force microscopic analysis of the influence of the molecular weight of poly-L-lysine on the size of polyelectrolyte complexes formed with DNA. *Gene Ther.*, 3: 269-273.

Wu, G.Y. and Wu, C.H. (1987). Receptor-mediated *in vitro* gene transformation by a soluble DNA carrier system. *J. Biol. Chem.*, 262: 4429-4432.

Wu, G.Y. and Wu, C.H. (1988). Receptor-mediated gene delivery and expression *in vivo*. *J. Biol. Chem.*, 263: 14621-14624.

Wu, G.Y., Wilson, J.M., Shalaby, F., Grossman, M., Shafritz, D.A. and Wu, C.H. (1991). Receptor-mediated gene delivery *in vivo*. Partial correction of genetic analbuminemia in Nagase rats. *J. Biol. Chem.*, 266: 14338-14342.

Wu, G.Y., Zhan, P., Sze, L.L., Rosenberg, A.R. and Wu, C.H. (1994). Incorporation of adenovirus into a ligand based DNA carrier system results in retention of original receptor specificity and enhances targeted gene expression. *J. Biol. Chem.*, 269: 11542-11546.

Yin, W. and Cheng, P.W. (1994). Lectin conjugate-directed gene transfer to airway epithelial cells. *Biochem. Biophys. Res. Commun.*, 205: 826-833.

Zanta, M.A., Boussif, O., Adib, A. and Behr, J.P. (1997). *In vitro* gene delivery to hepatocytes with galactosylated polyethylenimine. *Bioconjug. Chem.*, 8: 841-844.

Zanta, M.A., Belguise-Valladier, P. Behr, J.P. (1999). Gene delivery: A single nuclear localization signal peptide is sufficient to carry DNA to the cell nucleus. *Proc. Natl. Acad. Sci. USA*, 96: 91-96.

Zauner, W., Blaas, D., Kuechler, E. and Wagner, E. (1995). Rhinovirus-mediated endosomal release of transfection complexes. *J. Virol.*, 69: 1085-1092.

Zauner, W., Kichler, A., Schmidt, W., Sinski, A. and Wagner, E. (1996). Glycerol enhancement of ligand-polylysine/DNA transfection. *Biotechniques*, 20: 905-913.

Zauner, W., Kichler, A., Mechtler, K., Schmidt, W. and Wagner, E. (1997). Glycerol and polylysine synergize in their ability to rupture vesicular membranes and increase transferrin-polylysine mediated gene transfer. *Exp. Cell Res.*, 232: 137-145.

Zauner, W., Ogris, M. and Wagner, E. (1998). Polylysine-based transfection systems utilizing receptor-mediated delivery. *Adv. Drug Del. Rev.*, 30: 97-114.

9. SUMMARY — A goal in gene therapy is the specific delivery of genes to the surface of a target cell, followed by efficient uptake and transport across intracellular barriers into the nucleus of cells. Amongst other systems, DNA/polycation particles coated with cell-binding ("targeting") ligands such as transferrin, EGF, asialoglycoprotein receptor binding molecules, or various antibodies are being optimized for this purpose. To allow ionic binding to the DNA, in most cases ligands are covalently linked to the polycations polylysine or polyethylenimine (PEI). Ligand-polylysine mediated gene transfer is strongly dependent on additional endosomolytic agents to ensure release from the endosomes and can be improved by pH-specific endosomolytic peptides or inactivated viral particles. In contrast, ligand-PEI mediated gene transfer is far less dependent on the presence of endosomolytic agents. Coating of DNA complexes with polyethylene glycol (PEG) through covalent coupling to PEI strongly reduces plasma protein binding and stabilizes complexes in size. Intravenous injection of PEG-modified complexes into tumor-bearing mice results in significant gene expression in the tumor.

Key Words: gene transfer; transfection; polylysine; polyethylenimine; endosomes

Progress in Gene Therapy: Basic and clinical Frontiers, pp. 205-223
R. Bertolotti *et al.* (Eds)

Gene delivery by guanidinium-cholesterol cationic lipids: a 1999 update

Noufissa Oudrhiri, Rajen Ramasawmy, Jean-Pierre Vigneron[1], Michelle Hauchecorne, Renée Toury, S. Riquier[1], Eric Vivien, Michel Peuchmaur, Jean Navarro, Jean-Marie Lehn[1] and Pierre Lehn*

INSERM U.458, Hôpital R. Debré, 48 Bd Sérurier, 75019 Paris, and [1]Laboratoire Chimie des Interactions Moléculaires, Collège de France, Place M. Berthelot, 75005 Paris, France

Table of Contents

1. INTRODUCTION

Artificial self-assembling systems may represent an attractive alternative approach to viral vectors for human gene therapy. Indeed, synthetic gene

* Corresponding author. E-mail: plehn@infobiogen.fr

delivery systems are free of the problems associated with the use of recombinant viruses, including safety issues and practical issues relating to bulk production and quality control. Thus, the search for efficient and easy-to-use nonviral vector systems led to the development of different kinds of synthetic vectors (Behr, 1993; Felgner, 1990; Gao and Huang, 1995; Ledley, 1995; Michael and Curiel, 1994; Scheule and Cheng, 1996).

Gene therapy relies on efficient gene transfer via vectors that transport and introduce the therapeutic gene into the target cells. Thus, an artificial vector has firstly to be capable of binding (and also condensing and protecting) the DNA to be transferred, which usually is a plasmid containing an eukaryotic expression cassette. The synthetic vectors that are most actively investigated today can schematically be divided into 2 classes: cationic lipids and polycationic polymers. Formulation of DNA with such vectors results in the spontaneous formation of (often highly heterogeneous) DNA aggregates via a self-assembly process driven by electrostatic interactions between the positively charged vector and the negatively charged DNA. A nomenclature convention for synthetic gene delivery systems was recently proposed: cationic lipid/DNA complexes are called lipoplexes and cationic polymer/ DNA complexes are termed polyplexes (Felgner *et al.*, 1997). Complexes that contain both polycationic polymers and cationic lipids may be referred to as lipopolyplexes. The DNA complexes are characterized by their size and their charge ratio (positive charge equivalents/negative charge equivalents). Several steps need then to be performed by the DNA complexes in order to achieve efficient gene delivery and expression: binding to the target cell, internalization, endosomal escape, uncoating, nuclear transport and, finally, appropriate expression of the transgene (Perales *et al.*, 1996; Zabner *et al.*, 1995). Lipoplexes are thought to enter cells via a non-specific "zipper-like" endocytosis induced by an ionic attraction between residual positive charges on the complexes (DNA formulations with a charge ratio greater than 1) and negatively charged residues on the cell surface. However, fusion events at the plasma membrane could also be involved in lipofection.

Since the pioneering work of Felgner *et al.* and also of Behr *et al.*, different cationic liposome/lipid formulations have been described (Behr *et al.*, 1989; Felgner *et al.*, 1987; Felgner and Ringold, 1989). Cationic liposomes normally contain a cationic amphiphile and a helper lipid, which usually is dioleoylphosphatidylethanolamine (DOPE). DOPE can be required for the generation of stable cationic liposomes and also provides membrane fusion activity. All cationic lipids contain the following domains: a cationic head group, a spacer arm, a linker bond and a hydrophobic anchor (Gao and Huang, 1995). As cationic lipids have usually been synthesized on the basis of trial and error, there is little similarity in their structures. However, from the great amount of experience gained in the field, one can identify some structure-activity relationships that appear to be important for the design of efficient reagents. First, it should be stressed here that the cationic headgroup of all currently available lipids contains one or several amine functions with different degrees of substitution (ranging from primary amines to quaternary ammonium). Protonation of these various amine groups, which is pH-dependent, allows binding of the vector to the negatively charged DNA. It has been shown that cholesterol derivatives with tertiary amino groups are more efficient than their quaternary ammonium counterparts (Farhood *et al.*, 1992). Interestingly, it has been reported that lipids with multivalent head groups show higher transfection activities than monovalent lipids (Behr *et al.*, 1989; Gao and Huang, 1995). The spatial configuration of the head group also seems to play an important role, as reagents harboring a T-shaped head group were recently shown to be more active than their linear homologs (Lee *et al.*, 1996). As concerns the spacer arm, its length appears to be a critical factor for some compounds but not for others. Of note, liposomes containing cationic lipids with longer spacer arms showed increased interaction with mucosal tissues (Gao and Huang, 1995). As stated above, the linker bond is also an important parameter; carbamoyl and amide bonds offer the advantage of being both chemically stable and biodegradable. Finally, as concerns the hydrophobic anchor, cationic lipids can contain either a pair of aliphatic chains or a cholesterol ring (Felgner, 1990; Gao and Huang, 1991). Fatty chains (*e.g.* oleoyl C18), which provide the lipoplexes with membrane fluidity and high lipid mixing capacity, are often used. In contrast, a cholesterol ring offers rigidity. Most interestingly, it has been reported that cationic cholesterol-based liposomes display a high membrane fusion capacity while being stable in the presence of various substances (ions, mucus) (Gao and Huang, 1995; Leventis and Silvius, 1990). Of note, liposomes containing the cationic cholesterol derivative DC-chol were the first synthetic vectors used in clinical trials of gene therapy (Caplen *et al.*, 1995; Nabel *et al.*, 1993).

2. RATIONALE FOR AND FEATURES OF GUANIDINIUM-CHOLESTEROL REAGENTS FOR GENE TRANSFECTION

In view of developing novel and efficient cationic lipids, we took into account many points already discussed above, especially the following: 1) all currently available cationic lipids were amine-carrying compounds relying on the protonation of ammonium groups for DNA binding, 2) most cationic lipids were formulated as liposomes with the helper lipid DOPE, 3) a cationic cholesterol derivative (DC-chol) was already used in the clinical setting because of its efficiency and its safety. Thus, we decided to direct our efforts toward the design of cationic cholesterol derivatives characterized by novel cationic moieties and by a high versatility (*i.e.* reagents efficient with as well as without liposomal formulation).

We chose to synthesize and test guanidinium-cholesterol lipids, as such cationic cholesterol derivatives should combine the membrane compatible features of the cholesterol subunit (see above) and the favorable features of the guanidinium group for DNA binding (Oudrhiri *et al.*, 1998; Vigneron *et al.*, 1996). Indeed, the guanidinium function presents several interesting features: 1) being highly basic, a guanidinium group remains protonated over a much wider range of pH than an ammonium group and, therefore, DNA binding should be relatively insensitive to variations of pH during the in vitro generation of lipoplexes and their trafficking toward the nucleus of the target cells, 2) guanidinium groups form with the DNA phosphate anions parallel hydrogen bonds which provide binding strength, 3) the guanidinium group is also able to develop hydrogen bonding with nucleic bases, especially with guanine, 4) finally, guanidinium groups in arginine residues have a key role in DNA-binding proteins such as histones and protamines.

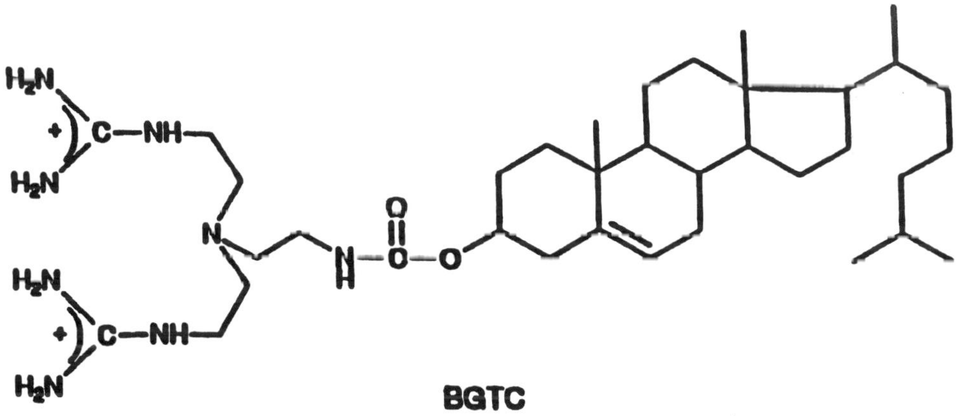

Figure 1. Structure of BGTC.

Two bis-guanidinium cholesterol derivatives were synthesized: BGSC (Bis-Guanidinium-Spermidine-Cholesterol) and BGTC (Bis-Guanidinium-Tren-Cholesterol). The structure of compound BGTC is indicated in Fig. 1. When taking into account several points discussed above in *Introduction*, one can note that BGTC has the following interesting features: 1) it is a multivalent (2 guanidinium groups providing DNA binding strength) T-shaped lipid, 2) it contains a carbamoyl bond known to be stable and biodegradable, 3) it contains a cholesterol ring providing the lipoplexes with rigidity, stability, and fusion activity, 4) finally, guanidinium functions and cholesterol are natural structures (Oudrhiri *et al.*, 1998). These 2 new cationic lipids could be obtained in good yields by a straightforward synthesis (Vigneron *et al.*, 1996). The purity of the final products was judged by 200 MHz ^{1}H NMR and by microanalysis. Of note, BGSC was found to be less soluble in aqueous medium than BGTC, probably because BGTC does also contain a tertiary amine; in addition, this tertiary amine of BGTC, which is situated between the positive guanidinium groups and probably has a lower pK, could also be able to buffer the acidic environment of endosomes and therefore protect the transferred DNA against lysosomal degradation as recently suggested for the polycationic polymer polyethylenimine (PEI) (Boussif *et al.*, 1995).

3. GUANIDINIUM-CHOLESTEROL LIPIDS ARE EFFICIENT FOR *IN VITRO* GENE TRANSFECTION INTO MAMMALIAN CELL LINES

As compound BGTC was found to be soluble in aqueous medium (by forming true micellar solutions), we first studied its usefulness for gene transfection by direct mixing of its micellar solution with the (luciferase reporter) plasmid-containing solution (Vigneron *et al.*, 1996). We especially investigated the influence of the charge ratio of the BGTC/DNA lipoplexes on transfection activity in 3 different easy-to-transfect mammalian cell lines (murine 3T3 fibroblasts, human epithelial HeLa cells and mouse neuro-blastoma NB2A cells). The transient transfection activity was highest when using strongly positive lipoplexes with a charge ratio of 6 to 8; this finding is in agreement with the literature, as lipoplexes are thought to interact with the negatively charged membrane of the target cells via their net positive charge (see above). Using similar transfection conditions, we could also satisfactorily transfect a variety of mammalian cell lines (human pulmonary A549 cells, simian COS cells, canine kidney MDCK cells, murine osteosarcoma ROS and pheochromocytoma PC12 cells...). These data demonstrate that reagent BGTC is efficient without liposomal formulation for gene transfection *in vitro*.

We next studied the *in vitro* transfection activity of compounds BGTC and BGSC formulated as cationic liposomes with DOPE, as: 1) BGSC does not yield micellar solutions in aqueous medium, 2) DOPE can help to stabilize cationic liposomes and provides fusion activity (see above) (Vigneron *et al.*, 1996). To prepare liposomes, a mixture of cationic lipid and DOPE (molar ratio 3: 2) was evaporated under vacuum and resuspended in Hepes buffer; the resulting dispersion was sonicated to form liposomes (with an average diameter of 50 nm by laser light scattering analysis). Both BGTC and BGSC cationic lipsomes gave satisfactorily transfection of various mammalian cell lines when using lipoplexes with a net positive charge (charge ratio ≈ 3); in addition, DOPE seemed also to facilitate gene transfection, probably because of its well-known fusogenic properties.

In conclusion, guanidinium-cholesterol lipids appear to be efficient and versatile reagents for gene transfection *in vitro* into mammalian cell lines.

4. BGTC LIPOPLEXES ARE EFFICIENT FOR GENE TRANSFECTION INTO PRIMARY AIRWAY EPITHELIUM CELLS *IN VITRO* OR *IN VIVO*

Gene transfer into the airway epithelium is a major goal for gene therapy, as it could offer treatment of both inherited (*e.g.* Cystic Fibrosis) and acquired (*e.g.* asthma) lung diseases (Alton and Geddes, 1995; Wilson, 1995). Thus, in view of lung-directed gene therapy, we have investigated the potential usefulness of the cationic lipid BGTC for gene transfection into primary human airway epithelium cells *in vitro* and into the mouse airway epithelium *in vivo* (Oudrhiri *et al.*, 1997).

Surface epithelial cells were isolated from human nasal polyps and cultured in a hormonally-defined serum-free medium. In this cell culture system, dissociated surface epithelial cells attach to a collagen substratum and form proliferative cell clusters containing the different epithelial cell types (differentiated ciliated cells, basal cells...). BGTC was used as a micellar solution for transfection of the reporter E.Coli LacZ gene into the primary cultures. Interestingly, X-Gal staining revealed a particular spatial distribution of the transfected cells, which were preferentially located in the peripheral areas of the cell clusters. We and others have also observed a similar distribution pattern when using other cationic lipids like Transfectam or Lipofectin (Fortunati *et al.*, 1996). Thus, we directed our efforts toward the identification of the biological features that may account for the higher tranfectability of the edge cells and found that it could be related to: 1) an increase in the accessibility for and the uptake of the lipoplexes, 2) a more active cell cycling status. Of note, elegant studies have indeed recently demonstrated that the surface of the edge cells is more negative than the surface of the central cells and that these edge cells are more cycling (Fasbender *et al.*, 1997; Matsui *et al.*, 1997). The role of mitosis in cationic lipid-mediated gene transfection had already previously been suggested, as it has been reported that transfected DNA does not easily enter the nuclear compartment during interphase.We are at present trying to better define the differentiation state of the transfected edge cells by studying the expression of the different cytokeratins (CK); indeed, CK are good markers of epithelial differentiation, *e.g.* CK14 is a basal cell specific marker and CK18 is a marker of fully differentiated ciliated cells. Biological studies have already suggested that these edge cells are poorly differentiated cells mimicking the regenerating cells present *in vivo* during the repair of the airway epithelium (Dupuit *et al.*, 1995; Plotkowski *et al.*, 1991). With regard to gene therapy of Cystic Fibrosis (CF), a key issue is however to target the mature fully-differentiated cells (notably ciliated cells) in the surface airway epithelium as these cells normally express CFTR. Thus, we are at present also investigating whether BGTC/DOPE liposomes can be used to transfer a CFTR plasmid into human (established and primary) CF cells expressing phenotypic markers of differentiated airway epithelial cells such as CK18 and polyglutamylated tubulins (specific of ciliated cells). Finally, it will also be important to examine the relevance of these *in vitro* findings for *in vivo* gene transfection.

Obtention of positive data not only with primary human airway cells *in vitro* but also in the airways of experimental animals *in vivo* should further strengthen the potential usefulness of reagent BGTC for lung-directed gene therapy. We have therefore also assessed the capacity of BGTC: DOPE liposomes for gene transfection into the mouse airways *in vivo* (Oudrhiri *et al.*, 1997). Positively charged lipoplexes containing a plasmid expressing a reporter gene (either the *E.Coli* LacZ gene or the *Photinus pyralis* luc gene) were delivered to the mouse airways by intratracheal instillation. The lungs and trachea were obtained for analysis from the treated animals at 48h after instillation. X-Gal positive cells could be detected in the airway epithelium, the transfected cells being mainly located in the trachea; such a distribution pattern could simply be related to the intratracheal instillation technique, as already suggested by others (Meyer *et al.*, 1995). Interestingly, mainly differentiated columnar cells were transfected in the surface epithelium; this observation is in agreement with data of the literature (Fortunati *et al.*, 1996). Most importantly, we also detected (both by X-Gal staining and by immunoperoxydase staining) transgene expression in submucosal glands; this is an important finding in the context of CF gene therapy, as it could then be required to efficiently transfect the submucosal glands which are the predominant site of CFTR expression in the respiratory tractus of man (Engelhardt *et al.*, 1992). Finally, we also could detect significant levels of luciferase activity in trachea homogenates of the treated mice, a result confirming unequivocally gene transfection. At present, studies in animal models (*e.g.* the CF mouse) should allow to still better assess the real usefulness of lipid BGTC for transfection of a therapeutic gene (*e.g.* CFTR).

In conclusion, as already stated above, positive data with primary human cells *in vitro* and in the mouse airways *in vivo* strongly support the potential of guanidinium-cholesterol cationic lipids for direct *in vivo* lung-directed gene therapy.

5. STRUCTURAL FEATURES OF SUPRAMOLECULAR ASSEMBLIES FORMED BY BGTC/DOPE/DNA AND BGTC/DNA LIPOPLEXES

In spite of being based on some rationale at the molecular level (see above), the development of these novel guanidinium-based synthetic vectors was nevertheless empirical as the mechanisms underlying cationic lipid-mediated gene transfer are still poorly understood. Thus, in order to get a better insight into the factors and processes involved, we studied (in collaboration with researchers at Rhône-Poulenc Rorer) the structural features of the supramolecular assemblies formed by BGTC/DOPE/DNA and BGTC/DNA lipoplexes (Pitard *et al.*, 1999). Indeed, because gene transfection by synthetic vectors involves the spontaneous *in vitro* formation of discrete DNA aggregates, the transfection efficiency of a given cationic lipid system depends not only on the properties (at the molecular level) of the cationic lipid, but also on the structural and functional properties of the self-assembled supramolecular assemblies. It should also be stressed here that the identification of the structural features of active lipoplexes should help in the design of improved guanidinium-based vectors.

Cryotransmission electron microscopy (cryoTEM) studies and small-angle X-ray scattering experiments were performed and indicated the presence of multilamellar domains with a regular spacing of 70 A and 68 A in BGTC/DOPE/DNA and BGTC/DNA aggregates, respectively (Pitard *et al.*, 1999). These results suggest that DNA condensation by multivalent guanidinium-cholesterol cationic lipids involves the formation of highly ordered multilamellar domains, the DNA being intercalated between the lipid bilayers. These results are also consistent with previous studies of the structural characteristics of lipoplexes resulting from the interaction of DNA with various cationic lipid formulations: lipopolyamine micelles, dioleoyl trimethylammonium propane/dioleoyl phosphatidylcholine (DOTAP/DOPC) liposomes and dioctadecyldiammonium bromide/cholesterol liposomes (Lasic *et al.*, 1997; Pitard *et al.*, 1997; Rädler *et al.*, 1997).

Conventional electron microscopy studies were performed on HeLa cells transfected with either unlabeled or gold-labeled plasmid DNA (to allow its identification) complexed with BGTC/DOPE liposomes or BGTC alone. The morphology of the internalized lipoplexes was in general quite similar to that of the lipoplexes imaged by cryoTEM. Thus, it is tempting to suggest that multilamellar lipoplexes may play an important role in gene transfection by BGTC/DOPE (molar ratio 3/2) liposomes and BGTC micelles. In addition, the presence of typical structures mainly in intracytoplasmic vesicles is consistent with electrostatic-driven endocytosis being the major route of DNA uptake during BGTC-mediated transfection. This is further supported by the observation of extracellular lipoplexes being engulfed by cellular digitations. This is also consistent with previous studies suggesting that lipoplexes are internalized by means of nonspecific endocytosis after binding of positively charged DNA complexes to anionic residues on the cell surface (Labat Moleur *et al.*, 1996; Wrobel and Collins, 1995; Zabner *et al.*, 1995; Zhou and Huang, 1994). Most importantly, as endocytosis leaves the DNA two membranes away from the nucleus, the internalized lipoplexes have still to carry out the following critical steps: endosomal escape, release into the cytosol, uncoating and nuclear uptake. The escape of BGTC/DOPE lipoplexes from the endosomes may be due to the fusogenic activity of DOPE and also to a "proton sponge" mechanism (buffering of the acidic environment of endosomes by the tertiary amine of BGTC leading to osmotic swelling and endosome disruption), as suggested for the lipopolyamine DOGS and the cationic polymer polyethylenimine PEI (Demeneix and Behr, 1996). Endosome buffering may even be a major mechanism of endosomal escape for BGTC/DNA lipoplexes which do not contain DOPE. Finally, in the context of BGTC-mediated transfection, it will be important to uncover and exploit the mechanisms of nuclear targeting and uptake. Interestingly, it has recently been shown that the efficacy of gene transfection can be greatly enhanced by using a linear expression cassette bearing a single nuclear localization signal (NLS) peptide (Zanta *et al.*, 1999).

Taken together, these studies highlight the importance of a « supramolecular galenics » approach to nonviral gene delivery. Ideally, such an approach would require determination of not only the initial structural features of the lipoplexes but also the precise structural changes that these lipoplexes undergo when they interact with the molecular machinery of the target cells during their trafficking toward the nucleus. It will obviously be difficult to develop such « supramolecular galenics » particularly because of the complexity of the different steps involved. Thus, one may predict that some empiricism will still persist when designing synthetic vectors for gene transfection.

6. TOWARDS IMPROVED SYNTHETIC VECTORS FOR HUMAN GENE THERAPY

Clearly, guanidinium-cholesterol lipids do constitute a novel and highly promising class of cationic lipids. However, as successful clinical application of direct *in vivo* lipofection will require an efficient, safe and pharmaceutically-acceptable vector, it will be important to: 1) get a better understanding of the mechanisms underlying lipofection, 2) develop further improved vectors, 3) work out optimal and pharmaceutically acceptable formulations, 4) and, finally, assess the *in vivo* toxicity of these novel synthetic vectors.

Several approaches can be used to investigate the factors and processes leading to transfection. We are at present synthesizing and testing new members of the family of guanidinium-based cationic lipids in order to get better insights into their structure-activity relationship. Biophysical and electron microscopy (EM) studies will help to elucidate the supramolecular structural features of the self-assembled lipoplexes. In ongoing morphological studies (EM, confocal immunofluorescence microscopy) we are investigating the intracellular trafficking of the internalized lipoplexes; we should thereby be able to contribute to a better understanding of the mechanisms of gene transfection by adding our own results on the nature and fate of BGTC lipoplexes to those already obtained by others when using "more classical" cationic lipids (Labat-Moleur *et al.*, 1996; Pitard *et al.*, 1997; Rädler *et al.*, 1997; Sternberg *et al.*, 1994; Szoka *et al.*, 1996; Zabner *et al.*, 1995). In other ongoing studies, we are evaluating the effects of various biological fluids (*e.g.* serum, pulmonary surfactant, human amniotic fluid) on gene transfection by lipid BGTC. Such transfection conditions aim to mimic *in vivo* conditions in man (respectively intravenous administration, in utero therapy and lung-directed therapy). It has been reported that these fluids can inhibit the transfection efficiency of some cationic lipids (Douar *et al.*, 1996; Duncan *et al.*, 1997; Tsan *et al.*, 1997). Interestingly, our preliminary results indicate that the *in vitro* transfection activity of lipid BGTC is not decreased in the presence of these various substances.

Attempts to develop improved vectors can be based on the design of new guanidinium-based cationic lipids and on the working out of optimal liposomal formulations (BGTC/DOPE molar ratio, use of other colipids than DOPE...). Most important, it should also be possible to design "modular" self-assembling systems. For example, several studies have recently reported a drastic improvement of lipofection by polycationic polymers, especially poly-L-lysine (Gao and Huang, 1996; Mack *et al.*, 1996). It has been suggested that polycationic polymers may play several roles, such as reducing the size of the DNA complexes (condensing reagents), allowing intracellular relocalization of the transferred DNA and even promoting nuclear trafficking (Pollard *et al.*, 1998). Broadly speaking, modular gene delivery systems can be based on a DNA core particle to which several components with various functions are anchored (for example, via their hydrocarbon domain) (Behr, 1993). A possible approach is to develop "artificial viruses", *i.e.* synthetic vectors incorporating functional elements mimicking viruses (Lehn *et al.*, 1998). Indeed, viruses are complex, precise and dynamic nucleic acid-containing structures that are able to efficiently infect their target cells via sophisticated mechanisms selected in the course of evolution.

Obviously, the design of such virus-like artificial systems relies on an in-depth understanding of the mechanisms underlying virus assembly and viral infection. The virus-mimicking elements incorporated in a modular artificial system could help the DNA complexes to successfully perform the various steps involved in gene tansfection, especially cell targeting, endosomal escape and nuclear transport. To date, some pionneering work has already been reported, but it is out of the scope of this article to discuss the various approaches that have been attempted (Lehn *et al.*, 1998). Of note, it should be stressed here that, alike viral infection which is a stepwise process involving conformational changes "priming" the viral particle to perform the next step, it could be required that the modular elements of virus-like synthetic vectors also work in a stepwise manner, *i.e.* in a chronological order. Such sophisticated gene delivery systems can be viewed as "programmed supramolecular systems" obtained via a defined plan, the information necessary for the assembly process to take place and the algorithm that it follows being stored in the components and operating via selective molecular recognition events (Lehn, 1993).

The development of pharmaceutical-grade formulations will also be important (Ledley, 1996). For example, the *in vitro* self-assembling process should allow to obtain, in a reproducible manner, homogenous and highly efficient lipoplexes. On the other hand, in the clinical trials already performed, plasmid DNA and cationic lipids were provided in separate vials and mixed at the bed site; a procedure based on the use of a single vial containing stable lipoplexes would be both more convenient and more acceptable. In order to develop gene -based medicines similar to current drug therapies, it will be important to work out stable and optimal formulations for systemic intravenous administration. As concerns lung-directed gene therapy, aerosol delivery needs to be investigated, its efficiency being dependent on the resistance of the lipoplexes to the shear forces involved in nebulization (Schwartz *et al.*, 1996). Optimal conditions for nebulization will probably have to be worked out, as considerable damage could occur to the lipoplexes.

Interestingly, it has recently been shown that BGTC/DOPE formulations have a degree of stability adequate to permit effective gene delivery by the aerosol route, the multilamellar structure of the lipoplexes being well maintained during nebulization; such formulations may therefore hold promise for successful clinical application of aerosol gene delivery (Densmore *et al.*, 1999).

7. CONCLUSIONS

Guanidinium-cholesterol cationic lipids do clearly constitute novel and highly promising reagents for gene delivery. The development of this novel class of synthetic vectors should allow to gain new insights into the mechanisms of gene transfection. Most important, it should also allow to design a broad range of improved artificial self-assembling systems for human gene therapy (Crystal, 1995; Ledley, 1995; Lehn *et al.*, 1998). Indeed, various gene-based medicines will need to be developed, as one might predict that there will be no ultimate "all purpose" vector, as each somatic target exhibits specific biological characteristics.

Acknowledgments

This work was supported by grants from the Association Française de Lutte contre la Mucoviscidose (AFLM) and the Association Française contre les Myopathies (AFM).

8. REFERENCES

Alton, E.W.F.W. and Geddes, D.M. (1995). Gene therapy for cysticfibrosis: a clinical perspective. *Gene Ther.*, 2: 88-95.

Behr, J.P. (1993). Synthetic gene transfer vectors. *Acc. Chem. Res.*, 26: 274-278.

Behr, J.P., Demeneix, B., Loeffler, J.P. and Perez-Mutul, J. (1989). Efficient gene transfer into mammalian primary endocrine cells with lipopolyamine-coated DNA. *Proc. Natl. Acad. Sci. USA*, 86: 6982-6986.

Boussif, O, Lezoualc'h, F., Santa, M., Mergny, M., Scherman, D., Demeneix, B. and Behr, J-P. (1995). A novel, versatile vector for gene and oligonucleotide transfer into cells in culture and in vivo: polyethylenimine. *Proc. Natl. Acad. Sci. USA*, 92: 7297-7303.

Caplen, N.J. Alton, E.W.F.W., Middleton, P.G., Dorin, J.R., Stevenson, B.J., Gao, X., Durham, S.R., Jeffery, P.K., Hodson, M.E., Coutelle, C., Huang, L., Porteous, D.J., Williamson, R. and Geddes, D.M. (1995). Liposome-mediated CFTR gene transfer to the nasal epithelium of patients with cystic fibrosis. *Nat. Med.*, 1: 39-46.

Crystal, R.G. (1995). The gene as the drug. *Nat. Med.*, 1: 15-17.

Demeneix, B. and Behr, J.P. (1996). The proton sponge: a trick the viruses did not exploit. In: *Artificial Self-Assembling Systems for Gene Delivery*, P.L. Felgner, M.J. Heller, P. Lehn, J.P. Behr and F.C. Szoka (Eds). Am. Chem. Soc. Washington, pp. 146-151.

Densmore, C.L., Giddings, T.H., Waldrep, J.C., Kinsey B. and Knight, V. (1999). Gene transfer by guanidinium cholesterol: dioleoylphosphatidyl-ethanolamine liposome-DNA complexes in aerosol. *J. Gene Med.*, 1: 251-264.

Douar, A.M., Themis, M., Sandig, V., Friedmann, T. and Coutelle, C. (1996). Effect of amniotic fluid on cationic lipid mediated transfection and retroviral infection. *Gene Ther.*, 3: 789-796.

Duncan, J.E., Whitsett, J.A. and Horowitz, A.D. (1997). Pulmonary surfactant inhibits cationic liposome-mediated gene delivery to respiratory epithelial cells in vitro. *Hum. Gene Ther.*, 8: 431-438.

Dupuit, F., Zahm, J.M., Pierrot, D., Brezillon, S., Bonnet, N., Imler, J.L., Pavirani, A. and Puchelle, E. (1995). Regenerating cells in human airway surface epithelium represent preferential targets for recombinant adenovirus. *Hum. Gene Ther.*, 6: 1185-1193.

Engelhardt, J., Yankaskas, J.R., Ernst, S., Yang, Y., Marino, C., Boucher, R., Cohn, J. and Wilson, J.M. (1992). Submucosal glands are the predominant site of CFTR expression in the human bronchus. *Nat. Genet.*, **2**: 240-248.

Farhood, H., Bottega, R., Epand, R.M. and Huang, L. (1992). Effect of cationic cholesterol derivatives on gene transfer and protein kinase C activity. *Biochim. Biophys. Acta*, 1111: 239-246.

Fasbender, A., Zabner, J., Zeiher, P.G. and Welsch, M.J. (1997). A low rate of cell proliferation and reduced DNA uptake limit cationic lipid-mediated gene transfer to primary cultures of ciliated human airway epithelia. *Gene Ther.*, 4: 1173-1180.

Felgner, P.L. (1990). Particulate systems and polymers for in vitro and in vivo delivery of polynucleotides. *Adv. Drug Deliv. Rev.*, 5: 163-187.

Felgner, P.L. and Ringold, G.M. (1989). Cationic liposome mediated transfection. *Nature*, 337: 387-388.

Felgner, P.L., Gadek, T.R., Holm, M., Roman, R., Chen, H.W., Wenz, M., Northrop, J.P., Ringold, G.M. and Danielsen, M. (1987). Lipofection: a highly efficient, lipid-mediated DNA transfection procedure. *Proc. Natl. Acad. Sci. USA*, 84: 7413-7417.

Felgner, P.L., Barenholz, Y., Behr, J.P., Cheng, S.H., Cullis, P., Huang, L., Jessee, J.A., Seymour, L., Szoka, F., Thierry, A.R., Wagner, E. and Wu, G. (1997). Nomenclature for synthetic gene delivery systems. *Hum. Gene Ther.*, 8: 511-512.

Fortunati, E., Bout, A., Zanta, M.A., Valerio, D. and Scarpa, M. (1996). In vitro and in vivo gene transfer to pulmonary cells mediated by cationic liposomes. *Biochim. Biophys. Acta*, 1306: 55-62.

Gao, W. and Huang, L. (1991). A novel cationic liposome reagent for efficient transfection in mammalian cells. *Biochem. Biophys. Res. Commun.*, 179: 280-285.

Gao, W. and Huang, L. (1995). Cationic liposome-mediated gene transfer. *Gene Ther.*, 2: 710-722.

Gao, W. and Huang, L. (1996). Potentiation of cationic liposome-mediated gene delivery by polycations. *Biochemistry*, 35: 1027-1036.

Labat-Moleur, F., Steffan, A.M., Brisson, C., Perron, H., Feugeas, O., Furstenberger, P., Oberling, F., Brambilla, E. and Behr, J.P. (1996). An electron microscopy study into the mechanism of gene transfer with lipopolyamines. *Gene Ther.* **3**, 1010-1017.

Lasic, D., Strey, H., Stuart, M., Podgornik, R. and Frederik, P. M. (1997). *J. Am. Chem. Soc.*, 119: 832-833.

Ledley, F.D. (1995). Nonviral gene therapy: the promise of genes as pharmaceutical products. *Hum. Gene Ther.*, 6: 1129-1144.

Ledley, F.D. (1996). Pharmaceutical approach to somatic gene therapy. *Pharm. Res.*, 13: 1595-1614.

Lee, E.R., Marshall, J., Siegel, C.S., Jiang, C., Yew, N.S., Nichols, M., Nietupski, J., Ziegler, R.J., Lane, M.B., Wang, K.X., Wan, N.C. Scheule, R.K., Harris, D.J, Smith, A.E. and Cheng, S.H. (1996). Detailed analysis of structures and formulations of cationic lipids for efficient gene transfer to the lung. *Hum. Gene Ther.*, 7: 1701-1717.

Lehn, J.M. (1993). Supramolecular chemistry-molecular information and the design of supramolecular materials. *Makromol. Chem., Macromol. Symp.*, 69: 1-17.

Lehn, P., Fabrega, S., Oudrhiri, N. and Navarro, J. (1998). Gene delivery systems: bridging the gap between recombinant viruses and artificial vectors. *Adv. Drug Deliv. Rev.*, 30: 5-11.

Leventis, R. and Silvius, R.J. (1990). Interactions of mammalian cells with lipid dispersions containing novel metabolizable cationic amphiphiles. *Biochim. Biophys. Acta*, 1023: 124-132.

Mack, K.D., Walzem, R.L., Lehmann-Bruinsma, K., Powell, J.S. and Zeldis, J.B. (1996). Polylysine enhances cationic liposome-mediated transfection of the hepatoblastoma cell line HepG2. *Biotechnol. Appl. Biochem.*, 23: 217-220.

Matsui, H., Johnson, L.G., Randell, S.H. and Boucher, R.C. (1997). Loss of binding and entry of liposome-DNA complexes decreases transfection efficiency in differentiated airway epithelial cells. *J. Biol. Chem.*, 272: 1117-1126.

Meyer, K.B., Thompson, M.M., Levy, M.Y., Barron, L.G. and Szoka, F.C. (1995). Intratracheal gene delivery to the mouse airways: characterization of plasmid DNA expression and pharmacokinetics. *Gene Ther.*, 2: 450-460.

Michael, S.I. and Curiel, D.T. (1994). Strategies to achieve targeted gene delivery via the receptor-mediated endocytosis pathway. *Gene Ther.*, 1: 223-232.

Nabel, G.J., Nabel, *E.G.*, Yang, Z.Y., Fox, B.A., Plautz, G.E., Gao, X., Huang, L., Shu, S., Gordon, D. and Chang, A.E. (1993). Direct gene transfer with DNA liposome complexes in melanoma: expression , biological activity, lack of toxicity in humans. *Proc. Natl. Acad. Sci. USA*, 90: 11307-11311.

Oudrhiri, N., Vigneron, J.P., Peuchmaur, M., Leclerc, T., Lehn, J.M. and Lehn, P.

(1997). Gene transfer by guanidinium-cholesterol cationic lipids into airway epithelial cells in vitro and in vivo. *Proc. Natl. Acad. Sci. USA*, 94: 1651-1656.

Oudrhiri, N., Vigneron, J.P., Hauchecorne, M., Toury, R., Lemoine, A.I., Peuchmaur, M., Navarro, J., Lehn, J.M. and Lehn, P. (1998). Guanidinium-cholesterol cationic lipids: novel reagents for gene transfection and perspectives for gene therapy. *Biogenic Amines*, 14: 537-552.

Perales, J.C., Molas, M. and Hanson, R.W. (1996). Advances in receptor-mediated gene transfer into cells in vitro and in vivo. In: *Artificial Self-Assembling Systems for Gene Delivery*, P.L. Felgner, M.J. Heller, P. Lehn, J.P. Behr and F.C. Szoka (Eds), Am. Chem. Soc., Washington, pp. 105-115.

Pitard, B., Aguerre, O., Airiau , M., Lachages, A.M., Bouknikachvili, T., Byk, G., Dubertret, C., Hervion, C., Scherman, D., Mayaux, J.F. and Crouzet, J. (1997). Virus-sized self-assembling lamellar complexes between plasmid DNA and cationic micelles promote gene transfer. *Proc. Natl. Acad. Sci. USA*, 94: 14412-14417.

Pitard, B., Oudrhiri, N., Vigneron, J.P., Hauchecorne, M., Aguerre, O., Toury, R., Airiau, M., Ramasawmy, R., Scherman, D., Crouzet, J., Lehn, J-M., and Lehn, P. (1999). Structural characteristics of supramolecular assemblies formed by guanidinium-cholesterol reagents for gene transfection. *Proc. Natl. Acad. Sci. USA*, 96: 2621-2626.

Plotkowski, M.C., Chevillard, M., Altemayer, D., Zahm, J.M., Colliot, G. and Puchelle, E. (1991). Differential adhesion of Pseudomonas aeruginosa to human respiratory epithelial cells in primary culture. *J. Clin. Invest.*, 87: 2018-2028.

Pollard, H., Remy, J.S., Loussouarn, G., Demolombe, S., Behr, J.P. and Escande, D. (1998). Polyethylenimine but not cationic lipids promotes transgene delivery to the nucleus in mammalian cells. *J. Biol. Chem.*, 273: 7507-7511.

Rädler, J.O., Koltover, I., Salditt, T. and Safinya, C.R. (1997). Structure of DNA-cationic liposome complexes: DNA intercalation in multilamellar membranes in distinct interhelical packing regimes. *Science*, 275: 810-814.

Scheule, R.K. and Cheng, S.H. (1996). Gene transfer into mammalian cells using synthetic cationic lipids. In: *Artificial Self-Assembling Systems for Gene Delivery,* P.L. Felgner, M.J. Heller, P. Lehn, J.P. Behr and F.C. Szoka (Eds), Am. Chem. Soc., Washington, pp 177-190.

Schwartz, L.A., Johnson, J.L., Black, M., Cheng, S.H., Hogan, M.E. and Waldrep, J.C. (1996). Delivery of DNA-cationic liposome complexes by small-particle aerosol. *Hum. Gene Ther.*, 7: 731-741.

Sternberg, B., Sorgi, F.L. and Huang, L. (1994). New structures in complex formation between DNA and cationic liposomes visualized by freeze-fracture electron microscopy. *FEBS Lett.*, 356: 361-366.

Szoka, F.C., Xu, Y. and Zelphati, O. (1996). How are nucleic acids released in cells from cationic lipid-nucleic acid complexes? *J. Liposome Res.*, 6: 567-587.

Tsan, M.F., Tsan, G.L. and White, J.E. (1997). Surfactant inhibits cationic liposome-mediated gene transfer. *Hum. Gene Ther.*, 8: 817-825.

Vigneron, J.P., Oudrhiri, N., Fauquet, M., Vergely, L., Bradley, J.C., Basseville, M., Lehn, P. and Lehn, J.M. (1996). Guanidinium-cholesterol cationic lipids: efficient vectors for the transfection of eukaryotic cells. *Proc. Natl. Acad. Sci. USA*, 93: 9682-9686.

Wilson, J.M. (1995). Gene therapy for cystic fibrosis: challenges and future directions. *J. Clin. Invest.*, 96: 2547-2554.

Wrobel, I. and Collins, D. (1995). Fusion of cationic liposomes with mammalian cells occurs after endocytosis. *Biochim. Biophys. Acta*, 1235: 296-304.

Zabner, J., Fasbender, A.J., Moninget, T., Poellinger, K.A. and Welsh, M.J. (1995). Cellular and molecular barriers to gene transfer by a cationic lipid. *J. Biol. Chem.*, 270: 18997-19007.

Zanta, M.A., Belguise-Valladier, P. and Behr, J. P. (1999). Gene delivery: a single nuclear localization signal peptide is sufficient to carry DNA to the cell nucleus. *Proc. Natl. Acad. Sci. USA*, 96: 91-96.

Zhou, X. and Huang, L. (1994). DNA transfection mediated by cationic liposomes containing lipopolylysine: characterization and mechanism of action. *Biochim. Biophys. Acta*, 1189: 195-203.

9. SUMMARY — Synthetic vectors are at present widely investigated as an alternative approach to recombinant viruses for gene transfer studies and gene therapy applications. Among these artificial self-assembling systems, cationic lipids are particularly attractive as it is possible to design and synthesize a great variety of reagents. Several amine-carrying cationic lipids have been shown to be efficient for gene transfection; moreover, some reagents (DC-Chol:DOPE, DOTAP...) have even already been used in clinical trials. Over the last several years, we have developed a novel class of cationic lipids: cholesterol derivatives characterized by polar head groups containing guanidinium functions. Such reagents combine the membrane compatible features of the cholesterol subunit and the favorable features of the guanidinium groups for DNA binding. We herein intend to summarize our work showing that these novel cationic lipids are efficient for gene transfection *in vitro* (into various mammalian cell lines and primary human airway cells) and also *in vivo* (into the mouse airway epithelium); we here also briefly outline the results of more recent work suggesting that gene transfection by multivalent guanidinium-cholesterol cationic lipids involves supramolecular bioassemblies characterized by highly ordered multilamellar domains. Taken together, our studies confirm the potential of cationic lipids for human gene therapy, especially lung-directed gene therapy for Cystic Fibrosis. Most importantly, our work also provides the basis for the design of improved synthetic gene delivery systems. In this forward-looking review, we will therefore also discuss some approaches to the remaining problems that need to be resolved in order to develop pharmaceutically-acceptable gene-based medicines.

Key Words: guanidinium; cholesterol; synthetic vectors; cationic lipids; gene therapy; transfection

Progress in Gene Therapy: Basic and Clinical Frontiers, pp. 225-243
R. Bertolotti *et al.* (Eds)

Progress of fusigenic viral-liposomes

Yasufumi Kaneda*, Hiroshi Nakasima, Naoki Tsuboniwa and Ryuichi Morishita

Division of Gene Therapy Science, Graduate School of Medicine, Osaka University, Suita, Osaka 565-0871, Japan

Table of Contents

* Corresponding author. E-mail: kaneday@gts.med.osaka-u.ac.jp

1. INTRODUCTION

To promote human gene therapy, development of more effective gene transfer vector systems is desired (Marshall, 1995). There are several barriers to be overcome in human gene therapy. First is improvement of gene transfer efficiency. Numerous viral and non-viral (synthetic) methods for gene transfer have been developed for human gene therapy (Mulligan, 1993; Ledley, 1995). Generally, viral methods are more efficient than non-viral methods for gene delivery to cells. However, viral vectors are safety hazards because of the co-introduction of essential genetic elements from the parent viruses, leaky expression of viral genes, immunogenicity, and alterations of host genomic structure. In general, non-viral vectors are less toxic and less immunogenic. However, most non-viral methods are less efficient for gene transfer, especially *in vivo*, compared to some of the viral vectors. Thus, both viral and non-viral vectors have limitations as well as advantages. Therefore, to develop an *in vivo* gene transfer vector with high efficiency and low toxicity, the limitations of one type of vector system should be compensated for by introducing the strengths of another type of system.

With this idea of compensation in mind, we developed a novel hybrid gene transfer vector by combining viral and non-viral vectors, and we constructed a fusigenic viral liposome with a fusigenic envelope derived from Hemagglutinating Virus of Japan (HVJ; Sendai virus) (Kaneda, 1998; Kaneda *et al.,* 1999). In this delivery system, DNA-loaded liposomes are fused with UV-inactivated HVJ to form the fusigenic viral-liposome, HVJ-liposome (400 to 500 nm in diameter). The advantage of fusion-mediated delivery is the protection of molecules in endosomes and lysosomes from degradation. DNA smaller than 100 kb can be incorporated and delivered to cells. RNA, oligonucleotides and drugs are also efficiently introduced into cells *in vitro* and *in vivo*. HVJ-liposomes have not been shown to induce significant cell damages *in vivo*. The repetitive transfection is successful *in vivo* because of the low immunogenicity of HVJ-liposomes (Hirano *et al.,*

1998). Numerous gene therapy strategies using this HVJ-liposome system have been successful (Dzau *et al.,* 1996; Kaneda *et al.,* 1999). Recently, we developed several types of HVJ-liposomes by altering the lipid components of the liposomes (Saeki *et al.,* 1997). HVJ-cationic liposomes facilitated efficient entrapment of DNA and yielded 100 to 800 times higher gene expression *in vitro* and were more useful for gene expression in restricted portions of organs and for gene therapy of disseminated cancers compared with the conventional HVJ-anionic liposome (Kaneda *et al.*, 1999). The novel anionic liposomes with a virus-mimicking lipid composition (HVJ-AVE liposome) increased transfection efficiency by 5-10 fold *in vivo*. Therefore, appropriate HVJ-liposomes should be used for different targets and objectives (Saeki *et al.,* 1997).

Another barrier for human gene therapy is long-term gene expression. The transient expression of genes in tissues and inability for targeted delivery are the main limitations of current HVJ-liposomes. The main reason for transient gene expression *in vivo* appears to be instability of the transgene in the cell nucleus. Southern blotting of the transgene in the liver revealed that the gene transferred to liver nucleus by HVJ-liposome existed extrachromosomally without integration into host genome and the episomal form of the transgene was degraded in the liver nucleus after 2 weeks. To achieve sustained gene expression, there are two distinct approaches. One is the integration of a transgene into the host genome. Insertion of a transgene into the genome of non-dividing cells has been problematic, and site-specific integration is desirable. Another approach is the stable retention of an episomal piece of DNA. Epstein-Barr virus (EBV) has been analyzed thoroughly in terms of its latent infection. The *cis*-acting oriP (the latent viral DNA replication origin) sequence and the *trans*-acting EBNA-1 (EBV nuclear antigen-1) are required for EBV latent-infection, which is characterized by autonomous replication and nuclear retention of the EBV genome in host cells (Reisman *et al.,* 1985; Sugden *et al.,* 1989; Wisokenski and Yates 1989; Yates *et al.,* 1985). However, this EBV vector can transfer genes dominantly to B lymphocytes, but hardly to other somatic cells. Thus, to augment long-term transgene expression in various animal organs, we constructed an EBV replicon-based plasmid, pEB, that contains oriP and EBNA-1, and coupled the replicon vector with HVJ-liposomes. Using this sytem, we succeeded in sustained and enhanced luciferase expression in several cell lines and in mouse liver (Saeki *et al.,* 1998).

However, for human gene therapy, the HVJ-liposome system should be improved to achieve a safe vector. From the viewpoint of safety, the limitation of the current HVJ-liposome system is that components other than fusion proteins are included in the vesicles, although the genome is inactivated by UV-irradiation. Several attempts have been made to construct a synthetic virosome derived from HVJ (Wu *et al.,* 1995; Ramani *et al.,* 1997 and 1998). Here, we also report the development of reconstituted fusion particles containing fusion proteins of HVJ, and we demonstrate gene delivery *in vitro* and *in vivo* using the reconstituted fusion vesicles.

2. MATERIALS AND METHODS

a) Preparation of EBV replicon vector

The *Cla*I-*Nar*I fragment of p205 which contains the oriP and triplet repeat (717bp)-deleted EBNA-1 sequences was isolated and the pEBc vector was constructed by cloning the fragment into the *Bgl*II site of pcDNA3 (Invitrogen) by blunt-end ligation. It has already been reported that this truncated EBNA-1 gave the most efficient replication of oriP-containing DNA (Yates *et al.,* 1985). Then, the EBV replicon vector, pEBAct, was constructed by replacing the CMV promotor of pEBc with the chicken β-actin promotor derived from pAct-CAT. The luciferase gene was isolated from the pGL2-promotor-vector (Promega) by *Hind*III and *Bam*HI digestion, and pEBActLuc was constructed by cloning the DNA at the *Hind*III and *Bam*HI sites of pEBAct. pActLuc was also constructed by cloning the luciferase gene at the *Hind*III and *Bam*HI sites of pAct-CAT.

b) Virus

HVJ, Z strain, was purified by differential centrifugation as described previously (Kaneda, 1994). The purified HVJ was resuspended in balanced salt solution (BSS: 137 mM NaCl, 5.4 mM KCl, 10 mM Tris-HCl, pH 7.5) and the virus titer was determined by measuring the absorbance at 540 nm. Optical density at 540 nm corresponds to 15,000 hemagglutinating units (HAU) per ml, which correlates with fusion activity.

c) Preparation of HVJ-liposome

Phosphatidylcholine, dioleoylphosphatidylethanolamine, sphingomyelin, phosphatidylserine, and cholesterol were mixed in a molar ratio of 13.3:13.3:13.3:10:50 for anionic type HVJ-AVE liposome. Phosphatidylcholine, dioleoylphosphatidylethanolamine, sphingomyelin, cholesterol and 3β[N-(N`,N`-dimethylaminoethane)-carbamoyl] (DC)-cholesterol were mixed in a molar ratio of 16.7:16.7:16.7:40:10 for cationic type HVJ-DC liposome. Dried lipid mixture was prepared and then hydrated in 200 μl of balanced salt solution (BSS; 137mM NaCl, 5.4mM KCl, 10mM Tris-HCl, pH 7.6) containing 200 μg of plasmid DNA. Liposomes were prepared by vortexing and extrusion. Purified HVJ (Z strain) was inactivated by UV irradiation (198 mjoule/mm2) just before use. The liposome suspension (2 ml, containing 9.75 mg lipid) was incubated with HVJ (15000 hemagglutinating units) for 10 min on ice and then for 1hr at 37 ℃ with gentle shaking. Free HVJ was removed from the HVJ-liposomes by sucrose density gradient centrifugation. HVJ-liposome was collected from the top of 30% sucrose layer, filled up to 1 ml by BSS and stored at 4 °C.

d) *In vitro* and *in vivo* transfection

Human embryonic kidney (HEK 293) cells or baby hamster kidney (BHK-21) cells (2 x 10^5 cells) were inoculated into a 100mm dish one day before transfection. One-hundred microliter of HVJ-DC liposome was added to one dish in the presence of Dulbecco's modified minimum essential medium supplemented with 10% fetal bovine serum (10% FBS-DMEM), incubated with cells for 1 hour at 37 °C, and then removed. Cells were cultured in 10% FBS-DMEM and harvested at various time points. To obtain stable transformants, HeLa-S3 were cultured in the presence of 500 μg/ml G418 (geneticin R; Gibco BRL) for three weeks after the transfer, fixed with 80% methanol and stained by 1% Giemsa in phosphate buffered saline (PBS).

For *in vivo* transfection, eight-week-old, male Balb/c mice were anesthetized by intraperitoneal injection of diluted pentobarbital (1 mg). The abdomen was opened with a middle-line incision, and a large lobe of the liver was exposed. Two-hundred μl of HVJ-AVE liposome in BSS with 1 mM CaC12 were injected into a lobe of the liver under the perisplanchnic membrane using a 1-ml syringe with a 26-gauge needle, and the incision was sutured. Mice were sacrificed on days 2, 7, 14, 21, 28 and 30 after the transfection.

e) Extraction of F and HN fusion proteins from HVJ

Nonidet P-40 (NP-40) and phenylmethylsulfonylfluoride (PMSF) dissolved in ethanol were added to 20 ml (1,750,000 HAU) of purified HVJ suspension at final concentrations of 0.5% and 2 mM, respectively. The mixture was incubated with rotation at 4 °C for 30 min. Then, the suspension was centrifuged at 100,000g for 75 min at 4 °C to remove insoluble proteins and viral genome (Uchida *et al.,* 1979). The supernatant was dialyzed for 3 days against 5 mM phosphate buffer (pH 6.0) to wash out the residual NP-40 and PMSF with the buffer being changed daily. The dialyzed solution was centrifuged at 100,000 g for 75 min at 4 °C to remove insoluble materials. The supernatant was applied to an ion-exchange column of CM-Sepharose CL6B (Pharmacia Fine Chemicals) equilibrated with 10 mM phosphate buffer (pH 5.2) containing 0.3 M sucrose and 1 mM KCl based on a previously described method (Yoshima *et al.,* 1981). Flow-through fraction and 0.2 M NaCl eluates were fractionated. Each fraction was subjected to sodium dodecyl sulfate-polyacrylamide gel electrophoresis (SDS-PAGE) to analyze protein components.

f) Insertion of F and HN proteins into liposomes

A lipid mixture of 3.56 mg phosphatidylcholine and 0.44 mg cholesterol was dissolved in chloroform, and the lipid solution was evaporated in a rotary-evaporator (Uchida *et al.,* 1979). The dried lipid mixture was dissolved thoroughly by vortexing in 2.0 ml (1.6 mg) of protein solution from the flow-through fraction containing 0.85% NP-40. Then, the solution was dialyzed against 10 mM phosphate buffer (pH 7.2) containing 0.3 M sucrose and 1 mM KCl to remove NP-40. Dialysis was performed for 6 days with daily buffer change. This dialyzed solution was applied to a column of agarose beads (Bio-Gel A-50m, BioRad Laboratories) equilibrated with 10 mM phosphate buffer (pH 5.2) containing 0.3 M sucrose and 1 mM KCl. Fractions over 1.5 optical density at 540 nm were collected as reconstituted fusion particles and fused with DNA-loaded liposomes prepared from 10 mg lipids as described below.

g) Preparation of the reconstituted fusion liposomes containing fluorescent isothiocyanate-labeled oligonucleotides

Fluorescent isothiocyanate (FITC)-oligonucleotides (ODNs) were incorporated into the liposome as described previously (Morishita, 1994). In brief, 10 mg of dried lipid mixture (phosphatidylserine, phosphatidylcholine and cholesterol) was hydrated in 200 μl of BSS containing 20 nmoles FITC-conjugated ODNs (5'-GAT-CCG-CGG-GAA-ATF-3'; Clontech Laboratories). FITC-ODN-loaded liposomes were prepared by vortexing and sonication. The liposomes incorporating FITC-ODN were incubated with the reconstituted fusion particles for 10 min at 4 °C and then for 60 min at 37 °C. The reconstituted fusion liposomes containing FITC-ODNs were purified by sucrose gradient ultracentrifugation as described previously (Kaneda, 1994).

h) Transfer of FITC-ODN to FL cells

Human amniotic FL cells were incubated in a 35 mm well with one-tenth of the reconstituted fusion liposomes with FITC-ODNs prepared as described above in 10% FBS-DMEM for 30 min at 37 °C. Then, the cells were washed with fresh medium to remove the liposomes and fixed with cold methanol containing 1% acetic acid at 4 ℃ for 5 min. After being washed with PBS, the nuclei were stained with 1 μg/ml Hoechst 33258 (Sigma Chemicals Inc.) for 5 min. Then, the cells receiving FITC-ODN were observed with a fluorescence microscope.

i) LacZ gene transfer to mouse muscle

The expression vector, pAct-LacZ-NII, was constructed as described previously (Saeki *et al.,* 1997). Eight-week-old, male C57BL/6 mice were anesthetized by intraperitoneal injection of diluted pentobarbital (1 mg). One-third of the reconstituted fusion liposomes containing 40 μg of pAct-LacZ-NII prepared as described above were injected directly into the quadriceps using a 1-ml syringe with a 26 gauge needle. On day 3 after the transfer, the muscle at the injected sites was isolated, fixed with 1% glutaraldehyde, immersed in 20% sucrose overnight, and the frozen sections (10 μm) were cut with a cryostat (Miles-Sankyo, Kanagawa, Japan). LacZ gene expression was detected as described elsewhere (Sanes *et al.,* 1986).

3. RESULTS

a) Sustained gene expression *in vitro* and *in vivo* by the EBV replicon vector coupled with HVJ-liposomes

Luciferase activity was measured in HEK 293 and BHK-21 cells at various time-points after gene transfer (Fig. 1 a and b). Luciferase gene expression by the introduction of pActLuc was highest on day 1 after the transfer, but then rapidly decreased. However, luciferase activity by the introduction of pEBActLuc was almost equal to that by pActLuc on day 1 in HEK 293 cells, and then increased and was maintained for at least 10 days

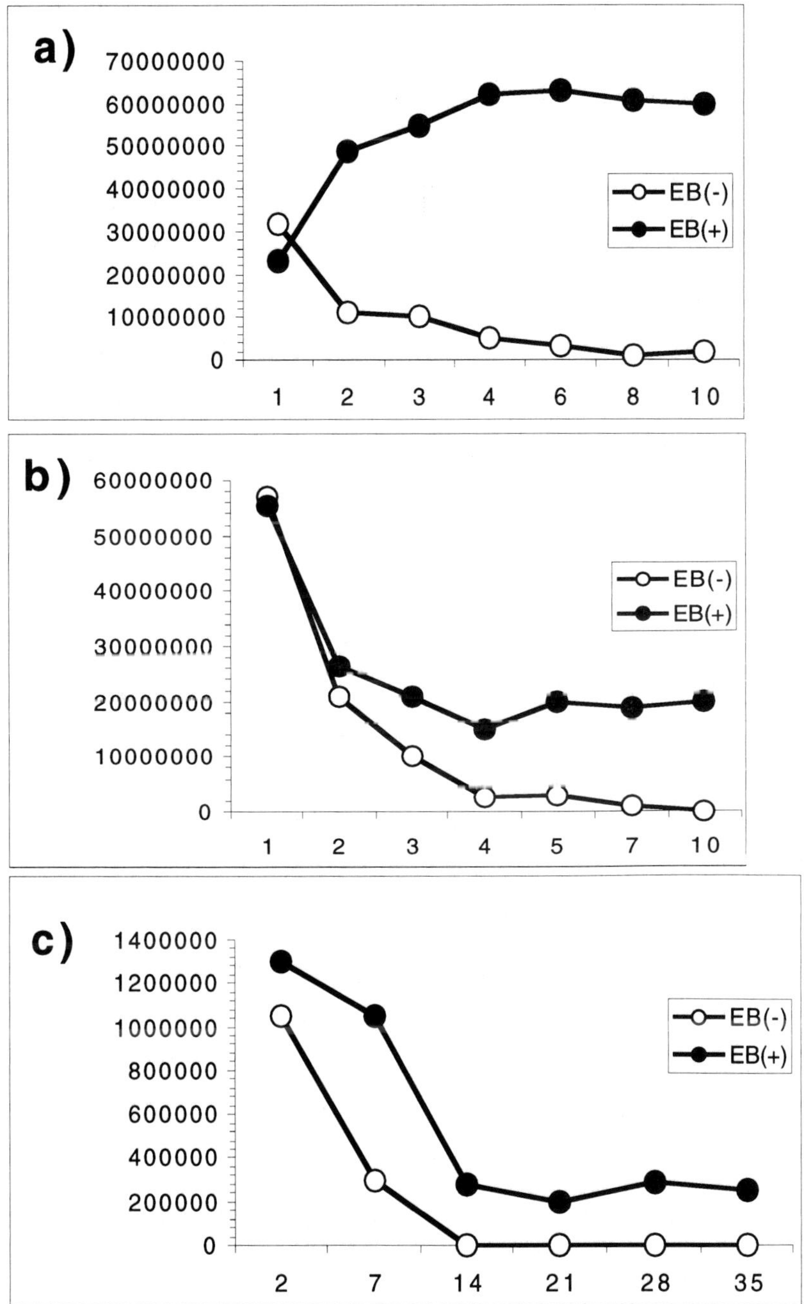

Figure 1. Luciferase activity in HEK 293 (a), BHK-21 (b) and mouse liver (c). The ordinate indicates light units per whole cells of one dish (a, b) or per liver (c) which were measured by a luminometer, and the abscissa represents the period (day) after transfer. Closed circle: luciferase activity upon transfer of the EBV replicon vector, pEBActLuc. Open circle : luciferase activity upon transfer of pActLuc without EBV sequences.

at 5-10 times higher level compared with that by pActLuc. Similar sustained gene expression was observed in human cervical cancer HeLa-S3 and human diploid fibroblast FS3 cells (Saeki *et al.,* 1998). Southern blot analysis revealed that in human cells transfected with the EBV replicon vector, the plasmid was episomally retained in the nucleus and that the amplification of the plasmid was detected probably due to the autonomous replication of the DNA (Saeki *et al.,* 1998). In BHK-21 cells, luciferase gene expression was highest on day 1 and then decreased for both plasmids, but it was maintained on and after day 3 post-transfection for pEBActLuc. Southern blot analysis of luciferase gene after gene transfer showed that the plasmid containing EBV sequences was retained extrachromosomally in the nucleus of BHK-21 cells, while the plasmid DNA without EBV sequences was lost between days 3 and 9 after transfer. About 100-150 copies of pEBActLuc were retained in the nucleus on day 9, decreasing between days 9 and 14 (Saeki *et al.,* 1998).

The EBV replicon vector was also very useful for obtaining stable transformants. Approximately 410 colonies resistant to G418 were obtained by the transfer of the pcEB vector, while only 10 colonies were resistant to G418 by the transfer of pcDNA3 (Saeki *et al.,* 1998).

Although we have succeeded in gene expression in many organs using HVJ-liposomes, the main limitation of the vector is transient gene expression *in vivo*. Thus, we attempted to prolong *in vivo* gene expression by combining the EBV replicon vector system with HVJ-AVE liposomes. HVJ-AVE liposomes incorporating pEBActLuc or pActLuc were directly injected under the perisplanchnic membrane of Balb/c mouse liver. Luciferase activity of the liver was assayed at various time points after the transfer (Fig. 1c). When pActLuc without EBV sequences was introduced, luciferase gene expression was detected on day 7, but rapidly decreased to insignificant levels by day 14. However, with pEBActLuc, luciferase gene expression also decreased between days 7 and 14, but the expression remained constant after day 14 for at least 35 days.

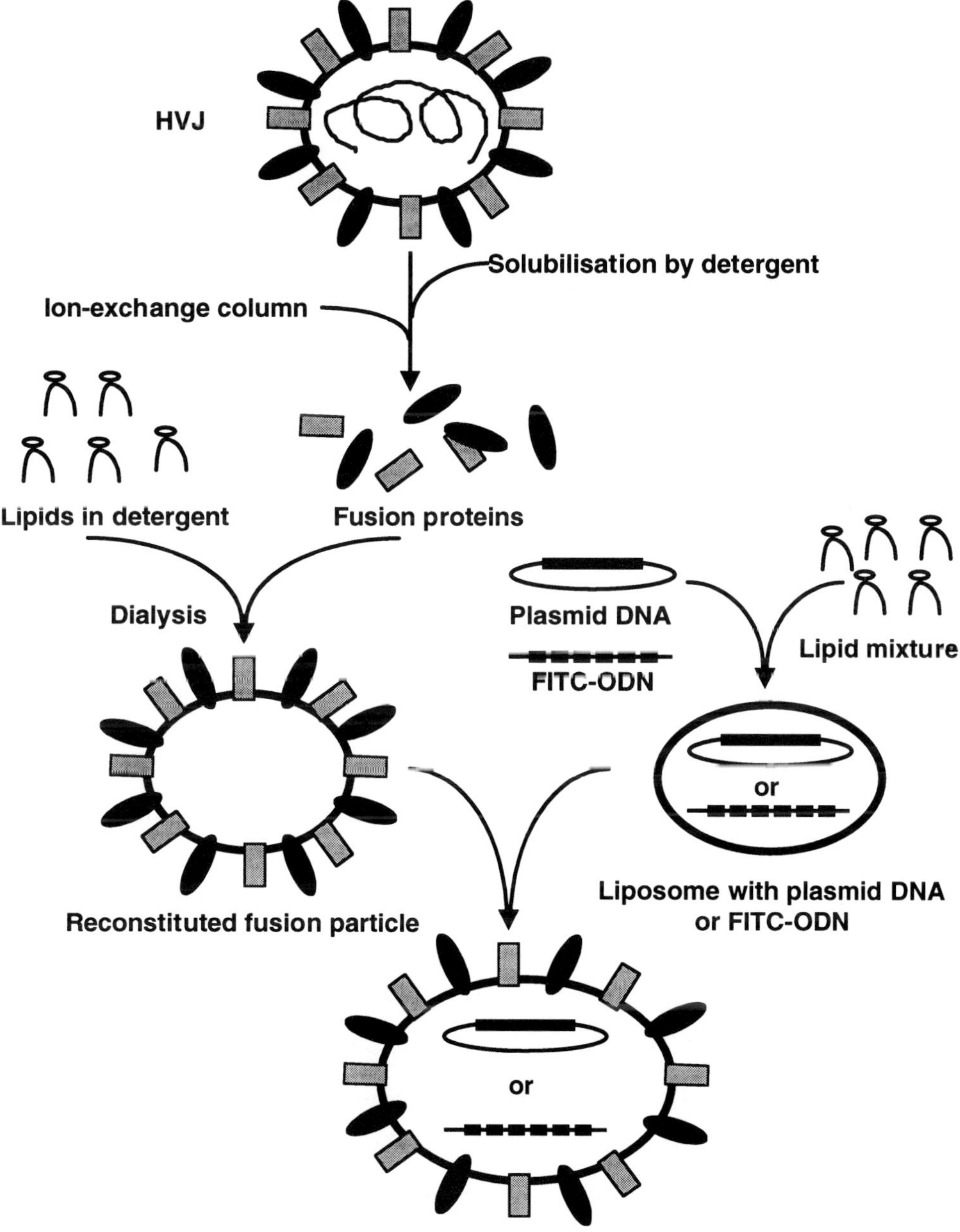

Figure 2. Preparation of reconstituted fusion liposomes. HVJ is solubilized and fusion proteins are isolated. The proteins are inserted into liposomes by dialysis to form reconstituted fusion particles. Liposomes trapping plasmid DNA or FITC-ODN are constructed and fused with reconstituted fusion particles. Reconstituted fusion liposomes with plasmid DNA or FITC-ODN are used as gene transfer vehicles *in vitro* and *in vivo*.

b) Reconstituted fusion liposomes for gene transfer *in vitro* and *in vivo*

Preparation of reconstituted fusion particles. The fusion proteins of HVJ were purified by applying detergent-lysed HVJ to ion-exchange column chromatography. The 52 kDa (F1) and 72 kDa (HN) proteins were dominantly eluted in the flow-through fractions (Suzuki *et al.*, 2000). The fusion proteins were mixed with NP-40 solubilized lipid mixture, and the liposomes were prepared by dialysis (Uchida *et al.,* 1979). The liposomes contained F1 and HN. However, we failed to trap DNA in the same liposomes by dialysis. The fusion particles containing DNA were constructed by incubating empty fusion particles with DNA-loaded liposomes prepared by a vortexing-sonication method. The schema of the construction is shown in Fig. 2.

Introduction of FITC-ODN into cultured cells by reconstituted fusion particles. Thirty minutes after the transfection, FITC-ODN were introduced into all the nuclei of FL cells (Fig. 3a). Without fusion particles, very faint signals were observed in FL cells (Fig. 3b). The efficiency of FITC-ODN introduction by the reconstituted liposomes was comparable to that by HVJ-liposomes using UV-irradiated whole HVJ virion. The empty reconstituted fusion particles were stored at 4 ^{0}C for 1 to 4 weeks, and the reconstituted fusion particles containing FITC-ODN were constructed at weekly intervals to introduce FITC-ODN into FL cells. FITC-ODN were introduced into more than 80% of the cells using the reconstituted fusion particles stored for 4 weeks (Suzuki *et al.*, 2000). However, when FITC-ODN-loaded HVJ-liposomes were prepared using HVJ that was stored at 4 ^{0}C for 1 to 4 weeks after UV-irradiation, the efficiency of the introduction of FITC-ODN into the nucleus decreased dramatically: it was reduced to less than 10% using HVJ-liposomes prepared with HVJ stored for only 1 week.

Gene expression in vitro and in vivo using the reconstituted fusion particles. Luciferase gene-loaded reconstituted liposomes were prepared as shown in Fig. 2 and added to HEK 293 cells. On day 1 after the transfer,

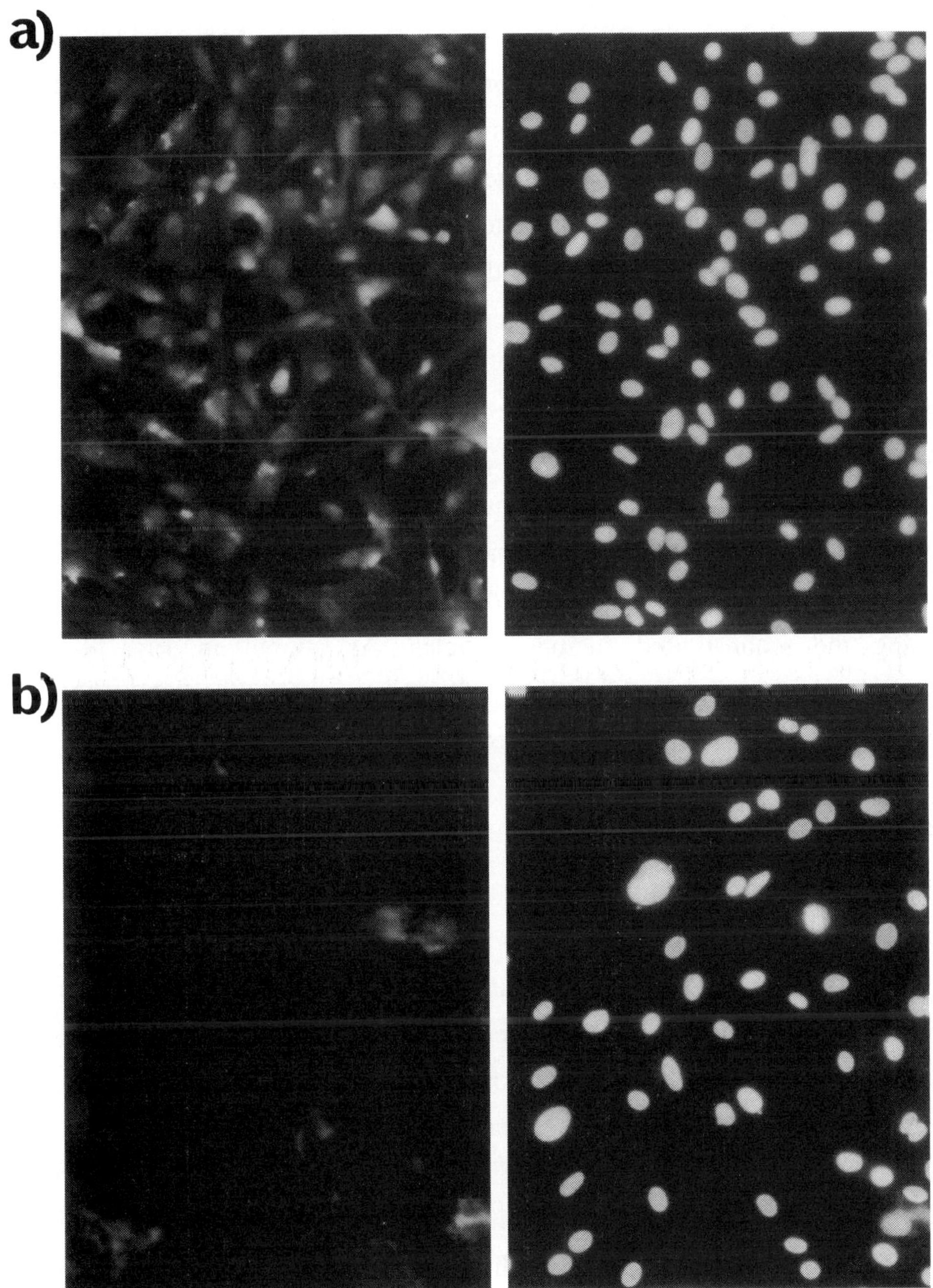

Figure 3. Transfer of FITC-ODNs to FL cells by the reconstituted fusion liposomes. At 30 min after the transfer, FITC fluorescence (left pictures) was detected in all the FL cells (a) but was absent when reconstituted fusion liposomes without FITC-ODNs were used (b). Right pictures: Hoechst staining to detect nuclei.

luciferase activity in the cells was measured as described previously (Saeki *et al.*, 1997). Luciferase gene expression with the reconstituted fusion liposomes was almost the same as that by HVJ-liposomes (Suzuki *et al.*, 2000). Both F and HN proteins are required for virus-cell fusion. Indeed, when the HN protein was inactivated by 3 mM dithiothreitol (DTT) before being incorporated into reconstituted vesicles as described previously (Bagai and Sankar, 1993), luciferase gene transfer was dramatically reduced. Briefly, fusion proteins were purified, exposed to HN-inactivating DTT treatment and used to prepare reconstituted liposomes. The resulting vesicles were able to transfer the luciferase gene to HEK 293 cells, however with a poor efficiency: luciferase activity was less than 10% of the level obtained with intact reconstituted fusion particles (Suzuki *et al.*, 2000).

LacZ gene was transferred directly to mouse skeletal muscle *in vivo* using the reconstituted fusion particles. As shown in Fig. 4, LacZ expression was observed in 40 to 50% of the muscle fibers. No significant toxicity was observed in mouse muscle.

4. DISCUSSION

a) Sustained gene expression by EBV replicon vector coupled with HVJ-liposomes

We showed that the combination of an EBV replicon vector with HVJ-liposomes resulted in sustained and enhanced gene expression both in various cultured cells and in mouse liver which are resistant to EBV infection. Since HVJ-liposomes can deliver genes to many other tissues including kidney, artery, lung, brain, heart, skin, muscle, eye, and articular cartilage, long-term gene expression will be induced in those tissues at high efficiency by the use of EBV replicon vectors. Sustained gene expression using the EBV replicon vector with HVJ-liposomes was also observed in other organs. Luciferase gene expression lasted more than 96 days in mouse muscle and more than 9 weeks in rat kidney, while without EBV replicon sequences, gene expression was not detected on day 28 in muscle and day 7

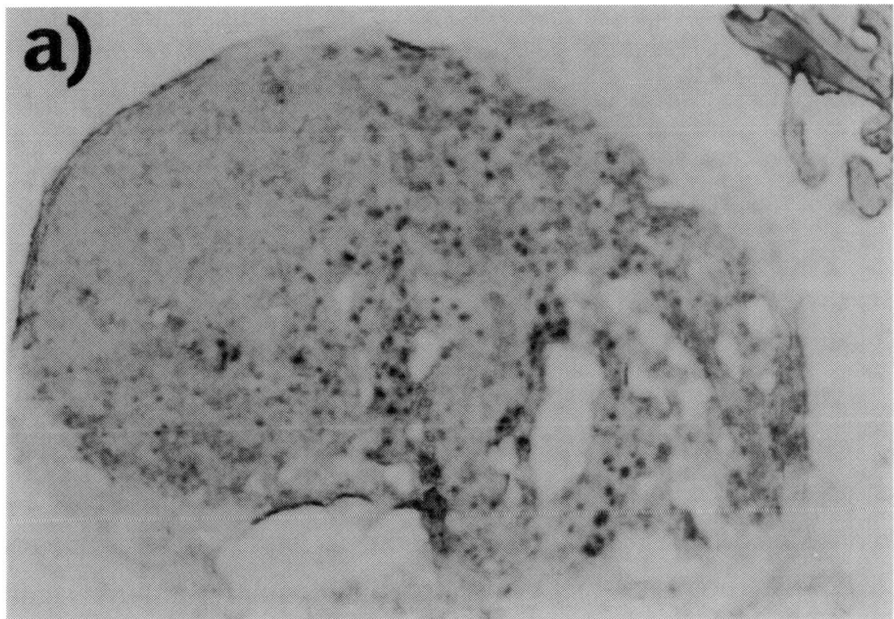

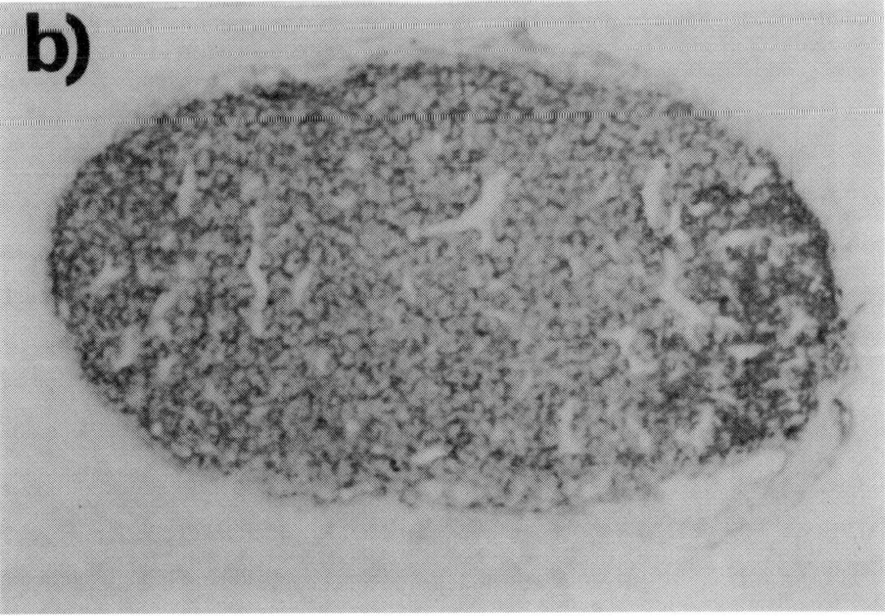

Figure 4. LacZ gene expression in mouse skeletal muscle. The reconstituted fusion liposomes containing pAct-LacZ-NII were injected into the quadriceps of C57BL/6 mice (a). As a negative control, empty reconstituted fusion liposomes were injected into the same muscle (b). On day 3 after the transfer, LacZ gene expression was examined in frozen sections of the quadriceps by X-gal staining. Transverse sections are presented.

in kidney (Kaneda *et al.,* unpublished results). Thus, the combination vector system of EBV components and HVJ-liposomes will provide us a powerful tool for future gene therapy as an artificial viral vector system.

One of the problems of EBV replicon vector plasmid is an unequal delivery of plasmid copies to daughter cells after cell division. Therefore, the current EBV replicon vector may be more beneficial for sustained gene expression in non-dividing cells. To realize an equal delivery of the plasmid after cell-division, the addition of centromere sequence will be desirable. This work will result in the construction of an artificial chromosome vector. Although the current artificial chromosome may be an ideal tool for the retention of genome of interest, it will be hard to deliver such a large-sized genome to the cytoplasm and to sort it to the nucleus. Further developments of EBV-based replicon vector systems may produce a novel artificial chromosome vector which is more practical for human gene therapy.

b) Reconstituted fusion liposomes for gene transfer *in vitro* and *in vivo*

We also demonstrated the construction of reconstituted fusion particles using fusion proteins of HVJ. The reconstituted fusion particles introduced both ODN and plasmid DNA into cells as efficiently as HVJ-liposomes based UV-inactivated HVJ whole virion, and also successfully transferred plasmid DNA to mouse muscle. The transfection efficiency with the reconstituted vesicles appeared to be comparable to that with HVJ-liposomes. The advantages of the reconstituted vesicles are their relative safety and stability. We found that the genome of HVJ was not detected in the reconstituted vesicles by RT-PCR (Kaneda, unpublished data). Therefore, there is no concern about viral replication after transfection. The reconstituted fusion liposomes retained significant fusion activity at 4 weeks after preparation, while UV-irradiated HVJ lost most of its fusion activity in one week. We speculate that fusion proteins may be also damaged by UV-irradiation. It is likely that fusion proteins isolated by detergent solubilization and column chromatography are less damaged than UV-irradiated proteins.

One limitation of the reconstituted vesicles is the difficulty of preparation. Two distinct vesicles are required for the construction of the reconstituted fusion liposomes for gene transfer. Both DNA and fusion proteins could not be incorporated into the same liposome by a detergent-solubilization and dialysis method. However, the reconstituted fusion particles could be stored more stably than UV-irradiated HVJ (Suzuki *et al.*, 2000). Therefore, the reconstituted fusion liposomes and DNA-loaded liposomes can be stored separately after preparation, and then both particles can be fused to construct the DNA transfer vehicle before transfection. Another difficulty of this reconstituted system is mass production. The present system to construct the reconstituted fusigenic vesicles requires about 20 times more HVJ than the method for conventional HVJ-liposomes based on UV-inactivated whole HVJ virion. To overcome this problem, we have attempted to produce fusion proteins of HVJ in bacteria (Nakasima and Kaneda, unpublushed data), although the production is so far inefficient. Moreover, recombinant F protein was fusion-inactive F0 form that required trypsinization to be converted into active F1 form. Thus, further examinations including the baculovirus system will be required for mass production of recombinant fusion proteins.

We suppose that the insertion of fusion proteins into the liposome membrane is randomly oriented when prepared by a detergent-solubilization and dialysis method. Therefore, the reconstituted fusion liposomes might present novel antigens on the cell surface. The immunogenicity of the reconstituted liposome system should be analyzed extensively after repetitive *in vivo* injection of the reconstituted vesicles as reported for our HVJ-liposome system (Hirano *et al.,* 1998).

Acknowledgments

This work was supported by grants from the Ministry of Public Welfare in Japan.

5. REFERENCES

Bagai, S. and Sarker, D.P. (1993). Targeted delivery of hygromycin B using reconstituted Sendai viral envelopes lacking hemagglutinin-neuraminidase. *FEBS Lett.*, 326: 183-188.

Dzau, V.J., Mann, M., Morishita, R. and Kaneda, Y. (1996). Fusigenic viral liposome for gene therapy in cardiovascular diseases. *Proc. Natl. Acad. Sci. USA*, 93: 11421-11425.

Hirano, T., Fujimoto, J., Ueki, T., Yamamoto, H., Iwasaki, T., Morishita, R., Sawa, Y., Kaneda, Y., Takahashi, H. and Okamoto, E. (1998). Persistent gene expression in rat liver *in vivo* by repetitive transfections using HVJ-liposome. *Gene Ther.*, 5: 459-464.

Kaneda, Y. (1994). Virus (Sendai virus envelopes) mediated gene transfer. In: *Cell Biology: A Laboratory Handbook*, J. E. Celis (Ed.), Academic Press Inc., Orlando, Florida, Vol. 3, pp. 50-57.

Kaneda, Y. (1998). Fusigenic Sendai-virus liposomes: a novel hybrid type liposome for gene therapy. *Biogenic Amines*, 14: 553-572.

Kaneda, Y., Saeki, Y. and Morishita, R. (1999). Gene therapy using HVJ-liposomes: the best of both worlds. *Mol. Med. Today*, 5: 298-303.

Ledley, F.D. (1995). Non-viral gene therapy; the promise of genes as pharmaceutical products. *Hum. Gene Ther.*, 6: 1129-1144.

Marshall, E. (1995). Gene therapy's growing pains. *Science*, 269 :1052-1055.

Morishita, R., Gibbons, G.H., Kaneda, Y., Ogihara, T. and Dzau, V.J. (1994). Pharmakokinetics of antisense oligonucleotides (cyclin B1 and cdc 2 kinase) in the vessel wall; enhanced therapeutic utility for restenosis by HVJ-liposome method. *Gene*, 149: 3-9.

Mulligan, R.C. (1993). The basic science of gene therapy. *Science*, 260: 926-932.

Ramani, K., Bora, R.S., Kumar, M., Tyagi, S.K. and Sarkar, D.P. (1997). Novel gene delivery to liver cells using engineered virosomes. *FEBS Lett.*, 404: 164-168.

Ramani, K., Hassan, Q., Venkaiah, B., Hasnain, S.E. and Sarkar, D.P. (1998). Site-specific gene delivery *in vivo* through engineered Sendai viral envelopes. *Proc. Natl. Acad. Sci. USA*, 95: 11886-11890.

Reisman, D., Yates, J. and Sugden, B. (1985). A putative origin of replication of plasmids derived from Epstein-Barr virus is composed of two cis-acting components. *Mol. Cell. Biol.*, 5: 1822-1832.

Saeki, Y., Matsumoto, N., Nakano, Y., Mori, M., Awai, K. and Kaneda, Y. (1997). Development and characterization of cationic liposomes conjugated with HVJ (Sendai virus): Reciprocal effect of cationic lipid for *in vitro* and *in vivo* gene transfer. *Hum. Gene Ther.*, 8: 1965-1972.

Saeki, Y., Wataya-Kaneda, M., Tanaka, K., and Kaneda, Y. (1998). Sustained transgene expression *in vitro* and *in vivo* using an Epstein-Barr virus replicon vector system combined with HVJ-liposomes. *Gene Ther.*, 5: 1031-1037.

Sanes, J.R., Rubenstein, J.L. and Nicolas, J.F. (1986). Use of a recombinant retrovirus to study post-implantation cell lineage in mouse embryo. *EMBO J.*, 5: 3133-3142.

Sugden, B. and Warren, N. (1989). A promotor of Epstein-Barr virus that can function during latent infection can be transactivated by EBNA-1, a viral protein required for viral DNA replication during latent infection. *J. Virol.*, 63: 2644-2649.

Suzuki, K., Nakasima, H., Morishita, R., Sawa, Y., Matsuda, H. and Kaneda, Y. (2000). Reconstitute fusion liposomes for gene transfer *in vitro* and *in vivo*. *Gene Ther. Regulation*, 1: 65-77.

Uchida, T., Kim., J., Yamaizumi, M., Miyake, Y. and Okada, Y. (1979). Reconstitution of lipid vesicles associated with HVJ (Sendai virus) spikes; purification and some properties of vesicles containing nontoxic fragement A of diphtheria toxin. *J. Cell Biol.*, 80: 10-20.

Wisokenski, D.A. and Yates, J. L. (1989). Multiple EBNA-1 binding sites are required to form an EBNA-1-dependent enhancer and to activate a minimal replication origin within oriP of Epstein-Barr virus. *J. Virol.*, 63: 2657-2666.

Wu, P., de Fiebre, C.M., Millard, W.J., Elmstrom, K., Gao, Y. and Meyer, E.M. (1995). Sendai virosomal infusion of an adeno-associated virus-derived construct containing neuropeptide Y into primary rat brain cultures. *Neuroscience Lett.*, 190: 73-76.

Yates, J.L., Warren, N. and Sugden, B. (1985). Stable replication of plasmids derived from Epstein-Barr virus in various mammalian cells. *Nature*, 313: 812-815.

Yoshima, H., Nakanishi, M., Okada, Y. and Kobata, A. (1981). Carbohydrate structures of HVJ (Sendai virus) glycoproteins. *J. Biol. Chem.*, 256: 5355-5361.

7. SUMMARY — We have developed HVJ (Hemagglutinating Virus of Japan; Sendai virus)-liposomes that are efficient *in vitro* and *in vivo* gene delivery vehicles using the fusion-mediated gene delivery. The HVJ-liposome was highly efficient for the introduction of oligonucleotides into cells *in vivo* as well as the transfer of genes less than 100 kbp without damaging cells. By coupling the Epstein-Barr (EB) virus replicon apparatus with HVJ-liposomes, transgene expression was sustained *in vitro* and *in vivo*. Most animal organs were found to be suitable targets for the fusigenic-viral liposomes, and numerous gene therapy strategies using this system were successful in animals. However, all the proteins and the genome of HVJ remain within the HVJ-liposomes, although replication of the viral genome is severely impaired by prior UV-irradiation. To construct more complete synthetic vehicles, we developed reconstituted fusion liposomes. Fusion proteins F1 and HN of HVJ were extracted by mild lysis of the viral particles and purified with ion-exchange column chromatography. Purified viral fusion proteins were inserted into liposome membranes by detergent-solubilization and dialysis to construct the reconstituted fusion particles. These particles retained fusion activity during more than 4 weeks. DNA-loaded liposomes, which were prepared by vortexing-sonication, were fused with the reconstituted fusion particles to deliver DNA to cells. Using the reconstituted vehicle, fluorescent isothiocyanate (FITC)-labeled oligonucleotides were introduced into 100% of the nuclei of human amniotic FL cells. Luciferase gene expression by these vehicles was almost the same as that by HVJ-liposomes based on UV-inactivated whole HVJ particles. LacZ gene was introduced into mouse skeletal muscle by the new vector, and 40 to 50% of the muscle fibers showed LacZ gene expression.

Key Words: gene transfer; viral-liposome; HVJ (Sendai virus); EBV replicon; synthetic vector

Gene Therapy:Basic and Clinical Frontiers, pp. 245-265
R. Bertolotti *et al.* (Eds)

Gene therapy for adenosine deaminase (ADA)-deficient severe combined immuno-deficiency (SCID)

Donald B. Kohn*

Division of Research Immunology /Bone Marrow Transplantation, Childrens Hospital Los Angeles, Departments of Pediatrics and Microbiology, University of Southern California School of Medicine, Los Angeles, CA 90027, USA

Table of Contents

1. SEVERE COMBINED IMMUNODEFICIENCY (SCID)

Severe combined immunodeficiency (SCID) is a clinical entity in which patients have profoundly defective immunity, with essentially no T lymphocyte function and minimal to absent B cell function (Weinberg and

* E-mail: dkohn@chla.usc.edu

Kohn, 1996). SCID patients typically present for medical attention between 3-9 months of age with multiple, recurrent severe infections with common childhood pathogens (encapsulated bacteria, respiratory syncytial virus, parainfluenza virus) or opportunistic organisms (Candida albicans, Pneumocystis carinii, cryptococcus). Related findings often include failure to thrive and wasting, persistent diarrhea, and delayed growth and dentition.

The clinical syndrome of SCID is due to a variety of specific genetic lesions, each of which impairs normal T cell development and function. The first specific genetic cause of SCID to be identified was deficiency of the enzyme adenosine deaminase (ADA) (Giblett *et al.*, 1972). It is estimated that ADA deficiency accounts for 20-25% of cases of SCID. Other known genetic lesions resulting in SCID include: mutations in the IL-2 receptor gamma chain (Yc), responsible for X-linked SCID (accounting for as much as 40% of cases of SCID), defects in Jak3 kinase and other components of T cell intracellular signaling pathways, the inability to produce interleukin-2 (IL-2), defects in components of the T cell antigen receptor complex, defects in the recombinase system which is responsible for rearrangement of immunoglobulin and T cell receptor genes, and impaired expression of HLA antigens (bare lymphocyte syndrome).

2. ADA DEFICIENCY

ADA acts on both adenosine and deoxyadenosine, producing inosine and deoxyinosine (Hershfield and Mitchell, 1995; Hirschhorn, 1995). In the absence of ADA, deoxyadenosine can be phosphorylated, particularly by lymphoid cells, which have high levels of the enzyme deoxycytidine kinase. The resulting dATP pool expansion has been shown to inhibit DNA replication and repair, and to induce apoptosis in immature thymocytes. In addition, deoxyadenosine inactivates the enzyme S-adenosylhomocysteine (SAH) hydrolase; SAH accumulation inhibits transmethylation reactions, which may contribute to immunodeficiency. It has been speculated that effects of adenosine acting through its various receptors may also play a role in pathogenesis.

3. BONE MARROW TRANSPLANTATION (BMT) FOR ADA DEFICIENCY SCID

Bone marrow transplantation was first successfully applied in the treatment of a patient with SCID in 1968 (Gatti *et al.*, 1968). ADA deficiency was recognized as a cause of SCID in 1972 and patients with ADA-deficient SCID were successfully transplanted at that time (Giblett *et al.*, 1972; Parkman *et al.*, 1975). For SCID patients with an HLA-matched sibling donor, BMT is the treatment of choice. Barring patients with severe pre-existing infections, greater than 90% of SCID patients (with or without ADA deficiency) can be cured with BMT from a matched sibling. Due to the lack of immune reactivity, a matched sibling BMT for SCID is accomplished quite easily; a modest amount of donor bone marrow (*e.g.* 5 x 10^7 cells/ kg) is taken directly from the donor and given to the SCID recipient without the use of cytoablation or immune suppression. The majority of patients treated by matched BMT will develop a normal immune system comprised of donor-derived T lymphocytes.

However, only 20% of SCID patients will have a matched sibling donor. An alternative treatment which has been developed is the use of haplo-identical (parental) bone marrow, depleted of mature T lymphocytes. The results with T cell-depleted BMT for SCID patients have gradually improved, so that 60-80% of patients will survive with gain of improved-to-normal immune function (Fisher *et al.*, 1990; O'Reilly *et al.*, 1989; Buckley *et al.*, 1999). Despite T cell depletion of the donor marrow, patients can suffer graft versus host (GvH) disease, requiring further immuno-suppressive therapy. Some patients will recover only moderate immune function, with B cell deficiency and hypogammaglobulinemia seen in up to one half of patients. Additionally, immune recovery is a slow process, taking 4-12 months, during which time the patients are vulnerable to opportunistic infections.

Buckley and co-workers have recently reported their results using T-depleted haplo-identical BMT to treat 77 SCID patients (Buckley *et al.*, 1999). Sixty of 77 survived (78%). Seventy-four of these received the transplants without the use of cytoablative therapy or GvH prophylaxis; 3 more received unrelated donor cord blood transplants, 2 with pre-conditioning and post-GvH prophylaxis after unsuccessful haplo-grafts. Nineteen required at least a second cell infusion, with a few subjects getting three or four transplants. Twenty-one of 60 were stated to have "some donor B cells", and 45 of the total 72 survivors continue to receive intravenous gammaglobulin. Specifically, 13 subjects had ADA-deficiency as the genetic basis of their SCID and 11 of these survived. Nine are stated to have received T-depleted haplo-grafts and, therefore, 4 of the ADA-deficient SCID subjects apparently received matched sibling transplants. Of 11 surviving ADA-deficient SCID, only 3 had evidence of donor B cell production and five were still receiving intravenous gammaglobulin. Nine children with ADA-deficiency received T- depleted haplo-grafts, with seven alive. Six of these have donor engraftment (66%) and one child with ADA deficiency is receiving PEG-ADA after rejection of two paternal T-depleted haplo-identical transplants.

Thus, while T cell-depleted BMT has become a useful treatment modality for ADA-deficient SCID patients, it is still beset by a number of serious limitations.

4. ADA ENZYME REPLACEMENT THERAPY

In the mid-1970's, it was observed that transfusion of packed red blood cells (RBC) from normal donors into ADA-deficient patients could cause a partial improvement in immune function, including a rise in the lymphocyte count (Polmar *et al.*, 1976). It was postulated that ADA within the donor RBC was acting to detoxify adenosine and deoxyadenosine metabolites, thereby protecting developing T lymphocytes. However, the responses to the RBC transfusions were only seen in a minority of patients and were typically modest in magnitude and duration.

Based upon the experience with RBC transfusions, a far more effective means of ADA enzyme replacement was developed. Bovine ADA may be isolated in relatively large quantities, but has an half-life of only minutes when injected parenterally and may be immunogenic upon repeated exposures. It had previously been shown that conjugation of proteins to polyethylene glycol (PEG) leads to greatly prolonged serum survival time and decreased immunogenicity. A pharmaceutical preparation of bovine ADA conjugated with polyethylene glycol (PEG-ADA, ADAGEN®) was approved by the USFDA in 1990 for treating ADA deficiency. It has been used to treat over 80 patients, primarily those considered to be too ill to safely undergo haplo-identical bone marrow transplantation (Hershfield, 1995a and 1995b). PEG-ADA avoids the risks of red cell transfusion, and the amount of ADA enzyme activity provided by one or two weekly intramuscular injections of PEG-ADA exceeds by as much as 10 fold the amount that can be supplied by repeated RBC transfusions. PEG-ADA has been well tolerated, with no reports of allergic reactions thus far.

Treatment with PEG-ADA consistently corrects metabolic abnormalities due to ADA deficiency. Lymphocyte counts and immune function begin to improve after 2 to 4 months of treatment. Most patients have done well clinically (overall mortality is <20%, and <12% in patients treated more than 6 months) (Hershfield 1995a). Recovery of immunologic function on PEG-ADA therapy has been variable, and about 20% of patients, primarily those with the most severely compromised immune function, have had minimal improvement in lymphocyte counts and function. The basis for this variability is unclear. One patient on PEG-ADA developed autoimmune hemolytic anemia, which resulted in her death. Autoimmune disorders are seen in some patients with incomplete immune function and may represent the presence of immune effector cells without an appropriate level of immune regulatory cells.

The other major limitation to PEG-ADA therapy is the high cost. Treatment of a child with weekly PEG-ADA injections may cost between $200,000-400,000 per year. Because PEG-ADA therapy is palliative and not curative, injections must be continued throughout the life of the patient. The high cost of this treatment may limit access and create severe financial burdens on patients and their families.

5. GENE THERAPY FOR ADA-DEFICIENT SCID

A new alternative therapy for ADA-deficient SCID has been developed, in which a normal ADA gene is put into cells of the patient (**gene therapy**). Studies performed at the NIH in the mid 1980's demonstrated that introducing a normal human ADA gene (using a retrovirus as the gene delivery vector) into cultured T lymphocytes from patients with ADA-deficient SCID resulted in the production of normal levels of ADA enzyme (Kantoff *et al.*, 1986). In cell culture, the "cured" cells were able to survive in the presence of levels of deoxyadenosine that were toxic to the parental ADA-deficient cells. Studies performed in Italy demonstrated that ADA-corrected human T lymphocytes survived significantly longer *in vivo* in immune deficient mice compared to ADA-deficient cells (Ferrari *et al.*, 1991).

a) T lymphocytes

Based upon these experimental results, a trial of gene therapy for ADA deficiency was performed with two patients at the NIH, beginning in 1990 (Culver *et al.*, 1991; Blaese *et al.*, 1995; Mullen *et al.*, 1996). Peripheral blood T lymphocytes were collected by leukopheresis, transduced in the laboratory with the ADA gene, expanded in numbers by stimulating with growth factors and then re-infused into the patients. The procedures of leukopheresis, transduction and reinfusion were repeated at 1-2 month intervals over the course of two years. Two children were treated by this method with no evidence of toxicity. The presence of cells containing the inserted ADA gene was documented in both patients, although one patient had a significantly greater number of cells transduced than the other. In at least one of the patients, ADA-transduced T lymphocytes have remained at a relatively stable level of 30-50% of total peripheral blood lymphocytes (PBL), more than six years after the last cell infusion; this observation shows that T lymphocytes subjected to *ex vivo* transduction are capable of extended *in vivo* survival. Both patients had significant increases in numbers of circulating T lymphocytes and some evidence of improved immune function. However, they remain on PEG-ADA enzyme replacement, making it difficult to attribute immunologic functions to the gene transfer. Recently, an ADA-deficient SCID patient in Japan was treated under this same protocol with good ADA gene transfer achieved (Onodera *et al.*, 1998a).

One drawback to gene therapy using T lymphocytes is that the immunologic repertoire of the transduced lymphocytes may be restricted to those specificities present in the population of cells which were treated; responses to newly encountered antigens may not be possible. Additionally, these cells may have a finite life-span (months to years), which may necessitate repeated ADA gene transfers to be performed to sustain a therapeutic level of transduced cells.

b) Hematopoietic stem cells

Insertion of the ADA gene into hematopoietic stem cells (HSC), rather than mature T lymphocytes, may produce a long-lived effect. HSC are cells which produce all blood cells by a complex process of proliferation and differentiation. HSC are long-lived and can function for the life of the individual after transplantation. For children and adults, HSC are found primarily in the bone marrow, but HSC are also present in the umbilical cord blood at birth. Genetic correction of a patient's HSC may provide a continuous, enduring source of gene corrected mature blood cells, including T lymphocytes.

From more than fifteen years of research on retroviral-mediated gene transfer into hematopoietic stem cells, it has been found that it is possible to routinely transduce the majority of stem cells in murine gene transfer/BMT models. However, work in large animal models (canine and rhesus) have achieved only 0.1-1.0% stem cell transduction (Kohn, 1995; Dunbar, 1996). Clinical trials of gene marking of autologous hematopoietic cells have also shown extremely limited abilities to transduce long-lived human stem cells, with marking of 0.1-1% of cells seen. The reasons for the greater difficulty in performing gene transfer into stem cells from the large animals and human subjects is unknown; possible explanations include lower numbers of receptors for the amphotropic virus, intracellular blocks to virus integration, or a lower percentage of stem cells in active cell cycle which can be transduced by retroviral vectors.

Because a small number of ADA-corrected T lymphocytes may have a selective survival advantage and lead to immune reconstitution, the current levels of gene transfer may be sufficient for clinical benefit. It is because of this latter presumed selective advantage that ADA deficiency is the focus of such intensive investigation.

Seven ADA-deficient SCID patients have now been treated by insertion of the ADA gene into their bone marrow cells (four in Italy and three in The Netherlands) and one of the patients who received ADA gene therapy via T lymphocytes has been treated by insertion of a different ADA vector into her G-CSF-mobilized peripheral blood stem cells. The first two patients treated in Italy have shown continued production of T lymphocytes containing the introduced ADA cDNA, with gradual withdrawal of PEG-ADA underway (Ferrari *et al.*, 1992; Bordignon *et al.*, 1993). Only low extents of gene transfer were achieved in the three patients treated in The Netherlands, with minimal evidence of *in vivo* production of transduced cells for less than six months (Hoogerbrugge *et al.*, 1992). No adverse reactions have been noted to date in these ADA deficient patients or any of the other patients who have received cells exposed to retroviral vectors.

c) Umbilical cord blood cells (UCBC)

An alternative to bone marrow as a source of HSC for gene therapy may be the umbilical blood. In the fetus, hematopoiesis occurs primarily in the liver. In the third trimester, HSC transit through the bloodstream from the liver to the bone marrow. At birth, the umbilical cord blood contains hematopoietic progenitor cells at approximately the same frequency as adult bone marrow. By the second day of life, the level of hematopoietic progenitor cells in the blood stream is reduced by several hundred-fold, to the level seen in adults. Thus, the cord blood represents a unique source of HSC which may be obtained without any invasive procedures.

Allogeneic HSC transfers have been performed in several hundred patients, using UCBC from HLA-matched siblings (Wagner *et al.*, 1995). The usual setting for this procedure has been in families with a child with a BMT-responsive malignancy in which an HLA-matched sibling is identified *in utero*. Collection of the fetal blood remaining in the umbilical cord yields between 50-200 ml of UCBC, which can be used for transplantation after high dose chemotherapy. Sustained engraftment with enduring hematologic reconstitution is routinely achieved with UCBC transplants, demonstrating the presence of long-lived HSC. More recently, the use of umbilical cord blood samples from matched or only partially-matched unrelated donors has been performed with significantly less graft versus host disease seen than would occur using bone marrow of a similar degree of histocompatibility (Kurtzberg *et al.*, 1995; Gluckman *et al.*, 1997). However, engraftment of allogeneic cells is likely to require imposition of cytoablative conditioning, even in SCID patients due to residual alloreactivity.

The use of autologous UCBC for gene therapy would obviate the need for performing a bone marrow aspiration procedure or serial leukopheresis to collect peripheral blood cells. We have performed ADA gene insertion into umbilical cord blood cells of three ADA-deficient neonates in may and june of 1993 (see section 6).

6. RESULTS OF PREVIOUS CHLA/NIH TRIAL OF GENE TRANSFER INTO UMBILICAL CORD BLOOD CD34+ CELLS FROM ADA-DEFICIENT SCID INFANTS

We have performed a clinical gene therapy study in which three infants diagnosed pre-natally with adenosine deaminase (ADA)-deficiency severe combined immunodeficiency (SCID) underwent ADA gene transfer (Kohn *et al.*, 1995). The Childrens Hospital Los Angeles (CHLA) group, in collaboration with Dr. Michael Blaese and colleagues at the NIH, developed an amendment to their protocol for gene therapy of ADA-deficient children to specify the use of CD34+ cells from newborn infant's umbilical cord blood. The protocol was reviewed by the governing IRB, as well as the NIH RAC and the CBER/FDA.

The three infants were born May 11, May 14 and June 7, 1993. Umbilical cord blood was obtained after clamping of the cord, ranging from 50-200 ml in volume. The cord blood was brought to our laboratory and processed to isolate CD34+ cells using the CellPro Ceprate device. Between 3-18 x 10^6 cells were obtained, with 32-62% CD34 positivity. The cells were transduced by three days culture in medium with interleukin-3, interleukin-6 and stem cell factor, with daily addition of retroviral vector supernatant (LASN; Hock *et al.*, 1989). The vector was designed and constructed by A. Dusty Miller at the Fred Hutchinson Cancer Center and produced at clinical grade by Genetic Therapy Inc., for Dr. Blaese's original trial. After transduction, the cells were extensively washed and infused intravenously into each child on their fourth day of life.

Each infant was immediately started on enzyme replacement therapy with PEG-ADA, 45-60 U/kg intramuscularly per week, divided twice weekly. From the PEG-ADA therapy, they had normalization of the levels of deoxyadenosine metabolites, which were elevated at birth, and improvements of their T cell numbers and function over the first few months. One patient was kept on a fixed dose of PEG-ADA so that the effective dose/kg decreased as he grew; the other two patients were maintained on a constant dose/kg (adjusted as they grew) for the first 18 months (through december 1994) and were then lowered to half this dose for the subsequent time (see below).

Peripheral blood samples were obtained at monthly intervals and the presence and frequency of gene-containing leukocytes was determined by semi-quantitative PCR. Over the first eighteen months, there was a relatively stable level of gene-containing cells in the peripheral blood mononuclear cell (PBMC) and granulocyte fractions, in the range of **1 positive cell per 10,000**. Bone marrow aspirates taken one year after treatment showed the presence of CD34+ cells and colony-forming progenitors containing the vector, at frequencies of approximately 1%. Expression of vector-specific transcripts was seen by RT-PCR of RNA from PBMC, and pooled CFU-GM grown from the marrow in the presence of G418 showed expression of enzymatically active ADA at or above normal levels.

Therefore, we concluded that CD34+ cells can be isolated from umbilical cord blood, be transduced *ex vivo* with retroviral vectors and will engraft for sustained production of gene-containing leukocytes. The relatively low frequency of transduced cells is in accord with theoretical predictions from previously described experimental studies of gene transfer/BMT in dogs and monkeys and in clinical marking studies which have used similar methods for retroviral-mediated gene transfer. In those studies, the highest levels of transduced cells which have been achieved have been 0.1-1%, despite elimination of the recipient's endogenous marrow by pre-treatment total-body irradiation. The major conclusion from those is that only a small fraction of reconstituting stem cells are transduced with the currently available retroviral vectors and methodologies. Our patients did not have their endogenous marrow eliminated and thus the treated CD34+ cells would be diluted by the non-transduced residual marrow. Thus, our measurements of 1/10,000 transduced cells may be the best that can be expected.

The reason that ADA deficiency has been the subject of so much of the initial efforts in clinical gene therapy is that there is the expectation that there would be a selective survival advantage to genetically corrected T cells. SCID patients undergoing transplant of bone marrow from HLA-matched siblings without prior cytoablation continue to produce their own granulocytes, RBC and platelets, but develop complete donor-derived T lymphoid systems. This consistent observation implies that genetically corrected T cells would have a similar survival advantage over uncorrected T cells and, with time, reconstitute the immune system with gene-containing cells.

It is likely that the PEG-ADA may blunt the selective advantage of transduced cells by detoxifying the deoxyadenosine metabolites which would otherwise eliminate non-transduced T cells. Therefore, after 18 months of observation, we reduced the dosages of PEG-ADA in each patient to 20-30 U/Kg/week (Kohn *et al.*, 1998). During this time, the subjects remained healthy, retained sub-normal but persistent blastogenic responses to mitogens and some antigens, and were moderately lymphopenic (absolute CD3 of 200-500/mm^3). Upon dosage reduction, each patient showed a 2-5-fold decrease in their T cell numbers and proliferative responses to phytohemagglutinin (PHA). However, over this same time, the frequency of gene-containing PBMC increased 30-100-fold in each patient. FACS sorting of PBMC into T cells, B cells and monocytes showed that the increase is selectively in the T cells, with between 1-10% of the T cell containing the vector, in contrast to the B cells, monocytes and granulocytes which remained in the 0.01% range (Table 1). **These findings support the postulated selective advantage for ADA-transduced T cells.**

Table 1. T lymphocyte counts and leukocyte gene frequency (4 years*)

Patient	PEG-ADA (U/kg)	Absolute numbers† CD3	CD4	Frequency of gene-containing cells PBMC‡	T cells	Mono‡	Gran‡
1	15	450	200	1%	3-10%	0.03%	0.03%
2	30	128	64	1%	1%	0.01%	0.01%
3	30	44	33	1%	10%	0.01%	0.03%

* May 1997

† absolute numbers of T lymphocytes per mm^3 of whole blood.

‡ PBMC: peripheral blood mononuclear cells; Mono: monocytes; Gran: granulocytes.

In may 1997, PEG-ADA enzyme replacement therapy was stopped completely in the one subject (#1, above) with the highest levels of gene marking and T lymphocyte numbers (Table 2). Over two months of observation, the absolute numbers of circulating CD3 and CD4 T lymphocytes decreased modestly (by 25%), the percentages of circulating T lymphocytes containing the introduced ADA gene increased to 30-100%, proliferative responses to PHA remained constant, but the proliferative response to tetanus toxoid was lost. Surprisingly, there were large decreases (over 100-fold) in the absolute numbers of B lymphocytes and natural killer (NK) cells. Because of the decline in these parameters of immune cell numbers and function, and development of oral thrush and a upper respiratory tract infection, the patient was re-started on PEG-ADA (45 U/kg/week).

We have measured expression of ADA by the vector in peripheral blood leukocytes, from a blood sample from subject # 1 taken 58 months after birth (march 1998; table 3). RT-PCR was used to quantify vector-derived transcripts. A standard curve was produced using Jurkat human T lymphocytes transduced with a single copy of the LASN vector which were serially diluted 10-fold with non-transduced Jurkat cells. We could detect

Table 2. Laboratory results before, during and after PEG-ADA cessation

	PEG-ADA	ADA*	dAXP*	PH†	Tetanus†	Absolute cell‡ numbers per mm^3					ALT*	Percentage¶ of gene-containing cells§			
	U/kg/wk	µmol/hr/ml	µmol/ml	cpm x 10^{-3}		CD3	CD4	CD8	CD19	16/56*	U/L	PBMC	Gran*	T cells	Mono*
Normal	--------->	*6-12*	*<0.01*	*>100*	*>5*	*>820*	*>440*	*>180*	*>50*	*>80*	*3-46*				
Month															
30	30	35.0	0.004	-	-	-	-	-	-	-	36	0.3	0.03	10	0.5
41	20	-	-	6.2	9.3	738	198	342	162	918	-	1.0	0.07	-	-
48	15 → 0	28.6	0.006	20.7	13.4	405	211	-	-	-	-	3.0	0.8	10	0.05
49	0	-	-	25.5	1.7	297	158	139	56	557	-	3.0	2.0	30	n.d.
49.5	0	0.33	0.712	32.7	0	370	139	218	9	35	244	1.0	0.08	-	-
50	0 → 45	0.26	0.272	-	-	-	291	313	22	29	71	-	-	-	-
51	45 → 40	119.5	-	-	-	419	987	239	987	37	-	-	-	-	-
53	40	36.2	0.000	-	-	-	-	-	-	-	-	1.0	0.01	5.0	0.003
54	40	-	-	65.9	12.4	329	145	-	-	-	-	-	-	-	-

* ADA: Plasma level of ADA enzymatic activity; dAXP: erythrocytic level of deoxyadenosine nucleotides; ALT: serum alanine aminotransferase; 16/56: CD16/56; Gran: granulocytes; Mono: monocytes.

† Blastogenic responses of PBMC to PHA (phytohemagglutinin) or tetanus toxoid as measured by ^{3}H-thymidine incorporation.

‡ Lineage-specific markers: CD3, CD4 & CD8 for T lymphocytes; CD19 for B lymphocytes; CD16 & CD56 for natural killer (NK) cells.

¶ The frequency of peripheral blood leukocytes containing the LASN vector was determined upon semi-quantitative PCR of cell genomic DNA extracts.

§ Peripheral blood leukocytes were separated into peripheral blood mononuclear cell (PBMC) and granulocyte (Gran) fractions by centrifugation on ficoll-hypaque gradients (Pharmacia). PBMC were further fractionated by labelling with fluorescent antibodies followed by sorting using a FACS Vantage (Becton-Dickinson, San Jose, CA). Monoclonal antibodies (Becton-Dickinson) used for sorting included an anti-human CD3 for T lymphocytes, and both an anti-human CD13 and an anti-human CD14 for monocytic cells.

the vector-specific transcripts from as few as 1/10,000 transduced Jurkat cells (0.01%) mixed with non-transduced Jurkat cells. Because Jurkat cells represent constitutively activated cells, expression from the MoMuLV LTR would be expected to be high.

Table 3. RT-PCR analysis of vector expression

Leukocyte fraction	Gene copy number	Gene expression level*
PBMC	1%	<0.01%
Granulocytes	0.03%	<0.01%
CD3+ lymphocytes	10%	0.01%
CD13/CD14+ monocytes	0.03%	<0.01%
PHA-treated PBMC (5 days)	N.D.	**0.3%**

* RT-PCR vs LASN Jurkat

Peripheral blood PBMC from the subject did not show detectable vector transcripts, nor did granulocytes or monocytes. Upon an approximate 10-fold enrichment of CD3+ T lymphocytes by FACS-sorting from the heterogenous PBMC fraction, LASN transcripts were detectable, at approximately the 0.01% level. The level of expression in the isolated primary T lymphocytes, normalized for a gene copy number of 10%, was 3 orders of magnitude lower than the signal from the standard curve mixture of Jurkat cells with a 10% copy number. Most interestingly, stimulation of the PBMC by growth for five days with PHA and IL-2 led to a readily detectable level of LASN transcripts (approximately equivalent to the signal from 0.3% + Jurkat cells).

These results show that there is a low level of expression *in vivo* in CD3+ T lymphocytes, most of which would be expected to be quiescent, but that expression is significantly up-regulated upon T lymphocyte stimulation. Similar observations have been made in one of the ADA-deficient patients from the first protocol in which mature T lymphocytes were transduced [and in human T lymphocytes derived in *scid*-Hu mice (Agarwal *et al.*, 1998)]. The low expression from the MoMuLV LTR in quiescent T lymphocytes may be responsible for the lack of significant immunologic benefits seen upon PEG-ADA withdrawal.

In summary, the trial of gene transfer into ADA-deficient CD34+ cells demonstrates:

1.) Long-lived hematopoietic progenitor (stem) cells among the CD34+ cells from umbilical cord blood can be transduced by retroviral vectors, engraft without prior administration of cytoablative conditioning, and produce gene-containing peripheral blood leukocytes for at least 6 years.

2.) There is selective accumulation of gene-containing T cells upon reduction of the PEG-ADA dosage.

3.) Expression by the LASN vector in resting CD3+ T lymphocytes is poor (<1% of that seen in Jurkat cells). Expression increases significantly upon stimulation of the T cells with PHA (to approximately the same level as in Jurkat cells).

4.) While there was no evidence that either B lymphocytes or NK cells containing the ADA gene selectively accumulate, both of these cell types were rapidly lost upon withdrawal of PEG-ADA.

5.) In one subject, complete cessation of PEG-ADA for two months did not demonstrate that he had protective immune function solely from gene-corrected cells. Thus, the low level of transduced progenitors present in this patient were not sufficient to sustain immune function following PEG-ADA withdrawal during the observation period. Potentially, higher frequencies of transduced progenitor cells may allow long-lasting, protective immune function independently of exogenous enzyme replacement.

7. FUTURE DIRECTIONS

Since these initial studies were performed, there has been significant progress in gene transfer technology. Recent incremental improvements in retroviral-mediated gene transfer into human hematopoietic stem cells (HSC) have been achieved, using GALV pseudotypes, "mobilized BM", recombinant fibronectin support, new cytokines (Flt-3 ligand, thrombopoietin), and manipulation of cell cycle kinetics (Kiem *et al.*, 1997; Dunbar *et al.*, 1996; Hanenberg *et al.*, 1996; Shah *et al.*, 1996; Dao *et al.*, 1997 and 1998). Combinations of these techniques have resulted in modest, yet significant, increases in gene marking in primate stem cell transplant models, *e.g.* 10%, up from the previous ceiling of 0.1-1.0% (Kiem *et al.*, 1998; Tisdale *et al.*, 1998). Additionally, modified transcriptional control elements, such as the MSCV and MND LTR, lead to improved gene expression (Hawley *et al.*, 1994; Onodera *et al.*, 1998b; Robbins *et al.*, 1998). A new set of clinical trials using these improved techniques may lead to more gene-transduced cells with better gene expression. This may then finally result in clear-cut evidence of clinical benefit for gene therapy of ADA-deficient SCID.

8. REFERENCES

Agarwal, M., Austin, T.W., Morel, F., Chen, J., Böhnlein, E. and Plavec, I. (1998). Scaffold attachment region-mediated enhancement of retroviral expression in primary T cells. *J. Virol.*, 72: 3720-3728.

Blaese, R.M., Culver, K.W., Miller, A.D., Carter, C.S., Fleisher, T., Clerici, M., Shearer, G., Chang, L., Chiang, Y., Tolstochev, P., Greenblatt, J.J., Rosenberg, S.A., Klein, H., Berger, M., Mullen, C.A., Ramsey, W.J., Muul, L., Morgan, R.A. and Anderson, W.F. (1995). T lymphocyte-directed gene therapy for ADA-SCID: Initial trial results after 4 years. *Science*, 270: 475-480.

Bordignon, C., Mavilio, F., Ferrari, G., Servida, P., Ugazio, A.G., Notarangelo, L.D., Gilboa, E., Rossini, S., O'Reilly, R.J., Smith, C.A. *et al.* (1993). Transfer of the ADA gene into bone marrow cells and peripheral blood lymphocytes for the treatment of patients affected by ADA-deficient SCID. *Hum. Gene Ther.*, 4: 513-520.

Buckley, R.H., Schiff, S.E., Schiff, R.I., Markert, M.L., Williams, L.W., Roberts, J.L., Myers, L.A. and Ward, F.E. (1999). Hematopoietic stem-cell transplantation for the treatment of severe combined immunodeficiency. *N. Engl. J. Med.*, 340: 508-516.

Culver, K.W., Anderson, W.F. and Blaese, R.M. (1991). Lymphocyte gene therapy. *Hum. Gene Ther.*, 2:107-109.

Dao, M.A., Hannum, C.H., Kohn, D.B. and Nolta, J.A. (1997). Flt3 ligand preserves the ability of human CD34+ progenitors to sustain long-term hematopoiesis in immune-deficient mice after ex vivo retroviral-mediated transduction. *Blood*, 89: 446-456.

Dao, M.A., Taylor, N. and Nolta, J.A. (1998). Reduction in levels of the cyclin-dependent kinase inhibitor p27(kip-1) coupled with transforming growth factor β neutralization induces cell-cycle entry and increases retroviral transduction of primitive human hematopoietic cells. *Proc. Natl. Acad. Sci. USA*, 95: 13006-13011.

Dunbar, C.E. (1996). Gene transfer to hematopoietic stem cells: implications for gene therapy of human disease. *Annu. Rev. Med.*, 47:11-20.

Dunbar, C.E., Seidel, N.E., Doren, S., Sellers, S., Cline, A.P., Metzger, M.E., Agricola, B.A., Donahue, R.E. and Bodine, D.M. (1996). Improved retroviral gene transfer into murine and rhesus peripheral blood or bone marrow repopulating cells primed in vivo with stem cell factor and granulocyte colony-stimulating factor. *Proc. Natl. Acad. Sci. USA*, 93: 11871-11876.

Ferrari, G., Rossini, S., Giavazzi, R., Maggioni, D., Nobili, N., Soldati, M., Ungers, G., Mavilio, F., Gilboa, E. and Bordignon, C. (1991). An in vivo model of somatic cell gene therapy for human severe combined immunodeficiency. *Science*, 251: 1363-1366.

Ferrari, G., Rossini, S., Nobili, N., Maggioni, D., Garofalo, A., Giavazzi, R., Mavilio, F. and Bordignon, C. (1992). Transfer of the ADA gene into human ADA-deficient T lymphocytes reconstitutes specific immune functions. *Blood*, 80:1120-1124.

Fischer, A., Landais, P., Friedrich, W., Morgan, G., Gerritsen, B., Fasth, A., Porta, F., Griscelli, C., Goldman, S.F., Levinsky, R. and Vossen, J. (1990). European experience of bone-marrow transplantation for severe combined immunodeficiency. *Lancet*, 336: 850-854.

Gatti, R.A., Meuwissen, H.J., Allen, H.D., Hong, R. and Good, R.A. (1968). Immunological reconstitution of sex-linked lymphopenic immunological deficiency. *Lancet*, 2:1366-1369.

Giblett, E.R., Anderson, J.E., Cohen, F., Pollara, B. and Meuwissen, H.J. (1972). Adenosine-deaminase deficiency in two patients with severely impaired cellular immunity. *Lancet*, 2:1067-1069.

Gluckman, E., Rocha, V., Boyer-Chammard, A., Locateli, F., Arcese, W., Pasquini, R., Ortega, J., Souillet, G., Ferreira, E., Laporte, J.P., Fernandez, M. and Chastang, C. (1997). Outcome of cord blood transplantation from related and unrelated donors. Eurocord Transplant Group and the European Blood and Marrow Transplantation Group. *N. Engl. J. Med.*, 337: 373-381.

Hanenberg, H., Xiao, X.L., Dilloo, D, Hashino. K., Kato, I. and Williams, D.A. (1996). Colocalization of retrovirus and target cells on specific fibronectin fragments increases genetic transduction of mammalian cells. *Nature Med.*, 2: 876-882.

Hawley, R.G., Lieu, F.H., Fong, A.Z. and Hawley, T.S. (1994). Versatile retroviral vectors for potential use in gene therapy. *Gene Ther.*, 1: 136-138.

Hershfield, M.S. (1995a). PEG-ADA replacement therapy for adenosine deaminase deficiency: an update after 8.5 years. *Clin. Immunol. Immunopathol.*, 76: S228-S232.

Hershfield, M.S. (1995b). PEG-ADA: an alternative to haploidentical bone marrow transplantation and an adjunct to gene therapy for adenosine deaminase deficiency. *Hum. Mutat.*, 5: 107-112.

Hershfield, M.S. and Mitchell, B.S. (1995). Immunodeficiency diseases caused by adenosine deaminase deficiency and purine nucleotide phosphorylase deficiency. In: *The*

Metabolic and Molecular Bases of Inherited Disease, Scriver, C.S., Beaudet, A.L., Sly, W.S. and Valle, D. (Eds.), McGraw-Hill, New York, pp. 1725-1768.

Hirschhorn, R. (1995). Adenosine deaminase deficiency: molecular basis and recent developments. *Clin. Immunol. Immunopathol.*, 76: S219-S227.

Hock, R.A., Miller, A.D. and Osborne, W.R. (1989). Expression of human adenosine deaminase from various strong promoters after gene transfer into human hematopoietic cell lines. *Blood*, 74: 876-881.

Hoogerbrugge, P.M., Vossen, J.M., van Beusechem, V.W. and Valerio, D. (1992). Treatment of patients with severe combined immunodeficiency due to adenosine deaminase (ADA) deficiency by autologous transplantation of genetically modified bone marrow cells. *Hum. Gene Ther.*, 3: 553-558.

Kantoff, P.W., Kohn, D.B., Mitsuya, H., Armentano, D., Sieberg, M., Zweibel, J.A, Eglitis, M.A., McLachlin, J.R., Wiginton, D.A., Hutton, J.J., Horowitz, S.D., Gilboa, E., Blaese, R.M. and Anderson, W.F. (1986). Correction of adenosine deaminase deficiency in human T and B cells using retroviral gene transfer. *Proc. Natl. Acad. Sci. USA*, 83:6563-6567.

Kiem, H.P., Heyward, S., Winkler, A., Potter, J., Allen, J.M., Miller, A.D. and Andrews, R.G. (1997). Gene transfer into marrow repopulating cells: comparison between amphotropic and gibbon ape leukemia virus pseudotyped retroviral vectors in a competitive repopulation assay in baboons. *Blood*, 90: 4638-4645.

Kiem, H.P., Andrews, R.G., Morris, J., Peterson, L., Heyward, S., Allen, J.M., Rasko, J.E.J., Potter, J. and Miller, A.D. (1998). Improved gene transfer into baboon marrow repopulating cells using recombinant human fibronectin fragment CH-296 in combination with interleukin-6, stem cell factor, FLT-3 ligand, and megakaryocyte growth and development factor. *Blood*, 92:1878-1886.

Kohn, D.B. (1995). The current status of gene therapy using hematopoietic stem cells. *Current Opinions in Pediatrics*, 7: 56-63.

Kohn, D.B., Weinberg, K.I., Nolta, J.A., Heiss, L.N., Lenarsky, C., Crooks, G.M., Hanley, M.E., Annett, G., Brooks, J.S., El-Khoureiy, A., Lawrence, K., Wells, S., Shaw, K., Moen, R.C., Bastian, J., Williams-Herman, D.E., Elder, M., Wara, D., Bowen, T., Hershfield, M.S., Mullen, C.A., Blaese, R.M. and Parkman, R. (1995). Engraftment of gene-modified cells from umbilical cord blood in neonates with adenosine deaminase deficiency. *Nature Med.*, 1: 1017-1026.

Kohn, D.B., Hershfield, M.S., Carbonaro, D., Shigeoka, A., Brooks, J., Smogorzewska, E.M., Barsky, L.W., Chan, R., Burotto. F., Annett, G., Nolta, J.A., Crooks, G., Kapoor, N., Elder, M., Wara, D., Bowen, T., Madsen, E., Snyder, F.F., Bastian, J., Muul, L., Blaese, R.M, Weinberg, K. and Parkman, R. (1998). T lymphocytes with a normal ADA gene accumulate after transplantation of transduced autologous umbilical cord blood CD34+ cells in ADA-deficient SCID neonates. *Nature Med.*, 4: 775-780.

Kurtzberg, J., Laughlin, M., Graham, M.L., Smith, C., Olson, J.F., Halperin, E.C., Ciocci, G., Carrier, C., Stevens, C.E. and Rubenstein, P. (1996). Placental blood as a source of hematopoietic stem cells for transplantation into unrelated recipients. *N. Engl. J. Med.*, 335: 157-166.

Mullen, C.A., Snitzer, K., Culver, K.W., Morgan, R., Anderson, W.F. and Blaese, R.M.

(1996). Molecular analysis of T lymphocyte-directed gene therapy for adenosine deaminase deficiency: Long-term expression in vivo of genes introduced with a retroviral vector. *Hum. Gene Ther.*, 7: 1123-1129.

Onodera, M., Ariga, T., Kawamura, N., Kobayashi, I., Otsu, M., Yamada, M., Tame, A., Furuta, H., Okano, M., Matsumoto, S., Kotani, H., McGarrity, G.J., Blaese, R.M. and Sakiyama, Y. (1998a). Successful peripheral T-lymphocyte-directed gene transfer for a patient with severe combined immune deficiency caused by adenosine deaminase deficiency. *Blood*, 91: 30-36.

Onodera, M., Nelson, D.M., Yachie, A., Jagadeesh, G.J., Bunnell, B.A., Morgan, R.A. and Blaese, R.M. (1998b). Development of improved adenosine deaminase retroviral vectors. *J. Virol.*, 72: 1769-1774.

O'Reilly, R.J., Keever, C.A., Small, T.N. and Brochstein, J. (1989). The use of HLA-non-identical T-cell depleted marrow transplants for correction of severe combined immunodeficiency disease. *Immunol. Rev.*, 1: 273-309.

Parkman, R., Gelfand, E.W., Rosen, F.S., Sanderson, A. and Hirschhorn, R. (1975). Severe combined immunodeficiency and adenosine deaminase. *N. Engl. J. Med.*, 292: 714-719.

Polmar, S.H., Stern, R.C., Schwartz, A.L., Wetzler, E.M., Chase, P.A. and Hirschhorn, R. (1976). Enzyme replacement therapy for adenosine deaminase deficiency and severe combined immunodeficiency. *N. Engl. J. Med.*, 295: 1337-1343.

Robbins, P.B., Skelton, D.M., Yu, X.J., Pepper, K.A. and Kohn DB. (1998). Consistent, persistent expression from modified retroviral vectors in murine hematopoietic stem cells. *Proc. Natl. Acad. Sci. USA*, 95: 10182-10187.

Shah, A.J., Smogorzewska, E.M., Hannum, C., Crooks, G.M. (1996). Flt3 ligand induces proliferation of quiescent human bone marrow CD34+CD38- cells and maintains progenitor cells in vitro. *Blood*, 87: 3563-3570.

Tisdale, J.F., Hanazono, Y., Sellers, S.E., Agricola, B.A., Metzger, M.E., Donahue, R.E. and Dunbar, C.E. (1998). Ex vivo expansion of genetically marked rhesus peripheral blood progenitor cells results in diminished long-term repopulating ability. *Blood*, 92: 1131-1141.

Wagner, J.E., Kernan, N.A., Steinbuch, M., Broxmeyer, H.E. and Gluckman, E. (1995). Allogeneic sibling umbilical-cord-blood transplantation in children with malignant and non-malignant disease. *Lancet*, 346: 214-219.

Weinberg, K.I. and Kohn, D.B. (1996). Gene therapy for congenital immunodeficiency disease. In: *Immunology and Allergy Clinics of North America, Vol. 16: Organ and Bone Marrow Transplantation*, W. Shearer and T. Fleisher (Eds.), W.B. Saunders, Philadelphia.

9. SUMMARY — Gene therapy is a novel treatment under investigation for congenital immune deficiencies, by transfer of the relevant gene in the hematopoietic stem cells of patients. Adenosine deaminase (ADA) deficiency is one cause of severe combined immunodeficiency (SCID) syndrome which has been approached by gene transfer. Clinical trials have been performed in which the human ADA cDNA was transferred into T lymphocytes or hematopoietic stem cells of ADA-deficient SCID patients. The initial studies demonstrated the safety of this approach, with some evidence of clinical effects. Therapeutic benefits from ADA gene transfer will require the use of more effective methods for gene transfer and expression. Gene therapy retains significant promise for the treatment of immune deficiencies, providing an alternative to bone marrow transplantation .

Key words: gene therapy; immune deficiency diseases; severe combined immune deficiency (SCID); adenosine deaminase; hematopoietic stem cells; retroviral vectors

Progress in Gene Therapy: Basic and Clinical Frontiers, pp. 267-288
R. Bertolotti *et al.* (Eds)

Transgenic rescue of tyrosine hydroxylase-deficient mice: application for generating animal models with catecholamine dysfunction

Kazuto Kobayashi* and Toshiharu Nagatsu[1]

Institute of Biomedical Sciences, Fukushima Medical University School of Medicine, Fukushima 960-1295, Japan, and [1]Institute for Comprehensive Medical Science, Graduate School of Medicine, Fujita Health University, Toyoake 470-0192, Japan

Table of Contents

1. INTRODUCTION

Catecholamines including dopamine (DA), noradrenaline (NA), and adrenaline (AD) are the principal neurotransmitters that mediate a variety of the central nervous system (CNS) and peripheral nervous system (PNS) functions, such as motor control, cognition, emotion, and autonomic regulation. Dysfunctions in catecholamine neurotransmission have been

---

* Corresponding author. E-mail: Kazuto@cc.fmu.ac.jp

implicated in some neurological and neuropsychiatric disorders including Parkinson's disease, manic-depressive illness, and schizophrenia (Nagatsu *et al.*, 1984; Davidson *et al.*, 1987). In addition, some automatic failures with primary deficits in the sympathetic neurons are accompanied by altered catecholamine metabolism, such as pure automatic failure and multiple system atrophy (Robertson *et al.*, 1993).

Tyrosine hydroxylase (TH) is the first and rate-limiting enzyme in the biosynthesis of catecholamines (Nagatsu *et al.*, 1964; Levitt *et al.*, 1965). Because of its key role in catecholamine metabolism, a number of researchers have suspected the possible involvement of the gene encoding TH in various hereditary disorders. Genetic studies of several families with manic-depressive illness or schizophrenia have suggested the possibility that the susceptibility factor in these diseases might be located around the TH gene locus (Leboyer *et al.*, 1993; Meloni *et al.*, 1995; Mallet, 1996). Also, a point mutation in the TH gene was found in a recessive form of hereditary progressive dystonia (Lüdecke *et al.*, 1994; Knappskog *et al.*, 1995). This finding is an inceptive case suggesting a specified TH mutation associated with the onset of a neurological disorder. In spite of little information about a clear relationship between mutations in the TH gene and the pathophysiology, the disorders caused by catecholaminergic defects are potential targets for gene therapy based on manipulation of TH gene expression. In particular, therapeutic approaches to treat Parkinson's disease with direct gene transfer and transplantation of artificially manipulated cells have provided new insights into gene therapy (Wolff *et al.*, 1989; Horellou *et al.*, 1990 and 1994; Le Gal La Salle *et al.*, 1993; Sabaté *et al.*, 1995).

In addition to development of novel therapeutic techniques, it is important to have animal models possessing altered catecholamine metabolism which mimics that of the pathological states. Recent advances in mouse genetic approaches provide an excellent system for creating these animal models by manipulating expression or activity of catecholamine-synthesizing enzymes. A series of technological advances have generated mice showing overproduction (Kaneda *et al.*, 1991; Kobayashi *et al.*, 1994), Misexpression (Cadd *et al.*, 1992; Kobayashi *et al.*, 1992), and deletion of the genes for these enzymes (Kobayashi *et al.*, 1995; Thomas *et al.*, 1995; Zhou and Palmiter, 1995; Zhou *et al.*, 1995; Nishii *et al.*, 1998).

In this review, we summarize genetic studies on the role of catecholamine neurotransmitters in the CNS and PNS functions. We describe 1) the mutant lethal phenotype of mice showing global catecholamine deficiency produced by targeted disruption of the TH gene and 2) the transgenic rescue of dopamine deficient mutants produced by transcriptional targeting of a tissue-specific TH minigene into adrenergic and noradrenergic cells. In addition, we indicate another transgenic rescue approach based on trans-complementation with human genomic DNA to create model animals in which the TH functions are derived from a transgene containing a whole human TH locus. This transgenic rescue approach provides a useful system to evaluate the mutations in the human TH gene that may be linked to neurological and neurospychiatric diseases.

2. TARGETED DISRUPTION OF THE TH GENE: CATECHOLAMINE DEFICIENCY

Since TH catalyzes the first step of catecholamine biosynthesis, we expected that the knockout of the TH gene would cause a depletion of the three kinds of catecholamines. We generated mutant mice lacking TH activity by the gene targeting approach utilizing embryonic stem cells (Kobayashi *et al.*, 1995). In our strategy of targeted disruption of the TH gene, we introduced a mutation into the middle portion of the TH structural gene by inserting a PGK-neo minigene with concomitant partial deletion of exons 7 and 8 (Fig. 1A). The mutation resulted in a loss of TH mRNA and enzyme activity in the homozygous mice (Fig. 1B and C). The levels of the three catecholamines in the tissues were severely reduced in the TH mutant mice as compared with those in the wild-type and heterozygous mice, although

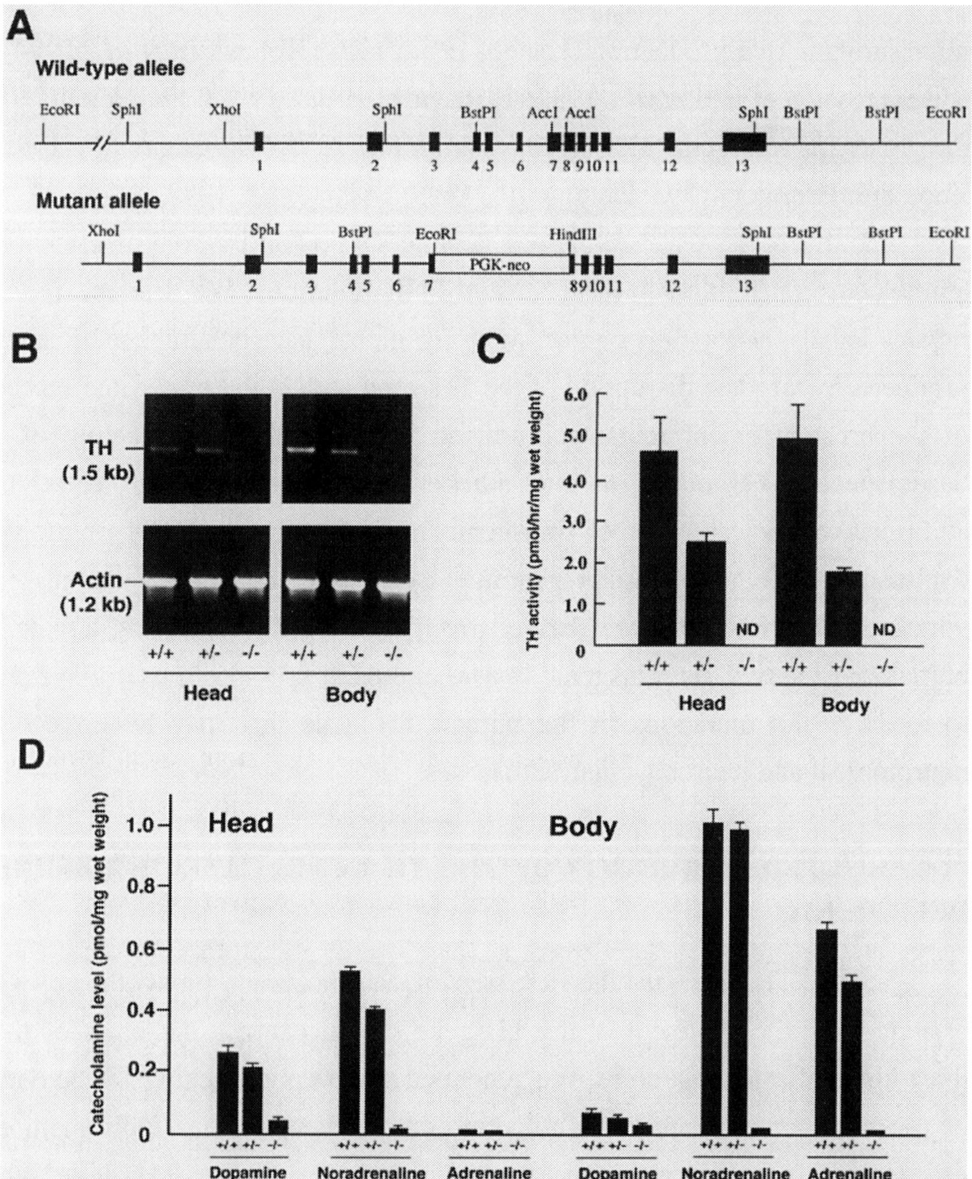

Figure 1. Generation of catecholamine-deficient mice by gene targeting of the mouse TH locus. A). Structures of wild type and mutant TH alleles. A small DNA fragment of the TH gene including exon 7 (3' portion) and exon 8 (5' portion) was deleted by exchanging with the PGK-neo gene cassette to generate null mutant animals. **B**) and **C**). Analysis of TH mRNA and enzyme activity in E12.5 embryos. Values are the mean ± SEM of the data from three to six embryos. The mutation led to complete loss of the TH mRNA and enzyme activity. **D**). Catecholamine contents of newborn mice. Values are the mean ± SEM of the data from three to six embryos. Loss of TH function in the mutant mice resulted in severe depletion of dopamine, noradrenaline, and adrenaline.

the extent of the decrease in dopamine levels was relatively moderate compared with that in noradrenaline and adrenaline levels (Fig. 1D). There seems to be a mechanism for dopamine synthesis bypassing the TH reaction; the small amount of dopamine accumulated in the mutants might be produced from L-DOPA that probaly originated from the maternal circulation. Recently, the involvement of tyrosinase, a key enzyme of melanin biosynthesis, in the production of a small amount of catecholamines has been suggested by Rios *et al.* (1999).

As a consequence of severe catecholamine depletion, the TH mutant mice showed lethality during the perinatal period (Fig. 2A). The cause of the lethality is likely because of cardiovascular dysfunctions in the mutant mice. One representative parameter of the defects was a significant decrease in basal heart beating rate in the mutants, which was recorded as surface electrocardiograms of the newborn mice (Fig. 2B). Other than cardiovascular regulation by catecholamines, it is known that they mediate metabolic responses in the developing embryos through modulation of carbohydrate and fatty acid mobilization (Jones, 1980; Silver and Edwards, 1980). In newborn animals, their action is also implicated in respiratory functions to prepare the lungs for ventilation after delivery (Olver, 1981; Faxelius *et al.*, 1983). However, at the time of birth, the surviving TH mutant mice could begin to breathe with no apparent histological abnormalities in their lung tissues. Also, the blood glucose levels in the newborn mutants were indistinguishable from the wild-type levels. In contrast to the cardiovascular dysfunctions observed in the mutants, metabolic and respiratory functions seemed to be normally maintained in spite of catechoamine depletion.

These phenotypic observations demonstrate that a crucial role of catecholamines during prenatal and neonatal stages is the control of cardiovascular functions, which ensures normal development and survival of the animals. Especially, during the perinatal development, circulating catecholamines derived from the adrenal glands are thought to be more important in physiological actions than the central and sympathetic neurotransmission, because these nervous systems are not fully functional due to immaturity at this stage (Seidler and Slotkin, 1985; Mills and Smith, 1986).

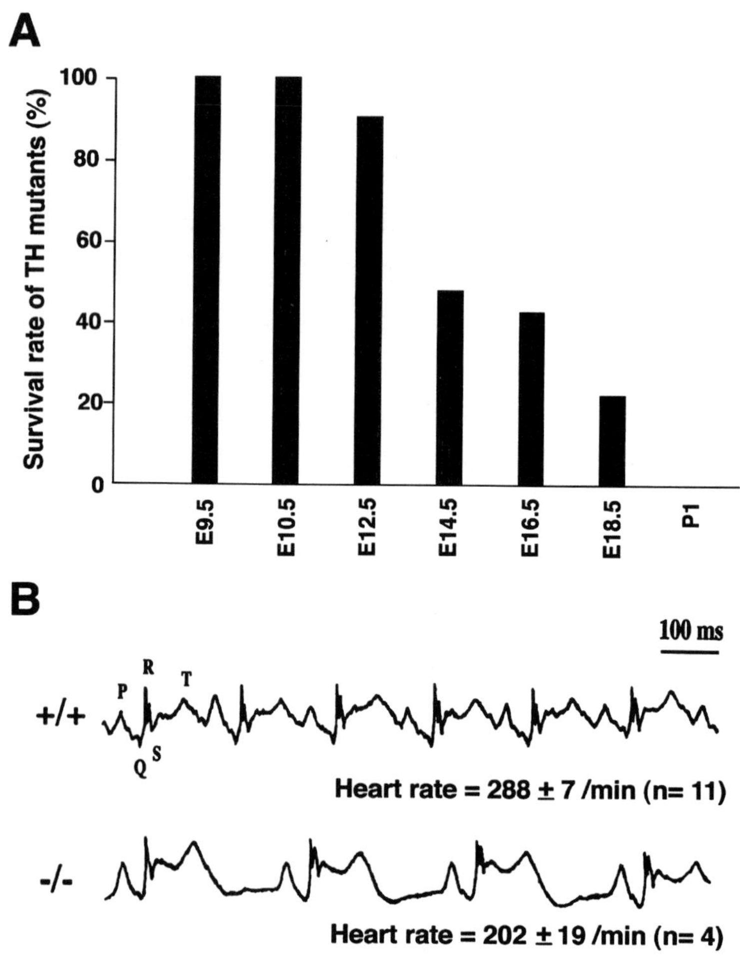

Figure 2. Perinatal lethality and cardiac dysfunction in mice lacking the TH gene. A) Survival rate of the TH mutant mice during development. Gross development of the mutants was normal up to E10.5, but the survival rate was gradually reduced during subsequent embryonic stages. The surviving mutant neonates delivered by caesarean section died soon after birth because of anoxia or during nursing by foster mothers until P1. **B)** Surface electrocardiography on the newborn mice. The wave morphology in the electrocardiograms was similar between the wild-type and mutant animals except for a slight elevation of the S-T segment in the mutants. However, there was a marked reduction in heart rate in the mutant mice as compared to the wild type rate, suggesting a defect in cardiac functions as a consequence of severe catecholamine depletion in the mutants.

3. TRANSGENIC RESCUE WITH A TH MINI-GENE: DOPAMINE DEFICIENCY

To generate mice deficient in dopamine synthesis only, we used the transgenic rescue approach (Nishii *et al.*, 1998). In mice with deficient TH genes, TH expression was rescued in noradrenergic and adrenergic cells upon transcriptional targeting with a human TH minigene that is driven by the tissue-specific promoter of the dopamine β-hydroxylase (DBH) gene (see Fig. 3A for experimental strategy). Indeed, transgenic mice were produced that carry a tissue-specific TH minigene, *i.e.* an artificial gene resulting from the insertion of a human TH cDNA into an expression cassette that is under the control of the 5'-upstream region of the human DBH gene (Fig. 3B). Transgenic offspring were mated with mice hetero-zygous for the TH mutation introduced by gene targeting. The offspring carrying the transgene and heterozygous mutation were further crossed with TH heterozygotes. At the next generation, we could obtain the rescued mutant mice that were homozygous for the TH mutation and had the transgene (Fig. 3C, see lane 6). The histological examination of the mouse embryos showed a loss of TH expression in the neuronal types that normally produce dopamine in the ventral midbrain (Fig. 3D). TH expression in noradrenergic and adrenergic cells was recovered in the mutant due to the tissue-specific expression of the transgene (Fig. 3D). Consistent with the TH expression pattern, dopamine accumulation was remarkably reduced in the forebrain, midbrain, hindbrain, and pituitary gland of the mutant (Fig. 3E), whereas noradrenaline and adrenaline accumulation was restored to the normal levels.

Dopamine-deficient mice grew normally up to around 10 days after birth, but at postnatal day 10-14 (P10-14) they were distinguishable from other littermates owing to their small size. During subsequent development, they further lost body weight and gradually weakened (see Fig. 4A). Finally, all of them died by P30. The brain weight of the mutant was reduced

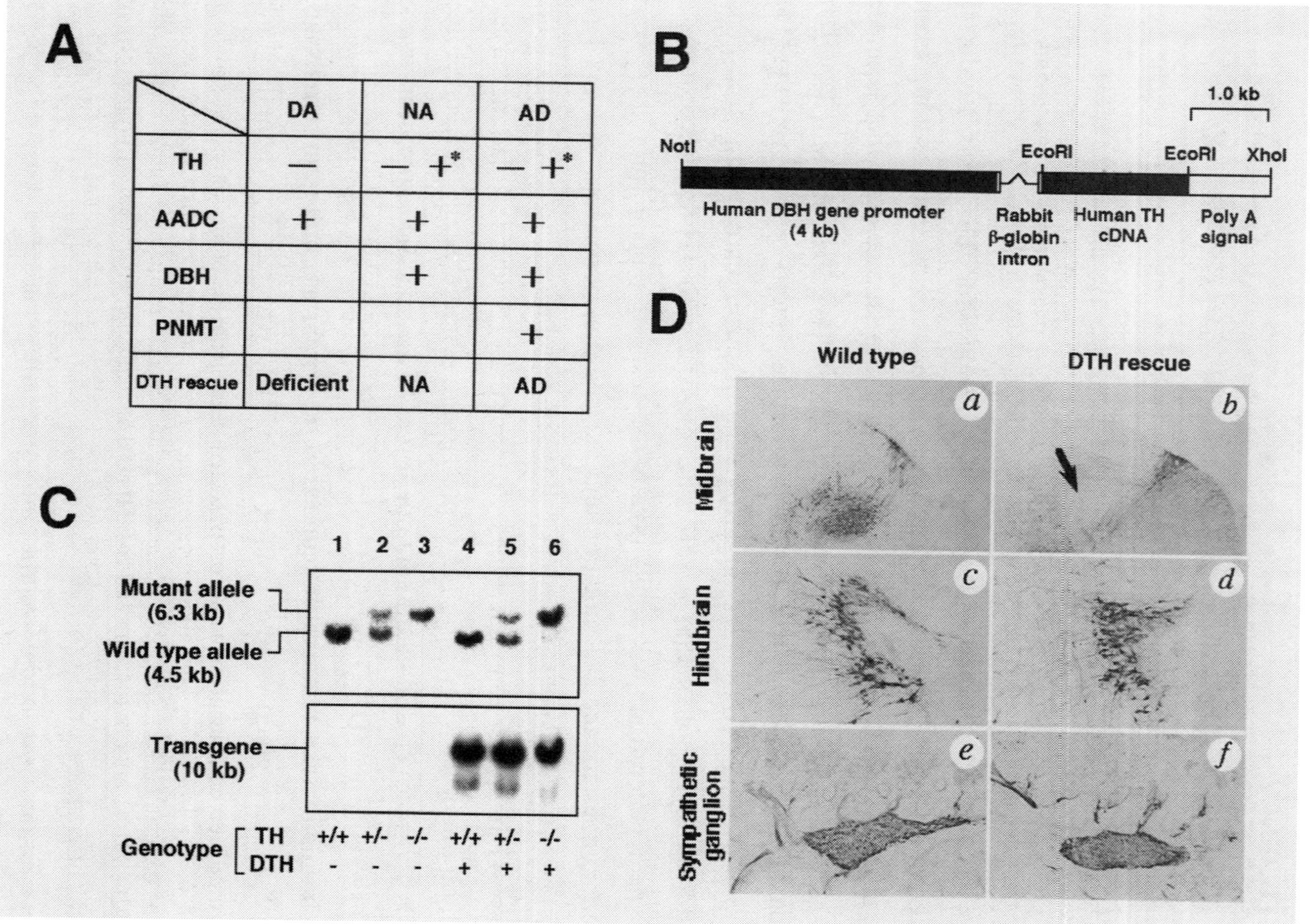
A
DA
NA
AD
TH — — +* — +*
AADC + + +
DBH + +
PNMT +
DTH rescue Deficient NA AD
B
NotI
EcoRI
EcoRI
XhoI
1.0 kb
Human DBH gene promoter (4 kb)
Rabbit β-globin intron
Human TH cDNA
Poly A signal
C
1 2 3 4 5 6
Mutant allele (6.3 kb)
Wild type allele (4.5 kb)
Transgene (10 kb)
Genotype
TH +/+ +/- -/- +/+ +/- -/-
DTH - - - + + +
D
Wild type
DTH rescue
Midbrain
Hindbrain
Sympathetic ganglion
a
b
c
d
e
f

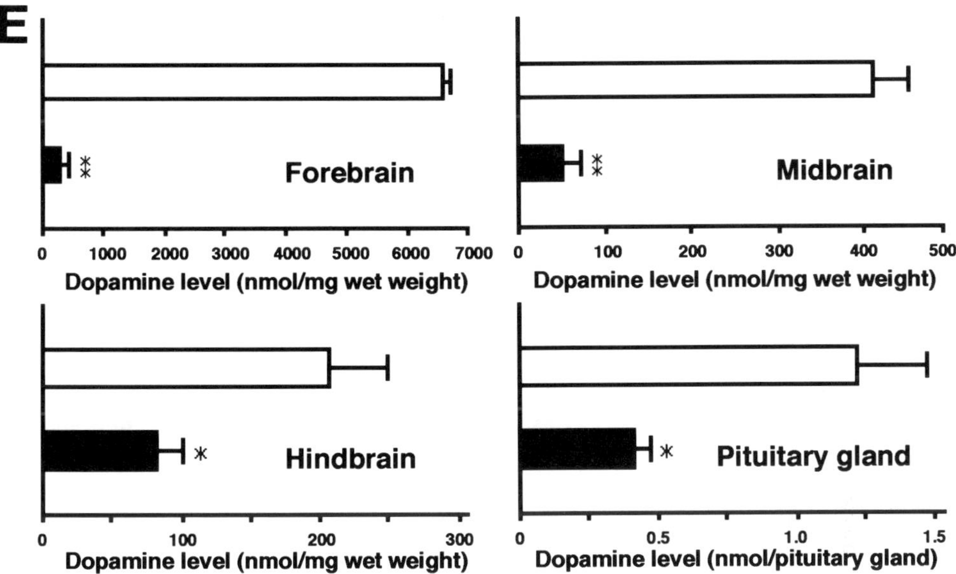

Figure 3. Generation of dopamine-deficient mice by the transgenic rescue approach. A) Experimental strategy. The expression pattern of catecholamine-synthesizing enzymes TH, aromatic L-amino acid decarboxylase (AADC), DBH, and phenylethanolamine N-methyltransferase (PNMT) — in catecholamine-producing cells (DA, NA and AD) of DTH-rescued mutant mice is shown. +: expression of AADC, DBH or PNMT in each cell type; -: deletion of endogenous TH expression due to the mutation; +*: expression of the TH transgene driven by the DBH promoter. Phenotypes of catecholamine-producing cells in the mutant are shown in the panel. **B)** In this strategy, transgenic mice were produced that carry a human TH cDNA downstream of the 5'-upstream region of the human DBH gene. Transgenic offspring were mated with mice heterozygous for the TH mutation introduced by gene targeting (see Fig. 1). The offspring carrying the transgene and heterozygous mutation were further crossed with TH heterozygotes. At the next generation, the rescued mutant mice were obtained that carried the homozygous TH mutation and the transgene. **C)** Genotyping of the second generation offspring: Southern blot analysis of tail DNA. The upper panel shows the genotype (+/+, +/- or -/-) of the mouse TH locus, and the lower panel shows the presence or absence of the transgene. The rescued mutant corresponded to the genotype of the individual shown in lane 6. **D)** Immunocytochemical localization of TH-expressing cells. Sagittal sections prepared from E14.5 embryos were stained with anti-TH antibodies. Light microscopic images of the midbrain, hindbrain and thoracic sympathetic ganglions of wild type and mutant mice are shown. In the mutant, TH expression was lacking only in the midbrain dopaminergic neurons (arrow) but was recovered in the noradrenergic and adrenergic cell types. **E)** Dopamine accumulation in tissues of juvenile mice. The dopamine levels in brain regions and pituitary glands of the wild type and mutant mice are shown. Values denote mean ± S.E. of the data obtained from six mice. Open columns: wild type; closed columns: mutant. The asterisks indicate a significant difference ($^*P < 0.05$; $^{**}P < 0.01$) from the wild type value (Student's *t* test).

to 87% of the wild type weight at P21, but the gross morphology of the mutant's brain appeared to be normal. The phenotypic abnormalities of the mutant revealed that dopamine is essential for animal development and survival during juvenile stage.

Dopamine-deficient mice exhibited impairment in motor control functions during juvenile stage past P14. First, the spontaneous locomotor activity of the mutant mice displayed a significant reduction relative to the wild type; the horizontal and vertical movements were decreased to about 43% and 30% of the controls, respectively (Fig. 4B). Second, the mutant mice exhibited the cataleptic behavior, which is defined as the absence of voluntary movement. When the mice were assayed by the parallel bar test, the mutant maintained a bizarre posture on the bar much longer than the wild type (Fig. 4C). Third, the mutant appeared to be insensitive to the methamphetamine treatment, which normally induces the hyperactivity of locomotion

Figure 4. Impaired motor and emotional learning in dopamine-deficient mice. A) Photograph of 3-week-old mice. Dopamine-deficient mice exhibited growth retardation until this stage. **B)** Spontaneous locomotor activity of 3-week-old mice. Locomotion activity (upper panel) and rearing activity (lower panel), defined as total number of beam breaks by horizontal and vertical movements, respectively, of the wild type and mutant mice are shown. Values represent mean ± S.E. of the data. Open columns: wild type (n = 13); closed columns: mutant (n = 12). The asterisks indicate a significant difference ($*P < 0.01$) from the wild type value (Student's *t* test). **C)** Cataleptic behavior of the mice monitored by the parallel bar test. Mice were forced to take a bizarre posture setting forepaws on a parallel bar and hindpaws on the floor. Mice were categorized into four groups (< 1 sec, 1-10 sec, 10-30 sec, and > 30 sec) depending on the degree of catalepsy. The percentage of the number of animals in each group is indicated (wild type, n= 26; mutant, n = 21). **D)** Methamphetamine-induced hyperactivity. Mice (3 weeks old) were treated with methamphetamine (3 mg/kg, s.c.) or saline, then locomotor activity was monitored for 20 min. Hyperactivity was evaluated as the relative ratio of the locomotor activity after treatment to the basal activity. Values are mean ± S.E. for methamphetamine injection (n = 6) and saline injection (n = 5). Open columns: wild type; closed columns: mutant. MAP: methamphetamine. Analysis of variance (ANOVA) indicated a significant interaction between genotype and treatment factors [$F(1, 10) = 11.93$, $P < 0.01$]. In the wild type, there was a significant difference between methamphetamine and saline treatment ($*P < 0.05$), but in the mutant the difference was not significant (NS) between the two kinds of treatment. The methamphetamine-induced hyperactivity in the mutant was significantly lower than that in the wild type ($^{\dagger}P < 0.05$).

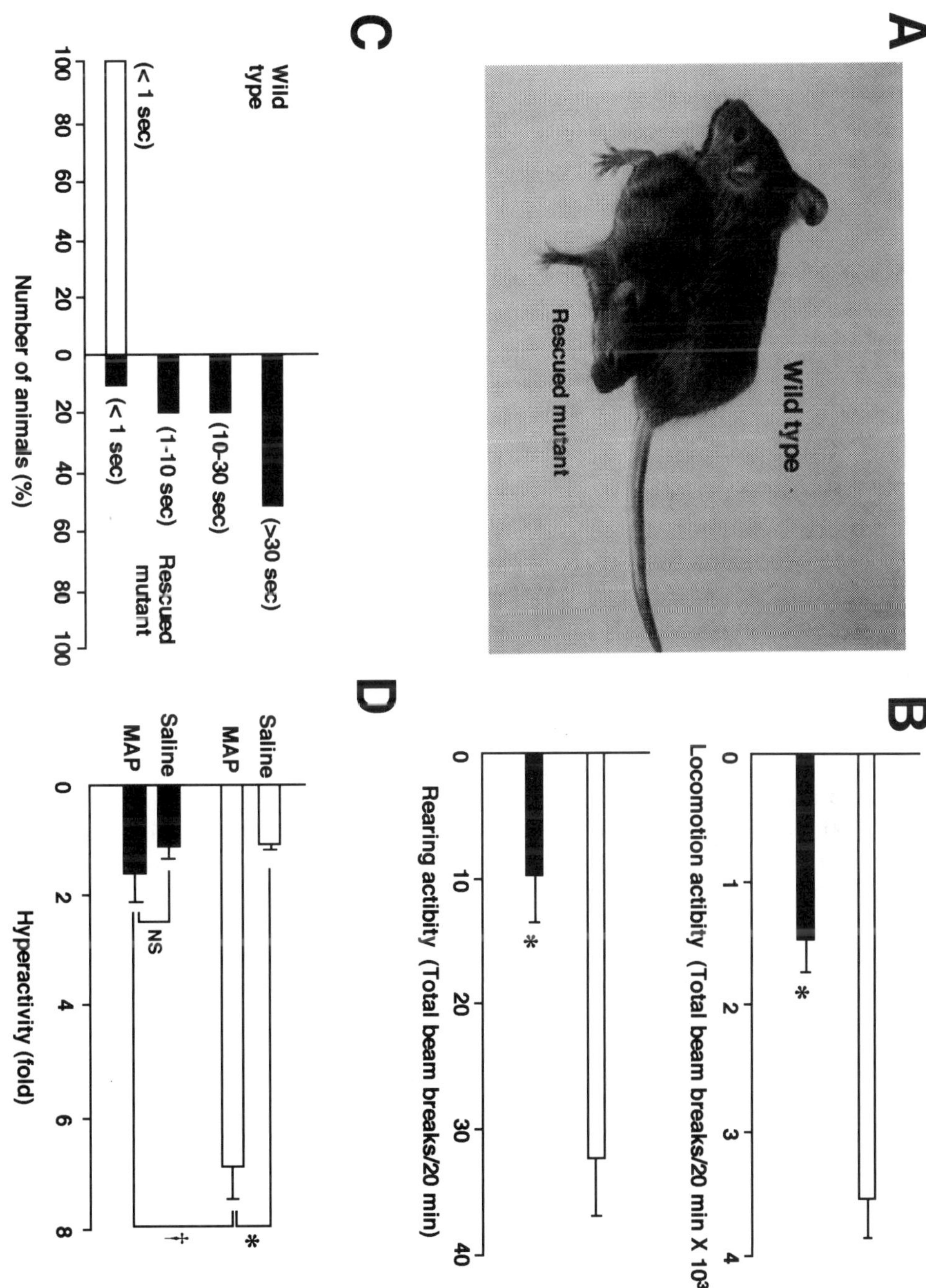
A
Wild type
Rescued mutant
B
Locomotion actibity (Total beam breaks/20 min X 10³)
Rearing actibity (Total beam breaks/20 min)
C
Wild type
(< 1 sec)
(>30 sec)
(10-30 sec)
(1-10 sec)
(< 1 sec)
Rescued mutant
Number of animals (%)
D
Saline
MAP
Saline
MAP
NS
Hyperactivity (fold)

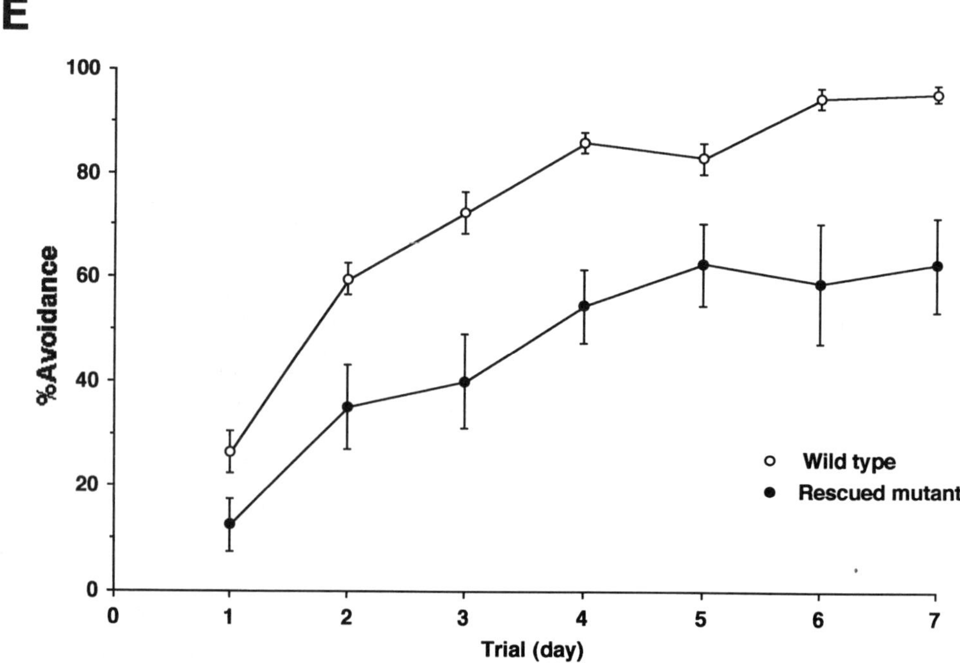

Figure 4 (*cont.*). Impaired motor and emotional learning in dopamine-deficient mice. E) Emotional learning evaluated by active avoidance task. Mice (3 weeks old) were tested for the acquisition of active avoidance response triggered by a tone stimulus. The shuttle box apparatus consisted of two compartments with a grid floor including one light and one dark chamber. The mice were given 5 sec of a tone stimulus in the dark compartment. Then an electric shock (0.2 mA, 5 sec) was delivered to the feet with a scambled shock generator. When the mouse moved into the light compartment before the onset of the footshock, such action was counted as an avoidance. Each animal was given 10 trials per day for 7 consecutive days. Percentage of successful avoidances in 10 trials per day is plotted. Values indicate mean ± S.E. of the data. Open circles: wild type (n = 10); closed circles: mutant (n= 6). ANOVA indicated significant main effects of genotypes [$F(1, 14) = 13.77$, $P < 0.01$] and trial days [$F(6, 84) = 22.04$, P , 0.01].

(Fig. 4D). Normally, these motor control functions are mediated by the nigrostriatal or mesocorticolimbic dopaminergic pathway (Lindvall and Björklund, 1983). The two pathways are assumed to participate in a neuronal network loop for modulating spontaneous and drug-induced locomotor activity (Kelly and Iversen, 1976; Koob *et al.*, 1978; Hu *et al.*, 1991). In contrast, it appears that the primary site of cataleptogenesis by dopamine receptor blockers is the caudate-putamen, the terminal region of the nigro-striatal pathway (Dunstan *et al.*, 1980). The abnormalities of motor responses in dopamine-deficient mice are considered to result from a combined loss of the actions through the two major dopaminergic pathways.

In addition to the motor dysfunctions, dopamine-deficient mice exhibited defects in a certain emotional learning paradigm. Active avoidance is the paradigm that monitors the performance to escape an aversive stimulus (a footshock) associated with a conditioned stimulus (a tone). The task requires the amygdala and its linking pathways. When the mice were explored to the tone-dependent active avoidance, the mutant learned the task much more slowly than the wild type (Fig. 4E). In the final session of the training, the success of avoidance in the mutant was only 60%. The difference in the shape of the learning curve between the two genotypes reveals that dopamine is essential for the acquisition of the active avoidance paradigm. In contrast, we found no difference in the escape latency, which defines the time taken between onset of the footshock and initiation of the response between the wild type and mutant mice. Also, there was no difference in the pain sensitivity measured by the tail-flick test between the two genotypes. Our observations suggest that the mutant mice can normally initiate movement in response to the footshock. Thus, the impairment in the active avoidance in the mutant seems to be attributable to deficits in the learning process. Recent studies with a different avoidance task using the lever press paradigm suggest that the avoidance response is disrupted by dopamine depletion in nucleus accumbens (McCullough *et al.*, 1993). The nucleus accumbens is reported to be an important route, in which the asociative information in the amygdala accesses to the emotional response (Everitt *et al.*, 1991). The abnormalities of emotional learning in dopamine-deficient mice could be explained predominantly by the deficits in the mesolimbic dopaminergic pathway projecting to the nucleus accumbens.

The behavioral abnormalities of the mutant mice demonstrate that dopamine plays an essential role in movement control and emotional learning during postnatal development through the nigrostriatal and mesocorticolimbic pathways. The dopaminergic cell groups become organized into a pattern similar to that seen in adult animals during the first postnatal week (Kalsbeek *et al.*, 1992). Also, TH activity and dopamine level in the brain show a progressive increase after birth until the third postnatal week (Kalaria and Prince, 1988; Giorgi *et al.*, 1987). The onset of the behavioral abnormalities agrees well with the period when the developmental changes in dopaminergic neurons exceedingly proceed.

4. TRANSGENIC RESCUE WITH HUMAN GENOMIC DNA: MODEL ANIMALS CARRYING THE HUMAN TH LOCUS

Gene expression patterns of human TH are divergent from those of mouse TH, particularly in the transcriptional and posttranscriptional control of the gene. For instance, alternative RNA splicing from a single TH gene generates mRNA diversity in humans (Grima *et al.*, 1987; Kaneda *et al.*, 1987; Le Bourdellés *et al.*, 1988; Kobayashi *et al.*, 1988; Dumas *et al.*, 1996), whereas the mouse TH gene is transcribed to produce only a single

Figure 5. Generation of model mice expressing the human TH enzyme. A) Genotyping of the F2 offspring at 4 weeks of age. By the transgenic rescue approach described in figure 3 legend, the transgene containing a whole human TH gene was introduced in the TH knockout mice. The genotyping of the mouse TH locus is shown in the upper panel. The lower panel shows the Southern blot for the human TH transgene integration. Patterns of the rescued mutant mice are shown in lanes 2 and 6. **B)** Detection of the human TH mRNA in the rescued mutants. Total RNA prepared from the wild type (lanes 3 and 6), homozygous mutants (lanes 4 and 7) and rescued mutant (lanes 5 and 8) mice were subjected to RT-PCR analysis with the mixed primer sets to detect the human and mouse TH mRNAs. The mouse TH cDNA (lane 1) and human TH cDNA (lane 2) were used as controls of the PCR reaction. As seen in lanes 5 and 8, the rescued mutant mice expressed only human TH mRNA in their brain and adrenal gland. **C)** TH enzyme activity in the tissues of the rescued mutant mice. Closed columns: wild type mice; hatched columns: transgenic mice; open columns: rescued mutant mice. Values are the means of the data from three animals. In the rescued mutant mice, TH activity was actually recovered to the level corresponding to the difference between the transgenic and wild–type levels in each tissue. **D)** Recovery of catecholamine accumulation in the rescued mutant mice. Closed columns: wild type mice; open columns: rescued mutant mice. Values are the means of the data from three animals. Catecholamine levels in the tissues of the rescued mutants are comparable to the corresponding wild type levels.

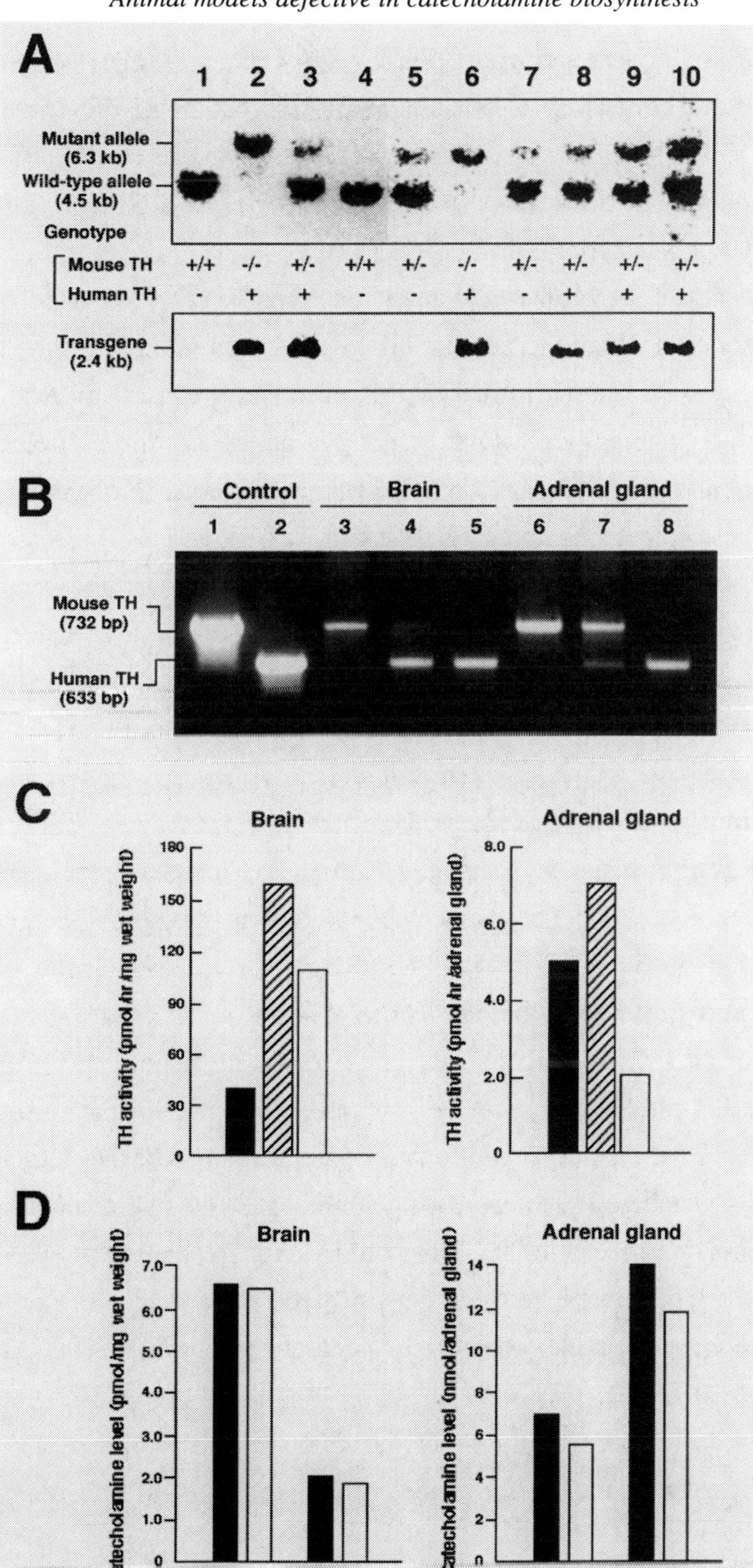
A
1 2 3 4 5 6 7 8 9 10
Mutant allele (6.3 kb)
Wild-type allele (4.5 kb)
Genotype
Mouse TH +/+ -/- +/- +/+ +/- -/- +/- +/- +/- +/-
Human TH - + + - - + - + + +
Transgene (2.4 kb)
B
Control Brain Adrenal gland
1 2 3 4 5 6 7 8
Mouse TH (732 bp)
Human TH (633 bp)
C
Brain
Adrenal gland
TH activity (pmol/hr/mg wet weight)
TH activity (pmol/hr/adrenal gland)
D
Brain
Adrenal gland
Catecholamine level (pmol/mg wet weight)
Catecholamine level (nmol/adrenal gland)
Dopamine Noradrenaline
Noradrenaline Adrenaline

mRNA species (Ichikawa *et al.*, 1991; Iwata *et al.*, 1992). Also, genomic DNA sequences conferring a tissue-specific expression to the human TH gene seem to be localized in different regions from the rodent TH gene (Kaneda *et al.*, 1991; Sasaoka *et al.*, 1992; Banerjee *et al.*, 1992 and 1994; Min *et al.*, 1994 and 1996; Kim *et al.*, 1997).

To evaluate the human TH gene functions *in vivo*, we need to have an animal model in which the human TH gene is expressed and thus able to replace the mouse TH functions. For this purpose, we used the transgenic rescue approach where mice with deficient TH genes are transcomplemented with human genomic DNA, *i.e.* a whole human TH locus (Kobayashi *et al.*, 1995). In these transgenic mice, the human TH enzyme is expressed in a tissue-specific manner (Kaneda *et al.*, 1991). The human TH locus was introduced into the TH knockout mice by the mating procedures described in the previous section. If the transgenic human TH locus could indeed correct the mutant phenotype of the homozygous mice, we should obtain viable rescued animals at the F2 generation. Genotyping of the weaned F2 offsprings indicated that the mice identified as homozygous for the TH mutation and comprising the transgenic human TH locus were viable (Fig. 5A, see lanes 2 and 6). These mice also failed to develop the abnormal cardiac response observed in the TH mutants at the time of birth. To determine the expression of human TH mRNA in the rescued mutant mice, we carried out RT-PCR analysis that could distinguish human TH mRNA from mouse TH mRNA by using the primer set specific for each sequence. As shown in Fig. 5B, a single band corresponding to the human TH sequence was visualized in the brain and adrenal gland of the rescued mutant animals. Also, in these animals, TH activity was recovered to the levels equivalent to the difference between the wild-type and transgenic levels (Fig. 5C), and the catecholamines were maintained at their normal levels in each tissue (Fig. 5D).

These results demonstrate that introduction of the human TH locus into homozygous mutant mice can correct the mutant phenotype by recovering catecholamine synthesis upon mediation of the human enzyme. In addition to the suppression of perinatal lethality and cardiac dysfunctions observed in the TH mutants, the rescued mutants remained healthy after the weaning, and their behavior was grossly normal. This approach enables us to introduce any mutation involved in expression of the human TH gene and regulation of TH enzyme activity. Analysis of the resulting phenotypic effects makes it possible to understand the physiological and behavioral significance of the mutations.

5. CONCLUSION

We described 1) lethal catecholamine deficiency in mutant mice generated by targeted disruption of the TH gene and 2) sublethal dopamine deficiency upon subsequent transgenic rescue with a tissue-specific TH minigene. These studies demonstrated an essential role of catecholamines in the physiological and behavioral functions of the developing animals. In addition, we showed that the transgenic rescue approach of mutant mice can be achieved with the human TH locus. Such a transcomplementation with human genomic DNA is thus useful to evaluate *in vivo* the functions of the human TH gene. Currently, efforts for identifying potential mutations in the TH gene are ongoing with respect to several neurobiological and neuropsychiatric disorders. It is important to characterize the relationship between the identified mutations and the pathogenesis of these disorders. The mouse genetic approach is a powerful technique to generate animal models with altered TH gene functions. Knowledge obtained from animal models provides thus an important basis for understanding the mechanisms linking the etiology of neurological and neuropsychiatric disorders. In addition, it should open new avenues for gene therapy of such nervous system disorders.

6. REFERENCES

Banerjee, S., Hoppe, P., Brilliant, M. and Chikaraishi, D.M. (1992). 5' flanking sequences of the rat tyrosine hydroxylase gene target accurate tissue-specific, developmental, and transsynaptic expression in transgenic mice. *J. Neurosci.*, 12: 4460-4467.

Banerjee, S.A., Roffler-Tarlov, S., Szabo, M., Frohman, L. and Chikaraishi, D.M. (1994). DNA regulatory sequences of the rat tyrosine hydroxylase gene direct correct catecholaminergic cell-type specificity of a human growth hormone reporter in the CNS of transgenic mice carrying a dwarf phenotype. *Mol. Brain Res.*, 24: 89-106.

Cadd, G.G., Hoyle, G.W., Quaife, C.J., Marck, B., Matsumoto, A.M., Brinster, R.L. and Palmiter, R.D. (1992). Alteration of neurotransmitter phenotype in noradrenergic neurons of transgenic mice. *Mol. Endocrinol.*, 6: 1951-1960.

Davidson, M., Keefe, R.S.E., Mohs, R.C., Siever, L.J., Losonczy, M.F., Horvath, T.B. and Davis, K.L. (1987). L-DOPA challenge and relapse in schizophrenia. *Am. J. Psychiatry*, 122: 509-522.

Dumas, S., Le Hir, H., Bodeau-Péan, S., Hirsch, E., Thermes, C. and Mallet, J. (1996). New species of human tyrosine hydroxylase mRNA are produced in variable amounts in adrenal medulla and are overexpressed in progressive supranuclear palsy. *J. Neurochem.*, 67: 19-25.

Dunstan, R., Broekkamp, C.L. and Lloyd, K.G. (1981). Involvement of caudate nucleus, amygdala or reticular formation in neuroleptic and narcotic catalepsy. *Pharmacol. Biochem. Behav.*, 14: 169-174.

Everitt, B.J., Morris, K.A., O'Brien, A. and Robbins, T.W. (1991). The basolateral amygdala-ventral striatal system and conditioned place preference: further evidence of limbic-striatal interactions underlying reward-related processes. *Neuroscience*, 42: 1-18.

Faxelius, G., Hägnevik, K., Lagercrantz, H., Lundell, B. and Irestedt, L. (1983). Catecholamine surge and lung function after delivery. *Arch. Dis. Child.*, 58: 262-266.

Giorgi, O., De Montis, G., Porceddu, M.L., Mele, S., Calderini, G., Toffano, G. and Biggio, G. (1987). Developmental and age-related changes in D1-dopamine receptors and dopamine content in the rat striatum. *Brain Res.*, 432: 283-290.

Grima, B., Lamouroux, A., Boni, C., Julien, J.-F., Javoy-Agid, F. and Mallet, J. (1987). A single human gene encoding multiple tyrosine hydroxylase with different predicted functional characteristics. *Nature*, 326: 707-711.

Horellou, P., Brundin, P., Kalén, P., Mallet, J. and Björklund, A. (1990). *In vivo* release of DOPA and dopamine from genetically engineered cells grafted to the denervated rat striatum. *Neuron*, 5: 393-402.

Horellou, P., Vigne, E., Castel, MN., Barnéoud, P., Colin, P., Perricaudet, M., Delaére, P. and Mallet, J. (1994). Direct intracerebral gene transfer of an adenoviral vector expres -sing tyrosine hydroxylase in a rat model of Parkinson's disease. *NeuroReport*, 6: 49-53

Hu, S.C., Chang, F.W., Sung, Y.J., Hsu, W.M. and Lee, E.H. (1991). Neurotoxic effects of 1-methyl-4-phenyl-1,2,3,6-tetrahydropyridine in the substantia nigra and the locus coeruleus in BALB/c mice. *J. Pharmacol. Exp. Ther.*, 259: 1379-1387.

Ichikawa, S., Sasaoka, T. and Nagatsu, T. (1991). Primary structure of mouse tyrosine hydroxylase deduced from its cDNA. *Biochem. Biophys. Res. Commun.*, 176: 1610-1616

Iwata, N., Kobayashi, K., Sasaoka, T., Hidaka, H. and Nagatsu, T. (1992). Structure of the mouse tyrosine hydroxylase gene. *Biochem. Biophys. Res. Commun.*, 182: 348-354

Jones, C.T. (1980). Circulating catecholamines in the fetus, their origin, actions and significance. In: *Biogenic Amines in Development,* Parvez, H. and Parvez, S. (Eds.), Elsevier/North-Holland Biomedical Press, Amsterdam, pp. 63-86.

Kalaria, R.N. and Prince, A.K. (1988). Neurochemical development of the striatum in a precocial (guinea pig) and an altricial (rat) species. *Int. J. Dev. Neurosci.*, 6: 161-166.

Kalsbeek, A., Voorn, P. and Buijs, R. (1992). Development of dopamine-containing systems in the CNS. In: *Handbook of Chemical Anatomy, Vol. 10, Ontogeny of Transmitters and Peptides in the CNS*, Björklund, A., Hökelt, T. and Tohyama, M. (Eds.), Elsevier Science Publishers B.V., Amsterdam, pp. 63-112.

Kaneda, N., Kobayashi, K., Ichinose, H., Kishi, F., Nakazawa, A., Kurosawa, Y., Fujita, K., and Nagatsu, T. (1987). Isolation of a novel cDNA clone for human tyrosine hydroxylase: Alternative RNA splicing produces four kinds of mRNA from a single gene. *Biochem. Biophys. Res. Commun.*, 146: 971-975.

Kaneda, N., Sasaoka, T., Kobayashi, K., Kiuchi, K., Nagatsu, I., Kurosawa, Y., Fujita, K., Yokoyama, M., Nomura, T., Katsuki, M., and Nagatsu, T. (1991). Tissue-specific and high-level expression of the human tyrosine hydroxylase gene in transgenic mice. *Neuron*, 6: 583-594.

Kelly, P.H. and Iversen, S.D. (1976). Selective 60HDA-induced destruction of mesolimbic dopamine neurons: abolition of psychostimulant-induced locomotor activity in rats. *Eur. J. Pharmacol.*, 40: 45-56.

Kim, S. J., Lee, J. W., Chun, H. S., Joh, T. H. and Son, J. H. (1997). Monitoring catecholamine differentiation in the embryonic brain and peripheral neurons using *E. coli lacZ* as a reporter gene. *Mol. Cells*, 7: 394-398.

Knappskog, P. M., Flatmark, T., Mallet, J., Ludecke, B. and Bartholomé, K. (1995). Recessively inherited L-DOPA-responsive dystonia caused by a point mutation (Q381K) in the tyrosine hydroxylase gene. *Hum. Mol. Genet.*, 4: 1209-1212.

Kobayashi, K., Kaneda, N., Ichinose, H., Kishi, F., Nakazawa, A., Kurosawa, Y., Fujita, K. and Nagatsu, T. (1988). Structure of the human tyrosine hydroxylase gene: Alternative splicing from a single gene accounts for generation of 4 mRNA types. *J. Biochem.*, 103: 907-912.

Kobayashi, K., Sasaoka, T., Morita, S., Nagatsu, I., Iguchi, A., Kurosawa, Y., Fujita, K., Nomura, T., Kimura, M., Katsuki, M. and Nagatsu, T. (1992) Genetic alteration of catecholamine specificity in transgenic mice. *Proc. Natl. Acad. Sci. USA*, 89: 1631-1635

Kobayashi, K., Morita, S., Mizuguchi, T., Sawada, H., Yamada, K., Nagatsu, I., Fujita, K. and Nagatsu, T. (1994). Functional and high level expression of human dopamine β-hydroxylase in transgenic mice. *J. Biol. Chem.*, 269: 29725-29731.

Kobayashi, K., Morita, S., Sawada, H., Mizuguchi, T., Yamada, K., Nagatsu, I., Hata, T., Watanabe, Y., Fujita, K., and Nagatsu, T. (1995). Targeted disruption of the tyrosine hydroxylase locus results in severe catecholamine depletion and perinatal lethality in mice. *J. Biol. Chem.*, 270: 27235-27243.

Koob, G.F., Riley, S.J., Smith, S.C. and Robbins, T.W. (1978). Effects of 6-hydroxydopamine lesions of the nucleus accumbens septi and olfactory tubercle on feeding, locomotor activity, and amphetamine anorexia in the rat. *J. Comp. Physiol. Psychol.*, 92: 917-927.

Le Bourdellés, B., Boularand, S., Boni, C., Horellou, P., Dumas, S., Grima, B. and Mallet, J. (1988). Analysis of the 5' region of the human tyrosine hydroxylase gene: combinatorial patterns of exon splicing generate multiple regulated tyrosine hydroxylase isoforms. *J. Neurochem.*, 50: 988-991.

Le Gal La Salle, G., Robert, J.J., Berrard, S., Ridoux, V., Stratford-Perricaudet, L.D., Pericaudet, M. and Mallet, J. (1993). An adenovirus vector for gene transfer into neurons and glia in the brain. *Science*, 259: 988-991.

Leboyer, M., Malafosse, A., Boularand, S., Campion, D., Gheysen, F., Samolyk, D., Henriksson, B., Denise, E., des Lauriers, A., Lepine, J-P., Zarifian, E., Clerget-Darpoux, F. and Mallet, J. (1993). Tyrosine hydroxylase polymorphisms associated with manic-depressive illness. *Lancet*, 335: 1219.

Levitt, M., Spector, S., Sjoerdsma, A. and Udenfriend, S. (1965). Elucidation of the rate-limiting step in norepinephrine biosynthesis in the perfused guinea-pig heart. *J. Pharmacol. Exp. Ther.*, 148: 1-8.

Lindvall, O. and Björklund, A. (1983). Dopamine- and norepinephrine-containing neuron systems: their anatomy in the rat brain. In: *Chemical Neuroanatomy*, Emson, P.C. (Ed.), Raven Press, New York, pp. 229-255.

Lüdecke, B., Dworniczak, B. and Bartholomé, K. (1994) A point mutation in the tyrosine hydroxylase gene associated with Segawa's syndrome. *Hum. Genet.*, 93: 123-125.

Mallet, J. (1996). Catecholamines: from gene regulation to neuropsychiatric disorders. *Trends Neurosci.*, 19: 191-196.

McCullough, L.D., Sokolowski, J.D. and Salamone, J.D. (1993). A neurochemical and behavioral investigation of the involvement of nucleus accumbens dopamine in instrumental avoidance. *Neuroscience*, 52: 919-925.

Meloni, R., Leboyer, M., Bellivier, F., Barbe, B., Samolyk, D., Alliaire, J. F. and Mallet, J. (1995). Association of manic-depressive illness with tyrosine hydroxylase microsatellite marker. *Lancet*, 345: 932.

Mills, E. and Smith, P. G. (1986) Mechanisms of adrenergic control of blood pressure in developing rats. *Am. J. Physiol.*, 250: R188-R192.

Min, N., Joh, T. H., Kim, K. S., Peng, C. and Son, J. H. (1994). 5' upstream DNA sequence of the rat tyrosin hydroxylase gene directs high-level and tissue-specific expression to catecholaminergic neurons in the central nervous system of transgenic mice. *Mol. Brain Res.*, 27: 281-289.

Min, N., Joh, T. H., Corp, E. S., Baker, H., Cubells, J. F. and Son, J. H. (1996). A transgenic mouse model to study transsynaptic regulation of tyrosine hydroxylase gene expression. *J. Neurochem.*, 67: 11-18.

Nagatsu, T., Levitt, M. and Udenfriend, S. (1964). Tyrosine hydroxylase: The initial step in norepinephrine biosynthesis. *J. Biol. Chem.*, 239: 2910-2917.

Nagatsu, T., Yamaguchi, T., Rahman, M. K., Trocewicz, J., Oka, K., Hirata, Y., Nagatsu, I., Narabayashi, H., Kondo, T. and Iizuka, R. (1984). Catecholamine-related enzymes and the biopterin cofactor in Parkinson's disease and related extrapyramidal diseases. In: *Advances in Neurology, Vol. 40*, Hassler, R.G. and Christ, J.F. (Eds.), Raven Press, New York, pp. 467-473.

Nishii, K., Matsushita, N., Sawada, H., Sano, H., Noda, Y., Mamiya, T., Nabeshima, T., Nagatsu, I., Hata, T., Kiuchi, K., Yoshizato, H., Nakashima, K., Nagatsu, T. and Kobayashi, K. (1998). Motor and learning dysfunction during postnatal development in mice defective in dopamine neuronal transmission. *J. Neurosci. Res.*, 54: 450-464.

Nakahara, D., Hashiguti, H., Kaneda, N., Sasaoka, T. and Nagatsu, T. (1993). Normalization of tyrosine hydroxylase activity *in vivo* in the striatum of transgenic mice carrying human tyrosine hydroxylase gene: A microdialysis study. *Neurosci. Lett.*, 158: 44-46.

Olver, R. E. (1981). Of labour and the lungs. *Arch. Dis. Child.*, 56: 659-662.

Rios, M., Habecker, B., Sasaoka, T., Eisenhofer, G., Tian, H., Landis, S., Chikaraishi, D. and Roffler-Tarlov, S. (1999). Catecholamine synthesis is mediated by tyrosinase in the absence of tyrosine hydroxylase. *J. Neurosci.*, 19: 3519-3526.

Robertson, D., Beck, C., Gary, T. and Picklo, M. (1993). Classification of autonomic disorders. *Int. Angiol.*, 12: 93-102.

Sabaté, O., Horellou, P., Vigne, E., Colin, P., Perricaudet, M., Buc-Caron, M.-H. and Mallet, J. (1995). Transplantation to the rat brain of human nueral progenitors that were genetically modified using adenoviruses. *Nature Genet.*, 9: 256-260.

Sasaoka, T., Kobayashi, K., Nagatsu, I., Takahashi, R., Kimura, M., Yokoyama, M., Nomura, T., Katsuki, M. and Nagatsu, T. (1992). Analysis of the human tyrosine hydroxylase promoter-chloramphenicol acetyltransferase chimeric gene expression in transgenic mice. *Mol. Brain Res.*, 16: 274-286.

Seidler, F.J. and Slotkin, T.A. (1985). Adrenomedullary function in the neonatal rat: Responses to acute hypoxia. *J. Physiol.*, 358: 1-16

Silver, M. and Edwards, A.V. (1980). The development of the sympatho-adrenal system with an assessment of the role of the adrenal medulla in the fetus and neonate. In: *Biogenic Amines in Development,* Parvez, H. and Parvez, S. (Eds), Elsevier/North-Holland Biomedical Press, Amsterdam, pp. 147-211.

Thomas, S.A., Matsumoto, A.M. and Palmiter, R.D. (1995). Noradrenaline is essential for mouse fetal development. *Nature*, 374: 643-646.

Wolff, J.A., Fisher, L.J., Xu, L., Jinnah, H.A., Langlais, P.J., Iuvone, P.M., O'Malley, K.L., Rosenberg, M.B., Shimohama, S., Friedmann, T. and Gage, F.H. (1989). Grafting fibroblasts genetically modified to produce L-dopa in a rat model of Parkinson disease. *Proc. Natl. Acad. Sci. USA*, 86: 9011-9014.

Zhou QY, Palmiter RD. (1995). Dopamine-deficient mice are severely hypoactive, adipsic, and aphagic. Cell, 83: 1197-1209.

Zhou, Q.-Y., Quaife, C.J., and Palmiter, R.D. (1995). Targeted disruption of the tyrosine hydroxylase gene reveals that catecholamines are required for mouse fetal development. *Nature*, 374: 640-643.

7. SUMMARY — Catecholamines (dopamine, noradrenaline and adrenaline) are the principal neurotransmitters that mediate a variety of CNS and PNS functions, such as motor control, cognition, emotion and autonomic regulation. Dysfunctions in catecholamine neuro-transmission are implicated in some neurological, neuropsychiatric and autonomic disorders. For a clearer understanding of these disorders, it is important to have animal models possessing altered catecholamine metabolism which mimics that of the pathological states. Mouse genetic approaches provide an excellent system for creating these animal models by manipulating expression or activity of catecholamine-synthesizing enzymes. In this review, we initially describe the mutant phenotype of mice carrying homozygous disruption of the gene encoding tyrosine hydroxylase (TH), which is the initial and rate-limiting enzyme in the catecholamine biosynthetic pathway. Disruption of the TH gene results in severe catecholamine depletion leading to cardiovascular failure during the embryonic or neonatal stage. Secondly, we describe the transgenic rescue of dopamine-deficient mice with a tissue-specific TH minigene. Transcriptional targeting of this minigene to noradrenergic and adrenergic neuronal cells restricts TH deficiency to dopaminergic neuronal type. Genetic depletion of dopamine causes an impairment in motor control and emotional learning at the juvenile stage. Finally, we indicate another transgenic rescue approach based on trans-complementation with human genomic DNA to create model animals in which TH functions rely on a whole human TH locus. This approach is useful to evaluate mutations in the human TH gene that may be linked to neurological and neuropsychiatric diseases. In addition, it should open new avenues for gene therapy of such nervous system disorders.

Key Words: catecholamine; dopamine; tyrosine hydroxylase; hereditary disorder; gene targeting; transgenic animal

Progress in Gene Therapy: Basic and Clinical Frontiers, pp. 289-306
R. Bertolotti *et al.* (Eds)

Adeno-associated virus (AAV) vectors and gene therapy of Parkinson's disease

Keiya Ozawa[1,*], Shin-ichi Muramatsu[2], Kunihiko Ikeguchi[2], Ken-ichi Fujimoto[2], Yang Shen[1, 2], Dong-Shen Fan[1,2,†], Hiroaki Mizukami[1], Masashi Urabe[1], Yutaka Hanazono[1], Akihiro Kume[1], Matsuo Ogawa[2] and Imaharu Nakano[2]

[1]*Division of Genetic Therapeutics, Center for Molecular Medicine, and* [2]*Department of Neurology, Jichi Medical School, Tochigi 329-0498, Japan*

Table of Contents

---

* Corresponding author. E-mail: kozawa@ms.jichi.ac.jp

† Present address: Department of Neurology, Third Hospital of Beijing Medical University, Beijing 10083, P.R. China

1. INTRODUCTION

Somatic gene therapy is an ideal approach to the treatment and correction of inherited disorders, and the concept that genetic manipulation might be used to treat disease is also applied to the treatment of acquired disorders, such as cancer and AIDS (acquired immunodeficiency syndrome). Accordingly, the types of disease under consideration for gene therapy are diverse, and many different treatment strategies are being investigated, each with an appropriate gene transfer system. During the past ten years, many clinical protocols of gene therapy and gene marking have been conducted, mainly in the United States. More than 3,000 patients have been transferred with marker or therapeutic genes in the world. Although most of the human gene therapy trials have failed to demonstrate clinical benefit, this kind of therapeutic modality is believed to be important to develop efficient ways for treatment of many diseases, including neurological disorders, in the near future.

Parkinson's disease (PD) seems to be one of the appropriate target diseases for gene therapy. PD is a common disorder characterized by progressive disturbance in the control of movement and has an age of onset between 40 and 70 years old, with the peak age of onset in the sixth decade. The symptoms result from a progressive loss of dopaminergic neurons in the substantia nigra. These neurons send dopamine down to their nerve endings that are located in the striatum. The severity of PD is proportional to the decline in the amount of dopamine in the striatum. Dopamine is produced in the midbrain from tyrosine, which is first hydroxylated and converted to l-dopa by tyrosine hydroxylase (TH) in the presence of tetrahydrobiopterin (BH4) as a cofactor. L-dopa is subsequently decarboxylated into dopamine by a second enzyme called aromatic l-amino acid decarboxylase (AADC) (Nagatsu, 1992). The BH4 is synthesized from guanosine triphosphate (GTP) in the TH-containing dopamine neurons where GTP cyclohydrolase I (GCH) catalyzes the first and rate-limiting step of BH4 biosynthesis. The GCH is considered to regulate the TH activity via regulation of BH4 biosynthesis and thus, controls indirectly dopamine production (Nagatsu *et al.*, 1997).

Patients with PD can be treated with oral administration of l-dopa, but, unfortunately, this therapy usually becomes less effective with progression of the disease and must often be discontinued due to the numerous serious side effects of the orally administered drug. Local production of dopamine in the striatum can be obtained by grafting of fetal mesencephalic neurons into the patient's brain using tissue from up to ten fetuses per patient (Freed *et al.*, 1992; Spencer *et al.*, 1992), but serious practical and ethical issues are associated with this approach. For example, a low transplant efficiency, the requirement for immunosuppression to prevent graft rejection, and the need for large amounts of human fetal tissue, have made this approach impractical for the majority of patients. Gene therapy offers a novel potential alternative to the treatment of PD. Initial gene therapy strategies for PD have focused on modifying cells *ex vivo* by delivering the gene for TH to these cells (Jiao *et al.*, 1993). The target cells used so far include fibroblasts, astrocytes and even neurons (Jiao *et al.*, 1994; Lundberg *et al.*, 1996; Martinez-Serrano *et al.*, 1997). Once implanted into the brain, these genetically modified cells produce l-dopa. A potentially simpler and more practical gene therapy approach involves delivering the gene directly into the brain (During *et al.*, 1994; Kaplitt *et al.*, 1994). A promising approach for achieving this is the use of a gene-delivery vehicle based on adeno-associated virus (AAV), a common virus that has never been associated with human disease. Kaplitt *et al.* have reported that AAV vectors can efficiently deliver TH gene into nerve cells following their direct injection into the rat striatum.

2. GENE TRANSFER VECTORS FOR THE NERVOUS SYSTEM

a) Representative viral vectors and their characteristics

To succeed in human gene therapy, we must overcome numerous fundamental technical issues. Especially, the development of efficient gene transfer systems is crucial to obtain therapeutic effects. A number of methods have been developed for introducing genes into living cells, including viral and non-viral vectors. The former is more commonly utilized for gene therapy or gene marking due to their relatively higher efficiencies of gene transfer. Several different viral vector systems are in use or under consideration for somatic gene transfer (Verma *et al.*, 1997). Among them, retroviral vectors based on the murine leukemia virus (MLV) are currently most extensively investigated as gene transfer vehicles and are employed in the majority of clinical trials of human gene therapy. However, since non-dividing cells cannot be transduced with retroviral vectors, this vector is not appropriate for *in situ* gene transfer into neurons. Herpes simplex virus (HSV) vectors (During *et al.*, 1994; Geller *et al.*, 1995) and adenoviral vectors (Le Gal La Salle *et al.*, 1993) are able to transduce neurons efficiently but their cytotoxicity may be problem. Because a large part of viral genes are remained in these vectors, their low-level expression causes some cytotoxic effects and induces immune reaction to transduced cells. In addition, transgene expression is transient in the case of adenovirus vectors. This vector is suitable for the purposes where transient expression of transduced genes is sufficient to get therapeutic effects. Therefore, the best application of adenovirus vectors may be immune-mediated gene therapy for cancer. AAV vectors and lentivirus vectors can transduce non-dividing neurons and long-term gene expression can be obtained. As for lentivirus vectors, human immunodeficiency virus (HIV) vectors are currently well investigated (Naldini *et al.*, 1996). It has been shown that HIV-based vectors could efficiently deliver genes *in vivo* into terminally differentiated neurons (Blömer *et al.*, 1997; Zufferey *et al.*, 1997). However, the safety issue of HIV vectors is serious problem when chronic diseases are target, and lentiviral vectors have not yet been used in human trials. On the other hand, AAV vectors, derived from non-pathogenic virus, possess several unique properties and are potentially useful gene transfer vehicles for gene therapy (Muzyczka, 1992; Kotin, 1994; Russell *et al.*, 1999; Snyder, 1999).

Therefore, AAV vectors are currently most appropriate for gene therapy of neurological diseases (Kaplitt *et al.*, 1994; Du *et al.*, 1996).

b) AAV vector system

AAV is naturally defective and requires coinfection with a helper virus such as adenovirus or herpes simplex virus to establish a productive infection. The AAV genome consists of 4681 nucleotides (nt) of single-stranded DNA (ssDNA), including two inverted terminal repeats (ITRs) of 145 nt long, which serve as an origin of replication and are also essential for packaging and rescue. The left half of the genome encodes nonstructural Rep proteins (Rep78, Rep68, Rep52, and Rep40) and the right half codes for structural Cap proteins (VP1, VP2, and VP3). The conventional method for AAV vector production is based on cotransfection of 293 cells with a helper plasmid expressing the viral genes, rep and cap, without ITRs and a vector plasmid consisting of a foreign gene flanked by ITRs and infection with a helper adenovirus. Recently, a very efficient helper-adenovirus-free system for AAV vector production has been developed (Matsushita *et al.*, 1998; Xiao *et al.*, 1998), in which 293 cells are co-transfected with the vector plasmid, the helper plasmid, and the plasmid containing adenovirus genes with helper function (E2A, E4, and VA) in place of helper adenovirus (E1 genes are contained in 293 cells). In this system, we can eliminate the contamination of most of the adenovirus-derived proteins. In addition, final yield of AAV vectors increases by omitting the heat-inactivation step for contaminated adenovirus. AAV particles are remarkably stable and are able to be purified by cesium chloride ultracentrifugation. However, it is still difficult to prepare clinical-grade AAV vectors on a large scale. Therefore, it is important to establish efficient AAV packaging cell lines for clinical use of AAV vectors (Inoue *et al.*, 1998; Ogasawara *et al.*, 1998; Ogasawara *et al.*, 1999; Grimm *et al.*, 1999).

The advantages of AAV vectors include lack of any associated disease with a wild-type virus, broad host cell range, integration of the gene into the host genome, and the ability to transduce non-dividing cells as well as dividing cells. The lack of pathogenicity seems to be the most important feature of AAV-based vectors, since the safety issue is crucial when gene therapy is applied to chronic diseases. AAV vectors are suitable for *in situ* transduction, and therefore, neurons (Kaplitt *et al.*, 1994; Du *et al.*, 1996; Wu *et al.*, 1998; Lo *et al.*, 1999), muscle cells (Kessler *et al.*, 1996; Fisher *et al.*, 1997; Herzog *et al.*, 1997 and 1999), and hepatocytes (Snyder *et al.*, 1997 and 1999) are considered to be appropriate target cells. Most part of the transferred genes with AAV vectors are initially present as an episomal form, and the transgenes are expressed after the conversion to double stranded DNA. Transgenes are believed to integrate gradually into the host cell genome, leading to their stable expression. Since the AAV vectors do not contain viral genes, immune response should not occur against the transduced cells. Cotransduction with a variety of AAV vectors would also be possible. This phenomenon is important, because the capacity of gene insertion is limited for small-sized AAV vectors (approximately, up to 4.5 kb of a fragment including promoter and polyadenylation signal sequences can be inserted between the ITRs).

The clinical trials using AAV vectors have already been initiated in patients with cystic fibrosis and hemophilia B, in which vectors expressing the therapeutic genes are injected into respiratory tracts and intramuscular sites, respectively.

c) Gene transfer into neurons using a AAV vector

To examine the efficiency of AAV vector-mediated gene transfer into neurons, primary cultured rat striatal cells were transduced with AAV vectors containing the LacZ gene (AAV-LacZ). When the LacZ expression by transduced cells was evaluated with X-gal staining, roughly 30% of striatal cells were positive for β-galactosidase activity, and about one week was required for maximum expression. The transduction efficiency was dose-dependent and time-dependent (Fan *et al.*, 1998b). The AAV-LacZ vector was also injected into the striatum of normal adult rats. Rats were anaesthetized with nembutal and placed in a stereotaxic instrument. The skull was exposed and 3 holes were drilled above the striatum. AAV vector solution was injected slowly using a Hamilton syringe into each of the three sites. Each injection was 2 μl (5 x 10^8 vector particles per μl) in volume. When the rats were sacrificed at 1 week postinjection, LacZ positive cells were clearly identified at the injection site within the striatum without any background staining using X-gal histochemistry. Most of the positive cells were neurons, based on the morphological criteria, although some of the positive cells were non-neuronal. We found that LacZ was expressed for several months in the striatum. There was no pathological change on microscopic examination in any animal (Fan *et al.*, 1998b).

3. DOPAMINE-SUPPLEMENT GENE THERAPY USING AAV VECTORS IN PARKINSONIAN RATS

a) *In vivo* double transduction using AAV-TH and AAV-AADC

Unlike l-dopa delivered orally, it is believed that l-dopa produced directly in the striatum is converted to dopamine within the striatum. However, the source and the site in the brain of the AADC has not yet been identified. The AADC activity within the striatum is considered to be very low (Jeager *et al.*, 1984; Tashiro *et al.*, 1989; Kang *et al.*, 1992). Therefore, one of the unsolved issues for gene therapy of PD is whether cells genetically engineered to produce TH would suffice for production of therapeutic levels of dopamine (Jinnah *et al.*, 1995). Expression of both TH and AADC directly in striatal cells may be a more straightforward approach to supply therapeutic levels of dopamine locally. *In vitro* studies using non-neuronal cells have already shown that the coexpression of TH and AADC augmented dopamine production (Kang *et al.*, 1993). As for *in vivo* studies, conflicting results have been reported (Wachtel *et al.*, 1997; Imaoka *et al.*, 1998).

In our study, two AAV vectors, one containing the TH gene (AAV-TH) (Kaneda *et al.*, 1987) and the second containing the AADC gene (AAV-AADC) (Ichinose *et al.*, 1989) were constructed, and *in vitro* cotransduction experiments were conducted using primary cultured rat striatal cells. Coexpression of TH and AADC in the same striatal cells was detected by dual immunofluorescent staining. Cotransduction efficiency was increased along with the increasing doses of AAV-TH/AAV-AADC vectors. The striatal cells cotransduced with AAV-TH/AAV AADC mixture synthesized dopamine more efficiently than the cells transduced with AAV-TH alone, when the intracellular l-dopa and dopamine were measured by HPLC (high-performance liquid chromatography).

To prepare a rodent model of PD, neurotoxic 6-hydroxydopamine (6-OHDA) was stereotaxically injected into the left medial forebrain bundle (MFB) to destroy the nigro-striatal pathway of dopaminergic neurons unilaterally. This MFB lesion causes the decrease of dopamine in the striatum like PD. The dopamine receptors of striatal cells in the lesioned side become hypersensitive to dopamine. After the intraperitoneal administration of dopamine agonist, apomorphine, unbalance of motor function is induced, and the rats show rotational behavior. The dopamine content in the striatum can be estimated by counting the rotation after apomorphine administration.

The AAV-TH and/or AAV-AADC vectors were then stereotaxically injected into the denervated striatum of 6-OHDA-lesioned rats. A behavioral test was conducted following the injection of AAV-TH/AAV-AADC mixture, AAV-TH alone, AAV-AADC alone, AAV-LacZ, or PBS, respectively. As a result, abnormal movement was significantly decreased in animals receiving AAV-TH alone or a combination of AAV-TH and AAV-AADC; animals receiving both vectors had better recovery than animals receiving only AAV-TH (Fan *et al.*, 1998b). To confirm that the synergistic effect on the behavior was caused by enhanced dopamine production *in vivo* with the combined use of TH and AADC genes, the striatal l-dopa and DOPAC (3,4-dihydroxyphenylacetic acid) levels were assayed by HPLC. DOPAC is one of the main dopamine metabolites, and is an indicator of dopamine production. In the lesioned rats injected with AAV-TH alone, a large amount of l-dopa was detected in the lesioned-side striatum, but the DOPAC level was low. On the other hand, when coinjected with AAV-TH and AAV-AADC vectors, a significantly larger amount of DOPAC was detected in the lesioned side striatum with low-level l-dopa, suggesting efficient conversion to dopamine. This finding also indicates that striatal AADC activity in 6-OHDA-lesioned rats is insufficient as in PD. Coexpression of TH and AADC in the same striatal cells was also confirmed with dual immuno-fluorescent staining. Long-term expression of transgenes can be expected when the genes are transferred into neurons using AAV vectors.

b) *In vivo* triple transduction using AAV-GCH, AAV-TH, and AAV-AADC

We are now interested in the therapeutic role of GCH, a critical enzyme for BH4 biosynthesis, because not only TH and AADC but also BH4 are considered to be insufficient in PD striatum (Mandel *et al.*, 1998; Szczypka *et al.*, 1999). A low level of endogenous BH4 does not yield sufficient TH activity. In addition, some investigators have reported that the more effective L-dopa production was observed in the 6-OHDA-lesioned rats by expression of TH and GCH (Bencsics *et al.*, 1996; Mandel *et al.*, 1998), although no effect on their behavior was found. A low AADC activity in the striatum may limit the conversion of l-dopa, although expression of TH and GCH in the 6-OHDA-lesioned striatum produced a higher level of l-dopa. We expect that the triple intrastriatal transduction with the TH, AADC, and GCH genes, which should cause the most efficient production of dopamine, would induce a better behavioral improvement than that with the combination of TH and GCH genes or TH and AADC genes in parkinsonian rats. In our unpublished observation (Shen *et al.*, manuscript in preparation), *in vitro* experiments showed that when 293 cells were cotransduced with fixed amounts of AAV-TH and AAV-AADC and an increasing amount of AAV-GCH, the intracellular dopamine level in the transduced cells was gradually elevated along with the increases of GCH activity and BH4 content. Furthermore, the triple transduction enhanced the BH4 and dopamine production in denervated striatum of parkinsonian rats and improved rotational behavior of the rats more efficiently, compared with the double transduction. This study would be valuable for developing more efficacious gene therapy strategies for PD. It is worthy of note that multiple genes can be delivered into neurons *in vivo* using separate AAV vectors. In addition, transgene expression and behavioral recovery persisted for at least 7 months after stereotaxic intrastriatal injection in parkinsonian rats.

4. PROTECTION OF DOPAMINERGIC NEURONS WITH AAV-GDNF

Although gene therapy may be effective to improve the clinical manifestations of PD, the treatments aimed at augmenting striatal dopamine production do not halt the progressive degeneration of nigral dopaminergic neurons. Therefore, another approach seems to be important to obtain the long-term therapeutic effects of gene therapy for PD. The identification of glial cell line-derived neurotrophic factor (GDNF), a potent neurotrophic factor for dopaminergic neurons, has attracted the attention of researchers as a promising therapeutic molecule for limiting the neuronal damage caused by PD (Unsicker *et al.*, 1996; Lin *et al.*, 1993). GDNF is known to promote the survival of cultured mesencephalic dopaminergic neurons. Although single or repeated injections of GDNF protein directly into the central nervous system were reported in some studies (Beck *et al.*, 1995; Sauer *et al.*, 1995; Tomac *et al.*, 1995; Gash *et al.*, 1996), continuous targeted delivery to specific neurons remains a challenge. Somatic gene therapy by local and stable gene transfer is a logical approach to meet this challenge. When Rat E14 mesencephalic cultures were transduced with an AAV vector containing the GDNF cDNA (Matsushita *et al.*, 1997) (AAV-GDNF), prevention of dopaminergic neuron death and prominent neurite outgrowth were observed *in vitro* (Fan *et al.*, 1998a).

Although the precise etiology of PD remains unknown, it is believed that the degeneration of nigrostriatal dopaminergic neurons relates to some neurotoxic factors. GDNF gene transfer may have therapeutic effects in patients with any type of PD by halting nigrostriatal degeneration. Promising data of GDNF gene therapy has already been reported using a 6 OHDA-lesioned rat model of PD (Bilang-Bleuel *et al.*, 1997; Choi-Lindberg *et al.*, 1997; Mandel *et al.*, 1997). The combination of dopamine-supplement gene therapy and GDNF gene therapy would be a logical approach for the treatment of PD.

5. FUTURE PERSPECTIVES

There are two major approaches to the treatment of PD; *i.e.* gene therapy and cell therapy. Cell-mediated gene therapy is a combination of both approaches, and is based on the amplification and genetic modification of cells *in vitro* prior to transplantation to the brain. Therefore, therapeutic genes can be introduced into the brain either by direct *in vivo* injection, or by *ex vivo* transduction. AAV vector would be the most appropriate vehicle for *in vivo* transduction. To determine if AAV vector-mediated gene transfer would be a feasible approach to the treatment of PD, primate model experiments are very important as a preclinical study. We are now generating a monkey model of PD using neurotoxin MPTP. The most critical difference between rats and monkeys is the size of striatum. To obtain positive results in monkeys, a sufficient number of striatal neurons have to be transduced (During *et al.*, 1998). Further investigation will be required to develop successful clinical gene therapy of PD using AAV vectors. In addition, safety issues of intrabrain injection of AAV vectors should be initially investigated in patients with life-threatening disease such as brain tumors, even if AAV vectors are believed to be non-pathogenic.

As for cell therapy, besides embryonic mesencephalic cells, other sources of catecholamine-producing cells, such as adrenal medulla, sympathetic ganglia, and carotid body, have also been considered potentially useful for cell replacement therapy of PD. Recently, it is reported that intrastriatal transplantation of carotid body cell aggregates in parkinsonian rats caused a dramatic behavioral recovery (Espejo *et al.*, 1998). The efficacy of carotid body autotransplants for the treatment of PD was then demonstrated in chronically MPTP-treated macaque monkeys (Luquin *et al.*, 1999). These studies suggest the feasibility of cell replacement therapy for PD. This line of research is now focusing on cell engineering that could make it possible to generate unlimited numbers of neurons suitable for transplantation in PD. Neural stem cells derived from the cerebellum of the mouse were shown to be induced to develop into dopaminergic neurons by the overexpression of the nuclear receptor Nurr1 and factors derived from local type 1 astrocytes (Wagner *et al.*, 1999). Neural progenitors may also be applied to cell-mediated therapy, since they are brain-derived and can be expanded and genetically modified *in vitro*. When transplanted to the brain, neural progenitors undergo glial and neuronal differentiation depending on local environment. A system for the regulated expression of a gene of therapeutic potential in human neural progenitors is also being developed (Corti *et al.*, 1999). In the future, dopaminergic neurons will be generated from human embryonic (pluripotent) stem cells for clinical use in cell replacement therapy of PD. Genetic manipulation will promote the conversion of human stem cells to dopaminergic neurons, and further augment the ability of dopamine production with an appropriate regulatory system.

At present, it is difficult to determine which strategy is superior as the treatment of PD, and therefore several approaches should be explored toward the realization of novel and efficient therapy of PD.

Acknowledgements

This study was conducted in collaboration with the group of Dr. Toshiharu Nagatsu (Institute for Comprehensive Medical Science, Fujita Health University, Aichi, Japan) and Avigen, Inc. (Alameda, California,

USA). This work was supported by a Grant-in-Aid for Scientific Research on Priority Areas from the Ministry of Education, Science, Sports and Culture of Japan, and a grant for Research on Human Genome and Gene Therapy from the Ministry of Health and Welfare of Japan.

6. REFERENCES

Beck, K.D., Valverde, J., Alexi, T., Poulsen, K, Moffat, B, Vandlen, R.A., Rosenthal, A. and Hefti, F. (1995). Mesencephalic dopaminergic neurons protected by GDNF from axotomy-induced degeneration in the adult brain. *Nature*, 373: 339-341.

Bencsics, C., Wachtel, S.R., Milstien, S., Hatakeyama, K., Becker, J.B. and Kang, U.J. (1996). Double transduction with GTP cyclohydrolase I and tyrosine hydroxylase is necessary for spontaneous synthesis of l-dopa by primary fibroblasts. *J. Neurosci.*, 16: 4449-4456.

Bilang-Bleuel, A., Revah, F., Colin, P., Locquet, I., Robert, J-J., Mallet, J. and Horellou, P. (1997). Intrastriatal injection of an adenoviral vector expressing glial-cell-line-derived neurotrophic factor prevents dopaminergic neuron degeneration and behavioral impairment in a rat model of Parkinson disease. *Proc. Natl. Acad. Sci. USA*, 94: 8818-8823.

Blömer, U., Naldini, L., Kafri, T., Trono, D., Verma, I.M. and Gage, F.H. (1997). Highly efficient and sustained gene transfer in adult neurons with a lentivirus vector. *J. Virol.*, 71: 6641-6649.

Choi-Lindberg, D.L., Lin, Q., Chang, Y-N., Chiang, Y.L., Hay, C.M., Mohajeri, H., Davidson, B.L. and Bohn, M.C. (1997). Dopaminergic neurons protected from degeneration by GDNF gene therapy. *Science*, 275: 838-841.

Corti, O., Sabaté, O., Horellou, P., Colin, P., Dumas, S., Buchet, D., Buc-Caron, M.H. and Mallet, J. (1999). A single adenovirus vector mediates doxycycline-controlled expression of tyrosine hydroxylase in brain grafts of human neural progenitors. *Nat. Biotechnol.*, 17: 349-354.

Du, B., Wu, P., Boldt-Houle, D.M., Terwilliger, E.F. (1996). Efficient transduction of human neurons with an adeno-associated virus vector. *Gene Ther.*, 3: 254-261.

During, M.J., Naegele, J.R., O'Malley, K.L., Geller, A.I. (1994). Long-term behavioral recovery in Parkinson's rats by an HSV vector expressing tyrosine hydroxylase. *Science*, 266: 1399-1403.

During, M.J., Samulski, R.J., Elsworth, J.D., Kaplitt, M.G., Leone, P., Xiao, X., Li, J., Freese, A., Taylor, J.R., Roth, R.H., Sladek, J.R., O'Malley, K.L. and Redmond, D.E. (1998). *In vivo* expression of therapeutic human genes for dopamine production in the caudates of MPTP-treated monkeys using an AAV vector. *Gene Ther.*, 5: 820-827.

Espejo, E.F., Montoro, R.J., Armengol, J.A. and Lopez-Barneo, J. (1998). Cellular and functional recovery of Parkinsonian rats after intrastriatal transplantation of carotid body cell aggregates. *Neuron*, 20: 197-206.

Fan, D., Ogawa, M., Ikeguchi, K., Fujimoto, K., Urabe, M., Kume, A., Nishizawa, M.,

Matsushita, N., Kiuchi, K., Ichinose, H., Nagatsu, T., Kurtzman, G.J., Nakano, I. and Ozawa, K. (1998a). Prevention of dopaminergic neuron death by adeno-associated virus vector-mediated GDNF gene transfer in rat mesencephalic cells in vitro. *Neurosci. Lett.*, 248: 61-64.

Fan, D., Ogawa, M., Fujimoto, K., Ikeguchi, K., Ogasawara, Y., Urabe, M., Nishizawa, M., Nakano, I., Yoshida, M., Nagatsu, I., Ichinose, H., Nagatsu, T., Kurtzman, G.J. and Ozawa, K. (1998b). Behavioral recovery in 6-hydroxydopamine-lesioned rats by cotransduction of striatum with tyrosine hydroxylase and aromatic l-amino acid decarboxylase genes using two separate adeno-associated virus vectors. *Hum. Gene Ther.*, 9: 2527-2535.

Fisher, K.J., Jooss, K., Alston, J., Yang, Y., Haecker, S.E., High, K., Pathak, R., Raper, S.E. and Wilson, J.M. (1997). Recombinant adeno-associated virus for muscle directed gene therapy. *Nat. Med.*, 3: 306 312.

Freed, C.R., Breeze, R.E., Rosenberg, N.L., Schneck, S.A., Kriek, E., Qi, J-X., Lone, T., Zhang, Y-B., Snyder, J.A., Wells, T.H., Ramig, L.O., Thompson, L., Mazziotta, J.C., Huang, S.C., Grafton, S.T., Brooks, D., Sawie, G., Schroter, G. and Ansari, A.A. (1992). Survival of implanted fetal dopamine cells and neurologic improvement 12 to 46 months after transplantation for Parkinson's disease. *N. Engl. J. Med.*, 327: 1549-1555.

Gash, D.M., Zhang, Z., Ovadia, A., Cass, W.A., Yi, A., Simmerman, L., Russell, D., Martin, D., Lapchak, P.A., Collins, F., Hoffer, B.J. and Gerhardt, D.A. (1996). Functional recovery in parkinsonian monkeys treated with GDNF. *Nature*, 380: 252-255.

Geller, A.I., During, M.J., Oh, Y.J., Freese, A. and O'Malley, K. (1995). An HSV-1 vector expressing tyrosine hydroxylase cause production and release of l-DOPA from cultured rat striatal cells. *J. Neurochem.*, 64: 487-496.

Grimm, D. and Kleinschmidt, J.A. (1999). Progress in adeno-associated virus type 2 vector production: promises and prospects for clinical use. *Hum. Gene Ther.*, 10: 2445-2450.

Herzog, R.W., Hagstrom, J.N., Kung, S.H., Tai, S.J., Wilson, J.M., Fisher, K.J. and High, K.A. (1997). Stable gene transfer and expression of human blood coagulation factor IX after intramuscular injection of recombinant adeno-associated virus. *Proc. Natl. Acad. Sci. USA*, 94: 5804-5809.

Herzog, R.W., Yang, E.Y., Couto, L.B., Hagstrom, J.N., Elwell, D., Fields, P.A., Burton, M., Bellinger, D.A., Read, M.S., Brinkhous, K.M., Podsakoff, G.M., Nichols, T.C., Kurtzman, G.J. and High, K. (1999). Long-term correction of canine hemophilia B by gene transfer of blood coagulation factor IX mediated by adeno-associated viral vector. *Nat. Med.*, 5: 56-63.

Ichinose, H., Kurosawa, Y., Titani, K., Fujita, K. and Nagatsu, T. (1989). Isolation and characterization of a cDNA clone encoding human aromatic l-amino acid decarboxylase. *Biochem. Biophys. Res. Commun.*, 164: 1024-1030.

Imaoka, T., Date, I., Ohmoto, T. and Nagatsu, T. (1998). Significant behavioral recovery in Parkinson's disease model by direct intracerebral gene transfer using continuous injection of a plasmid DNA-liposome complex. *Hum. Gene Ther.*, 9: 1093-1102.

Inoue, N. and Russell, D. (1998). Packaging cells based on inducible gene amplification for the production of adeno-associated virus vectors. *J. Virol.*, 72: 7024-7031.

Jeager, C.B., Ruggiero, D.A., Albert, V.R., Joh, T.H. and Reis, D.J. (1984). Immunocytochemical localization of aromatic l-amino acid decarboxylase. In: *Handbook of Chemical Neuroanatomy, Vol 2*, A. Björklund and T. Hokfelt (Eds.), Elsevier, Amsterdam, pp. 384-408.

Jiao, S., Gurevich,V. and Wolff, J.A. (1993). Long-term correction of rat model of Parkinson's disease by gene therapy. *Nature*, 362: 450-453.

Jiao, S., Hogan, K. and Wolff, J.A. (1994). Gene therapy for neuromuscular disorders. *Cur. Neurol.*, 14: 1-28.

Jinnah, H.A. and Friedmann, T. (1995). Gene therapy and the brain. *Bri. Med. Bull.*, 51: 138-148.

Kang, U.J., Park, D., Wessel, T., Baker, H. and Joh, T.H. (1992). DOPA-decarboxylation in the striata of rats with unilateral substantia nigra lesions. *Neurosci. Lett.*, 147: 53-57.

Kang, U.J., Fisher, L.J., Joh, T.H., O'Malley, K.L. and Gage, F.H. (1993). Regulation of dopamine production by genetically modified primary fibroblasts. *J. Neurosci.*, 13: 5203-5211.

Kaplitt, M.G., Leone, P., Samulski, R.J., Xiao, X., Pfaff, D.W., O'Malley, K.L. and During, M.J. (1994). Long-term gene expression and phenotypic correction using adeno-associated virus vectors in the mammalian brain. *Nat. Genet.*, 8: 148-154.

Kaneda, N., Kobayashi, K., Ichinose, H., Kishi, F., Nakazawa, A., Kurosawa, Y., Fujita, K. and Nagatsu, T. (1987). Isolation of a novel cDNA clone for human tyrosine hydroxylase: alternative RNA splicing produces four kinds of mRNA from a single gene. *Biochem. Biophys. Res. Commun.*, 146: 971-975.

Kessler, P.D., Podsakoff, G.M., Chen, X., McQuiston, S.A., Colosi, P.C., Matelis, L.A., Kurtzman, G.J. and Byrne, B.J. (1996). Gene delivery to skeletal muscle results in sustained expression and systemic delivery of a therapeutic protein. *Proc. Natl. Acad. Sci. USA*, 93: 14082-14087.

Kotin, R.M. (1994). Prospects for the use of adeno-associated virus as a vector for human gene therapy. *Hum. Gene Ther.*, 5: 793-801.

Le Gal La Salle, G., Robert, J.J., Bernard, S., Ridoux. V., Stratford-Perricaudet. L.D., Perricaudet, M. and Mallet, J. (1993). An adenovirus vector for gene transfer into neurons and glia in the brain. *Science*, 259: 988-990.

Lin, L-F.H., Doherty, D.H., Lile, J.D., Bektesh, S. and Collins, F. (1993). GDNF: a glial cell line-derived neurotrophic factor for midbrain dopaminergic neurons. *Science*, 260: 1130-1132.

Lo, W.D., Qu, G., Sferra, T.J., Clark, R., Chen, R. and Johnson, P.R. (1999). Adeno-associated virus mediated gene transfer to the brain: duration and modulation of expression. *Hum. Gene Ther.*, 10: 201-213.

Lundberg, C., Horellou, P., Mallet, J. and Björklund, A. (1996). Generation of DOPA-producing astrocytes by retroviral transduction of the human tyrosine hydroxylase gene: *in vitro* characterization and *in vivo* effects in the rat Parkinson model. *Exp. Neurol.*, 139: 39-53.

Luquin, M.R., Montoro, R.J., Guillen, J., Saldise, L., Insausti, R., Del Rio, J. and

Lopez-Barneo, J. (1999). Recovery of chronic parkinsonian monkeys by autotransplants of carotid body cell aggregates into putamen. *Neuron*, 22: 743-50.

Mandel, R.J., Spratt, S.K., Snyder, R.O. and Leff, S.E. (1997). Midbrain injection of recombinant adeno associated virus encoding rat glial cell line-derived neurotrophic factor protects nigral neurons in a progressive 6-hydroxydopamine-induced degeneration model of Parkinson's disease in rats. *Proc. Natl. Acad. Sci. USA*, 94: 14083-14088.

Mandel, R.J., Rendahl, K.G., Spratt, S.K., Snyder, R.O., Cohen, L.K. and Leff, S.E. (1998). Characterization of intrastriatal recombinant adeno-associated virus-mediated gene transfer of human tyrosine hydroxylase and human GTP-cyclohydrolase I in a rat model of Parkinson's disease. *J. Neurosci.*, 18: 4271-4284.

Martinez-Serrano, A., Lundberg, C. and Björklund, A. (1997). Use of conditionally immortalized neural progenitors for transplantation and gene transfer to the CNS. In: *Isolation, Characterization and Utilization of CNS Stem Cells*, F. Gage and Y. Christen (Eds.), Springer-Verlag, Berlin, pp 151-169.

Matsushita, N., Fujita, Y., Tanaka, N., Nagatsu, T. and Kiuchi, K. (1997). Cloning and structural organization of the gene encoding the mouse glial cell-line-derived neurotrophic factor, GDNF. *Gene*, 203: 149-157.

Matsushita, T., Elliger, S., Elliger, C., Podsakoff, G., Villarreal, L., Kurtzman, G.J., Iwaki, Y. and Colosi, P. (1998). Adeno-associated virus vectors can be efficiently produced without helper virus. *Gene Ther.*, 5: 938-945.

Muzyczka, N. (1992). Use of adeno-associated virus as a general transduction vector for mammalian cells. *Curr. Top. Microbiol. Immunol.*, 158: 97-129.

Nagatsu, T. (1992). Molecular biology of dopamine systems. In: *Controversies in the Treatment of Parkinson's Disease*, U.K. Rinne and N. Yanagisawa (Eds.), PMSI, Tokyo, pp 15-26.

Nagatsu, T. and Ichinose, H. (1997). GTP cyclohydrolase I gene, dystonia, juvenile parkinsonism, and Parkinson's disease. *J. Neural Transm.*, Suppl. 49: 203-209.

Naldini, L., Blömer, U., Gallay, P., Ory, D., Mulligan, R., Gage, F.H., Verma, I.M. and Trono, D. (1996). *In vivo* gene delivery and stable transduction of nondividing cells by a lentiviral vector. *Science*, 272: 263-267.

Ogasawara, Y., Urabe, M. and Ozawa, K. (1998). The use of heterologous promoters for adeno associated virus (AAV) protein expression in AAV vector production. *Microbiol. Immunol.*, 42: 177-185.

Ogasawara, Y., Urabe, M., Kogure, K., Kume, A., Colosi, P., Kurtzman, G.J. and Ozawa, K. (1999). Efficient production of adeno-associated virus vectors using split-type helper plasmids. *Jpn. J. Cancer Res.*, 90: 476-483.

Russell, D.W. and Kay, M.A. (1999). Adeno-associated virus vectors and hematology. *Blood*, 94: 864 874.

Sauer, H., Rosenblad, C. and Björklund, A. (1995). Glial cell line-derived neurotrophic factor but not transforming growth factor β3 prevents delayed degeneration of nigral dopaminergic neurons following striatal 6-hydroxydopamine lesion. *Proc. Natl. Acad. Sci. USA*, 92: 8935-8939.

Snyder, R.O. (1999). Adeno-associated virus-mediated gene delivery. *J. Gene Med.*, 1: 166-175.

Snyder, R.O., Miao, C.H., Patijn, G.A., Spratt, S.K., Danos, O., Nagy, D., Gown, A.M., Winther, B., Meuse, L., Cohen, L.K., Thompson, A.R. and Kay, M.A. (1997). Persistent and therapeutic concentrations of human factor IX in mice after hepatic gene transfer of recombinant AAV vectors. *Nat. Genet.*, 16: 270-276.

Snyder, R.O., Miao, C.H., Meuse, L., Tubb, J., Donahue, B.A., Lin, H-F., Stafford, D.W., Patel, S., Thompson, A.R., Nichols, T., Read, M.S., Bellinger, D.A., Brinkhous, K.M. and Kay, M.A. (1999). Correction of hemophilia B in canine and murine models using recombinant adeno-associated viral vectors. *Nat. Med.*, 5: 64-70.

Spencer, D.D., Robbins, R.J., Naftolin, F., Marek, K.L., Vollmer, T., Leranth, C., Roth, R.H., Price, L.H., Gjedde, A., Bunney, B.S., Sass, K.J., Elsworth, J.D., Kier, E.L., Makuch, R., Horrer, P.B. and Redmone, D.E. (1992). Unilateral transplantation of human fetal mesencephalic tissue into the caudate nucleus of patients with Parkinson's disease. *N. Engl. J. Med.*, 327: 1541-1548.

Szczypka, M.S., Mandel, R.J., Donahue, B.A., Snyder, R.O., Leff, S.E. and Palmiter, R.D. (1999). Viral gene delivery selectively restores feeding and prevents lethality of dopamine-deficient mice. *Neuron*, 22: 167-178.

Tashiro, T., Kaneko, T., Sugimoto, T., Nagatsu, I., Kikuchi, H. and Mizuno, N. (1989). Striatal neurons with aromatic l-amino acid decarboxylase-like immunoreactivity in the rat. *Neurosci. Lett.*, 100: 29-34.

Tomac, A., Lindqvist, E., Lin, L.-F.H., Ogren, S., Young, D., Hoffer, B.J. and Olson, L. (1995). Protection and repair of the nigrostriatal dopaminergic system by GDNF in vivo. *Nature*, 373: 335-339.

Unsicker, K. (1996). GDNF: a cytokine at the interface of TGF-β and neurotrophins. *Cell Tissue Res.*, 286: 175-178.

Verma, I.M. and Somia, N. (1997). Gene therapy - promises, problems and prospects. *Nature*, 389: 239-242.

Wachtel, S.R., Bencsics, C. and Kang, U.J. (1997). Role of aromatic l-amino acid decarboxylase for dopamine replacement by genetically modified fibroblasts in a rat model of Parkinson's disease. *J. Neurochem.*, 69: 2055-2063.

Wu, P., Phillips, M.I., Bui, J. and Terwilliger, E.R. (1998). Adeno-associated virus vector-mediated transgene integration into neurons and other nondividing cell targets. *J. Virol.*, 72: 5919-5926.

Xiao, X., Li, J. and Samulski, R.J. (1998). Production of high-titer recombinant adeno-associated virus vectors in the absence of helper adenovirus. *J. Virol.*, 72: 2224-2232.

Zufferey, R., Nagy, D., Mandel, R.J., Naldini, L. and Trono, D. (1997). Multiply attenuated lentiviral vector achieves efficient gene delivery in vivo. *Nat. Biotechnol.*, 15: 871-875.

6. SUMMARY — Parkinson's disease (PD) is characterized by the progressive loss of the dopaminergic neurons in the substantia nigra and a severe decrease in dopamine in the striatum. A promising approach to the gene therapy of PD is intrastriatal expression of enzymes required for dopamine production. Dopamine-synthesizing key enzymes are tyrosine hydroxylase (TH) and aromatic l-amino acid decarboxylase (AADC). In addition, since TH needs tetrahydrobiopterin (BH4) as a cofactor, GTP cyclohydrolase I (GCH), a critical enzyme for BH4 biosynthesis, is considered to regulate dopamine production indirectly. The most appropriate gene-delivery vehicle for neurons is an adeno-associated virus (AAV) vector, which has several unique characteristics, including lack of any associated disease with wild-type virus, the ability to transduce non dividing cells, and prolonged expression of the transgene. When TH and AADC genes were introduced into the striatum of the lesioned side using separate AAV vectors in parkinsonian rats (6-OHDA-lesioned rats), behavioral recovery was observed and the improvement was better than that with TH gene alone. Moreover, our preliminary data showed that triple transduction with AAV-TH, AAV-AADC, and AAV-GCH caused the most striking recovery. This finding is important, because not only TH and AADC but also BH4 are insufficient in PD striatum. Another strategy for gene therapy of PD is the protection of dopaminergic neurons in the substantia nigra using an AAV vector containing a glial cell line-derived neurotrophic factor (GDNF) gene. Combination of dopamine-supplement gene therapy and GDNF gene therapy would be a logical approach to the treatment of PD. On the other hand, several novel approaches to cell replacement therapy of PD are also explored using neural stem/progenitor cells or human embryonic stem cells, which would be amplified and genetically manipulated to generate dopaminergic neurons prior to transplantation. Further investigation would be needed to determine the most feasible approach to the novel and effective treatment of PD.

Key Words: AAV vector; gene therapy; Parkinson's disease; tyrosine hydroxylase (TH); aromatic l-amino acid decarboxylase (AADC); GTP cyclohydrolase I (GCH); glial cell line derived neurotrophic factor (GDNF)

Progress in Gene Therapy: Basic and Clinical Frontiers, pp. 307-320
R. Bertolotti *et al.* (Eds)

Bystander effect in suicide gene therapy of cancers

Marc Mesnil* and Hiroshi Yamasaki*

Unit of Multistage Carcinogenesis, International Agency for Research on Cancer, F-69372 Lyon cedex 08, France

Table of Contents

1. THE HSV-TK/GANCICLOVIR CYTOTOXICITY

The thymidine kinase (tk) protein encoded by the HSV-tk gene from the Herpes simplex virus (HSV) has a high affinity for a nucleoside used as the prodrug, an acyclic analogue of deoxyguanosine, ganciclovir [GCV or 9-(1,3-dihydroxy-2-propoxy-methyl)guanine], and phosphorylates it to a monophosphate form (GCV-MP). GCV-MP is further phosphorylated by

* Corresponding authors. E-mail: Marc.Mesnil@iarc.fr and Hiroshi.Yamasaki@iarc.fr

cellular kinases into its ultimate derivative, GCV-triphosphate (GCV-TP) which inhibits the DNA polymerase α and also prevents the elongation of DNA strands, leading the cells to an apoptotic death (Cheng *et al.*, 1983; Mar *et al.*, 1985; Smee *et al.*, 1985). Such a DNA-targeted gene therapy is particularly addressed against the replication of cancer cells. However, recent studies suggest that the high toxicity of GCV, compared to other substrates of HSV-tk as acyclovir or 1-β-D-arabinofuranosylthymine, is also due to the incorporation through the genome, and not only into early replicating regions, of GCV-MP (Rubsam *et al.*, 1998). Moreover, the inhibition of mitochondrial DNA polymerase may also explain a certain toxicity of phosphorylated GCV which has been described on quiescent cells (Wallace *et al.*, 1996; van der Erb *et al.*, 1998).

The cytotoxic effect of GCV on cells expressing HSV-tk was demonstrated more than a decade ago (Balzarini *et al.*, 1985; Nishiyama and Rapp, 1979) and its application to cancer therapy was first suggested by Moolten (1986) who transduced sarcoma cells with retroviral vectors carrying the HSV-tk gene and observed *in vivo* the prevention of the tumor growth by GCV treatment. However, and more importantly, this strategy also worked in a retro-active way: the resorption of induced tumors by introducing *in situ* the HSV-tk gene into the tumor cells (through intra-tumoral injections of retroviral vectors or cells producing the vectors), and treating them with GCV by intraperitoneal injections in the animals (Moolten and Wells, 1990).

2. THE BYSTANDER EFFECT

Through some experimental work, it became clear that this strategy lead to the regression of tumors even if the HSV-tk gene was introduced into a few tumor cells. Actually, the resorption of the tumors could be observed when only a small fraction (10%) of the tumor cells contained the suicide gene (Freeman *et al.*, 1993). This was initially observed in rats following the injection of 9L rat glioma cells into their brains. Five days later, cells producing the retroviral vector carrying the HSV-tk gene were injected at the same place in order to make the tumor cells transduced *in situ*. The subsequent intraperitoneal administrations of GCV cured most of the tumors (Culver *et al.*, 1992). This implies that cells that do not express the HSV-tk gene became sensitive to the GCV treatment. In other words, it implies that, in the presence of GCV, cells that produce HSV-tk are able to cause the death of neighboring cells that do not themselves produce the enzyme. This phenomenon has been termed the "bystander effect" and attracted considerable interest since, at a therapeutical point of view, it may complement the low efficiency of the gene transfer into the tumors.

3. THE BYSTANDER EFFECT HYPOTHESIS

Several mechanisms have been proposed to explain such a phenomenon. In particular, the involvement of immune responses (Freeman *et al.*, 1997) and blood vessel destruction have been suggested (Ram *et al.*, 1994). However, these events are probably a part of the phenomenon since the bystander effect mediated by HSV-tk/GCV strategy has been also observed *in vitro* where they do not occur. In such conditions, this phenomenon is dependent on the cell density. The emission of apoptotic bodies may be a possibility to transfer the toxic GCV metabolites from cells expressing the HSV-tk gene (tk^+ cells) to be phagocytized by neighbouring tk^- cells (Colombo *et al.*, 1995; Samejima and Meruelo, 1995). However, the necessity of a close contact between the two cell types supported the hypothesis that metabolic cooperation was mostly involved (Moolten, 1986). This was reinforced by the fact that the bystander effect cannot occur if the two cell types (tk^+ and tk^- cells) were separated by a permeable membrane (Freeman *et al.*, 1993) or through conditioned medium from tk^+ cells treated with GCV (Bi *et al.*, 1993).

The direct transfer of radiolabeled GCV from tk^+ cells to tk^- cells which were in close contact suggested the possibility of a transfer mediated by gap junctions. Indeed, such observations were very similar to the phenomenon already observed 30 years ago and described as the "metabolic cooperation". Cells deficient in hypoxanthine guanine phosphoribosyl transferase ($HPRT^-$ cells) could incorporate hypoxanthine only if they were in close contact with normal cells which are able to metabolize it (Hooper and Suback-Sharpe, 1981). The transfer of hypoxanthine from $HPRT^+$ cells to $HPRT^-$ cells was proven to be mediated by gap junctions (Cox *et al.*, 1970; Gilula *et al.*, 1972). Since hypoxanthine is a toxic metabolite, it became possible to use such a metabolic cooperation assay to estimate the cell-cell communication capacity by the extent of toxicity in the coculture. The intercellular transfer of this toxic metabolite was even called the "kiss of death".

4. THE GAP JUNCTIONS

Most of the cells in our body are able to collaborate metabolically by exchanging small fundamental and hydrophilic molecules (sugars, amino-acids, ions, nucleotides) from cytoplasm to cytoplasm through gap junctions. For instance, the gap junctional intercellular communication (GJIC) transfer of cAMP and also second messengers such as calcium and inositol triphosphate has been reported (Saez *et al.*, 1989). However, it has long been known that gap junctions can mediate cell-to-cell transfer of phosphorylated nucleotides (Subak-Sharpe *et al.*, 1969). This suggests that phosphorylated GCV may pass through the gap-junctional intercellular channels, an hypothesis even reinforced by the fact that such molecules have the required size (< 1,000Da) and properties to do so (Simpson *et al.*, 1977).

The gap-junctional communication is mediated by intercellular channels gathered into gap junction plaques which are made up of juxtaposed transmembrane hemichannels (connexons) provided by the adjacent cells. Each connexon is composed of six connexin protein subunits (Cx) forming a central pore through which the messengers are transmitted (for review see: Kumar and Gilula, 1996). The connexins are encoded by genes from a family containing at least thirteen members in mammals sharing a high sequence homology. The length of the carboxy terminal part of the connexins is variable, a characteristic which permits to identify them according to their molecular weight in kilodaltons, *i.e.* connexin 26 (Cx26), connexin 43 (Cx43), *etc*. All of these connexins are expressed differently in various tissues with a temporal specificity during development suggesting that they play an important role in cell differentiation and/or cell growth control (Bruzzone *et al.*, 1996; Yamasaki and Naus, 1996).

5. GAP JUNCTIONS AND THE BYSTANDER EFFECT

a) *In vitro* model studies of bystander effect

The first *in vitro* study to exploit GJIC ability to spread cancer therapeutic effect was conducted in our laboratory a decade ago (Yamasaki and Katoh, 1988). The study has demonstrated that GJIC diffusion of cytotoxic agents can be used to improve chemotherapy efficiency. We later demonstrated that gap junctions can also mediate *in vitro* a bystander effect of cancer gene therapy by using HeLa cells which exhibit a very poor gap-junctional intercellular communication capacity (GJIC). When HeLa cells were transfected with an expression vector carrying the HSV-tk gene (tk^+ cells), they became highly sensitive to GCV. If these cells were mixed with untransfected HeLa cells (1:1 ratio), GCV could not eliminate whole cell culture. As shown by Northern analysis, the tk^+ cells were eliminated and the surviving tk^- cells grew up to reconstitute a confluent layer of cells. However, the picture was quite different if we used HeLa cells which are able to communicate through gap junctions after transfection of a gene coding for a common connexin, Cx43. In those conditions, not only the tk^+ cells, but also the tk^- cells, were killed by GCV even if the culture contained only 10% tk^+ cells. This bystander effect, the transfer of the GCV toxicity to the resting 90% of the culture, was mediated by gap junctions since it was inhibited by either the lack of contact between the two cell types or by treating the cells with a long-term inhibitor of GJIC (Mesnil *et al.*, 1996). These data are supported by others reporting that GJIC is predictive of or correlated to the extent of the bystander effect *in vitro* whatever the origin of the cancer cell lines (Fick *et al.*, 1995; Elshami *et al.*, 1996, Yang *et al.*, 1998). Similar results were obtained with cells expressing other types of connexins such as Cx26 in HeLa cells (Mesnil *et al.*, 1997) or Cx37 in a mouse neuroblastoma cell line (Vrionis *et al.*, 1997), confirming that GJIC is involved in the bystander effect we observe *in vitro*. Our data as well as those from others did not prove that phosphorylated GCV passed from tk^+ cells to tk^- cells. However, a report has shown that cell lines exhibiting the bystander effect *in vitro* were able to transfer radiolabeled GCV between them (Ishii-Morita *et al.*, 1997). Such an intercellular transfer is probably mediated by gap junctions since most of the bystander effect we observed *in vitro* was enhanced by an efficient GJIC capacity.

b) Bystander effect *in vivo*

The bystander effect mediated by gap junctions that we reported *in vitro* has been also observed *in vivo*. The major conclusion from these *in vivo* studies was the same as *in vitro*, *i.e.* GJIC, as a mediator of the bystander effect, can compensate for a low number of tk^+ cells in propagating killing effect of GCV in an induced tumor. For instance, transfection of the Cx37 gene reduced up to 60% the weight of tumors induced by mouse neuroblastoma cells when 50% of the cells expressed the tk gene (Vrionis *et al.*, 1997). The Cx43 gene expression in rat C6 glioma cells induced a similar phenomenon in nude mice even when only 5% of the injected cells were tk^+ cells (Dilber *et al.*, 1997). However, the tumor suppressive role of endogenous Cx43, which is expressed in parental nontransfected C6 glioma cells, may have interferred with the observed bystander effect (Naus *et al.*, 1992). Moreover, in the above studies, the mice were killed soon after the end of GCV treatment, preventing estimation of the duration of the bystander effect mediated by connexins in down-regulating or preventing the growth of the induced tumors.

Since the immune system is considered to be involved significantly in the *in vivo* bystander effect, we used nude mice for estimating better the involvement of gap junctions in this phenomenon. In such a study, we found that the growth of 10^6 injected HeLa cells, which are highly tumorigenic, could be prevented by 10% tk^+ cells, only if they do express Cx43. These results suggest that Cx43 expression can efficiently induce the bystander effect, inhibiting the growth of tumors even when 90% of the injected cells were resistant to GCV, reproducing *in vivo* what we observed *in vitro*. Even if GCV is administered when the tumors are already palpable, their growth could be reduced by 66% or 77% (for mixtures containing 10 or 50% tk^+ cells, respectively) than that of tumors induced by communicating but tk^- cells (Duflot-Dancer *et al.*, 1998). The gap-junction dependent bystander effect was so efficient that some of the mice were still alive more than 2 months after the GCV treatment. In contrast, GCV decreased the size of the tumors in proportion to the percentage of tk^+ cells present in the original mixtures of HeLa cells, if they were $Cx43^-$. We think that these results are particularly significant, because we used immunodeficient mice. Indeed, it is possible that, for the same cancer cell line, almost total tumor regression is accomplished in immunocompetent mice contrary to immuno-compromised athymic nude mice (Gagandeep *et al.*, 1996). Therefore, in therapeutical approaches, GJIC and immunological responses may have cumulative implications leading to even increased bystander effect that could be split into local (through gap junctions) and distant (through the immune system) bystander effects (Wei *et al.*, 1998).

6. HOW TO INCREASE GJIC?

All the data we presented so far have shown that GJIC mediates a high bystander effect. Such a capacity to spread intracellularly produced toxic drug from cell to cell within a tumor mass helps to increase the efficiency of a suicide gene transfered in the tumor and makes feasible the use of GJIC for therapy of localized cancers (Ishii-Morita *et al.*, 1997). Unfortunately, most of the cancer cell lines we have tested exhibit a low GJIC capacity to such an extent that we considered it as a characteristic of cancer cells (Yamasaki and

Naus, 1996). The lack of GJIC can decrease the bystander effect and in turn an efficient eradication of the tumors even in immunocompetent animals may be hampered. In the few cases we could test, we indeed observed that the cancer cells which did not exhibit a subsequent bystander effect *in vivo* were GJIC deficient *in vitro*. This was the case for GJIC-deficient mammary tumor cells obtained from *neu*-transgenic mice; the transduction of the HSV-tk gene into 10% of the tumor cells did not produce a sufficient bystander effect to prevent the eradication of the tumors by GCV (Sacco *et al.*, 1996). A similar observation was made for colon cancer cells injected intra-peritoneally in rats. The bystander effect was so low, due to their poor GJIC capacity, that 75% of the injected cell population had to contain the transduced HSV-tk gene to see some curative effect by GCV treatment (Coll *et al.*, 1997). The low expression of Cx43 was also related to the limited by-stander effect observed *in vivo* for human medulloblastoma cells (Roselen *et al.*, 1998).

As a crucial parameter for the HSV-tk gene therapy, the analysis of GJIC in tumors would be predictive for an efficient bystander effect (Yang *et al.*, 1998). The *ex vivo* estimation of GJIC would not be easy to establish except for soft tissues such as liver (Krutovskikh and Yamasaki, 1995). Since the level of expression of connexins is not related to the GJIC ability, an indirect way for screening tumors susceptible to be treated by HSV-tk /GCV would be through the estimation of the correct immunolocalization of connexins at the cell-to-cell borders. Indeed, recently it was shown that a poor bystander effect was related, not to the level of expression of the connexins, but to their localization in the cytoplasm of the cells (McMasters *et al.*, 1998).

Our data suggest that GJIC is low not only in cancer cell lines but also in human tumors (Krutovskikh *et al.*, 1994). Therefore, an alternative way would be to increase *in situ* the GJIC capacity by specific chemical treatments. Some chemicals are known to have such significant effect *in vitro* such as cAMP (Mehta *et al.*, 1992), retinoic acids (Hossain and

Bertram, 1994), β-caroten. This effect may be cell-type specific, as for instance carotenoids increasing the dye-coupling of human dermal fibroblasts but not of human keratinocytes (Zhang *et al.*, 1995). It may even be connexin-type specific as, for example, glucocorticoids (dexamethasone and hydrocortisone) enhancing the GJIC capacity of primary rat hepatocytes by increasing the expression of Cx32 but not Cx43 (Ren *et al.*, 1994). Indeed, some data suggest that the artificial increase of GJIC (or connexin expression) by chemicals may lead to an enhanced bystander effect. This was observed both *in vitro* and *in vivo* by treating tumor cell lines with retinoids (Park *et al.*, 1997). Doubling the GJIC of Cx43-HeLa cells by cAMP also resulted in doubling the bystander effect (unpublished data).

However, more *in situ* studies have still to be performed on the specificity of chemicals to enhance GJIC before being applied to the suicide-gene therapy. Therefore, some research groups prefered to increase the GJIC capacity by transfering a connexin gene into the cancer cells as we have originally done (Mesnil *et al.*, 1996 and 1997). Some interesting results were obtained *in vitro* by treating hepatocellular cells with liposomes carrying either the HSV-tk or the Cx43 genes (Ghoumari *et al.*, 1998). Unfortunately, the *in vivo* application of this strategy was still limited by the gene transfer efficacy (Cirenei *et al.*, 1998).

7. CONCLUSIONS

GJIC modulates the extent of the bystander effect both *in vitro* and *in vivo*. The HSV-tk/GCV anti-cancer gene therapy may be more adapted for tumors in which the cells do have a basal level of GJIC. It is therefore crucial to establish whether some particular types of human tumors do exhibit such a capacity as that they become putative and immediate targets for such a strategy. We believe that this kind of therapy could be extended, in a next future, to GJIC-deficient tumors by enhancing artificially the *in situ* communication capacity of the tumor cells through appropriate chemical treatments or connexin gene transfer.

Acknowledgements

This work was supported by the National Cancer Institute of USA (Grant R01-CA-40534), the Association for International Cancer Research (Grant No. 96-27) and the Association pour la Recherche sur le Cancer.

8. REFERENCES

Balzarini, J., de Clercq, E., Verbruggen, A., Ayusawa, D. and Seno, T. (1985). Highly cytostatic activity of (E)-5-(2-bromovinyl)-2'-deoxyuridine derivatives for murine mammary carcinoma (FM3A) cells transformed with the herpes simplex virus type 1 thymidine kinase gene. *Mol. Pharmacol.*, 28: 581-587.

Bi, W.L., Parysek, L.M., Warnick, R. and Stambrook, P.J. (1993). In vitro evidence that metabolic cooperation is responsible for the bystander effect observed with HSV-tk retroviral gene therapy. *Hum. Gene Ther.*, 4: 725-731.

Bruzzone, R., White, T.W. and Paul, D.L. (1996). Connections with connexins: the molecular basis of direct intercellular signalling. *Eur. J. Biochem.*, 238: 1-27.

Cheng, Y., Grill, S.P., Dutschman, G.E., Nakayama, K. and Bastow, K.F. (1983). Metabolism of 9-(1,3-dihydroxy-2-propoxymethyl)guanine, a new anti-herpes virus compound, in Herpes simplex virus-infected cells. *J. Biol. Chem.*, 258: 12460-12464.

Cirenei, N., Colombo, B.M., Mesnil, M., Benedetti, S., Yamasaki, H. and Finocchiaro, G. (1998). In vitro and in vivo effects of retrovirus-mediated transfer of the connexin 43 gene in malignant gliomas: consequences for HSVtk/GCV anticancer gene therapy. *Gene Ther.*, 5: 1221-1226.

Coll, J-L., Mesnil, M., Lefebvre, M.F., Lancon, A. and Favrot, M.C. (1997). Long-term survival of immunocompetent rats with intraperitoneal colon carcinoma tumors using herpes simplex thymidine kinase/ganciclovir and IL-2 treatments. *Gene Ther.*, 4: 1160-1166.

Colombo, B.M., Benedetti, S., Ottolenghi, S., Poli, G., Pollo, B., Mora, M. and Finocchiaro, G. (1995). The "bystander effect": Association of U-87 cell death with ganciclovir-mediated apoptosis of nearby cells and lack of effect in athymic mice. *Hum. Gene Ther.*, 6: 763-772.

Cox, R.P., Krauss, M.R., Balis, M.E. and Dancis, J. (1970). Evidence for transfer of enzyme product as the basis of metabolic cooperation between tissue culture fibroblasts of Lessch-Nyhan disease and normal cells. *Proc. Natl. Acad. Sci. USA*, 67: 1573-1579

Culver, K.W., Ram, Z., Wallbridge, S., Ishii, H., Oldfield, E.H. and Blaese, R.M. (1992). In vivo gene transfer with retroviral vector-producer cells for treatment of experimental brain tumors. *Science*, 256: 1550-1552.

Dilber, M.S., Abedi, M.R., Christensson, B., Bjorkstrand, B., Kidder, G.M., Naus, C.C.G., Gahrton, G. and Smith, C.I. (1997). Gap junctions promote the bystander effect of herpes simplex virus thymidine kinase in vivo. *Cancer Res.*, 57: 1523-1528.

Duflot-Dancer, A., Piccoli, C., Rolland, A., Yamasaki, H. and Mesnil, M. (1998). Long-term connexin-mediated bystander effect in highly tumorigenic human cells in vivo in herpes simplex virus thymidine kinase/ganciclovir gene therapy. *Gene Ther.*, 5: 1372-1378.

Elshami, A.A., Albeda, S.M. Kaiser, L.R., Amin, K., Fishman, G., Spray, D.C., Kucharczuk, J.C., Zhang, H.B. and Saavedra, A., (1996). Gap junctions play a role in the "bystander effect" of the herpes simplex virus thymidine kinase/ganciclovir system in vitro. *Gene Ther.*, 3: 85-92.

Fick, J., Barker, II F., Dazin, P., Westphale, E.M., Beyer, E.C. and Israel, M.A. (1995). The extent of heterocellular communication mediated by gap junctions is predictive of bystander tumor cytotoxicity in vitro. *Proc. Natl. Acad. Sci. USA,* 92: 11071-11075.

Freeman, S.M., Abboud, C.N., Whartenby, K.A., Packman, C.H., Koeplin, D.S., Moolten, F.L. and Abraham, G.N. (1993). The bystander effect: tumor regression when a fraction of the tumor mass is genetically modified. *Cancer Res.,* 53: 5274-5283.

Freeman, S.M., Ramesh, R. and Marrogi, A.J. (1997). Immune system in suicide-gene therapy. *Lancet*, 349: 2-3.

Gagandeep, S., Brew, R., Green, B., Christmas, S.E., Klatzmann, D., Poston, G.J. and Kinsella, A.R. (1996). Prodrug-activated gene therapy: involvement of an immunological component in the "bystander effect". *Cancer Gene Ther.*, 3: 83-88.

Ghoumari, A.M., Mouawad, R., Zerrouqi, A., Nizard, C., Provost, N., Khayat, D., Naus, C.C.G. and Soubrane, C. (1998). Actions of HSVtk and connexin43 gene delivery on gap junctional communication and drug sensitization in hepatocellular carcinoma. *Gene Ther.*, 5: 1114-1121.

Gilula, N.B., Reeves, D.R. and Steinbach, A. (1972). Metabolic coupling, ionic coupling and cell contacts. *Nature*, 235: 262-265.

Hooper, M.L. and Suback-Sharpe, J.H. (1981). Metabolic cooperation between cells. *Int. Rev. Cytol.,* 69: 45-104.

Hossain, M.Z. and Bertram, J.S. (1994). Retinoids suppress proliferation, induce cell spreading, and up-regulate connexin43 expression only in postconfluent 10T1/2 cells: implications for the role of gap junctional communication. *Cell Growth Differ.*, 5: 1253-1261.

Ishii-Morita, H., Agbaria, R., Mullen, C.A., Hirano, H., Koeplin, D.A., Ram, Z., Oldfield, E.H., Johns, D.G. and Blaese, R.M. (1997). Mechanism of "bystander effect" killing in the herpes simplex thymidine kinase gene therapy model of cancer treatment. *Gene Ther.*, 4: 244-251.

Krutovskikh, V.A. and Yamasaki, H. (1995). Ex-vivo dye transfer assay as an approach to study gap junctional intercellular communication disorders in hepatocarcinogenesis. *Prog. Cell Res.*, 4: 93-97.

Kumar, N.M. and Gilula, N.B. (1996). The gap junction communication channel. *Cell,* 84: 381-388.

Mar, E.C., Chiou, J.F., Cheng, Y.C. and Huang, E.S. (1985). Inhibition of cellular DNA polymerase α and human cytomegalovirus-induced DNA polymerase by the triphosphates of 9-(2-hydroxyethoxymethyl) guanine and 9-(1,3-dihydroxy-2-propoxymethyl) guanine. *J. Virol.,* 53: 776-780.

McMasters, R.A., Saylors, R.L., R.L., Jones, K.E., Hendrix, M.E., Moyer, M.P. and Drake, R.R. (1998). Lack of bystander killing in herpes simplex virus thymidine kinase-transduced colon cell lines due to deficient connexin43 gap junction formation. *Hum. Gene Ther.*, 9: 2253-2261.

Mehta, P.P., Yamamoto, M. and Rose, B. (1992). Transcription of the gene for the gap junctional channel protein connexin 43 and expression of functional cell-to-cell channels are regulated by cAMP. *Mol. Biol. Cell*, 3: 839-850.

Mesnil, M., Piccoli, C., Tiraby, G., Willecke, K. and Yamasaki, H. (1996). Bystander killing of cancer cells by herpes simplex virus thymidine kinase gene is mediated by connexins. *Proc. Natl. Acad. Sci. USA*, 93: 1831-1835.

Mesnil, M., Piccoli, C. and Yamasaki, H. (1997). A tumor suppressor gene, Cx26, also mediates the bystander effect in HeLa cells. *Cancer Res.*, 57: 2929-2932.

Moolten, F.L. (1986). Tumor chemosensitivity conferred by inserted herpes thymidine kinase genes: paradigm for a prospective cancer control strategy. *Cancer Res.*, 46: 5276-5281.

Moolten, F.L. and Wells, J.M. (1990). Curability of tumors bearing herpes thymidine kinase genes transferred by retroviral vectors. *J. Natl Cancer Inst.*, 82: 297-300.

Naus, C.C.G., Elisevich, K., Zhu, D., Belliveau, D.J. and Del Maestro, R.F. (1992). In vivo growth of C6 glioma cells transfected with connexin 43 cDNA. *Cancer Res.*, 52: 4208-4213.

Nishiyama, Y. and Rapp, F. (1979). Anticellular effects of 9-(2-hydroxy-ethoxymethyl) guanine against herpes simplex virus-transformed cells. *J. Gen. Virol.*, 45: 227-230.

Park, J.Y., Elshami, A.A., Amin, K., Rizk, N., Kaiser, L.R. and Albelda, S.M. (1997). Retinoids augment the bystander effect in vitro and in vivo in herpes simplex virus thymidine kinase/ganciclovir-mediated gene therapy. *Gene Ther.*, 4: 909-917.

Ram, Z., Culver, K., Walbridge, S., Blaese, R. and Oldfield, E. (1993). In situ retroviral-mediated gene transfer for the treatment of brain tumors in rats. *Cancer Res.*, 53: 83-88

Ram, Z., Walbridge, S., Shawker, T., Culver, K.W., Blaese, R.M. and Oldfield, E.H. (1994). The effect of thymidine kinase transduction and ganciclovir therapy on tumor vasculature and growth of 9L gliomas in rats. *J. Neurosurg.*, 81: 256-260.

Ren, P., de Feijter, A.W., Paul, D.L. and Ruch, R.J. (1994). Enhancement of liver cell gap junction protein expression by glucocorticoids. *Carcinogenesis*, 15: 1807-1813.

Roselen, A., Frascella, E., di Francesco, C., Todesco, A., Petrone, M., Mehtali, M., Zacchello, F., Zanesco, L. and Scarpa, M. (1998). *In vitro* and *in vivo* antitumor effects of retrovirus-mediated herpes simplex thymidine kinase gene-transfer in human medulloblastoma. *Gene Ther.*, 5: 113-120.

Rubsam, L. R., Davidson, B. L. and Shewach, D. S. (1998). Superior cytotoxicity with ganciclovir compared with acyclovir and 1-β-D-arabinofuranosylthymine in Herpes simplex virus-thymidine kinase-expressing cells: A novel paradigm for cell killing. *Cancer Res.*, 58: 3873-3882.

Sacco, M.G., Benedetti, S., Duflot-Dancer, A., Mesnil, M., Bagnasco, L., Strina, D., Fasolo, V., Villa, A., Macchi, P., Faranda, S., Vezzoni, P. and Finocchiaro, G. (1996). Partial regression, yet incomplete eradication of mammary tumors in transgenic mice by retrovirally mediated HSVtk transfer in vivo. *Gene Ther.*, 3: 1151-1156.

Saez, J.C., Connor, J.A., Spray, D.C. and Bennett, M.W.L. (1989). Hepatocyte gap junctions are permeable to the second messenger, inositol 1,4,5-triphosphate, and to calcium ions. *Proc. Natl. Acad. Sci. USA*, 86: 2708-2712.

Samejima, Y. and Meruelo, D. (1995). "Bystander killing" induces apoptosis and is inhibited by forskolin. *Gene Ther.*, 2: 50-58.

Simpson, I., Rose, B. and Loewenstein, W.R. (1977). Size limit of molecules permeating the junctional membrane channels. *Science*, 195: 294-296.

Smee, D.F., Boehme, R., Chernow, M., Binko, B.P. and Matthews, TR (1985). Intracellular metabolism and enzymatic phosphorylation of 9-(1,3-dihydroxy-2-propoxymethyl)guanine and acyclovir in herpes simplex virus infected and uninfected cells. *Biochem. Pharmacol.*, 34: 1049-1056.

Subak-Sharpe, H., Burk, R.R. and Pitts, J.D. (1969). Metabolic co-operation between biochemically marked mammalian cells in tissue culture. *J. Cell Sci.*, 4: 353-367.

van der Erb, M.M., Cramer, S.J., Vergouwe, Y., Schagen, F.H., van Krieken, J.H., van der Erb, A.J., Rinkes, I.H., van de Velde, C.J. and Hoeben, R.C. (1998) Severe hepatic dysfunction after adenovirus-mediated transfer of the herpes simplex virus thymidine kinase gene and ganciclovir administration. *Gene Ther.*, 5: 451-458.

Vrionis, F.D., Spray, D.C., Cherington, V., Waltzman, M. Qi, P. and Wu, J. (1997). The bystander effect exerted by tumor cells expressing the herpes simplex virus thymidine kinase (HSVtk) gene is dependent on connexin expression and cell communication via gap junctions. *Gene Ther.*, 4: 577-585.

Wallace, H., MacLaren, K., Al-Shawi, R. and Bishop, J.O. (1996). Ganciclovir-induced ablation of non-proliferating thyrocytes expressing herpes thymidine kinase occurs by p53-independent apoptosis. *Oncogene*, 13: 55-61.

Wei, M.X., Bougnoux, P., Sacré-Salem, B., Peyrat, M.-B., Lhuillery, C., Salzmann, J-L. and Klatzmann, D. (1998). Suicide gene therapy of chemically induced mammary tumor in rat: efficacy and distant bystander effect. *Cancer Res.*, 58: 3529-3532.

Yamasaki, H. and Katoh, F. (1988). Novel method for selective killing of transformed rodent cells through intercellular communication, with possible therapeutic applications. *Cancer Res.*, 48: 3203-3207.

Yamasaki, H. and Naus, C.C.G. (1996). Role of connexin genes in growth control. *Carcinogenesis*, 17: 1199-1213.

Yang, L., Chiang, Y., Lenz, H.-J., Danenberg, K.D., Spears, C.P., Gordon, E.M., Anderson, W.F. and Parekh, D. (1998). Intercellular communication mediates the bystander effect during herpes simplex thymidine kinase/ganciclovir-based gene therapy of human gastrointestinal tumor cells. *Hum. Gene Ther.*, 9: 719-728.

Zhang, L.X., Acevedo, P., Guo, H. and Bertram, J.S. (1995). Upregulation of gap junctional communication and connexin 43 gene expression by carotenoids in human dermal fibroblasts but not in human keratinocytes. *Mol. Carcinog.*, 12: 50-58.

9. SUMMARY — Anti-tumoral suicide gene therapy consists of introduction into cancer cells of a gene capable of converting a non-toxic prodrug into a cytotoxic drug. Among others, the thymidine kinase gene from the Herpes simplex virus (HSV-tk) has been extensively and successfully used in some animal models of anti-cancer gene therapy. Since this gene cannot be easily introduced into the whole cell population of a tumor, the bystander effect exerted by gap junctional intercellular communication attracted much attention for complete elimination of cancer cells.

Key Words: anti-cancer gene therapy; bystander effect; gap junctions; connexins; thymidine kinase; ganciclovir

Progress in Gene Therapy: Basic and Clinical Frontiers, pp. 321-341
R. Bertolotti *et al.* (Eds)

Vaccination strategies for advanced melanoma

Yuansheng Sun and Dirk Schadendorf*

Clinical Cooperation Unit for Dermatooncology (DKFZ), Department of Dermatology, Clinics Mannheim, University of Heidelberg, Theodor kutzer Ufer 1, 68135 Mannhein, Germany

Table of Contents

1. MELANOMA AND TUMOR IMMUNOLOGY

Melanoma is a malignant tumor of neuroectodermal origin with an increasing incidence and mortality worldwide. It needs to be detected and eliminated early since advanced melanoma is highly resistant to conventional

* Corresponding author. E-mail: E-mail: d.schadendorf@dkfz-heidelberg.de
[†]GM-CSF: granulocyte-macrophage colony stimul. factor; IFN: interferon; IL: interleukine

therapies including surgery and chemotherapy (Ahmann *et al.*, 1989; Johnson *et al.*, 1995; Garbe, 1993). On the other hand, melanoma is supposed to be one of the most immunogenic tumors which is frequently demonstrated by an intense lymphocytic infiltration (Oettgen and Old, 1991; Parkinson *et al.*, 1992). This may also be responsible for the occasional occurence of spontaneous tumor regression and for concomitant destruction of melanocytes in benign lesions, leading to clinical phenomena such as halo nevi, uveitis and vitiligo in some patients with melanoma.

It is generally accepted that the spontaneous generation of cancer cells is a common event, and that the immune system assures a strict surveillance with the detection and elimination of these cells. In order to fight cancer, the idea of using the destructive power of immunological reactions is easily visualized in autoimmune diseases as well as in transplantation medicine. A number of clinical observations suggest the existence of a particularly vigorous occurring immune response to the melanoma tumors (Oettgen and Old, 1991; Parkinson *et al.*, 1992). And, there has been increasing evidence that T cells, particularly the cytotoxic T lymphocytes (CTLs) play a critical role in this immunological process. For instances, the magnitude of T lymphocytic tumor infiltration is an important prognostic parameter (Soong, 1992). Further, CTLs bearing a specific set of T-cell receptor (TCR) Vβ segments have been isolated from a case of regressing melanoma (Mackensen *et al.*, 1994), thus providing direct evidence for the concept of immunosurveillance. Another compelling evidence is the demonstration that adoptive transfer of *ex vivo* expanded tumor-infiltrating lymphocytes (TILs) along with IL-2 into the autologous patients resulted in impressive melanoma regressions (Kawakami *et al.*, 1994; Robbins *et al.*, 1994). In recent years, the availability and further characterization of such tumor-reactive CTL clones has led to the identification of several melanoma-associated antigens (Boon *et al.*, 1997; Rosenberg, 1999).

In order to fight the melanoma tumors, there are several different approaches currently being actively explored utilizing the host's immune system. These include 1) augmentation of the immunogenicity of tumor cells by genetical modification with cytokine genes, which is also known as "gene therapy", and 2) the use of T cell-defined melanoma-associated peptides, either directly for immunization (peptide vaccines) or after loading onto professional antigen-presenting cells (APCs) such as dendritic cells (DCs) (dendritic cell-based vaccines).

2. CYTOKINE GENE-MODIFIED TUMOR CELL VACCINES

a) Cytokine gene transfer: the rationale and preclinical results

In the past decades, the identification and cloning of cytokines provided an important set of tools for the manipulation of immunologic responses. For example, systemic infusions of IL-2 alone or in conjunction with lymphokine-activated killer (LAK) cells into autologous patients with advanced metastatic melanoma were shown to achieve comparable clinical responses as conventional therapics. However, this treatment regimen is sometimes associated with severe side effects (organ dysfunction or even death) related to high-dose cytokine administration (Rosenberg *et al.*, 1985 and 1989). Given that most cytokines act as local hormones whose accumulation in the serum is prevented by virtue of their short half-life, it was suggested that local and paracrine delivery of immunostimulatory cytokines by tumor cell itself may have a more pronounced therapeutic benefit while avoiding systemic toxicity. This mode of cytokine delivery can be achieved by introduction of cytokine genes into the tumor cells.

Starting from the pioneering work of Tepper and co-workers (1989) who showed that J558L plasmocytoma cells after transfection with IL-4-encoding gene were rejected, a large number of immunostimulatory cytokines have been tested in such a delivery mode in various animal tumor models (reviewed in: Vieweg and Gilboa, 1995; Schadendorf, 1997; Sun *et al.*, 1999). In most cases, the transduction of tumor cells with cytokine genes led to the rejection of the genetically modified cells by syngeneic hosts, suggesting the enhanced immunogenicity of transduced cells. Of more interest is that tumor cells transduced to express cytokines such as IL-2, IL-4, IL-7, IL-12, IFN-γ, TNF-α, or GM-CSF were capable of inducing a systemic anti-tumor immune memory: animals vaccinated with cytokine gene-transduced cells rejected a subsequent challenge of non-transduced (wild-type) tumor cells at a distant site, and in some cases eliminated a preexisting tumor. Because several studies have used poorly- or non-immunogenic tumor models, the protective anti-tumor immunity induced could thus be solely attributed to the transduced cytokine. Examinations of the cellular mechanisms underlying the protective anti-tumor immunity revealed the uniform dependence on T cells, mostly the subset of $CD8^+$ T cells and in part $CD4^+$ T subset (Vieweg and Gilboa, 1995; Schadendorf, 1997; Sun *et al.*, 1999). Probably, $CD8^+$ and $CD4^+$ T lymphocytes were activated directly by transfected tumor cells, or more likely indirectly *via* professional APCs such as DCs that can pick up tumor antigens (tumor-cell debris), process and finally present them to both class I-restricted $CD8^+$ T cells and class II-restricted $CD4^+$ T cells (Pardoll and Beckerleg, 1995).

In summary, animal studies have clearly demonstrated that genetic modification of tumor cells with cytokine genes can generate potent systemic anti-tumor immunity. However, no conclusive indication was provided as to which cytokine is overall the best in terms of increasing tumor immunogenicity, and certain cytokines are active in some, but not in all animal tumor models. Nevertheless, these preclinical studies have provided the regulatory, mechanistic, and functional principles for cytokine gene-modified tumor vaccines.

b) Cytokine gene-modified tumor cell vaccines: clinical phase I/II trials

The successful insertion of cytokine genes into human tumor cell lines and primary tumor explants has made cancer gene therapy realistic clinically. Since the first therapeutic experiments in 1990, more than 250 additional clinical gene trials were approved and more than 2000 patients treated worldwide by the end of 1996 (Marcel and Grausz, 1997; Schadendorf, 1998). Almost 25% of these studies have no therapeutic intent and are gene marking trials. The majority (60%) of the trials are aimed to treat cancer and the great majority of investigators used immunization strategies with cytokine-gene modified tumor cells. Melanoma is supposedly one of the most immunogenic human solid tumors, it was therefore a favorable target for gene-modified cancer vaccines (summarized in Schadendorf, 1997). In view of the ethic, phase I clinical trials with gene-modified tumor cell vaccines should be performed only in patients failing all current therapies. Thus, the major goal of these studies is to assess the safety of this innovative approach, to determine the toxicity caused by locally secreted cytokines, and to evaluate the immunological functions induced by these approaches.

Results of IL-7- or IL-12-secreting tumor cell vaccines in melanoma treatment. We have initiated two clinical phase I studies aiming at induction of T cell-mediated anti-tumor immune responses by immunizing advanced melanoma patients (Schadendorf *et al.*, 1995 and1996; Möller *et al.*, 1998; Sun *et al.*, 1998). Autologous melanoma cell lines established from metastases were transfected either with both chains of the IL-12-gene (Sun *et al.*, 1998) or with the IL-7-gene (Möller *et al.*, 1998). The transduced autologous tumor cells were subjected to irradiation, and were then injected subcutaneously (s.c.) into patients in weeks 1, 2, 3 and 6 with total numbers of 5×10^6 to 3×10^7. In parallel, DTH-reactivity and an extensive immunological monitoring including flow cytometry, NK- and LAK-activity as well as CTL analysis are performed. The cytokine gene transfer protocol uses a newly developed gene transfer technology which combines ballistic transfer of biological molecules and magnetic cell sorting (Schadendorf, 1998).

Evaluation of the first 10 patients immunized with autologous IL-7 gene-modified melanoma cells demonstrated the safety, the lack of toxicity and the feasibility of such an approach; however, no major clinical responses were achieved (Möller *et al.*, 1998). Four patients showed stable disease (SD) and 2 had a mixed response. Eight of 10 patients completed the initial three s.c. vaccinations and were eligible for immunological evaluation: four showing an increased NK response and seven showing an increased LAK response upon vaccination. In 3/7 patients, the frequency of tumor-reactive CTL precursors in peripheral blood increased between 2.6- to 28-fold after the third vaccination. Interestingly, all three patients with increased CTL responses were also clinical responders: two with a mixed response and one with a SD, implying a role of $CD8^+$ CTL in controlling tumor growth. The magnitude of the T-cell reactivity induced by the vaccine was found to be highly associated with the patient's Karnofsky-index and recall antigen skin reactivity (DTH reaction) before vaccination. This suggests that patients with minimal tumor load or minimal residual disease may preferentially benefit from such a tumor vaccine.

In a subsequent phase I study, 6 advanced melanoma patients were immunized with IL-12 gene-modified, irradiated, autologous tumor cells (Sun *et al.*, 1998). Clinically, there was no major toxicity except for mild fever and mild flu-like symptoms in some patients. All patients completed more than 4 vaccinations and were eligible for evaluation at week 5. No major clinical responses were achieved in any patient. Three patients showed SDs, two of which were alive for more than 10 months. One showed a minor clinical response (MR) with the regression of some cutaneous metastases over 3 months. Post-vaccination, two patients developed DTH reactivity against autologous melanoma cells, one showing a heavy infiltrate of $CD4^+$ and $CD8^+$ T-lymphocytes in regressing metastasis. Analysis of the frequency of tumor-reactive CTL precursors in peripheral blood revealed a significant increase (up to 15-fold) in 2 patients after immunization.

Results of GM-CSF- or IFN-γ-secreting tumor cell vaccines in melanoma therapy. Soiffer *et al.* (1998) have investigated the biologic activity of GM-CSF gene-modified melanoma vaccines in 29 patients with stage IV melanoma. Three successive patient cohorts were immunized intradermally (i.d.) and s.c. with 10^7 irradiated tumor cells (each treatment) administered at 28-, 14-, or 7-day intervals for a total of 3, 6, or 12 vaccinations. One partial response (PR), one mixed response, and 3 MRs were observed. Substantial erythema and induration were induced at vaccination sites in all patients and Grade 1 fatigue and nasal congestion were occasionally noted; however, no systemic toxicities were recorded. In all patients, an initially negative DTH reaction to irradiated, nontransfected, autologous tumor cells was converted to a strong response after several vaccinations. This conversion could be interpreted as an augmented antitumor cellular immunity towards original tumor. In addition, this immunization scheme also generated an enhanced anti-tumor humoral immune responses, as demonstrated by an increased anti-melanoma antibody response (IgG-type) in 7 patients examined. Immunohistological analysis revealed that distant metastases from 11 of 16 patients were largely infiltrated by both T cells ($CD4^+$ and $CD8^+$) and plasma cells after, but not before, vaccination, even though these anti-tumor immune responses failed to induce clinical tumor regressions in most cases.

In another phase I clinical trial, Abdel-Wahab *et al.* (1997) treated 20 stage III-IV melanoma patients with IFN-γ gene-modified irradiated autologous tumor vaccines, in escalating doses once every 2 weeks for a total of 6 injections (*i.e.* 2 vaccinations with 2 x 10^6, 6 x 10^6 and 18 x 10^6 each). Two patients showed a complete response and 2 additional patients had transient shrinkage of subcutaneous nodular disease. No side effects were noted. In this trial, the humoral immune responses of the immunized patients were intensively investigated: 8 of 13 patients evaluated showed an increasing anti-melanoma IgG titer during the course of immunization. IgG titers increased with the number of vaccinations and detected for up to 4 weeks following the last vaccination. The observations that the increased IgG response was dominated by IgG2 versus IgG1 isotypes in all 8 responders, thus suggest that T helper type 1 (Th1) cells were activated in these patients.

Results of IL-2 gene-transfected cell-based vaccines in treatment of melanoma. Because direct transduction of autologous tumors is highly individualized, expensive and labor intensive, simpler approaches that maintain the immunological activity of paracrine cytokine elaboration are also currently under clinical investigations; these include 1) to use standardi-zed gene-transduced allogeneic tumor cell lines as vaccines and 2) to admix tumor cells with a generic transduced bystander cells (*e.g.* fibroblasts). The former approach is based on the idea that some tumor rejection antigens are shared rather than unique, whereas the latter strategy takes advantage of the fact that the cytokine does not need to be produced by the tumor itself, thereby obviating the need for transduction of each patient' tumor cells.

Arienti *et al.* (1996) and Belli *et al.* (1997) have used an HLA-A2-matched, Melan-A/Mart-1$^+$, gp100$^+$, tyrosinase$^+$, allogeneic melanoma cell line engineered to secrete IL-2, to immunize 12 stage IV melanoma patients four times at doses of 5 or 15 x 10^7 cells. Both local and systemic toxicities were mild. Three out of 8 patients evaluated showed mixed clinical responses. Two patients showed increased reactivity of specific CTL directed against tyrosinase and gp100 melanoma-associated antigens in post-vaccination PBL. In additional two patients, the frequency of melanoma-specific CTL precursors in PBL was increased after vaccination. Veelken *et al.* (1997) applied a vaccine containing IL-2-secreting allogeneic fibroblasts and autologous tumor cells to 15 patients with advanced malignancies, including 6 melanoma patients. No major clinical response was observed, nor major side effects attributable to vaccines. In 2 melanoma patients, a dense infiltration of both CD4$^+$ and CD8$^+$ T cells at vaccination sites was demonstrated. Analysis of the variability of TCR in infiltrating lymphocytes obtained from vaccination sites or tumor sites of one patient revealed the identical V-D-J junctional sequence of CTL, indicating that the same CTL clone had infiltrated the tumor, circulated in the peripheral blood, and was amplified at the vaccination site (Mackensen *et al.*,1997).

Limitations and remaining questions. The above studies demonstrated the feasibility, apparent safety and low toxicity of cytokine gene-modified tumor cell vaccines, and suggested the potential of this approach to boost host anti-tumor immunity even in far advanced melanoma patients. However, it must be realized that *ex vivo* genetic modification for vaccination is still at the beginning of its development, and it is still far away from providing a curative T-cell response. A number of questions remain open which have only partly been addressed in animal studies:

1). The immunogenicity of gene-modified tumor cell preparations used for vaccinations might be crucially affected by irradiation or mitomycin treatment, since animal studies by Hock and co-workers (1993) suggested a dramatic decrease in protection after tumor cell inactivation.

2). The dose of gene-modified cells necessary for vaccinations is presently unclear.

3). The levels of cytokines produced by genetically altered tumor cells needed to effectively stimulate the immune system are not known. The first animal studies addressing this question suggest that different cytokines have different optimal concentration which further differ in the animal tumor model tested. Furthermore, in certain instances, a bell-shaped response curve has been observed with a dramatic decrease in vaccination efficacy after passing a certain cytokine level (Schmidt *et al.*, 1995).

4). The mechanisms involved in tumor suppression are not well understood possibly interfering with the vaccination effects at distant sites of the metastatic disease or in mounting an immune response including immunological memory.

5). The optimal immunization schedule and route have not been explored systematically.

6). The mechanism of action of gene-modified tumor cells in causing tumor immunity and tumor regression is still under discussion.

7). Furthermore, the evaluation of immunological endpoints (analysis of precursor frequency, CTL, DTH, *etc.*) is difficult to compare, since methodology, reliability and reproducibility to evaluate such parameters is still under discussion.

3. DEFINED ANTIGEN VACCINES

A number of antigens present on melanoma cells that could potentially serve as targets for the host T-cell responses have recently been identified (Boon *et al.*, 1997; Rosenberg, 1999). Although all of these antigens may not be present on every tumor cell, as the number of defined tumor antigens increases, it becomes progressively more likely that at least one relevant antigenic target can be identified for every patient's melanoma. Based on their expression pattern, these T cell-defined melanoma antigens can be categorized into three broad groups: 1) tissue-specific differentiation antigens expressed only in normal melanocytic cells and melanomas, such as tyrosinase, Melan-A/MART-1, gp100, TRP1 and TRP2; 2) tumor-specific shared antigens expressed in various types of human cancers but not in normal tissues except testis, as exemplified by MAGE, BAGE and GAGE gene family; and 3) tumor-specific unique antigens produced by point mutation in a gene that is ubiquitously expressed, including β-catenin, MUM-1 and CDK4. A summary of presently known T cell-recognized melanoma-associated peptides and their MHC restriction elements are listed in Table I. Since the first two classes of antigens are both expressed in the majority of melanoma, strategies using their antigenic peptides are currently being evaluated.

In a phase I clinical trial, Rosenberg and co-workers (1998) immunized HLA-A2$^+$ patients with stage IV metastatic melanoma using a modified peptide (g209-2M) derived from the gp100 melanoma differentiation antigen. Immunologic assays of peripheral blood mononuclear cells from patients after treatment showed that 91% of the patients were successfully immunized with this synthetic peptide. In another cohort of patients who received the peptide vaccine plus adjuvant IL-2, 13 of 31 patients (42%) had objective tumor regression (Rosenberg *et al.*, 1998). Similarly, Marchand *et*

Table 1. Melanoma-associated antigens recognized by human T lymphocytes[1]

Target Antigen	Restricting HLA molecule	Peptides sequence	Amino acid position
MAGE-1	HLA-A1	EADPTGHSY	161-169
	HLA-A3	SLFRAVITK	96-104
	HLA-A24	NYKHCFPEI	135-143
	HLA-A28	EVYDGREHSA	222-231
	HLA-B37	REPVTKAEML	127-136
	HLA-B53	DPARYEFLW	258-266
	HLA-Cw2	SAFPTTINF	62-70
	HLA-Cw3/-Cw16	SAYGEPRKL	230-238
	HLA-DR13	LLKYRAREPVTKAE	121-134
MAGE-2	HLA-A2	KMVELVHFL	112-120
		YLQLVFGIEV	157-166
	HLA-24	EYLQLVFGI	156-164
	HLA-B37	REPVTKAEML	127-136
	HLA-DR13	LLKYRAREPVTKAE	121-134
MAGE-3	HLA-A1	EVDPIGHLY	168-176
	HLA-A2	FLWGPRALV	271-279
		KVAELVHFL	112-120
	HLA-A24	IMPKAGLLI	195-203
		TFPDLESEF	97-105
	HLA-B37	REPVTKAEML	127-136
	HLA-B44	MEVDPIGHLY	167-176
	HLA-DR13	AELVHFLLLKYRAR	114-127
		LLKYRAREPVTKAE	121-134
	HLA-DR11	TSYVKVLHHMVKISG	281-295
MAGE-4	HLA-A2	GVYDGREHTV	230-239
MAGE-6	HLA-A34	MVKISGGPR	290-298
	HLA-B37	REPVTKAEML	127-136
	HLA-DR13	LLKYRAREPVTKAE	121-134
MAGE-10	HLA-A2.1	GLYDGMEHL	254-262
	HLA-B53	DPARYEFLW	290-298
MAGE-12	HLA-Cw7	VRIGHLYIL	170-178
	HLA-DR13	AELVHFLLLKYRAR	114-127
BAGE	HLA-Cw16	AARAVFLAL	2-10
RAGE	HLA-B7	SPSSNRIRNT	
GAGE-1, 2, 8	HLA-Cw6	YRPRPRRY	9-16

Table 1. (*Cont.*)

Target Antigen	Restricting HLA molecule	Peptides sequence	Amino acid position
GAGE-3 to 7	HLA-A29	YYWPRPRRY	10-18
NY-ESO-1/CAG-3	HLA-A2 (ORF-1)	QLSLLMWITQC	155-165
	HLA-A31 (ORF-1)	ASGPGGGAPR	53-62
	HLA-A31 (ORF-2)	LAAQERRYPR	
LAGE/CAMEL	HLA-A2 (ORF-2)	MLMAQEALAFL	
pMel-34/ Tyrosinase	HLA-A2	YM*D*GTMSQV	369-377
		YM*N*GTMSQV	369-377
		MLLAVLYCL	1-9
	HLA-A24	AFLPWHRLF	206-214
	HLA-B35	LPSSADVEF	312-320
	HLA-B44	SEIWRDIDF	192-200
	HLA-A1	KCDICTDEY	243-251
		SSDYVIPIGTY	146-156
	HLA-DR4	QNILLSNAPLGPQFP	56-70
		DYSYLQDSDPDSFQD	448-462
	HLA-DR15	FLLHHAFVDSIFEQWLQR- HRP	386-406
TRP-1/gp75	HLA-A31	MSLQRQFLR (ORF-3)	1-9
TRP-2	HLA-A*0201	SVYDFFVWL	180-188
	HLA-A*0201	SLHNLYHSFL	367-376
	HLA-A31, -A33	LLGPGRPYR	197-205
	HLA-CW8	ANDPIFVVL	387-395
TRP-2 (int2)	HLA-A68011/3301	EVISCKLIKR	222-231
pMel-17/gp100	HLA-A2	VLYRYGSFSV	476-485
		KTWGQYWQV	154-162
		YLEPGPVTA	280-288
		LLDGTATLRL	457-466
		SLADTNSLAV	570-579
		ITDQVPFSV	209-217
		RLMKQDFSV	619-627
		RLPRIFCSC	639-647
		(A)MLGTHTMEV	177(178)-186
	HLA-A3	LIYRRRLMK	614-622
		ALLAVGATK	17-25

	HLA-A3, -A11 HLA-Cw8 HLA-(DR1),DR4, (DR3)	(I)ALNFPGSQK SNDGPTLI WNRQLYPEWTEAQRLDV-YFFLPDHL	86(87)-95 71-78 44-59
gp100 (int-4)	HLA-A24		170-178
Melan-A/MART-1	HLA-A2	ILTVILGVL (E)AAGIGILTV GIGILTVL GILTVILGV	32-40 (26)27-35 29-36 31-38
	HLA-B45.1	AEEAAGIGIL(T)	24-33(34)
707-AP	HLA-A2	RVAALARDAP	
N-Acetyl-Gn-Transferase V	HLA-A2	VLPDVFIRC	38-46 (intron sequence)
p15	HLA-A24	AYGLDFYIL	
β-catenin	HLA-A24	SYLDSGIII*F*	29 37
SR-2	HLA-A3	KIFSEVT*L*K	
MUM-1	HLA-B44	EEKL*I*VVLF	782-808
Myosin class I	HLA-A3	*K*INKNPKYK	
CDK4	HLA-A2.1	A*C*DPHSGHFV	
TPI (Triosephosphate Isomerase)	HLA-DR1	GELIG*I*LNAAKVPAD	23-37
Fusion protein LDLR/FUT	HLA-DR1	(PVI)WRRAPA(PGA)	315-323
CDC27	HLA-DR4	FSWAMDLDPKGA	760-771

[1] Modified from Sun *et al.* (1999), *J. Mol. Med.*, 77: 593-608.

al. (1999) in another clinical trial examined therapeutic effect of a MAGE-3 peptide in HLA-A1$^+$ patients with metastatic melanoma. Of 25 immunized patients eligible for evaluation, 7 showed objective clinical responses, although in this study the effectors mediating tumor regression have yet to be identified. Taken together, these results suggest that peptide immunization may represent a potentially effective means for the treatment of melanoma.

4. DENDRITIC CELL-BASED VACCINES

Basis immunology demonstrated the pivotal role of dendritic cells (DCs) in generating an immune response. DCs are antigen presenting cells (APCs) specialized for the induction of a primary T cell response (Steinman, 1991). Evidence is mounting that DCs derived from bone marrow as well as from peripheral blood are responsible for initiating all T-cell responses *in vivo* (Grabbe *et al.*, 1995; Peters *et al.*, 1996), including both $CD4^+$ and $CD8^+$ T-cell responses, by their unique capability to present antigens to naive T-cells. Mouse studies have demonstrated the potent capacity of DCs to induce anti-tumor immunity (Zitvogel *et al.*, 1996; Celluzzi *et al.*, 1996; Paglia *et al.*, 1996; Porgador *et al.*, 1996; Mayordomo *et al.*, 1995 and 1996). Since DCs can now easily be generated from different sources including peripheral blood (Romani *et al.*, 1994; Sallusto and Lanzavecchia, 1994), these cells can be used for cancer vaccination purpose either *via* pulsing with peptides (Zitvogel *et al.*, 1996; Celluzzi *et al.*, 1996; Mayordomo *et al.*, 1995 and 1996) or after transfection with a tumor antigen (Alijagic *et al.*, 1995). The implication for vaccine design is that DCs would be more potent than tumor cells as immunogens. This way of treatment was pioneered in humans by Hsu and coworkers (1996) who treated four patients with B-cell lymphoma with DCs loaded with paraneoplastic immunoglobulins with some clinical success. Immature antigen-presenting cells were pulsed with MAGE-1 peptides, however, showing no clinical responses in 3 melanoma patients (Mukherji *et al.*, 1995).

In a clinical pilot study with 16 patients with advanced melanoma, DCs were generated in the presence of GM-CSF and IL-4 and were pulsed either with tumor cell lysate or a cocktail of peptides known to be recognized by CTLs, depending on the patient´s HLA haplotype. Multiple peptides (MAGE-1 and MAGE-3 for HLA-A1; gp100, tyrosinase, and Melan-A-derived peptides for HLA-A2) were used to diminish the chances of immune escape in a given patient. In 4/16 patients, tumor lysate was used instead of peptide as a source of tumor antigen. Keyhole limpet hemocyanin (KLH) was added as CD4 helper antigen and an immunological tracer molecule (Nestle *et al.*, 1998). DC vaccination was administrated on an outpatient basis and was well tolerated in all patients without significant toxicity. Occasionally, mild fever or swelling of the injected lymph node occurred lasting for 1-2 days. DC vaccination induced DTH-reactivity towards KLH in all patients, as well as a positive DTH-reaction to peptide-pulsed DC in 11/16 patients. Objective clinical responses were evident in 5 out of 16 evaluated patients (2 complete responses, 3 partial responses) with regression of metastases in various organs (skin, soft tissue, lung, pancreas) and one additional minor response. Clinical responsiveness seemed to correlate with DTH-reactivity to peptides used for vaccination and/or autologous tumor lysate. Furthermore, recruitment of peptide-specific CTL to the DTH challenge site was also demonstrated. Thus, antigen-specific anti-tumor immunity was induced during DC vaccination.

In conclusion, these data indicate that vaccination with autologous DCs generated from peripheral blood is a safe and promising approach to the treatment of metastatic melanoma (Nestle *et al.*, 1998). Further studies are necessary to demonstrate clinical effectiveness and impact on the survival of melanoma patients. A multicentre clinical phase II/III trial testing efficacy of DC vaccination in comparison to standard chemotherapy in advanced melanoma is underway.

Besides inducing anti-tumor immunity by DCs, there is also a chance of generating tumor tolerance possibly depending on the maturation stage of the DCs (Enk *et al.*, 1997). Therefore, it is highly critical to work with defined DC populations that are stable in their functional properties in vaccine settings. Methods to even improve efficacy of DC vaccination include improved DC generation protocols such as the inclusion of additional cytokines and prostaglandins (Jonuleit *et al.*, 1997), and the gene transfer of tumor antigens into DCs using DNA (Alijagic *et al.*, 1995; Bueler *et al.*, 1996; Perez-Diez *et al.*, 1998; Tuting *et al.*, 1998) or RNA (Boczkowski *et al.*, 1996; Nair *et al.*, 1998) in order to overcome HLA-restriction (associated with peptide-based approaches).

5. CONCLUSION AND PERSPECTIVES

Tumor immunology has made great progress in recent years. Recent animal studies have indicated that genetic modifications of tumor cells to express cytokines can enhance tumor immunogenicity. A long-lived systemic anti-tumor immunity can be generated *in vivo* by immunizations of animals with genetically-modified tumor cells secreting either IL-2, IL-4, IL-7, IL-12, IFN-γ, or GM-CSF. Based on these successful animal studies, a number of phase I clinical trials using cytokine-secreting tumor cell vaccines were initiated in recent years, predominantly in human melanoma. Some of them have been completed. At present, it seems safe to conclude that strategies involving cytokine-secreting tumor vaccines can influence the immunological tumor-host relationship; however, in most cases, the response seems not to be strong enough to completely eradicate the tumor in highly pretreated and far metastatically advanced patients. It must be realized that the application of gene therapy to the treatment of malignancies is still in the earliest stage of its development. The careful optimization of the most promising strategies, thoughtful selection of patient populations, and proper clinical trial design will accelerate identification of a reproducible clinical benefit from this strategy.

Besides the genetically modified tumor cell vaccines, other innovative approaches to melanoma vaccines are also being tested clinically, including peptide vaccines and dendritic cell vaccines. Both strategies have shown encouraging results, even though the number of patients treated in each clinical trial was small. It is believed that with the continuous increase in our knowledge regarding the interaction of the host immune response to vaccines and the effector mechanisms required to eradicate established tumor, active specific immunotherapy will eventually become a standard approach to treatment of melanomas as well as other malignancies.

Acknowledgements

The work was supported by the DFG and would not have been possible without the contributions of Drs. F.O. Nestle, M. Gillet, B. Wittig, P. Möller, B. Henz, and K. Jurgovsky, and without the excellent technical assistance of A. Sucker and T. Dorbic.

6. REFERENCES

Abdel-Wahab, Z., Weltz, C., Hester, D., Pickett, N., Vervaert, C., Barber, J.R., Jolly, D. and Seigler, H.F. (1997). A phase I clinical trial of immunotherapy with interferon-γ gene-modified autologous melanoma cells: monitoring the humoral immune response. *Cancer,* 80: 401-412.

Ahmann, D.L., Creagan, E.T., Hahn, R.G., Edmonson, J.H., Bisel, H.F. and Schaid, D.J. (1989). Complete responses and long-term survivals after systemic chemotherapy for patients with advanced malignant melanoma. *Cancer,* 63: 224-227.

Alijagic, S., Möller, P., Jurgovsky, K., Czarnetzki, B.M. and Schadendorf, D. (1995). Transfection of dendritic cells generated from peripheral blood with human tyrosinase induced specific T-cell activation. *Eur. J. Immunol.,* 25: 3100-3107.

Arienti, F., Sule-Suso, J., Belli, F., Macheroni, L., Rivoltini, L., Melani, C., Maio, M., Cascinelli, N., Colombo, M.P. and Parmiani, G. (1996). Limited antitumor T cell response in melanoma patients vaccinated with interleukin-2 gene-transduced allogeneic melanoma cells. *Hum. Gene Ther.,* 7: 1955-1963.

Belli, F., Arienti, F., Sule-Suso, J., Clemente, C., Mascheroni, L., Cattelan, A., Santantonio, C., Gallino, G.F., Melani, C., Rao, S., Colombo, M.P., Maio, M., Cascinelli, N. and Parmiani, G. (1997). Active immunization of metastatic melanoma patients with interleukin-2-transduced allogeneic melanoma cells: evaluation of efficacy and tolerability. *Cancer Immunol. Immunother.,* 44: 197-203.

Boczkowski, D., Nair, S.K., Synder, D. and Gilboa, E. (1996). Dendritic cells pulsed with RNA are potent antigen-presenting cells in vitro and in vivo. *J. Exp. Med.,* 184: 465-472.

Boon, T., Coulie, P.G. and Van den Eynde, B. (1997). Tumor antigens recognized by T cells. *Immunol. Today*, 18: 267-268.

Bueler, H. and Mulligan, R.C. (1996). Induction of antigen-specific tumor immunity by genetic and cellular vaccines against MAGE: enhanced tumor protection by coexpression of granulocyte colony-stimulating factor and B7-1. *Mol. Med.,* 2: 545-555.

Celluzzi, C.M., Mayordomo, J.I., Storkus, W.J., Lotze, M.T., Falo, L.D. Jr. (1996). Peptide-pulsed dendritic cells induce antigen-specific CTL-mediated protective tumor immunity. *J. Exp. Med.,* 183: 283-287.

Enk, A.H., Jonuleit, H., Saloga, J. and Knop, J. (1997) Dendritic cells as mediators of tumor-induced tolerance in metastatic melanoma. *Int. J. Cancer,* 73: 309-316.

Garbe, C. (1993). Chemotherapy and chemoimmunotherapy in disseminated malignant melanoma. *Melanoma Res.*, 3: 291-299.

Grabbe, S., Beissert, S., Schwarz, T. and Granstein, R.D. (1995). Dendritic cells as initiators of tumor immune responses: a possible strategy for immunotherapy? *Immunol. Today,* 16: 117-121.

Hock, H., Dorsch, M., Kunzendorf, U., Uberla, K., Qin, Z., Diamantstein, T. and Blankenstein, T. (1993). Vaccinations with tumor cells genetically engineered to produce different cytokines: effectivity not superior to a classical adjuvant. *Cancer Res.*, 53: 714-716.

Hsu, F.J., Benike, C., Fagnoni, F., Liles, T.M., Czerwinski, D., Taidi, B., Engleman, *E.G.* and Levy, R. (1996). Vaccination of patients with B-cell lymphoma using autologous antigen-pulsed dendritic cells. *Nat. Med.*, 2: 52-58.

Jonuleit, H., Kühn, U., Müller, G., Steinbrink, K., Paragnik, L., Schmitt, E., Knop, J. and Enk, A. (1997). Pro-inflammatory cytokines and prostaglandins induce maturation of potent immunostimulatory dendritic cells under fetal calf serum-free conditions. *Eur. J. Immunol.*, 27: 3135-3142.

Johnson, T.M., Smith, J.W., Nelson, B.R. and Chang, A. (1995). Current therapy for cutaneous melanoma. *J. Am. Acad. Dermatol.,* 32: 689-707.

Kawakami, Y., Eliyahu, S., Delgado, C.H., Robbins, P.F., Sakaguchi, K., Appella, E., Yannelli, J.R., Adema,G.J., Miki, T. and Rosenberg, S.A. (1994). Identification of a human melanoma antigen recognized by tumor-infiltrating lymphocytes associated with in vivo tumor rejection. *Proc. Natl. Acad. Sci. USA,* 91: 6458-6462.

Mackensen, A., Carcelain, G., Viel, S., Raynal, M.C., Michalaki, H., Triebel, F., Bosq, J. and Hercend, T. (1994). Direct evidence to support the immunosurveillance concept in a human regressive melanoma. *J. Clin. Invest.,* 93: 1391-1402.

Mackensen, A., Veelken, H., Lahn, M., Wittnebel, S., Becker, D., Kohler, G., Kulmburg, P., Brennscheidt, U., Rosenthal, F., Franke, B., Mertelsmann, R. and Lindemann, A (1997). Induction of tumor-specific cytotoxic T lymphocytes by immunization with autologous tumor cells and interleukin-2 gene transfected fibroblasts. *J. Mol. Med.*, 75: 290-296.

Marcel, T. and Grausz, J.D. (1997). The TMC worldwide gene therapy enrollment report, end 1996. *Hum. Gene Ther.,* 8: 775-800.

Marchand, M., van Baren, N., Weynants, P., Brichard, V., Dreno, B., Tessier, M.H., Rankin, E., Parmiani, G., Arienti, F., Humblet, Y., Bourlond, A., Vanwijck, R., Lienard, D., Beauduin, M., Dietrich, P.Y., Russo, V., Kerger, J., Masucci, G., Jager, E., De Greve, J., Atzpodien, J., Brasseur, F., Coulie, P.G., van der Bruggen, P., Boon, T. (1999). Tumor regressions observed in patients with metastatic melanoma treated with an antigenic peptide encoded by gene MAGE-3 and presented by HLA-A1. *Int. J. Cancer*, 80: 219-230.

Mayordomo, J.I., Zorina, T., Storkus, W.J., Celuzzi, C.M., Falo, L.D., Kast, W.M., Ildstad, S.T., DeLeo, A.B. and Lotze, M.T. (1995). Bone marrow-derived dendritic cells pulsed with synthetic tumor peptides elicit protective and therapeutic antitumor immunity. *Nature Med.,* 1: 1297-1302.

Mayordomo, J.I., Loftus, D.J., Sakamoto, H., De Cesare, S., Appasamy, R., Lotze, M., Storkus, W., Appella, E. and DeLeo, A.B. (1996). Therapy of murine tumors with p53 wild-type and mutant sequence peptide-based vaccines. *J. Exp. Med.,* 183: 1357-1365.

Möller, P., Sun,Y., Dorbic, T., Möller, H., Makki, A., Jurgovsky, K., Schrott. M., Henz, B.M., Wittig, B. and Schadendorf, D. (1998). Vaccination with IL-7 gene-modified autologous melanoma cells can enhance the anti-melanoma lytic activity in peripheral blood of patients with a good clinical performance status: a clinical phase I study. *Brit. J. Cancer,* 77: 1907-1916.

Mukherji, B. Chakraborty, N.G., Yamasaki, S., Okino, T., Yamase, H., Sporn, J.R., Kurtzman, S.K., Ergin, M.T., Ozols, J., Meehan, J. *et al.* (1995). Induction of antigen-specific cytolytic T cells in situ in human melanoma by immunization with synthetic peptide-pulsed autologous antigen presenting cells. *Proc. Natl. Acad. Sci. USA,* 92: 8078-8082.

Nair, S.K., Boczkowski, D., Morse, M., Cumming, R.I., Lyerly, H.K. and Gilboa, E. (1998). Induction of primary carcinoembryonic antigen (CEA)-specific cytotoxic T lymphocytes in vitro using human dendritic cells transfected with RNA. *Nat. Biotechnol.,* 16: 364-369.

Nestle, F.O., Alijagic, S., Gilliet, M., Sun, Y., Dummer, R., Burg, G. and Schadendorf, D. (1998). Vaccination of melanoma patients with peptide or tumor lysate-pulsed dendritic cells. *Nature Med.,* 4: 328-332.

Oettgen, H.F. and Old, L.J. (1991). The history of cancer immunotherapy. In: *Biologic Therapy of Cancer, Principles and Practice,* deVita, V. T., Hellman, S. and Rosenberg, S.A. (Eds.), J. B. Lippincott, pp. 87 ff.

Paglia, P., Chiodoni, C., Rodolfo, M. and Colombo, M.P. (1996). Murine dendritic cells loaded in vitro with soluble protein prime cytotoxic T lymphocytes against tumor antigen in vivo. *J. Exp. Med.,* 183: 317-322.

Pardoll, D.M. and Beckerleg, A.M. (1995). Exposing the immunology of naked DNA vaccines. *Immunity,* 3: 165-169.

Parkinson, D.R., Houghton, A.N., Hersey, P. and Borden, E.C. (1992). Biologic therapy for melanoma. In *Cutaneous Melanoma,* Balch, C.M., Houghton, A.N., Milton, G.W., Sober, A.J. and Soong, S.J. (Eds.), J.B. Lippincott Co, Philadelphia, p.522.

Perez-Diez, A., Butterfield, L.H., Li, L., Chakraborty, N.G., Economou, J. and Mukherji, B. (1998). Generation of $CD8^+$ and $CD4^+$ T-cell response to dendritic cells genetically engineered to express the MART-1/Melan-A gene. *Cancer Res.,* 58: 5305-5309.

Peters, J.H., Gieseler, R., Thiele, B. and Steinbach, F. (1996). Dendritic cells: from ontogenetic orphans to myelomonocytic descendants. *Immunol. Today,* 17: 273-278.

Porgador, A., Snyder, D. and Gilboa, E. (1996). Induction of antitumor immunity using bone marrow-generated dendritic cells. *J. Immunol.,* 156: 2918-2926.

Robbins, P.F., El-Gamil, M., Kawakami, Y. and Rosenberg, S.A. (1994). Recognition of tyrosinase by tumor-infiltrating lymphocytes from a patient responding to immunotherapy. *Cancer Res.,* 54: 3124-3126.

Romani, N., Gruner, S., Brang, D., Kämpgen, E., Lenz, A., Trockenbacher, B., Konwalinka, G., Fritsch, P.O., Steinman, R.M. and Schuler, G. (1994). Proliferating dendritic cell progenitors in human blood. *J. Exp. Med.,* 180: 83-93.

Rosenberg, S.A. (1999). A new era for cancer immunotherapy based on the genes that encode cancer antigens. *Immunity*, 10: 281-287.

Rosenberg, S.A., Lotze, M.T., Muul, L.M., Leitman, S., Chang, A.E., Ettinghausen, S.E., Matory, Y.L., Skibber, J.M., Shiloni, E., Vetto, J.T. *et al.* (1985). Observation on the systemic administration of autologous lymphokine-activated killer cells and recombinant interleukin-2 to patients with cancer. *N. Engl. J. Med.,* 313: 1485-1492.

Rosenberg, S.A., Lotze, M.T., Yang, J.C., Aebersold, P.M., Linehan, W.M., Seipp, C.A. and White, D.E. (1989). Experience with the use of high-dose interleukin-2 in the treatment of 652 cancer patients. *Ann. Surg.,* 210: 474-485.

Rosenberg, S.A., Yang, J.C., Schwartzentruber, D.J., Hwu, P., Marincola, F.M., Topalian, S., Restifo, N.P., Dudley, M.E., Schwarz, S.L., Spiess, P.J., Wunderlich, J.R., Parkhurst, M.R., Kawakami, Y., Seipp, C.A., Einhorn, J.H. and White, D.E. (1998). Immunologic and therapeutic evaluation of a synthetic peptide vaccine for the treatment of patients with metastatic melanoma. *Nature Med.,* 4: 321-327.

Sallusto, F. and Lanzavecchia, A. (1994). Efficient presentation of soluble antigen by cultured human dendritic cells is maintained by granulocyte/macrophage colony-stimulating factor plus interleukin-4 and downregulated by tumor necrosis factor-alpha. *J. Exp. Med.,* 179: 1109-1111.

Schadendorf, D. (1997). Cytokines, autologous cell immunostimulatory and gene therapy for cancer treatment. In: *Skin Immune System 2nd ed.*, Bos J.D. (Ed.), CRC Press Inc., Boca Raton, USA, pp. 657-669.

Schadendorf, D. (1998). *Ex vivo* cytokine gene transfer in melanoma cells by using particle bombardment. In: *Methods in Molecular Medicine "Gene Therapy of Cancer: Methods and protocols - Vol. II - clinical Protocols for Cancer Gene Therapy"*, Walther W. and Stein U. (Eds.), Humana Press Inc., Totowa, NJ, USA.

Schadendorf, D., Czarnetzki, B.M. and Wittig, B. (1995). Clinical Protocol - Interleukin-7-, interleukin-12-, and GM-CSF gene transfer in patients with metastatic melanoma. *J. Mol. Med.,* 73: 473-477.

Schadendorf, D., Henz, B.M. and Wittig, B. (1996). Interleukin 7 trials for melanoma treatment. *Mol. Med. Today,* 2: 143-144.

Schmidt, W., Schweighofer, T., Herbst, E., Maas, G., Berger, M., Schilcher, F., Schaffner, G. and Birnstiel, M.L. (1995). Cancer vaccines: The interleukin 2 dosage effect. *Proc. Nat. Acad. Sci. USA,* 92: 4711-4715.

Soiffer, R., Lynch, T., Mihm, M., Jung, K., Rhuda, C., Schmollinger, J.C., Hodi, F.S., Liebster, L., Lam, P., Mentzer, S., Singer, S., Tanabe, K., Cosimi, A.B., Duda, R., Sober, A., Bhan, A., Daley, J., Neuberg, D., Parry, G., Rokovich, J., Richards, L., Drayer, J., Berns, A., Clift, S., Cohen, L., Mulligan, R. and Dranoff, G. (1998). Vaccination with irradiated autologous melanoma cells engineered to secrete human granulocyte-macrophage colony-stimulating factor generates potent antitumor immunity in patients with metastatic melanoma. *Proc. Natl. Acad. Sci. USA*, 95: 13141-13146.

Soong, S.J. (1992). A computerized mathematical model and scoring system predicting outcome in patients with localized melanoma. In: *Cutaneous Melanoma*, Balch, C.M., Houghton, A.N., Milton, G.W., Sober, A.J. and Soong, S.J. (Eds.), J.B. Lippincott

Co, Philadelphia, pp. 200 ff.

Steinman, R.M. (1991). The dendritic cell system and its role in immunogenicity. *Annu. Rev. Immunol.*, 9: 271-296.

Sun, Y., Jurgovsky, K., Möller, P., Alijagic, S., Dorbic, T., Georgieva, J., Wittig, B. and Schadendorf, D. (1998). Vaccination with Il-12-gene modified autologous melanoma cells - preclinical results and a first clinical phase I study. *Gene Ther.*, 5: 481-490.

Sun, Y., Paschen, A. and Schadendorf, D. (1999). Cell-based vaccination against melanoma: background, preliminary results, and perspective. *J. Mol. Med.*, 77: 593-608.

Tepper, R.I., Pattengale, P.K. and Leder, P. (1989). Murine interleukin 4 displays potent anti-tumor activity in vivo. *Cell*, 57: 503-512.

Tuting, T., Wilson, C.C., Martin, D.M., Kasamon, Y.L., Rowles, J., Ma, D., Slingluff, C.L., Wagner, S.N., van der Bruggen, P., Baar, J., Lotze, M.T. and Storkus, W.J. (1998). Autologous human monocyte-derived dendritic cells genetically modified to express melanoma antigens elicit primary cytotoxic T cell responses in vitro: enhancement by cotransfection of genes encoding the Th1-biasing cytokines IL-12 and IFN-alpha. *J. Immunol.*, 160: 1139-1147.

Veelken, H., Mackensen, A., Lahn, M., Kohler, G., Becker, D., Franke, B., Brennscheidt, U., Kulmburg, P., Rosenthal, F.M., Keller, H., Hasse, J., Schultze-Seemann, W., Farthmann, E.H., Mertelsmann, R. and Lindemann, A. (1997). A phase-I clinical study of autologous tumor cells plus interleukin-2-gene-transfected allogeneic fibroblasts as a vaccine in patients with cancer. *Int. J. Cancer*, 70: 269-277.

Vieweg, J. and Gilboa, E. (1995). Considerations for the use of cytokine-secreting tumor cell preparations for cancer treatment. *Cancer Invest.*, 13: 193-201.

Zitvogel, L., Mayordomo, J.I., Tjandrawan, T., DeLeo, A.B., Clarke, M.R., Lotze, M.T. and Storkus, W.J. (1996). Therapy of murine tumors with tumor peptide-pulsed dendritic cells: dependence on T cells, B7 costimulation, and T helper cell 1-associated cytokines. *J. Exp. Med.*, 183: 87-97.

7. SUMMARY — Metastatic melanoma is presently incurable by conventional therapies including surgery and chemotherapy. Since melanoma is perceived as an immunogenic tumor and T lymphocytes play a crucial role in controlling tumor growth, considerable interest has arisen to develop and apply immunotherapeutic vaccines for treatment/prevention of the metastases. Vaccines consisting of tumor cells transduced with cytokine genes have been intensely investigated in the past decade and are currently being tested in numerous phase I clinical trials. The identification of melanoma-associated antigens and their peptide epitopes recognized by T cells in the context of MHC class I has allowed recent development of antigen-specific melanoma vaccines. Vaccinations with dendritic cells pulsed with or transduced with genes encoding tumor antigens are also a potent strategy to activate/induce antigen-specific T cell responses to the tumors. Both strategies are entering clinical testing as well. This chapter will review and discuss the foundations of and some clinical trials with these novel vaccines.

Key Words: melanoma; vaccines; gene therapy; dendritic cells; T lymphocytes; tumor antigens

Progress in Gene Therapy: Basic and Clinical Frontiers, pp. 343-360
R. Bertolotti *et al.* (Eds)

Apoptin® induces tumor-specific apoptosis: Potentials for a novel anti-tumor therapy

Mathieu H. M. Noteborn[1,2,*] and Alex J. Van der Eb[2]

[1]*Leadd BV, and* [2]*Department of Molecular Cell Biology, Leiden University, P.O. Box 9503, 2300 RA, Leiden, The Netherlands*

Table of Contents

* Corresponding author. E-mail: noteborn@leadd.nl

1. APOPTOSIS

Eukaryotic cells, when stimulated by certain signals, have the capacity to orderly enter a program leading to cell death. This process is called apoptosis and is a physiologically programmed form of cell death requiring *de novo* gene transcription and translation (Jacobson *et al.*, 1997; Chinnaiyan and Dixit, 1996). Apoptosis is characterized by shrinkage of cells, segmentation of the nucleus, and condensation and cleavage of DNA. The apoptotic remnants are enveloped by membranes and undergo rapid phagocytosis by neighboring cells. No cellular contents are released, so that inflammation reactions are avoided, in contrast to necrosis which constitutes a non-physiological process of cell death (Vaux and Strasser, 1996; Wyllie, 1995).

Apoptosis plays an important role in many natural processes, *e.g.* the formation of tissues and organs during embryogenesis, but also in the adult organism, in tissue renewal, in the regulation of the immune system, and the elimination of derailed cells which threaten to become tumors. In a variety of diseases, increased (*e.g.* AIDS) or decreased apoptosis (*e.g.* auto-immune disease and cancer) is involved (Duke *et al.*, 1996).

2. THE ROLE OF APOPTOSIS IN TUMOR FORMATION

Originally, it was thought that the development of tumors is caused mainly by enhanced cell proliferation. Recent research shows, however, that a decreased level of apoptosis also contributes to tumor formation (Duke *et al.*, 1996). Cells that have incurred damage in their DNA will have to repair this damage before they will divide, if possible. If they are not able to repair the damage, mutations may arise which may turn them into tumor cells. Alternatively, damaged cells may have sustained so much damage that they decide to undergo apoptosis and, by doing so, prevent tumor growth (Hunter, 1997; Paulovich *et al.*, 1997).

The tumor-suppressor protein p53 plays an important role in the prevention of oncogenic transformation. More than 90% of the individuals with a hereditary p53 defect develop sarcomas, breast tumors, adrenal carcinomas, *etc.* It is generally assumed that tumor suppression is due at least in part to p53-induced apoptosis and that, during tumor formation, a selection for p53 loss of function takes place (Smith and Fornace, 1995; Levine, 1997). Expression of the oncogene c-myc can induce either cell proliferation or apoptosis, depending on the cellular background. Over-expression of c-myc results in apoptosis in the presence of p53, whereas in its absence it leads to enhanced cell proliferation. Many tumors lack functional p53 and/or over-express c-myc (Packham and Cleveland, 1995; Canman and Kastan, 1995). Other tumor-suppressor genes, such as the retinoblastoma (Rb) gene, can similarly reduce tumor growth via induction of apoptosis (Vaux and Strasser, 1996). The anti-apoptotic protein Bcl-2, and some related proteins, *e.g.* Bcl-x-l, BAG-1, and Bak (Reed, 1997; Rao and White, 1997) are also involved in tumor development. Bcl-2 is over-expressed as a result of chromosomal translocations in various tumors, *e.g.* leukemia, lymphoma, and breast cancer. Another example is chronic myeloid leukemia caused by a translocation involving the *bcr* and *abl* genes. The resulting hybrid proteins have oncogenic activity and also a strong anti-apoptotic activity (Hunter, 1997).

Most of the DNA viruses that are known to date have oncogenic or transforming activity. Many of these viruses also harbor genes that prevent or inhibit apoptosis (Teodoro and Branton, 1997). For example, the human high risk papilloma viruses contain an early gene, E6, the product of which can suppress apoptosis by its ability to target p53 for rapid degradation. Epstein-Barr virus has 2 apoptosis-related proteins: BHRF1, which is a homologue of Bcl-2, and LMP-1, of which the product stimulate Bcl-2 expression.

3. APOPTOSIS PLAYS AN IMPORTANT ROLE IN CANCER THERAPY

Apoptosis is not only essential for tumor growth, but can also be exploited for the treatment of tumors. However, many tumor cells have a defect in the decision machinery for apoptosis, but an intact execution system. Such tumor cells will still die, if they are provided with an effective apoptotic signal (McDonnell *et al.*, 1995). Chemo-therapeutic agents normally induce apoptosis via activation of p53. However, more than 50% of the human tumors, including melanoma, lung cancer, or colon carcinoma do not contain functional p53. Patients with such tumors have a very low chance of responding to (chemo)-therapy (Levine, 1997; Lowe *et al.*, 1994). Likewise, over-expression of the anti-apoptosis proteins Bcl-2 or BCR-ABL negatively influences the chemo-therapeutic treatment of a large number of lymphomas (Rao and White, 1997; Martin and Green, 1994).

Considerable efforts are currently being invested in the development of new anti-tumor therapies, which are based on the induction of apoptosis and are not hampered by the lack of functional p53 and/or over-expression of anti-apoptotic genes. Some of these are based on the restoration of p53 expression, which will render tumor cells sensitive to anti-tumor drugs (Soengas *et al.*, 1999).

We have recently discovered a viral protein, Apoptin, which induces apoptosis specifically in tumor cells. Here, we will discuss its interactions with apoptosis-mediating (*e.g.* p53) and apoptosis-inhibiting proteins (*e.g.* Bcl-2), as well as a prototype anti-tumor therapy based on expression of Apoptin.

4. APOPTIN, A CHICKEN ANEMIA VIRUS (CAV)-DERIVED PROTEIN, INDUCES APOPTOSIS

CAV causes aplastic anemia in young chickens, due to destruction of erythroblastoid cells, and immunodeficiency due to severe depletion of thymocytes (reviewed in: Coombes and Crawford, 1996; Noteborn and Koch, 1995). The cause of the high susceptibility of young chickens, as opposed to older ones, is unknown. It may be due to the fact that the third wave of precursor cells that populate the thymus after hatching, are more resistant. This indicates that the cellular background specifies the susceptibility to CAV.

a) Chicken anemia virus induces apoptosis *in vivo* and *in vitro*

We found evidence that CAV induces apoptosis in the thymocytes of young chickens, as well as in transformed chicken and human cells under tissue-culture conditions (Noteborn *et al.*, 1993; Jeurissen *et al.*, 1992). DNA isolated from the infected cells shows the apoptosis-specific laddering pattern, which is not observed in DNA isolated from non-infected cells. Electron-microscopic analysis of infected thymocytes reveals the presence of cells containing condensed chromatin adjacent to the nuclear membrane, and apoptotic bodies in the cytoplasm of neighboring epithelial cells.

b) The CAV genome specifies a polycistronic mRNA encoding three proteins

The CAV genome, which was cloned in our laboratory, consists of a single-stranded circular DNA of 2319 nucleotides (Noteborn *et al.*, 1991). Upon infection of its target cells, it replicates via a double-stranded (ds) DNA intermediate. From this ds CAV molecule, a single polyadenylated polycistronic mRNA is transcribed (Noteborn and Koch, 1995; Noteborn *et al.*, 1992) which encodes three distinct proteins, VP1 (52 kDa), VP2 (24 kDa) and VP3 (14 kDa).

Since expression of VP3 alone induces apoptosis, this protein has been named Apoptin®. Soon after transfection, with a plasmid encoding Apoptin, the protein is dispersed throughout the nucleus and somewhat later the cell undergoes apoptosis. At this point in time, Apoptin aggregates and the cellular DNA condenses and/or is fragmented (Noteborn *et al.*, 1994). Apoptin is a 121-amino-acid protein, which does not resemble any other sequenced animal or viral protein. It carries several known protein domains, including two nuclear-localization signals, and a nuclear-export signal. Furthermore, it has regions rich in proline or basic amino acids and overall contains a high percentage of serine and threonine residues.

Zhuang *et al.* (1995) have provided evidence that the nuclear localization of Apoptin is important for its apoptotic activity. Deletion of one of the nuclear-localization signals significantly reduces the apoptotic activity of Apoptin, and causes partial translocation of Apoptin to the cytoplasm. Therefore, nuclear localization seems to be essential for Apoptin activity, as well as co-localization with chromatin. The basic regions of Apoptin may allow interaction with nucleic acids. The presence of Apoptin in the chromatin structure, and its high proline content, may cause a perturbation of the supercoil organization, which could result in apoptosis. Another possibility is that Apoptin acts as a transcriptional activator and/or inhibitor of genes, that directly bring about apoptosis (Noteborn *et al.*, 1998).

5. CELL DEATH INDUCED BY APOPTIN IS p53-INDEPENDENT AND IS NOT INHIBITED BY BCR-ABL

a) Apoptin induces p53-independent apoptosis

The apoptotic effect of Apoptin has been examined in many human and other mammalian tumorigenic/transformed cell lines of different origins, including breast and lung tumors, colon and cholangiocarcinomas, hepatoma, lymphoma, leukemia or neuroblastoma. In all these cell lines, Apoptin induced apoptosis. Although, the rate of Apoptin-induced apoptosis varies from one cell line to the other, it always affects 90-100% of the Apoptin-positive cells 6 days after transfection (Noteborn *et al.*, 1998a).

Many effectors and regulators of apoptotic pathways have been identified in various systems. Although more than one pathway per cell exists and cell specificity can be important, the principal steps are universal (White, 1996; Boise *et al.*, 1995). It is assumed that the apoptotic decision process is a cascade of events, followed by an execution stage. At each decision step after an apoptotic trigger, a go/no-go decision can be made. In many cases, apoptosis is triggered via activation of the tumor-suppressor protein p53 (Levine, 1997; Guchelaar *et al.*, 1997). The next step in the cascade might be controlled by antiapoptotic and proapoptotic Bcl-2 and/or Bcl-2-like proteins (Reed, 1997; Rao and White, 1997). This might subsequently result in activation of a cascade of specific cystein proteases (caspases), which is regarded as the point of no return after which the cells will rapidly undergo apoptosis (Nicholson and Thornberry, 1997).

To examine whether Apoptin requires expression of wild-type p53 to induce apoptosis, the human osteosarcoma cell lines U2OS (wild-type p53), Saos-2/ala143 (mutant p53), and Saos-2 (no p53) were transiently transfected with plasmid DNA encoding Apoptin (Zhuang *et al.*, 1995a). At various time points after transfection, the percentage of Apoptin-expressing p53-minus cells, in which the nuclei became apoptotic, was similar to that in U2OS cells containing wild-type p53. Six days after transfection, almost all Apoptin-positive cells had become apoptotic, whereas in the same cell cultures only 1-2% of the Apoptin-negative cells showed apoptotic features. Analysis of cellular DNA isolated from the p53-minus Saos-2 cells showed the expected oligonucleosomal DNA laddering pattern, which is the hallmark of apoptosis (Kerr *et al.*, 1994). The Apoptin-induced apoptosis does not seem to be slower in p53-minus Saos-2 cells than apoptosis induced by p53, which indicates that the distinct Apoptin pathway is as potent as the p53-regulated one.

Subsequently, we have shown that Apoptin can induce apoptosis in various other human or mammalian cells that do not express p53 (Noteborn *et al.*, 1998). Recently, we have found that various human cholangio-carcinoma-derived cell lines having non-functional or no p53 at all are very sensitive to Apoptin-induced apoptosis (Pietersen and Van Tongeren, unpublished results). Besides p53 gene deletion and point mutations, p53 can be inactivated by proteins encoded by DNA-tumor viruses. Expression of the adenovirus E1B 55K protein (Zhuang *et al.*, unpublished results) or of the SV40 large T antigen (Danen-van Oorschot *et al.*, 1997) do not interfere with apoptosis induced by Apoptin. All these data imply that Apoptin acts via a p53-independent apoptotic pathway.

b) Apoptin is not inhibited by BCR-ABL

Apoptin also induces apoptosis in human K562 cells derived from acute myeloid leukemias, which produce a chimeric BCR-ABL protein (Zhuang *et al.*, 1995). From a therapeutic viewpoint, these properties of Apoptin are important. Many tumors express BCR-ABL and/or lack p53, and thus are resistant to chemotherapeutic agents.

6. BCL-2 ENHANCES APOPTIN-INDUCED APOPTOSIS IN TRANSFORMED CELLS

In our laboratory and many others, Bcl-2 has been shown to inhibit p53-mediated apoptosis (Rao and White, 1997; Danen-van Oorschot *et al.*, 1997). Nevertheless, Apoptin could still induce apoptosis in human malignant blood cells expressing high levels of endogenous Bcl-2. In the lymphoblastoma-derived DoHH-2 which contains a high level of Bcl-2, Apoptin induced cell death even faster than in K562 cells, with a normal level of Bcl-2 (Zhuang *et al.*, 1995).

Therefore, we examined in more detail whether Bcl-2 can enhance Apoptin activity. To that end, human Saos-2 cells were transiently transfected with plasmids encoding Apoptin and/or Bcl-2. Several days after transfection, Saos-2 cells transfected with both Apoptin and Bcl-2 underwent apoptosis to a significantly higher level than cells expressing Apoptin alone (Danen-van Oorschot *et al.*, 1997a and 1998). Apparently, Bcl-2 accelerates Apoptin-induced apoptosis in transformed mammalian cells, which is surprising since Bcl-2 is known to inhibit apoptosis induced by the tumor-suppressor gene p53 (Rao and White, 1997).

In these experiments, the cells expressing Apoptin alone harbor the protein in their nuclei (Noteborn *et al.*, 1993). In transfected cells expressing both Apoptin and Bcl-2, Apoptin was also situated within the nucleus, and Bcl-2 in the cytoplasm, just as when they are over-expressed separately. These results indicate that Apoptin does not change the cytoplasmic localization of Bcl-2 or vice versa (Danen-van Oorschot *et al.*, 1997). Immunoprecipitation assays show that Bcl-2 does not co-precipitate with Apoptin, implying that no direct interaction between Apoptin and Bcl-2 exists (A. Den Hollander, unpublished data).

These data indicate that Bcl-2 indirectly stimulates the Apoptin-mediated reactions in tumor cells. It may be predicted that Apoptin should be especially effective in tumor cells over-expressing Bcl-2.

Caspases seem not to be involved in Apoptin-induced apoptosis

The cowpox-virus protein CrmA, which inhibits caspase 1 activity can inhibit p53-induced apoptosis, does not negatively influence Apoptin-induced apoptosis (Danen-van Oorschot, unpublished results). Similar results were obtained with modified peptides, known to specifically inhibit caspases, also those more downstream in the caspase cascade.

The fact that Apoptin acts independently of p53 and not via caspases (at least not those required for p53-induced apoptosis), that it is stimulated by the proto-oncogene Bcl-2 and not inhibited by other anti-apoptosis proteins such as BCR-ABL, suggests that Apoptin induces apoptosis via a unique and novel pathway.

7. APOPTIN DOES NOT INDUCE APOPTOSIS IN NORMAL NON-TRANSFORMED HUMAN CELLS

Apoptin can induce apoptosis in cell lines derived from a great variety of human tumors (Noteborn *et al.*, 1998a). On the other hand, Apoptin does not induce apoptosis in "normal" non-transformed human diploid cells, such as primary fibroblasts, keratinocytes, smooth muscle cells, T cells or endothelial cells. The possible cause for this phenomenon is that, in tumor cells, Apoptin is located in the nucleus, whereas in normal cells it was found to be present in the cytoplasm. Long-term expression of Apoptin in normal human fibroblasts revealed that it has no toxic or transforming activity in these cells and that Apoptin does not interfere with cell proliferation (Danen-van Oorschot *et al.*, 1997).

It is not yet known whether transformed cells, as opposed to normal cells, cannot recognize the nuclear export signal of Apoptin, or whether differential usage of the nuclear localization signals might be the explanation for the phenomena. An alternative explanation might be that Apoptin is differentially modified, such as phosphorylation, in tumorigenic/transformed cells versus normal cells. Transient transfection of normal human diploid cells with Apoptin gene and the SV40 transforming large-T antigen gene resulted in Apoptin-induced apoptosis. Immuno-fluorescence analysis showed that, in these cells, Apoptin had moved from the cytoplasm into the nucleus. Apparently, the expression of a transforming protein is sufficient to render normal cells susceptible to Apoptin, and the establishment of a stably transformed state is not required (Zhang *et al.*, unpublished results).

In view of the data obtained with co-expression of SV40 large-T antigen and Apoptin, one might assume that over-expression of the proto-oncogene Bcl-2 in "normal" human cells would also force Apoptin to enter the nucleus resulting in the induction of apoptosis. However, cells expressing Apoptin either alone or together with Bcl-2, do not undergo Apoptin-induced apoptosis (Danen-van Oorschot *et al.*, 1997a). Mitochondria are known to play a major role in Bcl-2-regulated apoptosis (Kroemer *et al.*, 1997). Staining of Apoptin-expressing cells with both Mito-Tracker dye, which specifically stains mitochondria, and a monoclonal antibody specific for Apoptin, however, shows that the majority of the Apoptin signal did not co-localize with the mitochondria (Danen-van Oorschot *et al.*, 1998). These results prove again that Bcl-2 and Apoptin do not affect each other's activity in normal cells, whereas they do so indirectly in transformed cells.

Apoptin induces apoptosis in UV-irradiated cells from individuals with hereditary cancer-prone syndromes

Noteborn *et al.* (1998b) demonstrated that UV-C-irradiated normal primary fibroblasts remain fully resistant to Apoptin, as could be expected since a single irradiation with UV will not cause oncogenic transformation of normal human diploid fibroblasts. Surprisingly, however, UV irradiation caused an apoptotic response to Apoptin in diploid fibroblasts from individuals with a hereditary cancer predisposition which is due to a germ-line mutation in a tumor-suppressor gene (Abrahams *et al.*, 1996). UV-C irradiation of cells from patients with Li-Fraumeni syndrome (p53+/–), dysplastic nevus syndrome (p16 –/–), or Lynch Type 2 syndrome (defect unknown), resulted in induction of Apoptin-induced apoptosis. As expected, these diploid non-transformed cells are resistant to Apoptin in the absence of UV irradiation. After transfection of untreated cells, Apoptin is found predominantly in the cytoplasm, whereas in UV-C-exposed cells derived from the individuals with hereditary cancer-prone syndromes it is present in the nucleus, causing rapid apoptosis. The induction of apoptosis by Apoptin in "cancer-prone" cells is UV-dose-dependent and is transient, just like many other UV-induced processes (Zhang *et al.*, 1999).

Loss of a tumor-suppressor gene apparently renders diploid cells susceptible to Apoptin-induced apoptosis, if they are exposed to a mutagenic radiation dose, which can be "handled" in diploid cells from healthy individuals.

8. APOPTIN IS A PROMISING ANTI-TUMOR AGENT

The findings that the Apoptin gene can specifically induce apoptosis in tumor cells, clearly make Apoptin a promising agent for anti-tumor therapy. Hence, we are attempting to develop strategies for gene therapy against cancer based on (viral) vectors expressing Apoptin.

The characterization of the Apoptin-induced apoptotic pathway has so far shown that it is specific for transformed cells and does not act according to known mechanisms. Cellular counterparts of Apoptin have so far not been discovered. Presently, we are isolating Apoptin-associating proteins by means of a yeast-2-hybrid system, which will hopefully help to elucidate the mechanism.

To investigate the potential of Apoptin for cancer gene therapy, the Apoptin gene has to be efficiently introduced into malignant cells *in vivo*. At present, the most efficient system to achieve this makes use of adenoviral vectors which have several advantages that make them particularly suitable for *in-vivo* gene transfer: recombinant adenoviral vectors (rAdVs) can be grown to high titers, have the capacity to transduce non-mitotic cells, and do not integrate their genomes into host-cell DNA (Graham *et al.,* 1977). Moreover, adenovirus vectors have already been applied for clinical gene-therapy trials.

Pietersen *et al.* (1999) generated and characterized an adenovirus vector, AdMLPvp3, for the expression of Apoptin. Infection with AdMLPvp3 of normal rat hepatocytes in cell culture did not cause more apoptosis than infection with a control Ad vector. In contrast, in the hepatoma cell line HepG2, infection with AdMLPvp3, but not with control vectors, led to a rapid cell loss. To characterize the nature of AdMLPvp3-induced cell death as apoptosis, the presence of DNA strand breaks was visualized with the aid of the enzyme terminal deoxynucleotidyl transferase and FITC-labeled dUTP (TUNEL assay).

Toxicity experiments in rats demonstrated that AdMLPvp3 could be safely administered intra-peritoneally, subcutaneously or intravenously. Repeated intravenous doses of AdMLPvp3 were also well tolerated, indicating that the Apoptin-expressing virus does not cause adverse effects (Pietersen *et al.*, 1999). The fact that transgenic-Apoptin mice have been obtained expressing the Apoptin gene in various tissues and organs under the control of a constitutive promoter, strengthens the low toxicity of Apoptin in normal healthy cells (Erkeland, unpublished publications).

Futher experiments showed that adenovirus-encoded Apoptin also has anti-cancer activity *in vivo*. One single intratumoral injection of AdMLPvp3 into a xenogeneic tumor (HepG2 cells in Balb/C$^{nu/nu}$ mice) resulted in a significant reduction of tumor growth (Pietersen *et al.*, unpublished results). Remarkably, the Apoptin-treated tumors showed a reduced vascularization, in comparison with the control groups which received intratumoral injection with control Ad vector. In a long-term experiment, we have found evidence that AdMLPvp3 has a significant tumor-growth reduction in comparison with a control adenovirus expressing LacZ (Pietersen, unpublished results).

The intrinsic tumor specificity of the Apoptin gene makes it a promising new tool for cancer gene therapy. So far, attempts by others to obtain tumor-cell specificity rely on targeting the vector to tumor cells, on tumor-specific expression of the transgene, or on local administration of the gene-transfer vehicle. To date, the applicability of the available systems is limited by their relatively low efficiency, or by insufficient specificity (Dachs *et al.*, 1997; Bui *et al.*, 1997; Roth and Christiano, 1997). Promising results have been obtained with the HSV-TK/gancyclovir combination in brain tumors (Moolten, 1986). The applicability of this system outside the central nervous system, however is hampered by toxic effects for certain normal tissues, *e.g.* the liver parenchyma (Van der Eb *et al.*, 1998; Brand *et al.*, 1997). Another approach is based on the reintroduction of tumor-suppressor genes, such as p53 or the Rb protein, into tumor cells in which the endogenous counterparts are affected (Hong *et al.*, 1996; Nielsen *et al.*, 1997).

The p53 approach combines two important features for cancer therapy: efficacy and specificity. For instance, studies with Ad-p53 show that re-expression of p53 in established tumors can induce apoptosis and tumor regression *in vivo* (Polyak *et al.*, 1996). Both preclinical and clinical studies suggest a low toxicity for non-transformed cells when they are forced to express exogenous p53 at levels lethal for tumor cells (Nielsen and

Maneval, 1998; Zhang *et al.*, 1995; Roth *et al.*, 1996). However, the expression of wild-type p53 is most effective in cells lacking functional p53. Tumor cells expressing wild-type p53 are only marginally affected and, at best, exhibit growth arrest (Polyak *et al.*, 1996). *In vitro*, Apoptin does not seem to discriminate between $p53^-$ and $p53^+$ cells (Zhuang *et al.*, 1995).

9. CONCLUDING REMARKS

Resistance to induction of apoptosis can play a major role in the induction of cancer. This can be caused by mutations in certain tumor-suppressor genes (p53) or oncogenes (*e.g.* Bcl-2), which result in inhibition of intrinsic programmed cell death pathways. These mutations also make many tumors resistant to apoptosis induced by cytotoxic agents.

Hence, new strategies for cancer therapy focus on by-passing the resistance to (chemo)therapy of tumors. In this report, we have provided evidence that Apoptin combines two valuable properties: specificity and efficacy. It induces apoptosis in tumor cells, but not in normal cells, and can exert its activity under conditions where most currently used anti-cancer agents fail. The first preclinical results are very encouraging and strengthens the specific anti-tumor potential of Apoptin-induced apoptosis.

Acknowledgements

This project was partially made possible by grants from The Netherlands Ministry of Economic Affairs, The Hague, The Netherlands.

10. REFERENCES

Abrahams, P.J., Houweling, A., Cornelissen-Steijger, P.D.M., Arwert, F., Menko, F.H., Pinedo, H.M., Terleth, C. and Van der Eb, A.J. (1996). Inheritance of abnormal expression of SOS-like response in Xeroderma pigmentosum and hereditary cancer-prone syndromes. *Cancer Res.*, 56: 2621-2625.

Boise, L.H., Gottschalk, A.R., Quintans, J. and Thompson, C.B. (1995). Bcl-2 and Bcl-2-related proteins in apoptosis regulation. *Curr. Top. Microbiol. Immunol.*, 200: 107-121.

Brand, K. Arnold, W., Bartels, T., Lieber, A., Kay, M.A., Strauss, M. and Dorken, B.

(1997). Liver-associated toxicity of the HSV-tk/GCV approach and adenoviral vectors. *Cancer Gene Ther.*, 4: 9-16.

Bui, L.A., Butterfield, L.H., Kim, J.Y., Ribas, A., Seu, P., Lau, R., Glaspy, J.A., McBride, W.H. and Economou, J.S. (1997). In-vivo therapy of hepatocellular carcinoma with a tumor-specific adenoviral vector expressing interleukin-2. *Hum. Gene Ther.,* 8: 2173-2182.

Canman, C.E. and Kastan, M.B. (1995). Induction of apoptosis by tumor suppressor genes and oncogenes. *Sem. Cancer. Biol.*, 6: 17-25.

Chinnaiyan, A.M. and Dixit, V. (1996). The cell-death machine. *Curr. Biol.*, 6: 555-562.

Coombes, A.L. and Crawford, G.R. (1996). Chicken anaemia virus: a short review. *World's Poultry Sci. J.*, 52: 267-277.

Dachs, G.U., Dougherty, G.J., Stratford, IJ. and Chaplin, D.J. (1997). Targeting gene therapy to cancer: a review. *Oncol. Res.*, 9: 313-325.

Danen-van Oorschot, A.A.A.M., Fischer, D.F., Grimbergen, J.M., Klein, B., Zhuang, S.-M., Falkenburg, J.H.F., Backendorf, C., Quax, P.H.A., Van der Eb, A.J. and Noteborn, M.H.M. (1997). Apoptin induces apoptosis in human transformed and malignant cells but not in normal cells. *Proc. Natl. Acad. Sci. USA*, 94: 5843-5847.

Danen-Van Oorschot, A.A.A.M., Den Hollander, A., Takayama, S., Reed, J., Van der Eb, A.J. and Noteborn, M.H.M. (1997a). BAG-1 inhibits p53-induced but not Apoptin-induced apoptosis. *Apoptosis*, 2: 395-402.

Danen-van Oorschot, A.A.A.M, Zhang, Y., Erkeland, S., Fischer, D.F., Van der Eb, A.J. and Noteborn, M.H.M. (1999). The effect of Bcl-2 on Apoptin in normal cells versus transformed human cells. *Leukemia*, 13: S1, S75-S77.

Duke R.C., Ojcius D.M., and Young J.D.-E. (1996). Cell suicide in health and disease. *Sci. American*, 274: 48-55.

Graham, F.L., Smiley, J., Russell, W.C. and Nairn, R. (1977). Characteristics of a human cell line transformed by DNA from human adenovirus type 5. *J. Gen. Virol.*, 36: 59-74.

Guchelaar, H.J., Vermes, A., Vermes, I. and Haanen, C. (1997). Apoptosis: molecular mechanisms and implications for cancer chemotherapy. *Pharm. World Sci.*, 19: 119-125.

Hunter T. (1997). Oncoprotein Networks. *Cell*, 88: 333-346.

Jacobson, M.D., Weil, M. and Raff, M. (1997). Programmed cell death in animal development. *Cell*, 88: 347-354.

Jeurissen, S.H.M., Wagenaar, F., Pol, J.M.A., Van der Eb, A.J. and Noteborn, M.H.M. (1992). Chicken anemia virus causes apoptosis of thymocytes after in-vivo infection and of cell lines after in-vitro infection. *J. Virol.*, 66: 7383-7388.

Kerr, J.F.R., Winterford, C.M. and Harmon, B.V. (1994). Apoptosis: Its significance in cancer and cancer therapy. *Cancer* , 73: 2013-2026.

Kroemer, G., Zamzani, N. and Susin, S.A. (1997). Mitochondrial control of apoptosis. *Immunol. Today*, 18: 44-51.

Levine, A.J. (1997). p53, the cellular gatekeeper for growth and division. *Cell*, 88: 323-331.

Lowe, S., Bodis, S., McClathey, A., Remington, L., Ruley, H.E., Fisher, D.E., Houseman, D.E. and Jacks, T. (1994). p53 status and the efficacy of cancer therapy *in vivo*. *Science*, 266: 807-810.

Martin, S.J. and Green, D.R. (1994). Apoptosis as a goal of cancer therapy. *Curr. Opin. Oncol.*, 6: 616-621.

McDonnell, T.J., Meyn, R.E. and Robertson, L.E. (1995). Implications of apoptotic cell death regulation in cancer therapy. *Sem. Cancer Biol.*, 6: 53-60.

Moolten, F. (1986). Tumor chemosensitivity conferred by inserted herpes thymidine kinase genes: paradigm for a prospective cancer control strategy. *Cancer Res.*, 46: 5276-5281.

Nicholson, D.W. and Thornberry, N.A. (1997). Caspases: killer proteases. *TIBS*, 22: 299-306.

Nielsen, L.L. and Maneval, D.C. (1998). p53 tumor suppressor gene therapy for cancer. *Cancer Gene Ther.*, 5: 52-63.

Nielsen, L.L., Dell, J., Maxwell, E., Armstrong, L,. Maneval, D., Catino, J.J. (1997). Efficacy of p53 adenovirus-mediated gene therapy against human breast cancer xenografts. *Cancer Gene Ther.*, 4: 129-138.

Noteborn, M.H.M. and Koch, G. (1995). Chicken anemia virus infection: molecular basis of pathogenicity. *Avian Pathol.*, 24: 11--31.

Noteborn, M.H.M., De Boer G.F., Van Roozelaar, D., Karreman, C., Kranenburg, O., Vos, J.G., Jeurissen, S.H.M., Hoeben, R.C., Zantema, A., Koch, G., Van Ormondt, H. and Van der Eb (1991). Characterization of cloned chicken anemia virus DNA that contains all elements for the infectious replication cycle. *J. Virol.*, 65: 3131-3139.

Noteborn, M.H.M., Kranenburg, O., Zantema, A., Koch, G., De Boer, G.F. and Van der Eb, A.J. (1992). Transcription of the chicken anemia virus (CAV) genome and synthesis of its 52-kDa protein. *Gene*, 118: 267-271.

Noteborn, M.H.M., Van der Eb, A.J., Koch, G. and Jeurissen, S.H.M. (1993). VP3 of the chicken anemia virus (CAV) causes apoptosis. In: *Vaccines 93: Modern approaches to new vaccines including prevention of AIDS*, Ginsberg, H.S., Brown, F., Chanock, R.M., Lerner, R.A. (Eds.), Cold Spring Harbor Laboratory Press, N.Y., pp 299-304.

Noteborn, M.H.M., Todd, D., Verschueren, C.A.J., De Gauw, H.W.F.M., Curran, W.L., Veldkamp, S., Douglas, A.J., McNulty, M.S., Van der Eb, A.J. and Koch, G. (1994). A single chicken anemia virus protein induces apoptosis. *J . Virol.*, 68: 346-351.

Noteborn, M.H.M., Danen-van Oorschot, A.A.A.M. and Van der Eb, A.J. (1998). Chicken anemia virus: Induction of apoptosis by a single protein of a single-stranded DNA virus. *Sem. Virol.*, 8: 497-504.

Noteborn, M.H.M., Danen-van Oorschot, A.A.A.M. and Van der Eb, A.J. (1998a). The Apoptin® gene of chicken anemia virus in the induction of apoptosis in human tumorigenic cells and in gene therapy of cancer. *Gene Therapy and Molecular Biology*, 1: 399-406.

Noteborn, M.H.M., Zhang, Y. and Van der Eb, A.J. (1998b). Apoptin specifically causes apoptosis in tumor cells and after UV-treatment in untransformed cells from cancer-prone individuals: A review. *Mutation. Res.*, 400: 447-456.

Packham, G. and Cleveland, J.L. (1995). c-Myc and apoptosis. *Biochim. Biophys. Acta*, 1242: 11-28.

Paulovich, A.G., Toczyscki, D.P. and Hartwell, L.H. (1997). When Checkpoints fail. *Cell*, 88: 315 321.

Pietersen, A.M., Van der Eb, M.M., Rademaker, H.J., Van den Wollenberg, D.J.M., Rabelink, M.J.W.E., Van Ormondt, H., Masman, D., Van de Velde, C.J.H., Van der Eb, A.J., Hoeben R.C. and Noteborn, M.H.M. (1999). Specific tumor-cell killing with adenovirus vectors containing the Apoptin gene. *Gene Ther.* 6: 882-892.

Polyak, K., Waldman, T., He, T.C., Kinzler, K.W. and Vogelstein, B. (1996). Genetic determinants of p53-induced apoptosis and growth arrest. *Genes Dev.*, 10: 1945-1952.

Rao, L. and White, E. (1997). Bcl-2 and the ICE family of apoptotic regulators: making a connection. *Curr. Opin. Genet. Dev.*, 7: 52-58.

Reed, J.C. (1997). Double identity for proteins of the Bcl-2 family. *Nature*, 387: 773-776.

Roth, J.A. and Christiano, R.J. (1997). Gene therapy for cancer: what have we done and where are we going? *J. Natl. Cancer Inst.*, 89: 21-39.

Roth, J.A., Nguyen, D., Lawrence, D.D., Kemp, B.L., Carrasco, C.H., Ferson, D.Z., Hong, W.K., Komaki, R., Lee, J.J., Nesbitt, J.C., Pisters, K.M., Putnam, J.B., Schea, R., Shin, D.M., Walsh, G.L., Dolormente, M.M., Han, C.I., Martin, F.D., Yen, N., Xu, K., Stephens, L.C., McDonnell, T.J., Mukhopadhyay, T. and Cai, D. (1996). Retrovirus-mediated wild-type p53 gene transfer to tumor of patients with lung cancer. *Nature Med.*, 2: 985-991.

Soengas, M.S., Alarcon, R.M., Yoshida, H., Giaccia, A.J., Hakem, R., Mak, W. and Lowe, S.W (1999). Apaf-1 and caspase-9 in p53-dependent apoptosis and tumor inhibition. *Nature*, 284: 156-159.

Smith, M.L. and Fornace Jr., A.J. (1995). Genomic instability and the role of p53 mutations in cancer cells. *Curr. Opin. Oncol.*, 7: 69-75.

Teodoro, J.G. and Branton, P.E. (1997). Regulation of apoptosis by viral gene products. *J. Virol.*, 71: 1739-1746.

Van der Eb, M.M., Cramer, S.J., Vergouwe, Y., Schagen, F.H.E., Van Krieken, J.H.J.M., Van der Eb, A.J., Borel Rinkes, I.H.M., Van de Velde, C.J.H. and Hoeben, R.C. (1998). Severe hepatic dysfunction after adenovirus-mediated transfer of the Herpes Simplex Virus thymidine-kinase gene and ganciclovir administration. *Gene Ther.*, 5: 451-458.

Vaux, D.L. and Strasser, A. (1996). The molecular biology of apoptosis. *Proc. Natl. Acad. Sci. USA*, 93: 2239-2244.

White, E. (1996). Life, death and the pursuit of apoptosis. *Genes Dev.*, 10: 1-15.

Wyllie, A.H. (1995). The genetic regulation of apoptosis. *Curr. Opin. Gen. Dev.*, 5: 97-104.

Xu, H.J., Zhou, Y., Seigne, J., Perng, G.S., Mixon, M., Zhang, C., Li, J., Benedict, W.F. and Hu, S.X. (1996). Enhanced tumor suppressor gene therapy via replication-deficient adenovirus vectors expressing an N-terminal truncated retinoblastoma protein. *Cancer Res.*, 56: 2245-2249.

Zhang, W.W., Alemany, R., Wang, J., Koch, P.E., Ordonez, N.G., and Roth, J.A.

(1995). Safety evaluation of Ad5CMV-p53 *in vitro* and *in vivo*. *Human Gene Ther.*, 6: 155-164.

Zhang, Y.-H., Abrahams, P., Van der Eb, and Noteborn, M.H.M. (1999). Viral protein Apoptin® induces apoptosis in UV-C-irradiated cells from individuals with various hereditary cancer-prone syndromes. *Cancer Res.*, 59: 3010-3015.

Zhuang, S.-M., Landegent, J.E., Verschueren, C.A.J., Falkenburg, J.H.F., Van Ormondt, H., Van der Eb, A.J. and Noteborn, M.H.M. (1995). Apoptin, a protein encoded by chicken anemia virus, induces cell death in various human hematologic malignant cells in vitro. *Leukemia*, 9 S1: 118-120.

Zhuang, S.-M., Shvarts, A., Van Ormondt, H., Jochemsen, A.-G., Van der Eb, A.J. and Noteborn, M.H.M. (1995a). Apoptin, a protein derived from chicken anemia virus, induces a p53-independent apoptosis in human osteosarcoma cells. *Cancer Res.*, 55: 486-489.

Zhuang, S.-M., Shvarts, A., Van Ormondt, H., Jochemsen, A.-G., Van der Eb, A.J. and Noteborn, M.H.M. (1995b). Differential sensitivity to Ad5 E1B-21kD and Bcl-2 proteins of Apoptin-induced versus p53-induced apoptosis. *Carcinogenesis*, 16: 2939-2944.

11. SUMMARY — Apoptosis is a programmed physiological process for eliminating superfluous, or altered cells. Apoptosis can be induced by many stimuli through different pathways, all of which seem to converge into one evolutionarily conserved process. Apoptosis plays an important role in tumor development, and can be exploited for therapeutic purposes. Many tumor cells have a disabling mutation in the decision-making machinery for apoptosis, but their execution system may still be intact. This means that tumor cells may die if they are provided with an appropriate apoptotic signal. Apoptin, a protein encoded by an avian virus, induces apoptosis in various cultured human tumorigenic and/or transformed cell lines. In such cells, Apoptin induces wild-type p53-independent apoptosis, which is not inhibited by BCR-ABL or Bcl-2. On the other hand, Apoptin is unable to induce apoptosis in normal non-tumorigenic human cells. In animal models, Apoptin appears to be a safe and efficient anti-tumor agent. The above data imply that Apoptin holds promise as a basis for anti-tumor therapy.

Key Words: anti-tumor therapy; Apoptin®; apoptosis; Bcl-2; gene therapy; p53

Progress in Gene Therapy: Basic and Clinical Frontiers, pp. 361-386
R. Bertolotti *et al.* (Eds)

Human DNA immunization with regulatory HIV-1 genes

Anne Kjerrström, Sandra A. Calarota and Britta Wahren*

Microbiology and Tumorbiology Center, Karolinska Institute, and Swedish Institute for Infectious Disease Control, Stockholm, Sweden

Table of Contents

---

* Corresponding author. E-mail: britta.wahren@smi.ki.se

1. ISOTYPIC REGULATORY GENES AS DNA VACCINES

HIV-1 encodes over 20 different mRNAs that are divided into three classes according to their degree of splicing (Cullen, 1998). The first genes that are transcribed are the regulatory genes that are needed for initiation of viral gene expression. These proteins are encoded by a group of ~1.8 kb mRNA. Human immunodeficiency virus type 1 (HIV-1) expresses several non-structural proteins, among them three regulatory proteins called Tat, Rev and Nef. These three proteins are expressed from the multiply spliced mRNAs that appear in the cytoplasm shortly after viral infection (Feinberg and Greene, 1992; Poeschla and Wong-Staal, 1994). The second class of singly spliced mRNA (~4.5 kb) encodes Vif, Vpr, Vpu, Env and a short form of the Tat protein. The third type of mRNA is unspliced and functions as the progeny genome. It also encodes virion structural proteins Gag and Pol.

Genetic vaccination appears to be a safe way to induce responses in animal models (for review, see: Ulmer, 1997; Wahren and Brytting, 1997) and in humans (Calarota *et al.,* 1998; McGregor *et al.,* 1998; Wang *et al.,* 1998). Immunization with cDNA encoding the influenza genes for nucleoprotein and hemagglutinin has resulted in complete protection against experimental disease (Montgomery *et al.,* 1993), but resulted in low immunogenicity in humans (Donelly *et al.,* 1999). Immunization of mice, rabbits and macaques with DNA encoding the HIV-1 envelope proteins and/or regulatory proteins have resulted in high levels of specific antibodies and CTL responses against infected cells (Okuda *et al.,* 1995; Wahren *et al.,* 1996; Hinkula *et al.,* 1997b; Putkonen *et al.,* 1998). Table 1 summarizes cytotoxic effector mechanisms that have been judged to be of benefit in HIV infection.

a) Intra- and extracellular Tat activities

The Tat protein is a 16 kD nuclear, trans-activating positive regulator that increases the processitivity of RNA polymerase II (RNAP II) activity

Table 1. Cellular immune responses in HIV-1 infection (mediated by CTL)*

In natural infection

Class I-restricted CTL (CD8+)

- unusually high frequency
- directed at multiple epitopes
- initial control of viral replication during primary infection
  - deterioration during disease progression

    Possible reasons

    - CTL exhaustion
    - lack of CD4+ Th cells
    - HIV infection of CTL
    - suppression of CTL activity by a soluble factor
- frequent cross-reactivity between different HIV-1 clades
- contribution to clearence of HIV-1 infection in HIV-1 exposed sero^{-} subjects

CTL evasion mechanisms used by HIV-1

- MHC class I downregulation
- amino acid sequence variation:
  - affecting MHC binding
  - affecting TCR recognition
- sequestration (in the brain)
- viral latency (provirus integrates but is not expressed)
- CD4 downregulation
  - affecting class II-restricted CTL own suggestion

Class II-restricted CTL (CD4+)

- in HIV-infected and gp160-immunized subjects

CD8+ anti-HIV

- inhibition of viral replication by soluble factors

Therapeutic implications of HIV-specific CTL

- adoptive transfer of CTL
- vaccines that induce CTL
  - recombinant proteins
  - naked DNA

Induction by prophylactic HIV-1 vaccines

- recombinant vaccinia alone or boosting with protein
- recombinant canarypox alone or boosting with protein

* Potentially protective cytolytic immune responses in HIV-1 infection. Effector mechanisms presented by: Autran *et al.*, 1996; Belshe *et al.*, 1998; Calarota *et al.*, 1998 and 1999; Corey *et al.*, 1998; Evans *et al.*, 1999; Fleury *et al.*, 1996; Koenig *et al.*, 1995; Koup, 1994; Kundu *et al.*, 1997; Heeney *et al.*, 1999; Littaua *et al.*, 1992; McMichael, 1998; McMichael and Walker, 1994; Ogg *et al.*, 1998; Rowland-Jones *et al.*, 1997; Sadat-Sowti *et al.*, 1991; Schwartz *et al.*, 1996; Shearer and Clerici, 1996; Stanhope, 1993; Walker *et al.*, 1986; Walker and Plata, 1990.

from the viral long terminal repeats (LTR) by a thousand fold that of the basal rate from the unaided promoter (Goldstein, 1996). Tat consists of six defined regions, one N-terminal, one cysteine-rich region containing 7 cysteine residues, one core region that is hydrophobic, one basic region that binds to the RNA target and contains 6 arginines and 2 lysines, and one glutamic acid region. The function of the second exon has not been described (Rana and Jeang, 1999). Directly after production, the Tat protein binds to an RNA secondary stem loop structure at the 5' end of all viral mRNAs called TAR (trans-activation response) element. The Tat-TAR interaction is mediated by the arginine rich motif present in the N-terminal of Tat. The arginine rich motif also interacts with a nuclear Tat-associated kinase (TAK), now identified as the cellular positive acting transcription elongation factor (P-TEFb) that is required for elongation of many genes. Purified TAK/P-TEFb hyperphosphorylates the RNAP II C-terminal domain (Parada and Roeder, 1999). The P-TEFb has many subunits, one is the cyclin dependent kinase 9 (CDK9) and another is the human cyclin T (CycT) which is a cyclin-related protein that has been found to interact with Tat (Wei *et al.,* 1998; Bieniasz *et al.,* 1998). Tat functions by recruiting already formed CycT-CDK9 complexes (P-TEFb) to RNAP II through enhancing the affinity and specificity of the Tat and TAR interactions. This interaction prevents premature transcription stops that would otherwise occur downstream of the HIV-1 LTR promotor. Viruses with defective Tat do not replicate efficiently.

Tat-TAR independent expression may however occur. Expression of Tat in a human cell line has led to over-expression of cellular genes encoding cytokines such as TNF-α and IL-2 (Biswas *et al.,* 1995; Ott *et al.,* 1997; Lafrenie *et al.,* 1997). The TNF-α cytokine binds to the viral NF-κB element in LTR, thereby allowing TAR independent expression of viral genes (Biswas *et al.,* 1995). IL-2 drives the immune response to a T-helper cell type 1 (Th1) directed cellular response, but Tat is seldom recognised by CTLs in HIV-1 infected individuals (Lamhamedi-Cherradi *et al.,* 1992).

New findings indicate that Tat may work as a β-chemokine interacting with the CCR2 and CCR3 receptors on monocytes. These receptors have key roles in chemotaxis (Albini *et al.*, 1998). Tat has also been shown to differentially induce chemokine receptors CXCR4, CCR5 and CCR3 expression in peripheral blood mononuclear cells. This phenomenon was correlated with Tat-enhanced infectivity of M- and T-tropic HIV-1 virus strains (Huang *et al.*, 1998; Albini *et al.*, 1998). The activity of Tat leads to expression also of Fas ligand, and therefore of cellular apoptosis (Li *et al.*, 1995).

Secreted Tat passes extracellularly to other cells and renders them susceptible to productive viral infection (Goldstein, 1996; Nagahara *et al.*, 1998). Thus, the release and secondary uptake of Tat can augment viral activation in latently infected cells.

b) Detection of Tat expression from Tat-encoding plasmids

An enzyme based ELISA was used to assay the amount of enzyme expressed from a CAT (chloramphenicol acetyl transferase) encoding plasmid (Kjerrström and Wahren, 1999). The plasmid, pNLCATw, carries the 5´and 3´ LTR from HIV-1 as a promoter and a poly A signal. The Tat protein binds the TAR element in the LTR and promotes CAT expression. A standard curve was fitted and the concentrations of Tat in the samples were calculated according to linear regression analysis.

c) Rev viral regulation

The Rev protein is a 19 kD nucleocytoplasmic phosphoprotein that regulates gene expression at a post-transcriptional level. Rev allows transport of the late mRNA classes from the nucleus to the cytoplasm (Kalland *et al.*, 1994; Miller and Sarver, 1997). This bidirectional transport is mediated by two functional regions in Rev. The N-terminal nuclear localisation signal (NLS) promotes binding to the Rev responsive element (RRE) present in the *env* region on all singly spliced and unspliced mRNA of HIV. Binding to this RRE that forms an RNA secondary stem loop structure, initiates multimerization of Rev on the target mRNA (Feinberg and Greene, 1992). The multimerisation completely masks the NLS. The C-terminal region of Rev is designated the nuclear export signal (NES) or activation domain, is highly conserved and contains several leucines. This region is responsible for the interaction of Rev with the cellular transport system. Mutation in this region has been associated with attenuated Rev function and asymptomatic infection (Hua *et al.,* 1996). In the absence of Rev, structural proteins are not made.

Rev interacts with several cellular cofactors. One is the human Rev interacting protein (hRIP)/Rab, which is required for efficient Rev response and for viral replication (Fritz *et al.,* 1995). This hRIP/Rab is predominantly found in the nucleoplasmic region and it specifically interacts with an element called the effector domain of viral mRNAs. The export of Rev itself proceeds through the nuclear pore complex and requires the NES along with the hRIP/Rab (Fritz *et al.,* 1995; Murphy and Wente, 1996).

Another cellular protein that has been described to interact with Rev is the eukaryotic initiation factor 5A (eIF-5A), a nucleoporin-like protein which normally interacts with hRIP/Rab and the nuclear pore-associated factor CRM1. It was speculated to have a function in the translocation of NES containing proteins through the nuclear pore complex. Mutations in eIF-5A have been shown to block Rev function *in vitro.* A cellular interaction partner of eIF-5A, the cellular ribosomal protein L5, enhances activity of Rev significantly (Schatz *et al.,* 1998). Thus, Rev positively regulates the expression of virion structural proteins, but negatively regulates the expression of the regulatory genes. Without Rev, Tat expression leads to unregulated positive feedback of the regulatory proteins (Malim *et al.,* 1988).

d) Detection of Rev expression from Rev-encoding plasmids

An antibody based ELISA was used to assay the amount of Gag (p24) protein expressed from a Gag-encoding plasmid, pNLgagSty330 (Kjerrström and Wahren, 1999). This plasmid carries the RRE sequence, the *gag* gene and the HIV-1 5´ and 3´ LTRs as promoter and as polyA signal. This results in Tat dependent transcription activation and Rev dependent transport of Gag-encoding mRNAs from the cell nucleus to the cytoplasm (Schwartz *et al.*, 1991). The assay thus includes cotransfection with three plasmids, one encoding the Tat protein, one the Rev protein and one the Gag protein. The pNLgagSty330 was also used alone as a negative control (0.5 μg) for unspecific Gag production. A standard curve was drawn and the p24 concentrations that indirectly determine Rev concentration of the samples were calculated according to linear regression analysis.

e) The Nef protein

The third regulatory protein, Nef, is a 27kD-myristoylated protein, and is found only in primate lentiviruses. Nef down-regulates both the CD4 receptor and the Major Histocompatibility Complex (MHC) class I from the surface of infected cells (Aiken *et al.*, 1996; Cohen *et al.*, 1999; Harris 1998; Schwartz *et al.*, 1996). HLA A and B but not HLA C or E are down regulated by Nef (Cohen *et al.*, 1999). This down-regulation prevents both superinfection and premature cell death of the infected cells by reducing the epitope density on their surface. It partially allows evasion of MHC-regulated lysis. NK lysis also appears to be blocked *in vitro*, since the major inhibitors of NK-lysis, HLA class C and E molecules, are not down regulated.

The CD4 receptor is attached to p56Lck at the plasma membrane. Activation from T-cell receptors initiates CD4 recruitment to clathrin-coated pits. Clathrin is recruited by adaptor protein 2 (AP2) to the cell surface. CD4 is internalised and normally recycled back to the surface. Nef disrupts the CD4 p56 Lck complex, allowing CD4 internalisation but not recycling. Then Nef associates with other proteins that are believed to be AP-1 and 3, to redirect CD4 transport to late endosomes and lysosomes (Kim *et al.*, 1999; Piguet and Trono, 1999). Nef recognises the SH3 domain of the Hck and Lyn non-receptor protein tyrosine kinases via a conserved PxxP motif in the core domain of Nef (Khan *et al.,* 1998; Piguet and Trono, 1999). This is believed to disturb normal T-cell signalling.

The main effect of external Nef secretion would be to activate T-cells, permitting secondary viral infection. This in turn allows virus to establish a secondary pool of latently infected cells large enough to evade the primary CTL immune response, but also to maintain a high level of chronic infection (Montagnier, 1995). Control of viral expression during the initial viral infection should be a key point for intervention against chronic viremia and clinical progression in HIV infection.

f) Detection of Nef expression from Nef-encoding plasmids

Protein expression from the *nef* genes was confirmed by Western blot and by immunofluorescence on HeLa cells transfected with Nef-encoding plasmids (Kjerrström and Wahren, 1999). The amount of *nef* DNA used varied from 0.4 to 0.8 μg. For transfection, 15000 cells were seeded per well in a 96-well plate or 200,000 cells per well in a 6-well plate. Immunofluorescence was performed by fixing the cells in acetone and after the cells were completely dried, incubating them at 37 ^{0}C for 1 hour with mouse monoclonal antibodies (kindly provided by Drs K. Krohn and V. Ovod) against Nef. The cells were incubated with FITC-conjugated mouse monoclonal antibodies (Sigma) and read by light microscopy.

For Western blot, the cells were lyzed and separated on an SDS gel. The proteins were transferred to nitrocellulose and incubated with anti-Nef monoclonal antibodies followed by incubation with goat-anti mouse IgG. The blots were developed with 4CN and photographed.

g) The use of a genetic handle for vaccine construction

For evaluation of immunogenicity of the viral genes on a similar background, we have constructed a genetic handle that has been used as a backbone for our genetic immunogens. The handle is a pUC8-derived plasmid comprising the major immediate early (MIE) promoter from human cytomegalovirus (CMV) and the polyadenylation signal from human papilloma virus type 16 (HPV 16, Hinkula *et al.,* 1997a and 1997b). In the first clinical trial on HIV-1 infected patients (Calarota *et al.,* 1998), the plasmids contained an ampicillin resistance gene. In the new generation of plasmids, we have exchanged this gene for a kanamycin resistance gene by homologous recombination (Bubeck *et al.,* 1993). Ampicillin is a widely used antibiotic and is known to occasionally cause unwanted allergic reactions. To eliminate the risk of such reactions, the less used antibiotic kanamycin has been used as selective marker during production of the plasmids. Homologous recombination occurs naturally between genetic elements when overlapping complementary sequences interact. To construct new plasmids, the gene of interest was amplified by polymerase chain reaction (PCR) using primers that were partly homologous to sequences in a linearised, purified plasmid vector. The two fragments were then mixed in equal amounts and transformed by heat shock into *E. coli DH5α*.

In the first and second generation of genetic vaccines, we used the *nef* gene from the HIV-1 subtype B strain HXB3 while the *rev* gene (Mermer *et al.,* 1990) and *tat* gene derived from the HXB2 strain. In the third generation of vaccines, we used viral genes derived from primary MT-2 negative and MT-2 positive viruses (Karlsson *et al.,* 1994) directly amplified from lymphocytes of HIV-1 infected patients.

h) Immunogenicity in animals

Ten 12-week-old (C57BL/6 x DBA/2)F1 ($H\text{-}2^{b/d}$) mice were immunised with the Nef-encoding plasmids three times with 50 μg doses. ELISA and B-cell stimulation *in vitro* measured humoral responses. Cellular responses were measured by T-cell proliferation assays and cytokine secretion assays (Hinkula *et al.,* 1997b).

2. EXPRESSION OF PATIENT-DERIVED GENE CONSTRUCTS

The primary HIV-1 isolates derive from five patients infected with non-syncytium forming viruses. Five Tat-encoding plasmids were constructed, each patient represented by one plasmid, pCMV*tat*K1-5. Figure 1 shows the quantity of proteins expressed. Protein expression from patient-derived plasmid clones was at least 20 % lower than that from the HXB2 derived clone HCMV*tat*. One of the patient-derived clones, pCMVtatK2, had at least 20 % lower Tat expression than the other four primary clones. This plasmid had a glutamic acid in amino acid position number 50 instead of a lysine, which the other four plasmids had. The amino acid sequence of the HXB2 clone was identical to four of the patient-derived clones. Substituting the ampicillin resistance gene for a kanamycin resistance gene in the HXB2 Tat-encoding plasmids resulted in a 20 % lower protein expression.

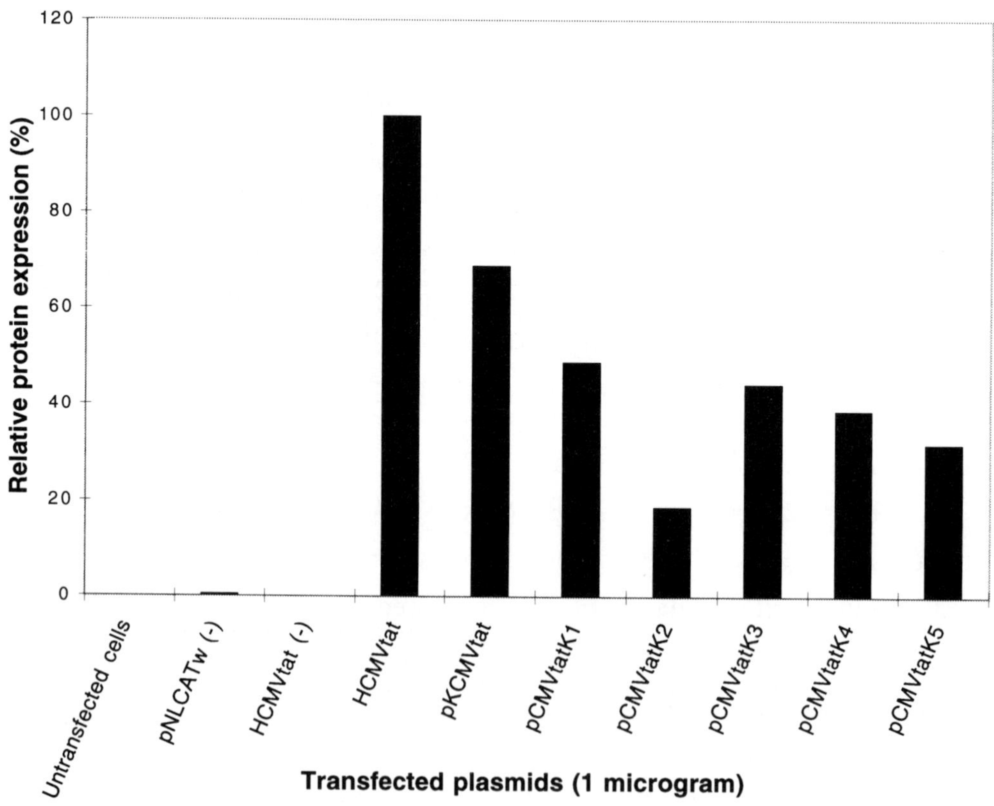

Figure 1. Relative Tat expression from the Tat-encoding plasmids in transfected HeLa cells after 24 hours. Protein expression was compared to that from HCMV*tat* that has established immunogenicity in HIV-1 infected individuals.

Five Rev-encoding plasmids were constructed from the same patients. Rev protein expression capacity was lowered by around 50% in plasmids encoding patient derived genes (Fig. 2). Compared to the HXB2 gene, the patient derived genes contained differences at both nucleotide and amino acid sequence levels. Sequence analysis revealed no nucleotide differences between the patient derived clones. No difference was seen between plasmids carrying ampicillin or kanamycin resistance genes.

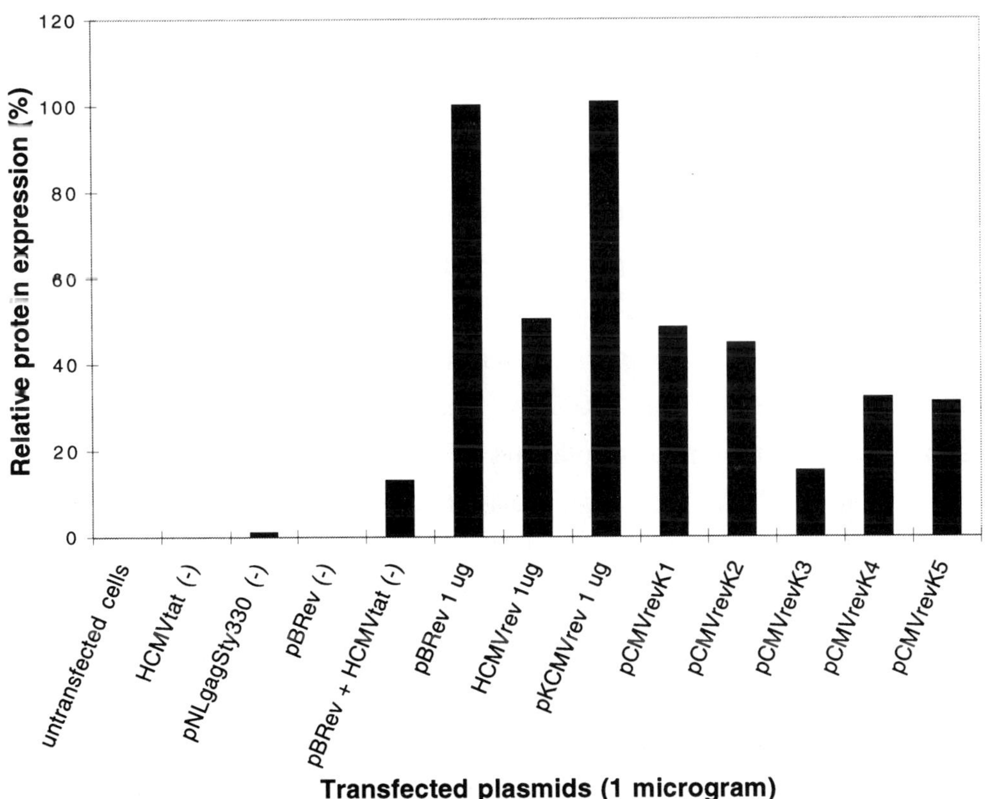

Figure 2. Relative Rev expression from the Rev-encoding plasmids in transfected HeLa cells after 24 hours. Protein expression was compared to that of pBRev which was used as positive control.

Three Nef-encoding plasmids were constructed from patient-derived genes of one HIV-1 infected individual. All three clones were shown to be sequence identical. Compared to the HXB3 gene, differences were found both in the nucleic acid and amino acid sequence. Protein expression was confirmed both by Western blot and immunofluorescence but not quantified. The three plasmids carrying patient-derived genes were used in an animal model, comparing the immunogenicity of these clones with that of the plasmid carrying the HXB3 *nef* gene. Both groups of plasmids evoked specific cellular and humoral immune responses in mice (Hinkula *et al.*, 1997b; Kjerrström, unpublished). Epitope mapping comparing consensus with primary immunizing genes showed similar results with strong T-cell epitopes in the C- and N-terminal ends of the Nef protein. B-cell epitopes were seen in the C- and N-terminals and in the central part of the protein.

3. CLINICAL TRIALS

We aimed to induce HIV-specific immune responses by means of DNA immunization with regulatory genes in asymptomatic HIV-infected individuals. The study was designed to identify immunogenicity and side effects in humans. Three plasmids with ampicillin resistance encoding Nef, Rev and Tat were given. 100 μg of each DNA were repeatedly given as single or as a combination of plasmids (Fig. 3).

Our data show that single plasmid immunizations with Nef or Rev or Tat cDNA induced HIV-specific CTL memory cell responses in immuno-deficient individuals (Fig. 3; Calarota *et al.*, 1998 and 1999).

When regulatory HIV-1 genes were given in separate plasmids, HIV-specific CTL responses were induced in eight patients, and these responses efficiently lysed target cells infected with recombinant vaccinia virus vectors expressing the same protein encoded by the gene used in the immunization. T helper cell responses were seen in all immunized individuals except one.

Figure 3. HIV-specific cellular immune responses in an asymptomatic HIV-1 infected patient (naive to antiretroviral therapy) immunized with a DNA vaccine encoding the HIV-1 Nef protein (first three immunizations) followed by three immunizations with a combination of HIV-1 *nef*, *rev* and *tat* genes. Cells for cytolysis were infected by pseudotyped HIV-1 (HIV-1/MuLV) or with modified vaccinia Ankara carrying *nef*, *rev* or *tat* genes (Calarota *et al.*, 1998 and 1999).

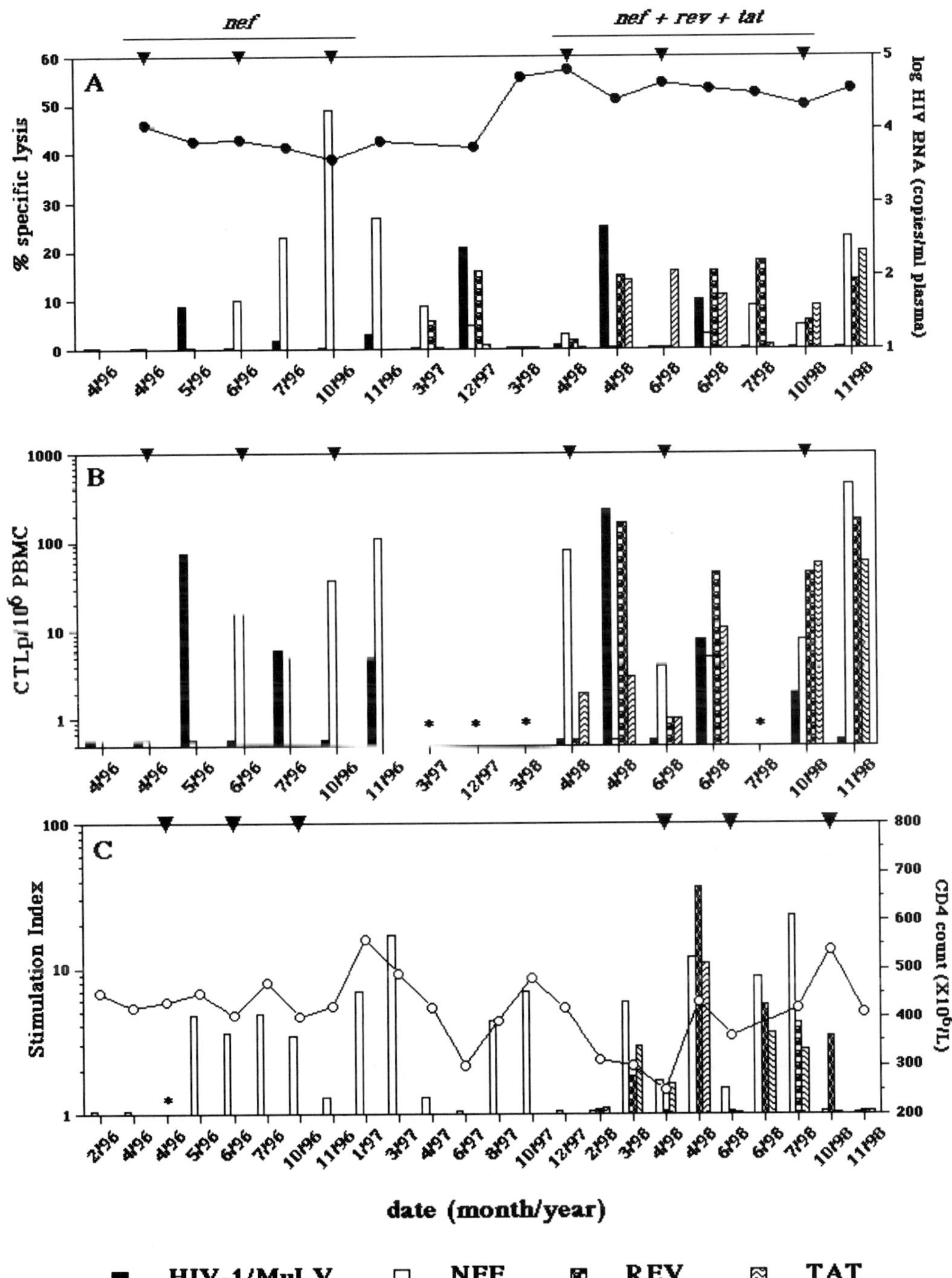

A) CTL responses (effector to target ratio 6:1) and viral load (●). **B**) Frequencies of CTL precursor cells. **C**) Lymphocyte proliferative responses and CD4 counts (○); ∗, not tested. DNA immunizations are indicated (▼) at the top of the figure.

Extending those findings, patients were subsequently immunized with a combination of all three plasmids. It is of interest to note that the T-helper cell, the CTL and the CTL precursor cell populations were stable or augmented in activity by these immunizations.

The data presented demonstrates that it is feasible to elicit virus CTL and T helper responses in humans that become capable of recognizing cells presenting HIV viral peptides. The immune responses occurred without highly active anti-retroviral treatment (HAART). It was reported to occur also with HAART (Calarota *et al.,* 1999). In natural chronic infection, there has been no recovery of HIV-specific T-cell immunity following HAART only (Kalams *et al.,* 1999; Leandersson *et al.,* 1998).

It has been suggested that the immune responses induced by immunizations with a combination of plasmid DNAs might be reduced due to interference at the level of antigen production or antigen competition. However, our data suggest that the combination of plasmids was more effective than the individual components in inducing cytolytic responses to whole HIV-1 gene products.

4. CYTOTOXICITY INDUCED BY DNA

To control and prevent viral infections, a strong cellular response is needed in addition to a specific humoral response. The peptides of regulatory proteins are presented together with MHC class I molecules like other peptides and may serve to induce CTL responses. In primary infection, such exposure can lead to elimination of infected cells before viral production can take place. It is of utmost importance to define the immunogenicity of the regulatory proteins, and to determine their role in inhibition of viral replication.

a) Tat immunity

Tat is present as a transactivator within the infected cell, and is also secreted from the infected cells. It can, at least *in vitro*, be shown to have effects on cellular regulation in non-infected cells as well as of trans-activation in latently infected cells. Both B-cell and CTL activities are important for immunoprotection. The major B-cell determinants are located in the cystein-rich region of Tat, in the core region, and in a less prominent region in the N-terminal part (Hinkula *et al.,* 1997b). Cellular reactivities were located mainly to the cystein-rich domain (Blazevic *et al.,* 1993). Only around 40 % of HIV-infected individuals naturally respond by developing CTL against Tat (Lamhamedi-Cherradi *et al.,* 1992).

Van Baalen *et al.* (1997) have shown that both Tat- and Rev-specific CTL frequencies inversely correlate with rapid progression to AIDS. Tat CTL reactivities were recently described to occur after plasmid immunization of humans (Calarota *et al.,* 1998). Our patient data show very small variations in Tat sequences of the five HIV-infected individuals. Immunizations by Tat constructs therefore seem to mediate new immunities. The immunogenic regions of Tat are conserved and thus a large variation between strains is not expected.

Antibodies towards Tat have been shown to cause significant reduction and consistent delay in viral replication *in vitro*. There has also been speculations on whether Tat can block cytotoxicity of NK cells. Tat would block the release of serine esterases that are responsible for the killing by inhibiting the rise of intracellular calcium that occurs when the target cell is recognised by the effector cell. It may prevent calcium influx by blocking the L-type calcium channels (Zocchi *et al.*, 1998).

By immunization with native Tat protein derived from HIV-1, Ensoli and collaborators were able to reduce or inhibit primary SHIV replication in macaques (Cafaro *et al.*, 1999). These findings have led to renewed interest to use Tat or inactivated Tat toxoid as an immunogen (Wallace *et al.*, 1999).

b) Rev immunity

The capacity of cDNA plasmids containing the regulatory gene *rev* to induce immune responses was studied in immunocompetent and huPBL repopulated SCID mice (Wahren *et al.,* 1996). IgG responses during immunization were noted in almost all mice after 2-3 injections. Rev expression as measured by our quantitative assay appeared to be rapid and strong from all primary *rev* genes. The arginine rich region of Rev contributes to the nucleic acid binding, and the C-terminal induces Th and antibody reactivity (Hinkula *et al.,* 1997b). It was also demonstrated that *rev* in plasmids could induce new CTL in HIV-infected individuals (Calarota *et al.,* 1998).

Rev, like Tat, is not well recognised by specific CTLs in the infected patient. Only around 10–40% of HIV-infected individuals respond to these two antigens. In spite of its relatively small size, HIV-1 Rev is rich in potential HLA-A2 binding regions (Wahren *et al.,* 1996). Most of these are relatively conserved between different virus isolates. Several synthetic Rev peptides showed a significantly increased surface expression of HLA class I molecules. Of special interest was that some of the best binding peptides had amino acid sequences of native Rev protein from different HIV-1 isolates.

c) Nef immune responses

The protein Nef has several disabilitating effects on immunologically active cells. CD4 molecules are downregulated by binding of the C-terminal of Nef to a dileucine motif (Schwartz *et al.,* 1996; Iafrate *et al.,* 1997). MHC molecules are rapidly endocytosed and degraded. The endocytosis is dependent on the SH3 binding motif of Nef and a cluster of acidic aminoacids. Cytotoxic cells were inefficient in killing HIV-infected cells if the infected cells expressed the viral gene product Nef. Primary cells infected with a *nef* molecular clone had a decreased expression of MHC class I molecules, while its effects on cells of the myelopoietic lineage have been questioned. Primary lymphoid cells infected with a virus clone containing defective Nef did not down regulate MHC class I and were more efficiently killed (Collins *et al.,* 1998).

Nef appears to have a specific disease-inducing property in mice transgenic for expression of this protein (Hanna *et al.,* 1998). We noted, however, no clinical effects in mice by injecting large amounts of plasmid DNA synthesizing Nef, as determined by antibody production. The weanling mice grew as fast as their non-injected littermates, and no clinical or pathological effects were noted (Hinkula and Wahren, unpublished). For short-term synthesis thus, these deleterious effects of the regulatory genes are not noticed, possibly due to apoptosis and disappearance of transiently transfected cells (Li *et al.,* 1995).

Around 60-70 % of infected individuals naturally develop both antibody responses as well as potent CTL activity to Nef (Choppin *et al.,* 1991; Riviere *et al.,* 1994; Rowland-Jones *et al.,* 1997; Tähtinen *et al.,* 1992). CTLs specific for Nef are among the strongest seen against HIV-1 antigens, and studies of a group Gambian women that, despite reoccurring exposure to HIV, remain seronegative, have revealed high levels of Nef specific CTLs (Rowland-Jones *et al.,* 1995). Virus isolates from long term non progressor (LTNP) patients may have deleted *nef* genes (Deacon *et al.,* 1995; Mariani *et al.,* 1996; Hua *et al.,* 1996). Thus, Nef plays an important role in determining the outcome of disease.

It is thus of importance to consider all activities of Nef in immunization with plasmid DNA. Such immunizations were shown in experimental animals and in humans to increase the Nef-specific cytotoxic levels of CD8-associated killing. It is however feasible that the presence of a functional *nef* gene in an attenuated vaccine or in a plasmid mixture may partially down-regulate MHC class I molecules like in natural infection, so that also CTL reactitivites to other HIV antigens are decreased.

A question is then how the exceptionally vigorous CTL that occurs already early after primary infection is obtained *in vivo*. The Nef protein is highly immunogenic, carrying a multitude of B-cell and T-cell epitopes (Hadida *et al.,* 1992; Ranki *et al.,* 1994). One hypothesis is that CTLs may be generated but not effective, another that the CTL may be directed to peptides presented on uninfected APCs. It is also possible that NK cells which recognize the absence of MHC class I molecules may kill a higher proportion of cells than previously estimated.

d). Highly active anti-retroviral treatment (HAART)

A finding of interest was observed in those patients who started antiretroviral treatment during the study. We detected moderate HIV-specific CTL responses when they were immunized with the combination of genes. Our previous data showed that after immunization with a single gene followed by the initiation of HAART, these responses decreased (Calarota *et al.,* 1999). It has been suggested that the loss of HIV-1 effector CTLs after the initiation of HAART is due to changes in the CD4+ T-cell population or redistribution between blood and tissues (Ogg *et al.,* 1999). However, it appears that the combination of plasmids in a vaccine can induce CTL responses to at least two of the individual components of the new immunogen.

e) Routes of administration

The DNA dose and the routes of administration are current problems that remain to be solved. They may, in each case, be dependent on the encoded antigen. The efficacy of protein expression from the plasmids is therefore very important, and we have developed techniques to measure this in viral genes derived from primary viral isolates. The gene constructs derived directly from infected patients and their functional and immunogenic properties were measured.

5. FUTURE ORIENTATIONS

Further work must be devoted to optimizing the constructs that carry the various genes. It would be preferable to keep the genetic content rather than the gene itself to a minimum, in order to diminish the potential risks of integration of the plasmid content. The viral cytomegalovirus promoter has been shown to be useful, perhaps since it contains several immune enhancing CpG motifs (Roman *et al.,* 1997). The part played by tissue-specific elements has not been sufficiently well explored, although there have been some reports that show no special benefits. It is also not clear whether signal sequences for excretion of these proteins are suitable for all genes or just for some. Recently, the first experiences with genetic vaccination in man were published (Calarota *et al.,* 1998; McGregor *et al.,* 1998; Wang *et al.,* 1998). They demonstrated the feasibility of immunization in humans using regulatory or structural viral genes and sporozoan genes. They gave rise to dose dependent and gene dependent T helper cell responses, cytolytic responses and antibody responses, all that are desirable properties against intracellular parasites.

6. REFERENCES

Aiken, C., Krause, L., Chen, Y.L. and Trono, D. (1996) Mutational analysis of HIV-1 Nef: identification of two mutants that are temperature sensitive for CD4 down regulation. *Virology*, 217, 293-300.

Albini, A., Ferrini, S., Benelli, R., Sforzini, S., Giunciuglio, D., Grazia-Aluigi, M., Proudfoot, A.E.I., Alouani, S., Wells, T.N.C., Mariani, G., Rabin, R.L., Farber, J.M. and Noonan, D. (1998). HIV-1 Tat protein mimicry of chemokines. *Proc. Natl. Acad. Sci. USA*, 95: 13153-13158.

Autran, B., Hadida, F. and Haas, G. (1996). Evolution and plasticity of CTL responses against HIV. *Curr. Opin. Immunol.,* 8: 546–553.

Belshe, R.B., Gorse, G.J., Mulligan, M.J., Evans, T.G., Keefer, M.C., Excler, J-L., Duliege, A-M., Tartaglia, J., Cox, W.I., McNamara, J., Hwang, K.L., Bradney, A., Montefiori, D. and Weinhold, K.J. (1998). Induction of immune responses to HIV-1 by canarypox virus (ALVAC) HIV-1 and gp120 SF-2 recombinant vaccines in uninfected volunteers. *AIDS*, 12: 2407 – 2415.

Bieniasz, P.D., Grdina, T.A., Bogerd, H.P. and Cullen, B.R. (1998). Recruitment of a protein complex containing Tat and cyclin T1 to TAR governs the species specificity of HIV-1 Tat. *EMBO J.*, 17: 7056–7065.

Biswas, D.K., Salas, T.R., Wang, F., Ahlers, C.M., Dezube, B.J. and Pardee, A.B.A. (1995). Tat-induced auto-up-regulatory loop for superactivation of the Human Immunodeficiency Virus type 1 promoter. (1995). *J. Virol.*, 69: 7437-7444.

Blazevic, V., Ranki, A., Mattinen, S., Valle, S.L., Koskimies, S., Jung, G. and Krohn, K.J. (1993). Helper T-cell recognition of HIV-1 Tat synthetic peptides. *J. Acquir. Immune Defic. Syndr.,* 6: 881-890.

Bubeck, B., Winkler, M. and Bautsch, W. (1993). Rapid homologous recombination in vivo. *Nucleic Acids Res.,* 21: 3601-3602.

Cafaro, A., Caputo, A., Fracasso, C., Maggiorella, M.T., Goletti, D., Baroncelli, S., Pace M., Sernicola, L., Koanga-Mogtomo, M.L., Betti, M., Borsetti, A., Belli, R., Åkerblom, L., Corrias, F., Butto, S., Heeney, J., Verani, P., Titti, F. And Ensoli, B. (1999). Control of SHIV-89.6P-infection of cynomologus monkeys by HIV-1 Tat protein vaccine. *Nat. Med.,* 6: 643–650.

Calarota, S., Bratt, G., Nordlund, S., Hinkula, J., Leandersson, A-C., Sandström, E. and Wahren, B. (1998). Cellular cytotoxic response induced by DNA vaccination in HIV-1 infected patients. *J Lancet,* 351**:** 1320-1325.

Calarota, S., Leandersson, A-C., Bratt, G., Hinkula, J., Klinman, D.M., Weinhold, K.J., Sandström, E. and Wahren, B. (1999). Immune responses in asymptomatic HIV-1 infected patients after HIV-DNA immunization followed by highly active antiretroviral treatment. *J. Immunol.*, 163: 2330-2338.

Choppin J, Martinon F, Connan F, Gomard E and Levy J-P. (1991). HLA-binding regions of HIV-1 proteins. *J. Immunol.*, 147: 569-574.

Cohen, G.B., Gandhi, R.T., Davis, D.M., Mandelboim, O., Chen, B.K., Strominger, J.L. and Baltimore, D. (1999). The selective downregulation of class I major histocompatibility complex proteins by HIV-1 protects HIV-infected cells from NK cells. *Immunity*, 10: 661–671.

Collins, K., Chen, B., Kalams, S., Walker, B., and Baltimore, D. (1998). HIV-1 nef protein protects infected primary cells against killing by cytotoxic T lymphocytes. *Nature*, 391: 397-401.

Corey, L., McElrath, M.J., Weinhold, K., Mattews, T., Stablein, D., Graham, B., Keefer, M., Schwartz, D., Gorse, G. and the AIDS Evaluation Group. (1998). Cytotoxic T cell and neutralizing antibody responses to human immunodeficiency virus type 1 envelope with a combination vaccine regimen. *J. Infect. Dis.,* 177**:** 301–309.

Cullen B.R. (1998). Posttranscriptional regulation by the HIV-1 Rev protein. *Seminars in Virology*, 8: 327–334.

Deacon, N.J., Tsykin, A., Solomon, A., Smith, K., Ludford-Menting, M., Hooker, D.J., McPhee, D.A., Greenway, A.L., Ellet, A., Chatfield, C., Lawson, V.A., Crowe, S., Maerz, A., Sonza, S., Learmont, J., Sullivan, J.S., Cunningham, A., Dwyer, D., Dowton, D. and Mills, J. (1995). Genomic structure of an attenuated quasispecies of HIV-1 from a blood transfusion donor and recipients. *Science*, 270: 988-991.

Donelly, J., Ulmer, J., Liu, M., Evans, R., Boslego, J., Hastings, J., Chan, I., Mazurick, C., Thaler, S., Clements-Mann, M-L., Eichelberger, M. And Harro, C. (1999). Adjuvants for DNA vaccines. In: *DNA vaccines: Mechanisms and manipulating antigen processing,* Keystone Symp. Mol. Cell. Biol., Abstract p. 28.

Evans, T.G., Keefer, M.C., Weinhold, K.J., Wolff, M., Montefiori, D., Gorse, G.J., Graham, B.S., McElrath, M.J., Clements-Mann, M.L., Mulligan, M.J., Fast, P., Walker, M.C., Excler, J.L., Duliege, A.M. and Tartaglia, J. (1999). A canary pox vaccine expressing multiple human immunodeficiency virus type 1 genes given alone or with rgp120 elicits broad and durable CD8+ cytotoxic T lymphocyte responses in seronegative volunteers. *J. Infect. Dis.,* 180: 290–298.

Feinberg, M. B. and Greene, W. C. (1992). Molecular insights into immunodeficiency virus type 1 pathogenesis. *Curr. Opin. Immunol.,* 4: 466-474.

Fleury, B., Janvier, G., Pialoux, G., Buseyne, F., Robertson, M.N., Tartaglia, J., Paoletti, E., Kieny, M.P., Excler, J.L., and Riviere, Y. (1996). Memory cytotoxic T lymphocyte responses in human immunodeficiency virus type 1 (HIV-1)-negative volunteers immunized with a recombinant canarypox expressing gp160 of HIV-1 and boosted with a recombinant gp160. *J. Infect. Dis.*, 174: 734 738.

Fritz, C.C., Zapp, M.L. and Green, M.R. (1995). A human nucleoporin-like protein that specifically interacts with HIV Rev. *Nature,* 376**:** 530-533.

Goldstein, G. (1996). HIV Tat protein as a potential AIDS vaccine. *Nat. Med.,* 9: 960-964.

Hadida, F., Parrot, A., Kieny, M-P., Sadat-Sowti, B., Mayaud, C., Debre, P. and Autran, B. (1992). Carboxy-terminal and central regions of Human Immunodeficiency virus-1 Nef recognized by cytotoxic T lymphocytes from lymphoid organs. *J. Clin. Invest.,* **89**: 53-60.

Hanna, Z., Kay, D., Rebai, N., Guimond, A., Jothy, S. and Jolicoeur, P. (1998). Nef harbors a major determinant of pathogenicity for an AIDS-like disease induced by HIV-1 in transgenic mice. (1998), *Cell,* 95: 163-175.

Harris, M. (1998). From negative factor to a critical role in virus pathogenesis: the changing fortunes of Nef. *J. Gen. Virol.,* 77: 2379–2392.

Heeney, J.L., Beverly, P., McMichael, A., Shearer, G., Strominger, J., Wahren, B., Weber, J. and Gotch, F. (1999). Immune correlates of protection from HIV and AIDS more answers but yet more questions. *Immunol. Today,* 20: 247–251.

Hinkula, J., Lundholm, P. and Wahren, B. (1997a). Nucleic acid vaccination with HIV regulatory genes: a combination of HIV-1 genes in separate plasmids induces strong immune responses. *Vaccine,* 15: 874- 878.

Hinkula, J., Svanholm, C., Schwartz, S., Lundholm, P., Brytting, M., Engström, G., Benthin, E., Glaser, H., Sutter, G., Kohleisen, B., Erfle, V., Okuda, K., Wigzell, H., and Wahren, B. (1997b). Recognition of prominent viral epitopes induced by immunization with human immunodeficiency virus type 1 regulatory genes. *J. Virol.,* 71: 5528–5539.

Hua, J., Caffrey, J.J., Cullen, B.R. (1996). Functional consequences of natural sequence variation in the activation domain of HIV-1 Rev. *Virology,* 222: 423-429.

Huang, L., Bosch, I., Hofmann, W., Sodroski, J. and Pardee, A.B. (1998). Tat protein induces human immunodeficiency virus type 1 (HIV-1) coreceptors and promotes infection with both macrophage-tropic and T-lymphotropic HIV-1 strains. *J. Virol.,* 72: 8952-8960.

Iafrate AJ, Bronson S and Skowronski J. (1997). Separable functions of Nef disrupt two aspects of T cell receptor machinery: CD4 expresson and CD3 signalling. *EMBO J.,* 16: 673-684.

Kahn, I.H., Sawai, E.T., Antonio, E., Weber, C.J., Mandell, C.P., Montbriand, P. and Luciw, P.A. (1998). Role of the SH3-ligand domain of simian immunodeficiency virus Nef in interaction with Nef-associated kinase and simian AIDS in rhesus macaques. *J. Virol.,* 72: 5820-5830.

Kalams, S.A., Goulder, P.J., Shea, A.K., Jones, N.G., Trocha, A.K., Ogg, G.S., Walker, B.D. (1999). Levels of human immunodeficiency virus type 1-specific cytotoxic T-lymphocyte effector and memory responses decline after suppression of viremia with highly active antiretroviral therapy. *J. Virol.*, 73: 6721-6728.

Kalland, K-H., Szilvay, A.M., Brokstad, K.A., Saetrevik, W. And Haukenes, G. (1994). The human immunodeficiency virus type 1 Rev protein shuttles between the cytoplasm and nuclear compartments. *Mol. Cell. Biol.*, 14: 7436-7444.

Karlsson, A., Parsmyr, K., Sandström, E., Fenyö, E.M. and Albert, J. (1994). MT-1 cell tropism as prognostic marker for disease progression in human immunodeficiency virus type 1 infection. *J. Clin. Microbiol.*, 32: 364-370.

Kin Y.H., Chang, S.H., Kwon, J.H. and Rhee, S.S. (1999). HIV-1 Nef plays an essential role in two independent processes in CD4 down-regulation: Dissociation of the CD4 p56 ^{lck} complex and targeting of CD4 to lysosomes. *Virology*, 257: 208-219.

Kjerrström, A. and Wahren, B. (1999). Expression of HIV regulatory DNA vaccine constructs. *Biogenic amines*, 15: 93-112.

Koenig, S., Conley, A.J., Brewah, Y.A., Jones, G.M., Leath, S., Boots, L.J., Davey, V., Pantaleo, G., Demarest, J.F., Carter, C., Wannebo, C., Yanelli, J.R., Rosenberg, S.A. and Lane, H.C. (1995). Transfer of HIV-1-specific cytotoxic T lymphocytes to an AIDS patient leads to selection for mutant HIV variants and subsequent disease progression. *Nat. Med.*, 1: 330-336.

Koup, R.A. (1994). Virus escape from CTL recognition. *J. Exp. Med.*, 180**:** 779-782.

Kundu, S.K., Katzenstein,D., Valentine, F.T., Spino, C., Efron, B. And Merigan, T.C. (1997). Effect of therapeutic immunization with recombinant gp160 HIV-1 vaccine on HIV-1 proviral DNA and plasma RNA: relationship to cellular immune responses. *J. Acquir. Immune Defic. Syndr. Hum. Retrovirol.*, 15**:** 269-274.

Lafrenie, R. M., Wahl, L. M., Epstein, J. S., Yamada, K. M., Dhawan, S. (1997). Activation of monocytes by HIV-Tat treatment is mediated by cytokine expression. *J. Immunol.*, 159: 4077-4083.

Lamhamedi-Cherradi, S., Culmann-Penciolelli, B., Guy, B., Kiény, M-P., Dreyfus, F., Saimont, A-G., Sereni, D., Sicard, D., Lévy, J-P. and Gomard, E. (1992). Qualitative and quantitative analysis of human cytotoxic responses to HIV-1 proteins. *AIDS*, 6: 1249-1258.

Leandersson, A-C., Bratt, G., Hinkula, J., Gilljam, G., Cochaux, P., Samson, M., Sandström, E. And Wahren, B. (1998). Induction of specific T-cell responses in HIV infection. *AIDS,* 12: 157-66.

Li, C.J., Friedman, D.J., Wang, C., Metelev, V., Pardee, A.B. (1995). Induction of apoptosis in uninfected lymphocytes by HIV-1 Tat protein. *Science*, 268: 429-431.

Littaua, R.A., Oldstone, M.B.A., Takeda, A. And Ennis, F.A. (1992). A CD4+ cytotoxic T-lymphocyte clone to a conserved epitope on human immunodeficiency virus type 1 p24 : cytotoxic activity and secretion of interleukin-2 and interleukin-6. *J. Virol.,* 66: 608-611.

MacGregor, R.R., Boyer, J.D., Ugen, K.E., Lacy, K.E., Gluckman, S.J., Bagarazzi, M.L., Chattergoon, M.A., Baine, Y., Higgins, T.J., Ciccarelli., R.B., Coney, L.R., Ginsberg, R.S. and Weiner, D.B. (1998). First human trial of a DNA-based vaccine for treatment of human immunodeficiency virus type 1 infection: safety and host response. *J. Infect. Dis.,* 178: 92-100

Malim, M.H., Hauber, J., Fenrick, R and Cullen, B.R. (1988). Immunodeficiency virus *rev trans*-activator modulates the expression of the viral regulatory genes. *Nature*, 335: 181-183.

Mariani, R., Kirchhoff, F., Greenough, T.C., Sullivan, J.L., Desrosier, R.C. and Skowronski, J. (1996). High frequency of defective nef alleles in a long-term survivor with non-progressive HIV-infection. *J. Virol.*, 70: 7752-7764.

McMichael, A. (1998). T cell responses and viral escape. *Cell,* 93: 673-676.

McMichael, A.J. and Walker, B.D. (1994). Cytotoxic T lymphocyte epitopes: implications for HIV vaccines. *AIDS*, 8 (Suppl 1): S155-S173.

Mermer, B., Felber, B.K., Campbell, M. and Pavlakis, G.N. (1990). Identification of trans-dominant HIV-1 rev protein mutants by direct transfer of bacterially produced proteins into human cells. *Nucleic Acids Res.,* 18: 2037-2044.

Miller, R.H. and Sarver, N. (1997). HIV accessory proteins as therapeutic targets. *Nat. Med.,* 3: 389-394.

Montagnier, L. (1995). Nef Vaccination against HIV disease. *Lancet,* 346: 1170.

Montgomery, D.L., Shiver, J.W., Leander, K R., Perry, H.C., Friedman, A., Martinez, D., Ulmer, J.B., Donelly, J.J., and Liu, M.A. (1993). Heterologous and homologous protection against Influenza A by DNA vaccination: Optimization of DNA vectors. *DNA Cell Biol.*, 12: 777-783.

Murphy, R. and Wente, S.R. (1996). An RNA-export mediator with an essential nuclear export signal. *Nature,* 383: 357-360.

Nagahara, H., Vocero-Akbani, A.M., Snyder, E.L., Ho, A., Latham, D.G., Lissy, N.A., Becker-Hapak, M., Ezhevsky, S.A. and Dowdy, S.F. (1998). Transduction of full-length TAT fusion proteins into mammalian cells: TAT-p27Kip1 induces cell migration. *Nat. Med.,* 12: 1449-1452.

Ogg, G.S., Jin, X., Bonhoeffer, S., Rod Dunbar, P., Nowak, M.A., Monard, S., Segal, J.P., Cao, Y., Rowland-Jones, S.L., Cerundolo, V., Hurley, A., Markowitz, M., Ho, D.D., Nixon, D.F. and McMichael, A.J. (1998). Quantification of HIV-1-specific cytotoxic T lymphocytes and plasma viral load of viral RNA. *Science,* 279: 2103 2106.

Ogg, G.S., Jin, X., Bonhoeffer, S., Moss, P., Nowak, M.A., Monard, S., Segal, J.P., Cao, Y., Rowland-Jones, S.L., Hurley, A., Markowitz, M., Ho, D.D., McMichael, A.J. and Nixon, D.F. (1999). Decay kinetics of human immunodeficiency virus-specific effector cytotoxic T lymphocytes after combination antiretroviral theraphy. *J. Virol.,* 73: 797-800.

Okuda, K., Bukawa, H., Hamajima, K., Kawamoto, S., Sekigawa, K-I., Yamada, Y., Tanaka, S-I., Ishii, N., Aoki, I., Nakamura, M., Yamamoto, H., Cullen, B. R., and Fukushima, J. (1995). Induction of potent humoral and cell-mediated immune responses following direct injection of DNA encoding the HIV Type 1 env and rev gene products. *AIDS Res. Hum. Retroviruses,* 11: 933-943.

Ott, M., Emiliani, S., Van Lint, C., Herbein, G., Lovett, G., Chirmule, N., McCloskey, T., Pahwa, S. and Verdin, E. (1997). Immune hyperactivation of HIV-1-infected T cells mediated by Tat and the CD28 pathway. *Science*, 275: 1481-1485.

Parada, C.A. and Roeder, R.G. (1999). A novel RNA polymerase II-containing complex potentiates Tat-enhanced HIV-1 transcription. *EMBO J.,* 18: 3688-3701.

Piguet, V. and Trono, D. (1999). The Nef protein of primate lentiviruses. *Rev. Med. Virol.*, 9: 111-120.

Poeschla, E. and Wong-Staal, F. (1994). Molecular biology of HIV: challenges for the second decade. *AIDS Res. Hum. Retroviruses,* 10: 111-112.

Putkonen, P., Quesada-Rolander, M., Leandersson, A.-C., Schwartz, S., Thorstensson, R., Okuda, K., Wahren, B. and Hinkula, J. (1998). Immune responses but no protection against SHIV by gene-gun delivery of HIV-1 DNA followed by recombinnant subunit protein boosts. *Virology*, 250: 293-301.

Rana, T.M. and Jeang, K.T. (1999). Biochemical and functional interactions between HIV-1 Tat protein and TAR RNA. *Arch Biochem Biophys.,* 365: 175-185.

Ranki, A., Lagerstedt, A., Ovod, V., Aavik, E. and Krohn, K. (1994). Expression kinetics and subcellular localization of HIV-1 regulatory proteins nef, tat and rev in acutely and chronically infected lymphoid cells. *Arch. Virol..*, 139: 365-378.

Riviere, Y., Robertson, M.N. and Buseyne, F. (1994). Cytotoxic T lymphocytes in human immunodeficiency virus infection: regulator genes. *Curr. Top. Microbiol. Immunol.*, 189: 65-74.

Roman, M., Martin-Orozco, E., Goodman, J.S., Nguyen, M.D., Sato, Y., Ronaghy, A., Kornbluth, R.S., Richman, D.D., Carson, D.A. and Raz, E. (1997). Immunostimulatory DNA sequences function as T helper-1-promoting adjuvants. *Nat. Med.,* 3: 849-854.

Rowland-Jones, S., Sutton, J., Ariyoshi, K., Dong, T., Gotch, F., McAdam, S., Whitby, D., Sabally, S., Gallimore, A. and Corrah, T. (1995). HIV-specific cytotoxic T-cells in HIV-exposed but uninfected Gambian woman. *Nat. Med.,* 1: 59-64.

Rowland-Jones, S., Tan, R. and McMichael, A. (1997). Role of cellular immunity in protection against HIV infection. *Adv. Immunol.*, 65: 277-346.

Sadat-Sowti, B., Debré, P., Idziorek, T., Guillon, J-M., Hadida, F., Okzenhendler, E., Katlama, C., Mayaud, C. and Autran, B. (1991). A lectin-binding soluble factor released by CD8+CD57+ lymphocytes from AIDS patients inhibits T cell cytotoxicity. *Eur. J. Immunol.,* 21: 737-741.

Schatz, O., Oft, M., Dascher, C., Schebesta, M., Rosorius, O., Jaksche, H., Dobrovnik, M., Bevec, D., and Hauber, J. (1998). Interaction of the HIV-1 Rev cofactor eukaryotic initiation factor 5A with ribosomal protein L5. *Proc. Natl. Acad. Sci. USA.,* 95: 1607-1612.

Schwartz, O., Marechal, V., LeGall, S., Lemonnier, F. and Heard, J. M. (1996). Endocytosis of major histocompatibility complex class I molecules is induced by the HIV-1 Nef protein. *Nat. Med.,* 217: 293-300.

Schwartz, S., Felber, B., Pavlakis, G. (1991). Expression of human immunodeficiency virus type-1 Vif and Vpr mRNA is Rev-dependent and regulated by splicing. *Virology*, 183: 677-686.

Shearer, G.M. and Clerici, M. (1996). Protective immunity against HIV infection : has nature done the experiment for us? *Immunol. Today,* 17: 21-24.

Stanhope, P.E., Clements, M.L. and Siliciano, R.F. (1993). Human CD4+ cytolytic T lymphocyte responses to a human immunodeficiency virus type 1 gp160 subunit vaccine. *J. Infect. Dis.*, 168**:** 92-100.

Tähtinen M, Gombert F, Hyytinen E-R, Jung G, Ranki A and Krohn K. (1992). Fine specificity of the B-cell epitopes recognized in HIV-1 Nef by human sera. *Virology*, 187: 156 164.

Ulmer, J. B. (1997). Elegantly presented DNA Vaccines. *Nat. Biotechnol.*, 15: 842.

Van Baalen, C.A., Pontesilli, O.P., Huisman, R.C., Geretti, A.M., Klein, M.R., deWolf, F., Miedema, F., Gruters, R.A. and Osterhaus, A.D. (1997). Human Immunodeficiency Virus type 1 Rev- and Tat-specific cytotoxic T lymphocyte frequencies inversly correlate with rapid progression to AIDS. *J. Gen. Virol.*, 78**:** 1913-1918.

Wahren, B. and Brytting, M. (1997). DNA increases the potency of vaccination against infectious disesases. *Curr. Opin. Chem. Biol.*, 1: 183-189.

Wahren, B., Brytting, M., Engström, G., Hinkula, J., Levi, M., Ståhle, E.L., Ohlsson, S., Rude´n, U., Schwartz, S. (1996). Immune responses to the HIV rev regulatory gene. *Antibiot. Chemother.*, 48: 105-112.

Walker, B.D. and Plata, F. (1990). Cytotoxic T lymphocytes against HIV. *AIDS*, 4: 177-184.

Walker, C.M., Moody, D.J., Stites, D.P. and Levy, J.A. (1986). CD8+ lymphocytes can control HIV infection *in vitro* by suppressing virus replication. *Science*, 234: 1563-1566.

Wallace, M., Trivedi, P., Yin, C., Schmidt, D., Salvato, M.S. and Pauza, C.D. (1999). Mechanisms controlling virus replication and lymphocyte loss during acute SIV or SHIV infection in rhesus macaques. *J. Hum. Virol.*, 2: 177.

Wang, R., Doolan, D.L., Le, T.P., Hedstrom, R.C., Coonan, K., Charoenvit, Y., Jones, T.R., Hobart, P., Margalith, M., Ng, J., Weiss, W.R., Sedegah, M., de Taisne, C., Norman, J.A. and Hoffman, S.L. (1998). Induction of antigen-specific cytotoxic T lymphocytes in humans by a malaria DNA vaccine. *Science*; 282**:** 476-480.

Wei, P., Garber, M.E., Fang, S.M., Fischer, W.H. and Jones, K.A. (1998). A novel CDK9-associated C-type cyclin interacts directly with HIV-1 Tat and mediates its high-affinity, loop-specific binding to TAR RNA. *Cell*, 92: 451-462.

Zocchi, M.R., Rubartelli, A., Morgavi, P. And Poggi, A. (1998). HIV-1 Tat inhibits human natural killing cell function by blocking L – type calcium channels. (1998). *J. Immunol.*, 161: 2938-2943.

7. SUMMARY — Viral regulatory sequences may be exploited as immunogens for prophylactic and therapeutic vaccination. Deleted forms of HIV-1 regulatory *nef*, *tat* and *rev* genes have been associated with less virulent human infections. During the natural infection, they have unwanted virus-activating effects, are subjected to a low interstrain variation, and are poorly immunogenic. They are thus of interest for immunization approaches. The construction of primary HIV DNA plasmids and their capacity to produce functional and immunogenic proteins were recently demonstrated. HIV regulatory proteins carry several amino acid stretches with good binding capacity to HLA class I molecules, which induce cytotoxicity T lymphocyte (CTL) responses. Also, antibody responses are induced, which may serve to inhibit regulatory molecules that have intercellular trafficking.

We have shown that DNA immunization with HIV-1 regulatory genes can induce specific cellular responses in asymptomatic HIV-1 infected patients with no apparent side effects. The next goal will be to analyse the immunogenicity of these genes in uninfected individuals and to study the cross-reactivity of isotypic regulatory genes from clinical samples of rapid progressors (RP) and long term non progressors (LTNP).

Key Words: Gene vaccination; immunotherapy; regulatory genes; immunogenes

Progress in Gene Therapy: Basic and Clinical Frontiers, pp. 387-432
R. Bertolotti *et al.* (Eds)

Alphavirus vectors: from RNA replicons to suicidal DNA vaccines

Cristian Smerdou[1] and Peter Liljeström[1,2,*]

[1] *Microbiology and Tumor Biology Center, Karolinska Institutet, Box 280, S-17177, Stockholm, Sweden, and* [2]*Department of Vaccine Research, Swedish Institute for Infectious Disease Control, S-17182 Solna, Sweden*

Table of Contents

---

* Corresponding author. E-mail: Cristian.Smerdou@mtc.ki.se

1. INTRODUCTION

Alphaviruses are enveloped positive strand RNA viruses that have been extensively studied (for a review see Strauss and Strauss, 1994). They include viruses with a very wide geographic distribution, which have been classified in three main groups on the basis of their sequence data. The main representative viruses of each group are the Sindbis virus (SIN), the Semliki Forest virus (SFV) and the Venezuelan Equine Encephalitis virus (VEE). These viruses are arthropod borne, with mosquitoes as the usual vector, and can be transmitted to birds and mammals, including humans. Although SIN and SFV are considered avirulent to humans, VEE can cause serious human illness. Several features have made alphaviruses especially attractive for the development of expression systems: 1) they have a very broad host range being able to infect cells of many different origins, such as insect, reptilian, avian or mammalian cells; 2) they can express very high levels of the structural proteins in the infected cells, a mechanism which has been utilized for expression of heterologous proteins at similar high levels; 3) cDNAs of their relative small genomes (around 11-12 kb) allow easy manipulation *in vitro*, and 4) the naked genomic RNA is infectious, being able to give rise to a productive infection when transfected into cells. Expression vectors have been developed for each of the three representative alphaviruses, SIN, SFV and VEE (Liljeström and Garoff, 1991; Pushko *et al.*, 1997; Xiong *et al.*, 1989). These vectors can be used for gene delivery into cells both *in vitro* or *in vivo*, and have shown a great potential in a variety of applications. The vectors can be delivered as recombinant viral particles, naked RNA or DNA. In the latter case, the replicon vector is transcribed in the nucleus of the transfected cell from an RNA polymerase II promoter. These delivery systems and their applications will be widely discussed in this chapter.

2. THE LIFE CYCLE OF ALPHAVIRUSES

The alphavirus particles contain a single positive strand RNA molecule associated to 240 subunits of the capsid protein, and surrounded by a lipid bilayer containing 240 heterodimers of the spike proteins E1 and E2 (Strauss and Strauss, 1994). When the virus enters the cell, through interaction with one or more receptors on the cell surface, the genomic RNA is released into the cytoplasm where the viral replication takes place. The genomic RNA, which is capped and polyadenylated, can function as a mRNA. The 5' two-thirds of this RNA is first translated to produce the nonstructural proteins (nsp) 1-4 constituting the viral replicase (Fig. 1). This enzyme copies the genome into negative strands, which serve as templates for the production of new positive strand genomes. The replicase also utilizes an internal promoter on the minus strand (subgenomic promoter) to produce a subgenomic RNA that corresponds to the last third of the virus genome. This subgenomic RNA encodes the structural proteins of the virus. These proteins are synthesized as a polyprotein precursor (NH_2-C-p62-6K-E1-COOH) which is cotranslationally cleaved by the capsid proteinase and signal peptidase to produce the capsid protein (C) and the envelope proteins p62, 6K and E1. In some alphaviruses, such as SFV and SIN, a sequence at the 5' end of the capsid gene functions as a translational enhancer, providing high level production of the structural proteins (Frolov and Schlesinger, 1994a and 1996; Sjöberg and Garoff, 1996; Sjöberg *et al.*, 1994). The C protein associates with new viral genomes to form cytoplasmic nucleocapsid structures, while the spike proteins are translocated to the endoplasmic reticulum (ER), where p62 and E1 dimerize. These heterodimers are routed to the cell surface where budding occurs through an interaction between the nucleocapsid and the spike proteins (Lee *et al.*, 1996; Lopez *et al.*, 1994; Skoging *et al.*, 1996; Suomalainen *et al.*, 1992; Zhao *et al.*, 1994). During transport to the cell surface, p62 is cleaved to its mature form E2 by a host cell protease. This cleavage is necessary for the infectivity of the particles (Jain *et al.*, 1991; Lobigs and Garoff, 1990; Salminen *et al.*, 1992) by allowing E1 to dissociate from E2 and rearrange into a trimeric complex which catalyzes membrane fusion (Bron *et al.*, 1993; Wahlberg *et al.*, 1992). A few hours after infection, the alphavirus induces a strong shut-off

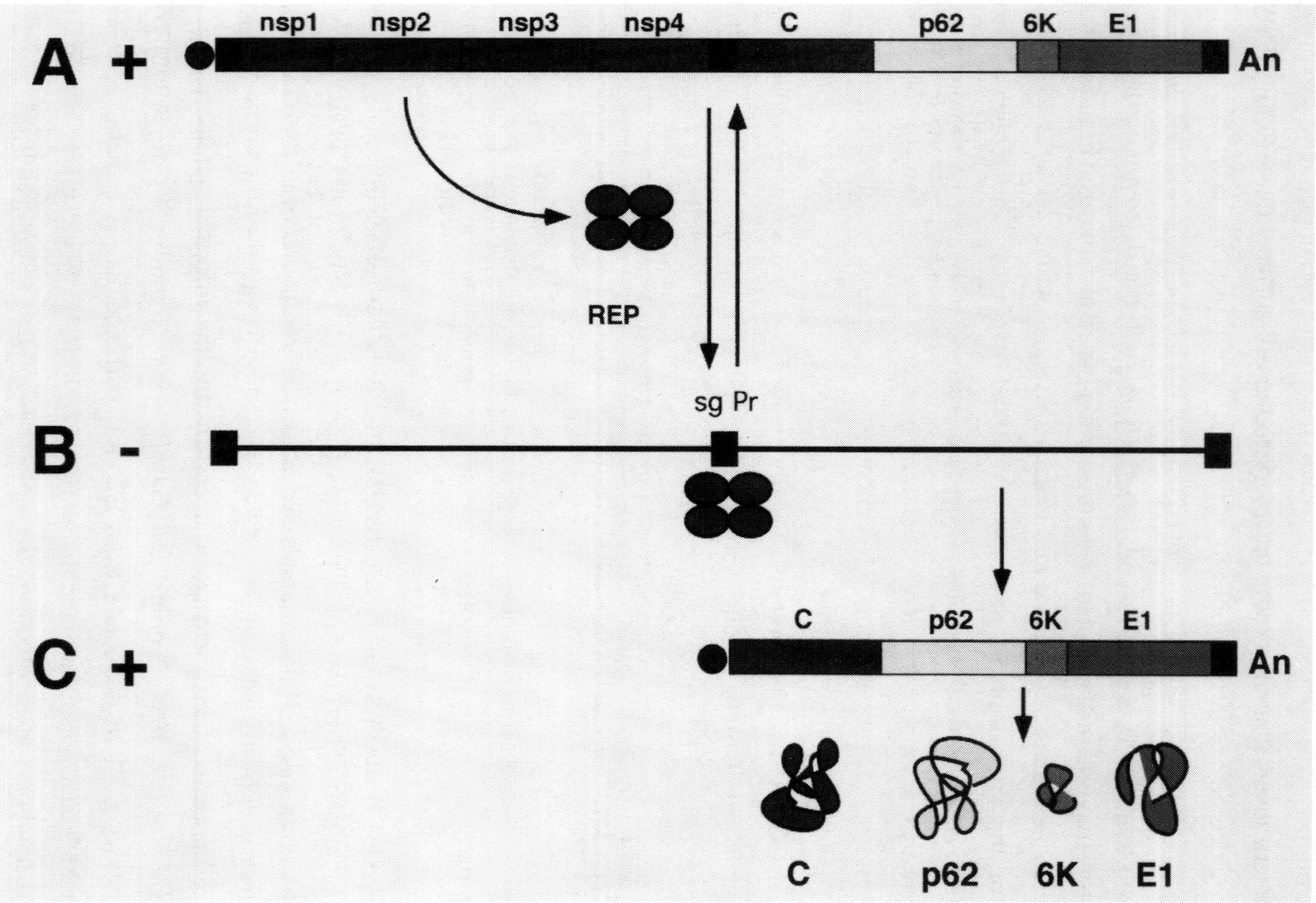
A +
nsp1
nsp2
nsp3
nsp4
C
p62
6K
E1
An
REP
sg Pr
B -
C +
C
p62
6K
E1
An
C
p62
6K
E1

of the host cell protein synthesis in such a way that only the structural proteins of the virus are being synthesized. These can constitute up to 50% of the total protein content of the infected cell. The viral infection also induces a cytopathic effect that leads to cell death by apoptosis in 48-72 hrs.

3. ALPHAVIRUS EXPRESSION SYSTEMS

The first step in the development of an alphavirus expression system was the cloning of the RNA genome in the form of a cDNA in a plasmid under the control of a prokaryotic promoter, such as SP6 or T7. By using SP6/T7 polymerase, the genomic RNA can be transcribed *in vitro* and used to transfect cells by different methods, including DEAE dextran, cationic lipid or electroporation. This last procedure has shown to be the most efficient, allowing 100% transfection of cells (Liljeström *et al.*, 1991). cDNA clones have been obtained for several alphaviruses, including SFV (Liljestrom *et al.*, 1991), SIN (Rice *et al.*, 1987), VEE (Davis *et al.*, 1989), and Ross River virus (RRV) (Kuhn *et al.*, 1991). The availability of such cDNA clones has made it possible to study multiple viral functions through reverse genetics, including the localization of the minimal replication and packaging sequences. In order to express foreign proteins in the context of an alphavirus genome, two different types of replicon vectors have been developed: replication-packaging proficient and suicidal vectors (Fig. 2).

Figure 1. Replication in alphavirus. The alphavirus genome is a single positive strand RNA molecule of approximately 11-12 kb (A). The nsp 1-4 genes encoding the viral replicase (REP) are located at the 5'-end of the genome and will be translated in the cytoplasm of the infected/transfected cell to produce this enzyme. REP will recognize specific sequences at the 3' end of the genomic RNA and synthesize a complementary strand of negative polarity (B). This RNA will also be copied by REP to produce more positive genomic RNA that can be encapsidated into viral particles. The replicase also recognizes a subgenomic promoter (sg Pr) sequence in the negative RNA from which it synthesizes a smaller positive subgenomic RNA (C). This subgenomic RNA will be translated into a polypeptide that is processed into the different structural proteins of the virus (capsid, p62, 6K and E1). The small circle at the 5' end of the positive RNAs represents the CAP; black squares at the 5’ and 3’ ends of the RNAs, are sequences necessary for RNA replication; nsp, nonstructural proteins; An, polyA; C, capsid.

a) Replication-packaging proficient vectors.

These vectors contain the complete alphavirus sequence including the 5' and 3' sequences necessary for replication, the replicase genes and the subgenomic promoter followed by the structural genes. The expression of the foreign gene is achieved by the inclusion of a second subgenomic promoter followed by the heterologous sequence in a nonessential region of the genome (Fig. 2A and 2B). These extra sequences can be added in a position 3' of the structural genes or inserted at the end of the replicase genes (before the viral subgenomic promoter). When this type of RNA is transfected into cells, both the heterologous protein and the viral structural proteins will be made. The latter will package the amplified recombinant genome into viral particles that can propagate. This type of strategy has been used to construct SIN (Hahn *et al.*, 1992; Raju and Huang, 1991) and VEE vectors (Davis *et al.*, 1996). In general, these vectors show stability problems after several passages, probably due to restrictions in the packaging capacity of the particles. Vectors containing inserts located 5' of the structural genes have usually shown greater stability than those in which the insertion was placed at the 3' end (Pugachev *et al.*, 1995). Although these vectors have shown high levels of foreign protein expression and are easy to produce, their use *in vivo* presents important biosafety limitations due to their infectious nature. A variant of these replication-packaging proficient vectors that has been used to express small peptides or short relevant epitopes is the so-called chimeric vectors. In these vectors, the foreign sequence is expressed on the surface of the viral proteins E2 or E3, taking advantage of the existence of some permissive sites in these proteins where short sequences can be inserted without disrupting their folding and functionality (London *et al.*, 1992).

b) Suicidal vectors

In order to render the alphavirus expression vectors non-infectious, the genes coding for the structural proteins were deleted and substituted by a polylinker where the foreign sequence can be cloned downstream of the subgenomic promoter (Fig. 2C and 2D). RNA transcribed from this type of vector can be transfected into cells where only the replicase and the heterologous protein or RNA will be made (Fig. 3A). Since no structural proteins are synthesized, viral particles will not be generated. These vectors have been developed for SIN (Xiong *et al.*, 1989), SFV (Liljeström and Garoff, 1991), and VEE (Pushko *et al.*, 1997). High level of foreign protein

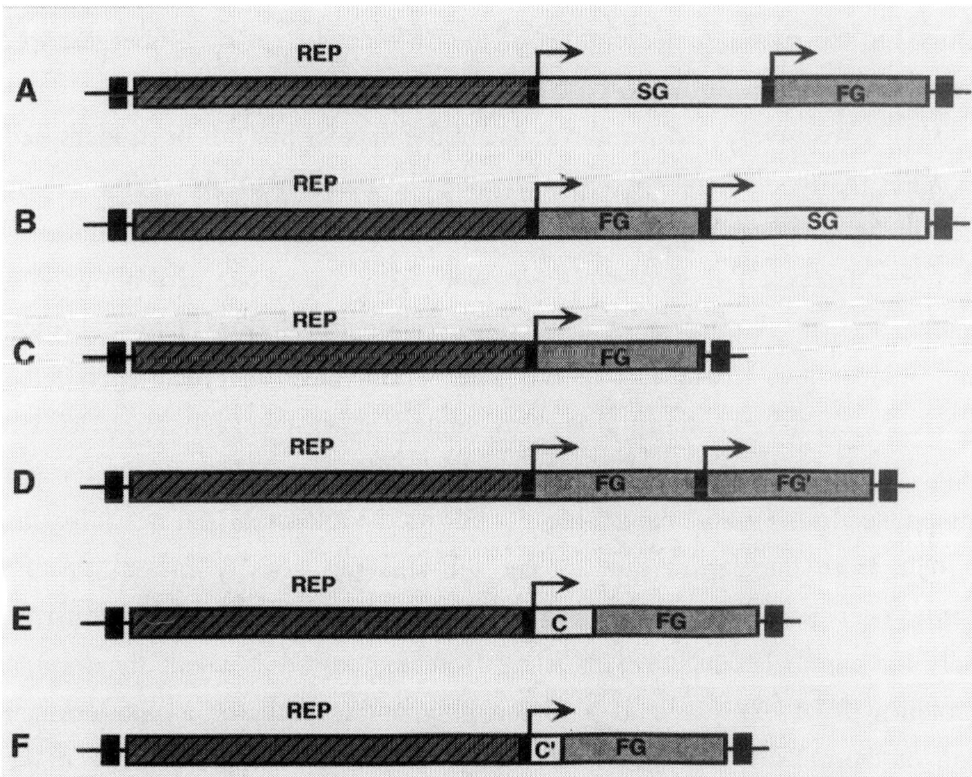

Figure 2. Alphavirus vectors. A and B, alphavirus replication–packaging proficient vectors with a second subgenomic promoter located 3'or 5' of the structural genes, respectively; C and D, alphavirus suicidal vectors containing one or two foreign genes, respectively; E and F, enhanced alphavirus suicidal vectors containing a fusion of the foreign gene with the whole viral capsid gene or the 5' translation enhancer region of this gene, respectively. REP, genes coding for the viral replicase; SG, genes coding for the viral structural proteins; FG, foreign gene; C, capsid gene; C', 5' end of the capsid gene containing the translation enhancer; black squares with horizontal arrows, subgenomic promoters; black squares at the ends of the vectors, 5' and 3' sequences necessary for vector RNA replication.

expression has been reported with this type of vectors, reaching 60-100 pg protein/cell when *LacZ* was used as a reporter gene (Berglund *et al.*, 1998; Bredenbeek *et al.*, 1993; Liljeström and Garoff, 1991). However, this level of expression can be increased around ten times by making an in-frame fusion of the gene of interest at the 3' end of the viral capsid gene (Fig. 2E and 2F) (Berglund *et al.*, 1998; Frolov and Schlesinger, 1994a; Sjöberg *et al.*, 1994). Further characterization showed that the 5'end of the capsid gene was enough to provide the translation enhancing effect. This region has been limited to the first 102 nucleotides of the SFV capsid gene (Sjöberg *et al.*, 1994) or the first 226 nt of the SIN capsid gene (Frolov and Schlesinger, 1994a), respectively. No similar translation enhancer has yet been identified in VEE. A hypothetical model to explain the mechanism of action of the translation enhancer suggests the formation of a stable hairpin structure in this area that could slow down the movement of the ribosomes, allowing a more efficient recruitment of low concentration translation factors (Frolov and Schlesinger, 1994a and 1996). This would have an effect only if the RNA containing the hairpin is very abundant in the cell, such as the subgenomic RNA in the transfected/infected cells. Expression of the foreign gene fused to the whole capsid usually allows the release of the heterologous protein from the capsid due to the self-cleaving activity present in the carboxy-terminal domain of this protein (Sjöberg *et al.*, 1994). However, if only the minimal translation enhancer from the capsid is used, the foreign protein will be expressed as a fusion polypeptide with the amino-terminal end of the capsid. One possible strategy to provide cleavage between the capsid enhancer and the foreign protein is to use the sequence of the 2A autoprotease from foot-and-mouth disease virus (FMDV) as a linker. This protease will cleave in *cis* at its carboxy-terminal end, allowing the release of a foreign protein containing only one residual proline at the amino-terminal end (Ryan and Drew, 1994; Ryan *et al.*, 1991; Smerdou and Liljeström, 1999).

4. PACKAGING OF ALPHAVIRUS REPLICONS

In order to increase the efficiency in the delivery of the alphavirus suicidal RNA into the cells, several strategies have been developed that allow encapsidation of these RNA molecules into viral particles.

a) Non packagable alphavirus helper RNA.

To package the alphavirus recombinant replicon into viral particles it is necessary to cotransfect the cells with a helper RNA that can provide the structural proteins of the virus in *trans* (Fig. 3B). The helper RNA contains the 5'and 3' sequences necessary for replication and the viral structural genes downstream of the subgenomic promoter. However, in the helper RNA, most of the replicase sequence has been deleted, including the packaging signals which are located in this area of the genome (Frolova *et al.*, 1997; White *et al.*, 1998). When cells are cotransfected with the vector RNA carrying the foreign gene and the helper RNA, the replicase encoded by the former will amplify both vector and helper RNAs. The replicase will also make two subgenomic RNAs that will be translated to produce the foreign and the structural proteins. The capsid protein will form nucleocapsids by interacting with the packaging signal present only in the vector RNA and viral particles will be generated at the cell surface containing this RNA. The particles generated in this way are suicidal since they contain a genome that, while able to replicate, cannot result in a productive infection. Such helper systems have been developed for SFV (Liljeström and Garoff, 1991), SIN (Bredenbeek *et al.*, 1993), and VEE (Pushko *et al.*, 1997) allowing production of high titer stocks of alphavirus recombinant particles, usually in the range of 10^9 particles per 10^6 transfected cells. However, these packaging systems have the drawback that recombination between the helper and vector RNAs can take place during the packaging reactions producing wild type genomes that can be incorporated into viral particles and thus further propagated. Recombination can take place at a relative high frequency in alphaviruses (Berglund *et al.*, 1993; Haijou *et al.*, 1996; Hill *et*

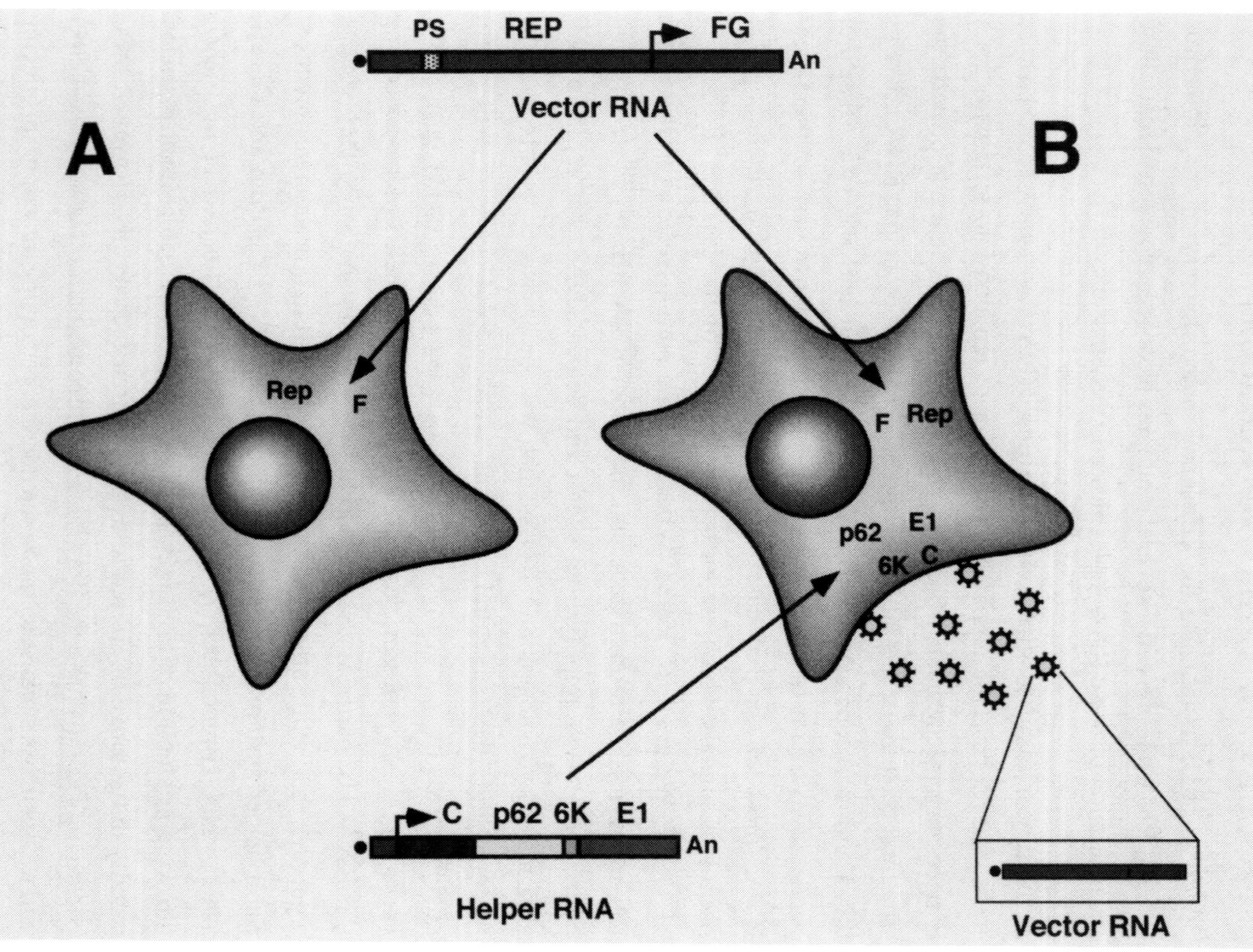
A
B
PS
REP
FG
An
Vector RNA
Rep
F
Rep
F
p62
6K
C
E1
C
p62 6K E1
An
Helper RNA
Vector RNA

al., 1997; Raju *et al.*, 1995; Schlesinger and Weiss, 1994; Weiss and Schlesinger, 1991) and replication proficient virus (RPV) have been detected with the three alphavirus helper systems described here. The frequency of recombination has been estimated to be around 10^{-6} or less (Berglund *et al.*, 1993). Due to the higher pathogenicity of VEE in humans, some additional attenuating mutations were introduced in the E2 and E1 genes present in the VEE helper RNA, so that any potential recombinant would have a good chance of being attenuated (Davis *et al.*, 1996 and 1991).

b) Packaging of alphavirus conditionally infectious particles

In order to reduce the amount of RPVs in the packaging system, a new helper RNA was developed for SFV in which a mutation abolishing intracellular cleavage of p62 was introduced, rendering the particles conditionally infectious (helper-2) (Berglund *et al.*, 1993). The non-infectious particles can be artificially "activated" by incubation with chymotrypsin *in vitro*. Using this packaging strategy, RPVs have not been detected to date, increasing considerably the biosafety of the expression system in SFV. One drawback of this helper system is the necessity of

Figure 3. Alphavirus suicidal RNA expression and packaging. A). Alphavirus suicidal vector RNA (vector RNA) can be synthesized *in vitro* and introduced into eukaryotic cells by lipofection or electroporation. Once inside the cell, the replicase (Rep) will be made. This enzyme will amplify the vector RNA and also synthesize a small subgenomic RNA from which the foreign gene product (F) will be translated at high levels. No viral particles are produced with this system, since the viral structural genes are absent. **B).** Packaging of alphavirus recombinant RNA into viral particles with one single helper RNA. The structural proteins of the virus can be provided in *trans* from a second RNA (helper RNA) that can also be synthesized *in vitro*. When cells are cotransfected with the vector RNA and the helper RNA, the replicase will be synthesized from the first one and amplify both RNA species. The replicase will also synthesize subgenomic RNAs from both vector and helper RNAs. The helper subgenomic RNA will be translated to produce the viral structural proteins: capsid (C), p62 (E2 precursor), 6K and E1. The specific interaction of the capsid protein with the packaging signal (PS) present in the vector RNA will produce nucleocapsids that will bud at the cell surface by interacting with the viral spike proteins (E2, E1 and 6K). The viral particles produced in this way are able to infect cells but cannot propagate since the genes coding for the structural proteins are not present in the encapsidated RNA. FG: foreign gene; black circle at the left of the RNAs: CAP; An: polyA

activating the particles with chymotrypsin, which makes their use less convenient. However, recombination between the vector and the helper RNAs is not prohibited per se, and generation of RPVs would also require reversions or suppressor mutations able to regenerate a functional p62 cleavage site. Such mutations have indeed been found, which either regenerated the p62 cleavage site or restored infectivity without cleavage of p62, probably through conformational changes of the protein (Tubulekas and Liljeström, 1998). New helpers were constructed containing additional nucleotide changes to reduce the frequency of reversion, but in all cases, new suppressors were found in a different domain of the protein which lead to extensive attenuation (Tubulekas and Liljeström, 1998). Although the probability of simultaneously having recombination and suppressor mutations in this system is very low, it cannot be completely excluded.

c) Split helper systems

An alternative strategy to increase the biosafety of the alphaviruses packaging system has been to reduce the frequency of recombination by splitting the helper into two independent molecules. This strategy, which has been used in SIN (Frolov *et al.*, 1997), SFV (Smerdou and Liljeström, 1999) and VEE (Pushko *et al.*, 1997), is based on expressing the capsid and the spike proteins from two different helpers (Fig. 4). With this strategy, cells have to be transfected with three different RNAs (the vector replicon and the two helper RNAs) and the generation of wild type genomes would require two independent recombination events between the three RNA molecules. This reduces considerably the probability of RPV generation. While the expression of the capsid protein by itself does not affect its expression level, the expression of the spikes without the capsid would reduce considerably their level of expression in SIN and SFV. Two different strategies have been used to address this problem. In the SIN model, the translation enhancer was provided by fusing the spikes with a heterologous capsid protein from RRV, in which the RNA binding domain had been deleted (Fig. 4A) (Frolov *et al.*, 1997). This mutated RRV capsid maintained the self-cleaving domain and was able to cleave itself off the spikes. Trans-

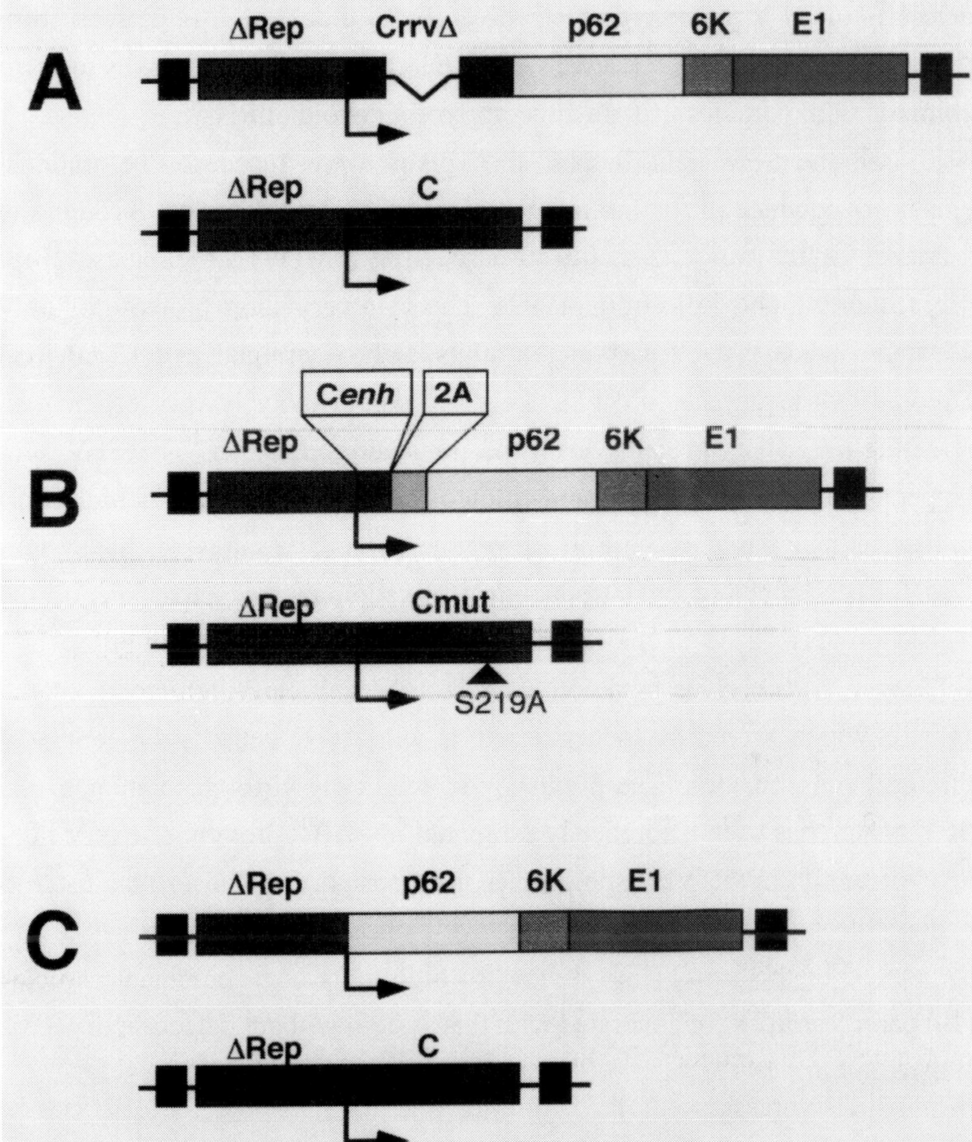

Figure 4. Alphavirus split helper RNAs for packaging of RNA vector replicons based on SIN (A), SFV (B), or VEE (C). ΔRep, viral replicase genes containing a large deletion that includes the packaging signals; CrrvΔ, gene coding for the Ross River virus capsid protein containing a deletion that includes the RNA binding motif; C*enh*, translation enhancer region of SFV capsid; 2A, 2A autoprotease from foot and mouth disease virus; Cmut, SFV capsid gene containing the mutation Ser219→Ala; C, p62, 6K, E1, genes coding for the structural proteins of each of the respective viruses.

spikes. Transfection with the two SIN helper RNAs and SIN vector RNA led to the production of high titer stocks (1-2x10^9 particles/ml) with no detection of RPVs. One possible problem with this system is derived from the fact that the mutated RRV capsid can interact with the SIN spikes to form chimeric viral particles that are not completely devoid of RNA.

In the SFV split helper, the spikes were fused to the minimal enhancer sequence of the homologous capsid (102 nucleotides) through the sequence of the foot-and-mouth disease virus (FMDV) 2A protease (Fig. 4B) (Smerdou and Liljeström, 1999). This is a very short protease of only 17 amino acids that cleaves in *cis* at its carboxyterminal end (Ryan and Drew, 1994; Ryan *et al.*, 1991). This cleavage activity is very efficient in releasing the spikes proteins from the translation enhancer. The SFV split helper system also allows the generation of high stock titers (10^9particles/ml) of recombinant particles with no RPV generation. The safety of this system was further increased by introducing a mutation in the active site of the capsid protease gene in the helper coding for this protein (Smerdou and Liljeström, 1999). This mutation abolished the self-cleaving activity of the capsid, which would be necessary if a wild type genome is generated through recombination. The frequency of wild type virus generation in the SFV system has been theoretically estimated as $<10^{-17}$ providing a very high level of safety. In the VEE split helper, no translation enhancer was used to express the spike polyprotein at high level, since it may not be necessary in the context of this virus (Fig. 4C). Also high titers of recombinant particles (10^8 particles/ml) were generated with this system without detection of RPVs (Pushko *et al.*, 1997).

d) Packaging cell lines

A most recent strategy to package alphavirus vector replicons is based on the use of stable cell lines able to produce the viral structural proteins upon transfection or infection with a functional replicon (Polo *et al.*, 1999). These cell lines are obtained by stably transforming cells with a plasmid that contains the alphavirus helper sequences downstream of an RNA polymerase-II promoter. In such a cell line, the helper RNA will be transcribed in the nucleus of the cell and transported to the cytoplasm in a constitutive way. However, the structural proteins will not be produced in the absence of the viral replicase, since they are translated from the subgenomic RNA that is produced by the replicase. Upon transfection of the cells with a functional replicon, the vector-encoded replicase will amplify the vector and helper RNAs and synthesize the subgenomic RNA, allowing the structural proteins to be made. The recombinant particles generated in this way will be propagated in this cell line since all the cells can provide the structural proteins. Several packaging cell lines have been generated for SIN, which include the use of one or two helpers (Polo *et al.*, 1999). In the latter case, the split helper approach described above for SIN was used. Upon infection of this cell line at a low multiplicity of infection (moi 0.2), recombinant particles were produced with titers of $0.5\text{-}1 \times 10^7$ particles/ml 48 hours post infection, and RPVs could not be found. Higher particle titers were obtained in packaging cell lines containing a single helper expressing all the structural proteins, but this was accompanied by the presence of RPVs. These packaging cell lines were also able to package SFV replicons with similar efficiency.

5. HETEROLOGOUS PROTEIN EXPRESSION USING ALPHAVIRUS VECTORS

The alphavirus expression systems have been used to express foreign proteins for different applications. These include protein purification and characterization, functional studies of proteins through mutagenesis, determination of cellular localization of proteins, virus assembly, control of viral infections in insects, vaccination and gene therapy. Functional studies have included the expression and characterization of different receptors such as the human neurokinin-1 and 2 receptors (Lundström *et al.*, 1995a, 1994 and 1995b), human dopamine D3 receptor (Lundström *et al.*, 1995a; Lundström and Turpin, 1996), ligand-gated ion channels (Lundström *et al.*, 1997), the human gonadotropin releasing hormone receptor (Marheineke *et al.*, 1998) and several other G protein-coupled receptors (Monastyrskaia *et al.*, 1999; Scheer *et al.*, 1999). Cellular localization studies have included the infection of neurons with recombinant SFV particles to determine the polarized distribution of heterologous proteins such as rab8 and VIP21 (de Hoop *et al.*, 1994; Olkkonen *et al.*, 1993), as well as the intracellular localization of rat odorant receptors in several mammalian cell lines (Monastyrskaia *et al.*, 1999). SIN based vectors have also been used to produce a functional monoclonal antibody specific for SIN E2 protein through cotransfection of BHK cells with two SIN vectors expressing the light and heavy chain of the antibody, respectively (Liang *et al.*, 1997).

Several viral proteins have been expressed and characterized using alphavirus vectors. Some examples include the HIV-1 envelope proteins (Paul *et al.*, 1993; Verrier *et al.*, 1997), hepatitis C virus envelope glycoproteins (Dubuisson *et al.*, 1994), Japanese encephalitis virus structural proteins (Pugachev *et al.*, 1995), rubella virus structural proteins (Chen *et al.*, 1995), the capsid protein of hepatitis E virus (Torresi *et al.*, 1999 and 1997), replication proteins from human cytomegalovirus (McCue and Anders, 1998), LaCrosse virus envelope proteins (Kamrud *et al.*, 1998), the NS3-NS4A protease from hepatitis C virus (Filocamo *et al.*, 1999), and the nonstructural protein (NSs) from Rift Valley fever virus (Yadani *et al.*, 1999). Many other viral proteins have also been expressed in alphavirus for vaccination studies, as it will be discussed with more detail below. Assembly studies of viral proteins expressed from an alphavirus vector have included the human papilloma virus type 16 capsid protein (Heino *et al.*, 1995), capsid proteins from rotaviruses (Nilsson *et al.*, 1998), Moloney leukemia virus envelope proteins (Opstelten *et al.*, 1998), and mutant capsid proteins from SFV (Skoging and Liljeström, 1998; Skoging *et al.*, 1996; Suomalainen *et al.*, 1992).

An interesting set of studies has focused on the use of alphavirus replicons to control viral infection in mosquitoes to prevent disease transmission to humans. These studies were based on the ability of SIN vectors to express heterologous proteins or RNAs in mosquito cells and transduced mosquitoes (Higgs *et al.*, 1995; Kamrud *et al.*, 1995; Olson *et al.*, 1992 and 1994; Seabaugh *et al.*, 1998). SIN vectors expressing antisense RNAs to different genes of viruses transmitted by mosquitoes have been used to transduce these insects (Higgs *et al.*, 1998; Olson *et al.*, 1996; Powers *et al.*, 1994; Rayms-Keller *et al.*, 1995). Transduction of mosquitoes with a SIN vector expressing an antisense RNA targeted to the premembrane coding region of DEN-2 inhibited replication of this virus in the salivary glands and the mosquitoes were not able to transmit the virus (Olson *et al.*, 1996). A similar result was obtained when transducing mosquitoes with SIN expressing antisense RNAs to the premembrane (prM) or polymerase (NS5) coding regions of a vaccine strain of yellow fever virus (Higgs *et al.*, 1998). The use of alphavirus vectors in gene therapy applications will be more extensively discussed at the end of this chapter.

6. VACCINATION WITH ALPHAVIRUS RECOMBINANT PARTICLES

Recombinant alphavirus particles have shown to be very effective in inducing long lasting humoral and cellular immune responses against heterologous antigens in different animal models. The immune responses generated with this type of particles have conferred protection against different pathogens, such as viruses and parasites. Initial studies involved vaccination with recombinant replication-packaging proficient vectors which were able to propagate in the vaccinated animal (Caley *et al.*, 1997; Davis *et al.*, 1996; Hahn *et al.*, 1992; London *et al.*, 1992; Pugachev *et al.*, 1995; Tsuji *et al.*, 1998). While these vectors were efficient in inducing immune responses against heterologous antigens, their use as vaccines cannot be considered due to their propagating nature. Other problems associated with the use of replication-packaging proficient viruses are related to the low stability of the insert, and to the generation of strong immune responses against the vector proteins, which can prevent their use in repetitive immunizations.

On the other hand, also suicide recombinant alphavirus particles have been able to induce strong immune responses against a variety of antigens with very low immune response against the vector itself (Berglund *et al.*, 1999; Pushko *et al.*, 1997). Conditionally infectious SFV particles expressing the nucleoprotein of influenza virus (SFV-NP) or β-galactosidase (SFV-LacZ) have been used to extensively characterize the immune responses generated with this type of particles in mice (Berglund *et al.*, 1999; Zhou *et al.*, 1994 and 1995). SFV-NP particles were extremely efficient in inducing cellular immune responses shown by the fact that as few as 100 particles inoculated intravenously generated a strong CD8+ cytotoxic T-cell (CTL) response. With higher doses of particles (10^6), mice developed CTL responses comparable to those in mice infected with influenza virus. In these animals, the responses were long lasting and CTL memory persisted over one year, even when the mice had received only one single immunization. The fact that good CTL responses can be induced with a small number of particles suggests that important antigen presenting cells (APCs) become directly infected by the recombinant SFV or that crosspriming is highly efficient. Antibodies were also efficiently induced with SFV recombinant particles. These antibody responses were long lasting and persisted for over 1.5 years. IgG2a antibodies were predominantly induced, indicating a $CD4^+$ Th-1 type immune response. SFV–NP particles combined with SFV particles expressing hemagglutinin from influenza virus (HA) were able to confer total protection in mice when the animals were challenged with the homologous influenza virus. A similar level of protection against influenza was obtained in mice immunized with VEE particles expressing HA (Davis *et al.*, 1996; Pushko *et al.*, 1997). Mucosal immunity has also been induced in mice inoculated intranasally with recombinant SFV particles (Berglund *et al.*, 1999; Fleeton *et al.*, 1999b;

Malone *et al.*, 1997) or subcutanously with recombinant VEE particles (Caley *et al.*, 1997; Davis *et al.*, 1996). VEE based vectors have been reported to replicate in the lymph nodes draining the site of inoculation where high level synthesis of a heterologous antigen can result in efficient immunization (Gleiser *et al.*, 1962; Grieder *et al.*, 1995; Jackson *et al.*, 1991). Other studies in which suicide alphavirus recombinant particles have been successfully used to induce immune responses in mice include SFV particles expressing proteins prM and E of the flavivirus Murray Valley encephalitis virus (MVE) (Colombage *et al.*, 1998), HIV-1 envelope glycoproteins (Brand *et al.*, 1998), prME and NS1 proteins from louping ill virus (LIV) (Fleeton *et al.*, 1999a and 1999b), and respiratory syncytial virus glycoproteins (Chen and Liljeström, unpublished results). SIN particles expressing the glycoprotein B of herpes simplex virus (Polo *et al.*, 1999), and VEE particles expressing Lassa virus nucleocapsid (Pushko *et al.*, 1997) were also able to induce immune responses in mice.

The high immunogenicity of the suicide alphavirus particles together with their high level of safety make them good candidates for human clinical applications. Vaccination trials in nonhuman primates have been done involving vaccination of monkeys with SFV or VEE recombinant particles. SFV recombinant particles expressing the envelope protein of simian immunodeficiency virus PBj14 (SIV-PBj14) were able to induce antibodies in pigtail macaques (Mossman *et al.*, 1996). Upon challenge with SIV-PBj14, the vaccinated animals were protected from acute disease and the viral load was reduced as compared to control animals. In a similar study, cynomolgus macaques were immunized with SFV particles expressing gp160 from HIV-1 (Berglund *et al.*, 1997). Humoral and cellular responses and a reduced viral load could be detected in these animals, but not in unimmunized controls, after challenge with a chimeric simian-human immunodeficiency virus (SHIV), indicating that a specific immune response had been primed. Priming of the immune response has also been shown by the immunization of rhesus monkeys with SFV particles expressing SIV

Pr56^{gag} and HIV-1 Env derivatives (Notka *et al.*, 1999). In these animals, the immune response against these antigens was enhanced when the animals were boosted with purified chimeric SHIV virus-like particles expressed from baculovirus, as compared to monkeys that had not been prevaccinated with SFV recombinant particles. A complete protection against Marburg virus (MBGV) was achieved in cynomolgus macaques with VEE particles expressing the glycoprotein of MBGV (Hevey *et al.*, 1998).

7. GENE TRANSFER USING ALPHAVIRUS SELF-AMPLIFYING NAKED RNA

Gene transfer or vaccination with naked RNA has not been widely used, mainly due to the instability of this type of molecules, as well as to the high cost of synthesizing RNA *in vitro*. If compared to DNA vectors, naked RNA has the advantage of a higher safety level, due to the elimination of chromosomal integration risk, potentially present when DNA vectors are used *in vivo*. Transfection of cells with conventional mRNA coding for a foreign antigen requires the use of very large amounts of RNA in order to achieve high level expression of the recombinant protein. In contrast, the use of self-amplifying RNA derived from alphavirus vectors could provide a high level of expression in the transfected cells with less RNA. As shown by Johanning *et al.*, a recombinant SIN RNA carrying the luciferase gene can express up to 1000-fold more protein than a conventional luciferase mRNA in BHK cells transfected *in vitro* (Johanning *et al.*, 1995). Delivery of this RNA vector *in vivo* by intramuscular injection in the tongue provided an expression level which was 24-fold higher than that observed with a conventional mRNA. The duration of the expression was also more prolonged, being at least 5-fold longer in the case of the SIN RNA vector. This expression, however, was already undetectable 28 days after the injection. The reasons for the loss of expression could include RNA degradation in transfected cells, induction of immune responses against these cells, or cytopathic effects produced by the self-replicative vector. Apoptosis induction in cells transfected with RNA replicons derived from alphavirus has been described for SIN and SFV, being responsible for the transient nature of these vectors (Frolov and Schlesinger, 1994b; Glasgow *et al.*, 1998). Recently, it was shown that cells transfected with SFV replicons underwent p53-independent apoptosis 48 h after transfection, where the replicase function was necessary for the cytopathic effect (Glasgow *et al.*, 1998).

Several studies have utilized recombinant alphavirus RNA for vaccination. In one of these studies, immunization of mice with 10 μg of recombinant SFV RNA expressing influenza virus NP induced specific antibodies to this protein (Zhou *et al.*, 1994). The antibody level induced with such an RNA vector was lower than that obtained when the same RNA was packaged into SFV viral particles, probably due to the higher efficiency of entry into the cells of the particles. The antibody response against the heterologous antigen induced with the SFV-RNA was long lasting, even with one single immunization, as shown by the fact that the antibody titers did not decrease 4 months after the vaccination. Similar results have been reported in mice immunized with recombinant SFV RNA expressing HIV protein gp160 (Brand *et al.*, 1998). In this case, the SFV RNA vector showed to be more effective than a conventional DNA vaccine encoding the same antigen in inducing antibodies specific for the recombinant protein. As described before, the immune response was higher when the mice were inoculated with recombinant SFV particles carrying the SFV HIVgp160 RNA, confirming that efficient delivery of the RNA into the cell is a crucial factor. Immunization of mice with SFV RNA expressing the HIV-1 envelope protein has also been used to generate monoclonal antibodies against this protein (Giraud *et al.*, 1999). In another study, mice immunized with recombinant SFV RNA expressing the hemagglutinin (HA) of influenza virus and subsequently challenged with the homologous influenza strain, were able to completely clear the virus from the lungs and to partially reduce the amount of virus in the trachea, which indicates that immunization with SFV RNA can control the spread of disease (Dalemans *et al.*, 1995). SFV recombinant RNA expressing the flavivirus LIV prME or NS1 proteins has also been able to induce protective immunity against LIV infection in mice (Fleeton *et al.*, in preparation).

8. ALPHAVIRUS DNA/RNA LAYERED SYSTEMS

One potential problem associated with the *in vivo* use of naked RNA for gene delivery or vaccination is the variability due to degradation of this type of molecule. Vectors based on DNA, on the other hand, are highly stable and easily produced with low cost, making them more suitable for clinical applications. A new generation of DNA vectors based on alphavirus replicons have recently been developed. These vectors are based on plasmid DNAs in which the alphavirus replicon sequence is placed downstream of an eukaryotic promoter. When the alphavirus plasmid enters a cell, the vector RNA will be transcribed in the nucleus by the host cell RNA polymerase II and then transported to the cytoplasm, where the replicase is produced by translation of the first part of the RNA. This replicase subsequently initiates the amplification of the vector RNA, and the synthesis of the subgenomic RNA from which the foreign gene will be expressed (Fig. 5). One self-amplifying alphavirus DNA vector was constructed by placing the SIN replicon sequence downstream of the Rous sarcoma virus (RSV) LTR promoter (Herweijer *et al.*, 1995). When expression of a recombinant antigen such as luciferase was measured in rat myotubes transfected *in vitro*

Figure 5. Alphavirus DNA/RNA layered vectors. Alphavirus DNA vectors contain the alphavirus vector sequence downstream of an eukaryotic promoter such as the CMV promoter (CMV Pr). When transfected into cells these plasmids can be transported into the nucleus where the cellular RNA polymerase II (RNA pol II) will transcribe the vector RNA from the CMV Pr. This RNA will be transported to the cytoplasm where the replicase (REP) will be made. REP will amplify the vector RNA and synthesize a subgenomic RNA from which expression of the foreign gene (FG) will take place at high levels. T/T, transcription termination signals; nsp, nonstructural proteins; An, polyA.

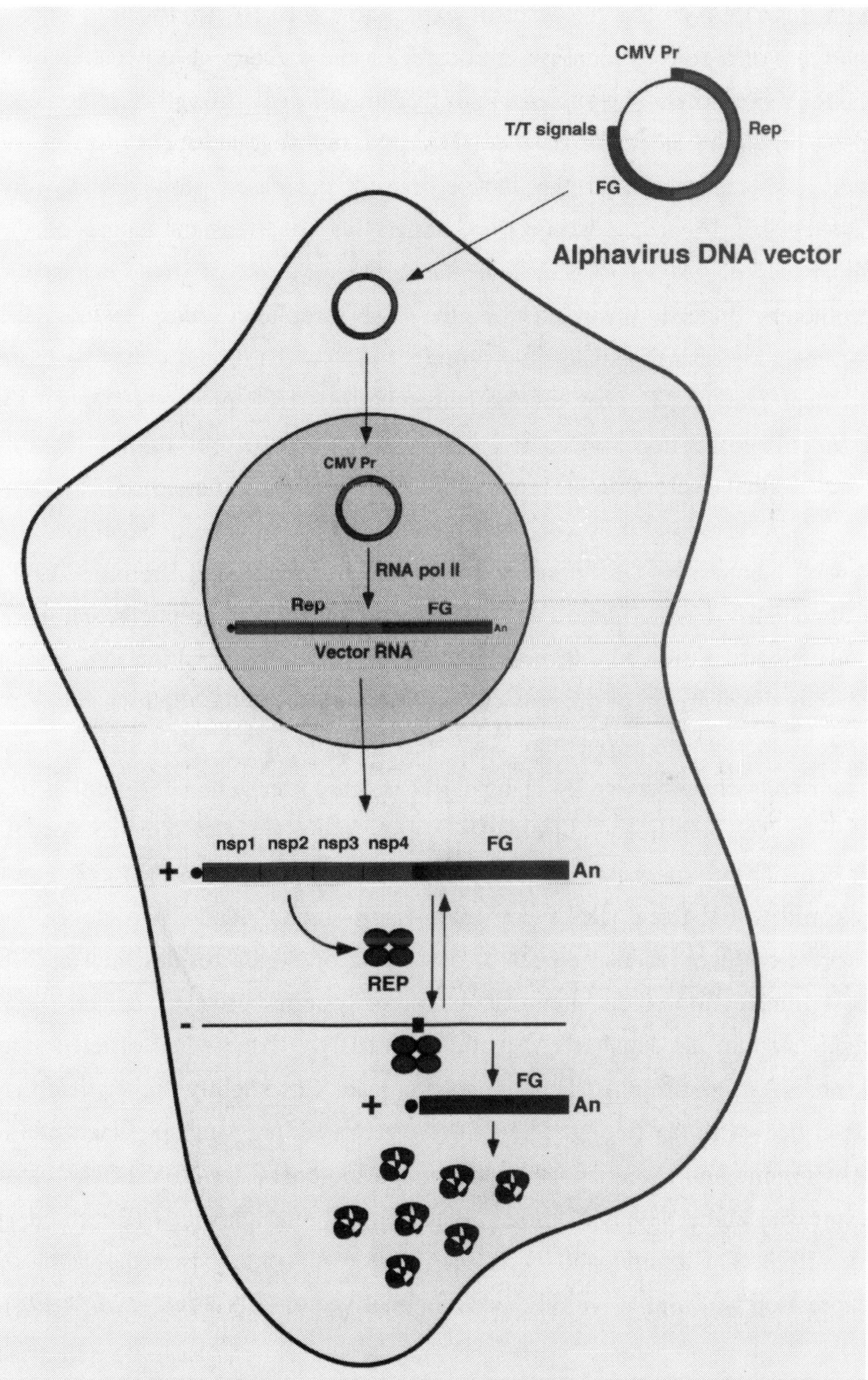
CMV Pr
T/T signals
Rep
FG
Alphavirus DNA vector
CMV Pr
RNA pol II
Rep
FG
An
Vector RNA
nsp1 nsp2 nsp3 nsp4
FG
+
An
REP
-
FG
+
An

with this DNA vector, the level of expression was 10- to 240-fold higher than that of a conventional plasmid carrying the same gene downstream of the RSV promoter. The expression of luciferase *in vivo* from this vector was also shown by injection of the DNA into mouse quadriceps. Levels of luciferase expression in the mouse muscle decreased slowly and were undetectable 16 days postinjection, underlying the transient nature of the alphavirus DNA vectors. When the efficiency of different eukaryotic promoters placed upstream of the SIN replicon was tested, the cytomegalovirus immediate early promoter (CMV IE) turned out to be more efficient than the Moloney murine leukemia virus LTR promoter (Dubensky *et al.*, 1996). The expression levels were 2 to 10-fold higher than in conventional expression plasmids, depending on the promoter used.

DNA expression vectors based on SFV have been developed by several laboratories (Berglund *et al.*, 1998; DiCiommo and Bremner, 1998; Kohno *et al.*, 1998). These DNA vectors contain the SFV replicon sequence downstream of the CMV IE promoter, and an SV40 transcription termination cassette downstream of the poly (dA-dT) at the end of the replicon. Foreign protein expression from the SFV recombinant DNA vector has been extensively characterized by using *LacZ* as a reporter gene (Berglund *et al.*, 1998). No significant differences were found in the kinetics of β-galactosidase expression in transfected cells between the SFV-LacZ DNA plasmid, SFV-LacZ RNA or SFV-LacZ recombinant particles. The expression level reached a peak of 50- 60 pg of β-galactosidase/cell at 24h post transfection, indicating that all three systems used to deliver SFV replicons into the cells (RNA, particles and DNA) have equal efficiency to express foreign proteins. This expression level was slightly higher than that described with a similar SFV DNA vector expressing β-galactosidase (DiCiommo and Bremner, 1998). Although in general the SFV DNA vector expressed higher levels of protein than conventional plasmids (Berglund *et al.*, 1998; DiCiommo and Bremner, 1998), in some cases the level of expression was higher with the conventional vector (Berglund *et al.*, 1998).

The inclusion of the SFV translation enhancer sequences (first 144 nucleotides of the capsid gene) upstream and in frame with the reporter gene provided a considerably higher expression level (Berglund *et al.*, 1998) as it had been observed in the RNA vector. In this way, an expression peak of 355 pg of β-galactosidase/cell was obtained 24 hours after transfection, which is almost 10-fold higher than the expression level observed without enhancer. The transient nature of the SFV-derived plasmids was indicated by a strong cytopathic effect in the transfected cells 72 h after transfection (Berglund *et al.*, 1998; Kohno *et al.*, 1998). This cytopathic effect was very similar in cells transfected with SFV RNA or infected with particles, but did not occur when a conventional CMV-LacZ plasmid was used.

9. SUICIDAL DNA VACCINES BASED ON ALPHAVIRUSES

Vaccination with naked DNA has been shown to be an efficient method for inducing long-lasting immune responses and protective immunity against a variety of pathogens in several animal models. A draw-back that has been proposed for DNA vaccines is the risk of chromosomal integration, which could potentially give rise to tumor development. However, evidence of DNA integration into the host chromosome could not be found when PCR-based studies were done in mice immunized with high doses of plasmid DNA (Nichols *et al.*, 1995). This event would have a theoretical probability at least 3-fold lower than the spontaneous mutation rate in the host chromosome, suggesting that the risk of integration does not constitute an important problem *in vivo*. Another problem that has been associated with DNA vaccines is the induction of tolerance due to long-term expression of antigen at low levels in the organism (Mor *et al.*, 1996; Wolff *et al.*, 1992). The use of alphavirus DNA-based vectors for vaccination could theoretically avoid both of these problems since: 1) high level of foreign protein can be expressed from these vectors, and 2) cell death will be induced in transfected cells after a few days.

A SIN plasmid expressing hepatitis B virus *c* or *e* genes was able to induce specific antibodies against the foreign antigens in mice inoculated intramuscularly (Dubensky *et al.*, 1996). In other studies, the induction of immune responses in mice with alphavirus DNA vectors has been extensively characterized. These include the use of SIN derived DNA vectors expressing the glycoprotein B (gB) of herpes simplex virus (HSV-1) (Hariharan *et al.*, 1998), as well as SFV-derived plasmids expressing β-galactosidase or the influenza virus antigens NP or HA (Berglund *et al.*, 1998). The transcription of SIN and SFV recombinant replicons in these plasmids was driven by the CMV IE promoter. In both cases, the immune responses induced with these vectors were compared to those generated with conventional DNA vaccines. Although the level of recombinant expression *in vitro* was similar between conventional plasmids and the alphavirus-derived vectors, the latter always showed to be superior in inducing both humoral and cellular responses. In order to induce the same level of responsiveness in the inoculated animals, up to 1000-fold more of the conventional DNA was needed. The inclusion of the capsid translation enhancer in the SFV DNA vector expressing β-galactosidase increased the level of the immune response when doses lower than 1 μg of vector were used, suggesting that an increase in the level of antigen expression has an additive effect in the immune response in the context of the SFV vector. An additional advantage of the alphavirus based DNA vectors was that they were able to induce immune responses in a larger number of animals as compared to the conventional vectors. The alphavirus DNA vectors are not expected to transfect cells more efficiently than conventional DNA vectors (probably even less efficiently due to their larger sizes) and therefore, the larger number of responders could only be explained if the alphavirus vectors have a higher probability to establish transfection due to their self-amplifiable nature.

The induction of immune responses with the SFV-LacZ DNA vector was also compared to those generated with SFV-LacZ RNA and recombinant SFV-LacZ particles. Although no major differences were observed between naked DNA and RNA SFV-LacZ vectors, SFV-LacZ recombinant particles were able to induce a higher antibody response in mice (Berglund *et al.*, 1998). CTL responses induced with recombinant SFV particles carrying the influenza virus NP gene and SFV-DNA encoding the same antigen were, however, very similar. The higher humoral immune response observed with particles could be associated with a higher efficiency of infection as compared to transfection. A total higher amount of antigen would thus be produced with the particles, elevating the antibody immune response. Induction of cellular immunity is, on the other hand, dependent on antigen presentation by a small number of APCs and small amounts of naked alphavirus DNA could be enough to induce a strong T-cell response. This hypothesis is supported by the fact that only 100 infectious units of recombinant SFV particles are sufficient to induce strong CTL responses against a heterologous antigen (Zhou *et al.*, 1995).

Protection against viral challenge was also induced in both systems. Mice immunized with 30 μg of the SIN-HSV-gB plasmid were completely protected when challenged with HSV-1 (Hariharan *et al.*, 1998). A level of 80% protection was maintained when the animals were vaccinated with only 0.1 μg of SIN DNA, while 3 to 30 μg DNA of the conventional vector was needed to achieve the same level of protection. The immunization of mice with a mixture of SFV plasmids encoding the NP and HA influenza virus antigens was also able to provide partial protection against influenza challenge (Berglund *et al.*, 1998), although the level of protection was higher when using recombinant particles.

In spite of one single report in which an alphavirus DNA vector based on SIN expressing encephalomyocarditis virus *VP1* gene was unable to induce a measurable immune response in mice (Lee *et al.*, 1997), it can be concluded that alphavirus DNA-based vectors induce stronger humoral and cellular immune responses than conventional DNA vaccines. One question that should be addressed is why alphavirus DNA vectors are more efficient than conventional plasmids in inducing immune responses when they do not differ significantly in the level of heterologous protein expression. One of the reasons could be the induction of cytokines by the self-replicating vectors. Both α and β interferons are induced following infection with RNA viruses, including alphaviruses, (Brandt *et al.*, 1994; Vilcek and Sen, 1996) and this could also be the case in cells transfected with the alphavirus-based plasmids. The self-amplifying process of the replicon transcribed in the cell mimics that of the wild-type virus and this would result in the induction of interferon (IFN). This molecule can activate natural killer cells and plasma IFN would be induced (Gates *et al.*, 1984; Kaluza *et al.*, 1987), including IFN-γ (Donnelly *et al.*, 1997), which can activate macrophages and stimulate differentiation of Th0 cells to become inflammatory CD4 T-cells (Th1). As it was shown in mice immunized with SFV alphavirus recombinant particles, the antibodies induced by SIN and SFV DNA vectors were found to be mainly of the IgG2a isotype, indicating that a CD4+ Th1 type cell-mediated immune response had taken place. Additionally, it may be that the induction of cell death by apoptosis could enhance antigenic crosspriming.

10. THE USE OF ALPHAVIRUS VECTORS IN GENE THERAPY

Several alphavirus recombinant vectors with potential antiviral and antitumoral applications have been described. Antiviral approaches have included the use of a SIN vector expressing a single chain antibody directed against a neutralizing epitope in the envelope protein of louping ill virus (LIV) (Jiang *et al.*, 1995). This vector was able to reduce significantly the infectivity of LIV in cells expressing the antibody. In another study, a functional hammerhead ribozyme specific for the U5 region of the HIV-1 long terminal repeat was expressed from a SFV vector (Smith *et al.*, 1997). Anti-tumoral strategies have been based on the expression of tumor antigens or cytokines from alphavirus vectors. Colmenero and coworkers showed that mice inoculated with SFV suicide particles expressing p815, a weak tumor rejection antigen from mouse mastocytoma, were protected from tumor challenge (Colmenero *et al.*, 1999). The intratumoral injection of B16 murine melanomas with SFV expressing mouse IL-12 genes was also able to inhibit tumor growth probably through inhibition of neovascularization within the tumor (Asselin-Paturel *et al.*, 1998; Zhang *et al.*, 1997). However, the use of alphavirus vectors in certain gene therapy applications requires some improvements in the present technology. One is the need to target the vectors to specific cells or tissues, and the other is the development of vectors able to maintain a more prolonged expression of the therapeutic product in the transfected/infected cells. Both of these problems have been recently addressed and will be discussed below.

a) Targeting alphavirus vectors

Alphavirus vectors lack target-cell specificity. The construction of targeted vectors requires both the elimination of the endogenous high-range tropism as well as the introduction of a novel tropism. One possible strategy to redirect the alphavirus tropism is the generation of "minimal" viruses by encapsidating the recombinant alphavirus genome inside an envelope containing a nonrelated viral protein involved in receptor binding and entry. This type of minimal viruses has been generated by transfecting cells with a SFV recombinant RNA vector expressing the vesicular stomatitis virus glycoprotein (VSV-G) (Rolls *et al.*, 1994). These minimal viruses were able to propagate and could be neutralized with specific anti-VSV serum, indicating that the infection process was probably mediated by the interaction with the specific VSV-G receptor. By including a second subgenomic promoter followed by a heterologous gene in the SFV-VSV-G vector, these minimal viruses were able to express foreign proteins at high levels (Rolls *et al.*, 1996). A similar result has been obtained by expressing the murine leukemia virus (MLV) envelope protein with a truncated cytoplasmic tail in an SFV vector. Minimal viruses were produced in cells transfected with this vector that contained the recombinant SFV genome and were able to infect specifically cells bearing the ecotropic MLV receptor (Lebedeva *et al.*, 1997).

A different group of studies has focused on the modification of the alphavirus spike proteins, to target the viral particles to specific cells. By random insertion mutation in the p62 gene, several SIN mutants with impaired binding to vertebrate cells were obtained (Dubuisson and Rice, 1993), some of which exhibited normal particle assembly, release and penetration. The insertion of foreign sequences in permissive sites in the p62 glycoprotein has recently led to the development of SIN virus vectors that can target specific cells. Alphavirus vectors that were able to target human choriocarcinoma cells were constructed by insertion of the alpha- and beta-hCG genes in the SIN envelope sequence (Sawai and Meruelo, 1998). These vectors had minimal infectivity in cells not containing the LH/CG receptors on their surface. A more general approach used by Ohno and coworkers was based on the insertion of two IgG binding domains of protein A from *Staphylococcus aureus* in the E2 region of SIN (Ohno *et al.*, 1997). The chimeric particles obtained in this way had minimal infectivities against BHK and several human cell lines, but when used in conjunction with monoclonal antibodies specific for cell surface antigens, they were able to infect specific cells lines. These chimeric particles could be targeted virtually to any cell type, provided that a specific antibody is available. However, the infectivity mediated by the antibody was not as efficient as that mediated by normal spike-receptor interaction, which could limit the possible applications of these vectors. Using these chimeric particles, a SIN recombinant RNA expressing herpes simplex virus thymidine kinase was specifically introduced in several human tumor cell lines, which after infection showed an increased sensitivity to the prodrug ganciclovir (Ijima *et al.*, 1999).

The targeting of the alphavirus nucleic acid vectors offers additional difficulties. Several strategies could be approached to deliver specifically the alphaviral RNA/DNA to the target cells including complexing the alphavirus nucleic acid with specific monoclonal antibodies (mAbs) or using virosomes containing specific proteins in the surface. The first approach has been used to deliver a SIN recombinant RNA to specific cancer cells *in vitro* in a mAb dose-dependent manner (Sawai *et al.*, 1998). In this case, the SIN RNA was biotinylated and coupled to the specific antibody through a streptavidin-protein A fusion protein. Another possibility would be the use of specific promoters in such a way that the amplification process would take place only in certain cells or at a certain time. This could also include the use of externally inducible promoters.

b) Non-cytopathic alphavirus vectors

Alphavirus vectors induce cell death in most vertebrate cells within 2-3 days after infection/transfection. The cytopathic effect has been shown to be mediated by apoptosis (Berglund *et al.*, 1998; Frolov and Schlesinger, 1994b; Glasgow *et al.*, 1997 and 1998; Sammin *et al.*, 1999). Although this transient nature of the vectors does not represent a problem in applications such as vaccination, for gene therapy purposes a more prolonged expression of the therapeutic product is usually desired. Basically, two different strategies have been used to eliminate the cytopathic effects induced by the alphavirus replication and render the vectors persistent in the infected cells. The first one is based on the expression of antiapoptotic genes, such as the proto-oncogen *bcl-2* (for a review see Griffin and Hardwick, 1997). This gene encodes an inner 26 kD mitochodrial membrane protein that blocks programmed cell death in neurons. A long term persistent infection could be established with wild type SIN or SFV in rat prostatic adenocarcinoma (AT-3) or mice neuroblastoma N18 cells transfected with *bcl-2* (Levine *et al.*, 1993; Liao *et al.*, 1997; Scallan *et al.*, 1997). In addition, the expression of bcl-2 from a replication-packaging proficient SIN vector significantly reduced the mortality rate in neonatal mice inoculated with this recombinant virus, as compared with control SIN vectors expressing non relevant antigens (Levine *et al.*, 1996). A similar result was obtained in animals inoculated with a SIN recombinant virus expressing beclin, a protein that interacts with bcl-2 in mammalian cells (Liang *et al.*, 1998). The cleavage of bcl-2 by caspases has been shown to inhibit the anti-apoptotic activity of this protein (Cheng *et al.*, 1997). In certain cases, alphaviruses could trigger a bcl-2 caspase mediated inactivation in order to evade the death protection provided by this protein (Grandgirard *et al.*, 1998). In this situation, the expression of bcl-2 cleavage resistant mutants could provide increase protection to alphavirus induced apoptosis (Cheng *et al.*, 1997).

The second strategy to render alphavirus vectors noncytopathic has been the isolation and cloning of mutant viruses able to establish a persistent infection in mammalian cells. In preliminary studies, one of these mutants (Sin-1) was isolated from cells infected with SIN heavily enriched with defective-interfering (DI) particles (Weiss *et al.*, 1980). Sin-1 contained a mutation in the nsp2 gene (Pro726 → Ser) that was responsible for the capacity to establish persistent infection (Dryga *et al.*, 1997). In more recent studies, SIN mutant replicons able to establish persistent replication have been isolated by expressing the puromycin N-acetyltranferase gene (*pac*) from the subgenomic promoter in the SIN expression vector. Cells transfected with recombinant SIN-*pac* were selected in the presence of puromycin and mutant replicons rescued from the surviving cells (Agapov *et al.*, 1998; Frolov *et al.*, 1999). These mutants contained single amino acid changes in nsp2, either in the same position as in Sin-1 (Pro726 → Leu) or in a different position (Asn779 → Lys). Additional SIN replicons engineered with different amino acids at position 726 in nsp2 allowed to establish a direct correlation between the level of viral RNA replication and cytopathogenicity (Frolov *et al.*, 1999). These noncytopathic vectors have been used to express different proteins *in vitro*, such as β-galactosidase, green fluorescent protein, T7 RNA polymerase, CD46 (measles virus receptor), yellow fever virus prM, E, and NS1 proteins etc (Agapov *et al.*, 1998; Lindenbach and Rice, 1997; Olivo *et al.*, 1998). The expression level reached with these vectors was only 4% of that obtained with their cytopathic counterparts, although it could be increased by expressing the foreign gene from a DI RNA cotransfected into the cells. An additional problem with these noncytopathic vectors is that they show certain instability, leading to a progressive loss of foreign gene expression after several passages, which could limit the duration of expression *in vivo*. A DNA version of a SIN attenuated vector has also been constructed by cloning the replicon sequence downstream of the RSV LTR promoter (Agapov *et al.*, 1998).

11. OTHER APPLICATIONS OF THE ALPHAVIRUS EXPRESSION SYSTEM

a) Packaging retroviral particles with SFV

Retrovirus particles are usually produced by transfecting a retrovirus recombinant proviral DNA into a packaging cell line able to provide the retroviral structural proteins in *trans*. Li and Garoff have developed a new strategy to package retroviral particles based on the SFV expression vector (Li and Garoff, 1996). In this system, cells are cotransfected with three different recombinant SFV RNAs which carry the *gag-pol*, *env* and a recombinant retrovirus genome from Moloney leukemia virus (MLV), respectively. Both the MLV genome and structural proteins are produced in the cytoplasm of the transfected cells from the different SFV vectors. Retroviral particles containing the MLV genome are produced in these cells with titers of 10^7 particles/10^6 transfected cells and with no detection of RPVs. Both particles carrying the ecotropic or the amphotropic envelope could be generated by using different SFV-*env* vectors, which makes the system specially attractive for the purpose of human gene therapy. This system has also been used to package retroviral genomes carrying heterologous genes containing introns, which can increase the expression levels of the foreign gene (Li and Garoff, 1998). This is not possible by using conventional retrovirus packaging cell lines, since in these cells retrovirus genomes are generated in the nucleus and are subjected to splicing. An alternative way to use the SFV vector to produce retroviral particles has been to transfect retroviral packaging cell lines with only one SFV vector carrying a retroviral recombinant genome (Wahlfors and Morgan, 1999; Wahlfors *et al.*, 1997). In this case, the retroviral structural proteins were provided in *trans* by the host cell.

b) Alphavirus stable cell lines for virus detection

The replication strategy of alphaviruses has been used to develop several cell lines that can be used for detection of homologous or heterologous viruses. A cell line that can be used for titration of SIN was developed by Olivo and coworkers (Olivo *et al.*, 1994). In this cell line, the RSV-LTR promoter controls the transcription of a SIN defective RNA carrying the luciferase gene downstream of the subgenomic promoter (SIN-*luc*). Although the SIN-*luc* RNA is produced constitutively in these cells, expression of luciferase only takes place when the cells are infected with SIN particles carrying a functional replicase. Cell lines for detection of herpesviruses have been developed more recently (Ivanova *et al.*, 1999). In this case, the cells were stably transformed with a SIN-LacZ recombinant replicon cDNA placed downstream of a herpes simplex virus (HSV) early promoter. Since transcription from this promoter requires the expression of HSV immediate early genes, the SIN-LacZ RNA will not be normally produced. However, upon infection of these cells with HSV, SIN-LacZ RNA will be transcribed, producing the alphavirus replicase that will amplify the replicon and synthesize a subgenomic RNA, allowing the synthesis of β-galactosidase. These cell lines were improved by transformation with a SIN helper cDNA that provides the SIN structural proteins upon HSV infection. In this way, the SIN-LacZ RNA can be packaged into particles and the signal amplified. A similar cell line that allowed detection of human cytomegalovirus (HCMV) was also developed (Ivanova *et al.*, 1999).

Acknowledgements

We thank the Swedish Medical Research Council, the Swedish Council for Engineering Sciences, the Swedish Cancer Research Foundation and the European Union Biotechnology Program for support.

12. REFERENCES

Agapov, E.V., Frolov, I., Lindebach, B.D., Prágal, B.M., Schlesinger, S. and Rice, C.M. (1998). Noncytopathic Sindbis virus RNA vectors for heterologous gene expression. *Proc. Natl. Acad. Sci. USA*, 95: 12989-12994.

Asselin-Paturel, C., Lassau, N., Guinebretiere, J.M., Zhang, J., Gay, F., Bex, F., Hallez, S., Leclere, J., Peronneau, P., Mami-Chouaib, F. and Chouaib, S. (1998). Transfer of the murine interleukin-12 gene in vivo by a Semliki Forest virus vector induces B16 tumor regression through inhibition of tumor blood vessel formation monitored by Doppler ultrasonography. *Gene Ther.*, 5: 606-615.

Berglund, P., Sjöberg, M., Garoff, H., Atkins, G. J., Sheahan, B. J. and Liljeström, P. (1993). Semliki Forest virus expression system: production of conditionally infectious recombinant particles. *Bio/Technology*, 11: 916-920.

Berglund, P., Quesada-Rholander, M., Putkonen, P., Biberfeld, G., Thorstensson, R. and Liljeström, P. (1997). Outcome of immunization of cynomolgus monkeys with recombinant Semliki Forest virus encoding human immunodeficiency virus type 1 envelope protein and challenge with a high dose of SHIV-4 virus. *AIDS Res. Hum. Retrovir.*, 13: 1487-1495.

Berglund, P., Smerdou, C., Fleeton, M.N., Tubulekas, I. and Liljeström, P. (1998). Enhancing immune responses using suicidal DNA vaccines. *Nat. Biotech.*, 16: 562-565.

Berglund, P., Fleeton, M.N., Smerdou, C. and Liljeström, P. (1999). Immunization with recombinant Semliki Forest virus induces protection against influenza challenge in mice. *Vaccine*, 17: 497-507.

Brand, D., Lemiale, F., Turbica, I., Buzelay, L., Brunet, S. and Barin, F. (1998). Comparative analysis of humoral immune responses to HIV type 1 envelope glycoproteins in mice immunized with a DNA vaccine, recombinant Semliki Forest virus RNA, or recombinant Semliki Forest virus particles. *AIDS Res. Hum. Retrovir.*, 14: 1369-1377.

Brandt, E.R., Linnane, A.W. and Devenish, R.J. (1994). Expression of IFN A genes in subpopulations of peripheral blood cells. *Brit. J. Haematol.*, 86: 717-725.

Bredenbeek, P.J., Frolov, I., Rice, C.M. and Schlesinger, S. (1993). Sindbis virus expression vectors: packaging of RNA replicons by using defective helper RNAs. *J. Virol.*, 67: 6439-6446.

Bron, R., Wahlberg, J.M., Garoff, H. and Wilschut, J. (1993). Membrane fusion of Semliki Forest virus in a model system: correlation between fusion kinetics and structural changes in the envelope glycoprotein. *EMBO J.*, 12: 693-701.

Caley, I.J., Betts, M.R., Irlbeck, D.M., Davis, N.L., Swanstrom, R., Frelinger, J.A. and Johnston, R.E. (1997). Humoral, mucosal, and cellular immunity in response to a human immunodeficiency virus type 1 immunogen expressed by a Venezuelan equine encephalitis virus vaccine vector. *J. Virol.*, 71: 3031-3038.

Chen, J.P., Miller, D., Katow, S. and Frey, T.K. (1995). Expression of the rubella virus structural proteins by an infectious Sindbis virus vector. *Archiv. Virol.*, 140: 2075-2084.

Cheng, E.H., Kirsch, D.G., Clem, R.J., Ravi, R., Kastan, M.B., Bedi, A., Ueno, K. and Hardwick, J.M. (1997). Conversion of Bcl-2 to a Bax-like death effector by caspases. *Science*, 278: 1966-1968.

Colmenero, P., Liljeström, P. and Jondal, M. (1999). Induction of P815 tumor immunity by recombinant Semliki Forest virus expressing the P1A gene. *Gene Ther.*, 6: 1728-1733.

Colombage, G., Hall, R., Pavy, M. and Lobigs, M. (1998). DNA-based and alphavirus-vectored immunisation with prM and E proteins elicits long-lived and protective immunity against the flavivirus Murray Valley encephalitis virus. *Virology*, 250: 151-163.

Dalemans, W., Delers, A., Delmelle, C., Denamur, F., Meykens, R., Thiriart, C., Veenstra, S., Francotte, M., Bruck, C. and Cohen, J. (1995). Protection against homologous influenza challenge by genetic immunization with SFV-RNA encoding Flu-HA. *Ann. N. Y. Acad. Sci.*, 772: 255-256.

Davis, N.L., Willis, L.V., Smith, J.F. and Johnston, R.E. (1989). In vitro synthesis of infectious Venezuelan equine encephalitis virus RNA from a cDNA clone: Analysis of a viable deletion mutant. *Virology*, 171: 189-204.

Davis, N.L., Powell, N., Greenwald, G.F., Willis, L.V., Johnson, B.J.B., Smith, J.F. and Jonhston, R.E. (1991). Attenuating mutations in the E2 glycoprotein gene of Venezuelan equine encephalitis virus: Construction of single and multiple mutants in a full-length cDNA clone. *Virology*, 183: 20-31.

Davis, N.L., Brown, K.W. and Johnston, R.E. (1996). A viral vaccine vector that expresses foreign genes in lymph nodes and protects against mucosal challenge. *J. Virol.*, 70: 3781-3787.

de Hoop, M.J., Olkkonen, V.M., Ikonen, E., Williamson, E., von Poser, C., Meyn, L. and Dotti, C.G. (1994). Semliki Forest virus as a tool for protein expression in cultured rat hippocampal neurons. *Gene Ther.*, 1: S28-31.

DiCiommo, D.P. and Bremner, R. (1998). Rapid, high level protein production using DNA-based Semliki Forest virus vectors. *J. Biol. Chem.*, 273: 18060-18066.

Donnelly, S.M., Sheahan, B.J. and Atkins, G. J. (1997). Long-term effects of Semliki Forest virus infection in the mouse central nervous system. *Neuropathol. Appl. Neurobiol.*, 23: 235-241.

Dryga, S.A., Dryga, O.A. and Schlesinger, S. (1997). Identification of mutations in a Sindbis virus variant able to establish persistent infection in BHK cells: the importance of a mutation in the nsP2 gene. *Virology*, 228: 74-83.

Dubensky, T.W.Jr., Driver, D.A., Polo, J.M., Belli, B.A., Latham, E.M., Ibanez, C.E., Chada, S., Brumm, D., Banks, T.A., Mento, S.J., Jolly, D.J. and Chang, S.M.

(1996). Sindbis virus DNA-based expression vectors: utility for in vitro and in vivo gene transfer. *J. Virol.*, 70: 508-519.

Dubuisson, J. and Rice, C.M. (1993). Sindbis virus attachment: isolation and characterization of mutants with impaired binding to vertebrate cells. *J. Virol.*, 67: 3363-3374.

Dubuisson, J., Hsu, H.H., Cheung, R.C., Greenberg, H.B., Russell, D.G. and Rice, C.M. (1994). Formation and intracellular localization of hepatitis C virus envelope glycoprotein complexes expressed by recombinant vaccinia and Sindbis viruses. *J. Virol.*, 68: 6147-6160.

Filocamo, G., Pacini, L., Nardi, C., Bartholomew, L., Scaturro, M., Delmastro, P., Tramontano, A., De Francesco, R. and Migliaccio, G. (1999). Selection of functional variants of the NS3-NS4A protease of hepatitis C virus by using chimeric Sindbis viruses. *J. Virol.*, 73: 561-575.

Fleeton, M.N., Liljeström, P., Sheahan, B.J. and Atkins, G.J. (1999a). Protective responses induced in mice by louping ill virus prME and NS1 antigens expressed from Semliki Forest virus vectors. *J. Gen. Virol.*, In press.

Fleeton, M.N., Sheahan, B.J., Gould, E.A., Atkins, G.J. and Liljeström, P. (1999b). Recombinant Semliki Forest virus particles encoding the prME or NS1 proteins of louping ill virus protect mice from lethal challenge. *J. Gen. Virol.*, 80: 1189-1198.

Frolov, I. and Schlesinger, S. (1994a). Translation of Sindbis virus mRNA: effects of sequences downstream of the initiating codon. *J. Virol.*, 68: 8111-8117.

Frolov, I. and Schlesinger, S. (1994b). Comparison of the effects of Sindbis virus and Sindbis virus replicons on host cell protein synthesis and cytopathogenicity in BHK cells. *J. Virol.*, 68: 1721-1727.

Frolov, I. and Schlesinger, S. (1996). Translation of Sindbis virus mRNA: analysis of sequences downstream of the initiating AUG codon that enhance translation. *J. Virol.*, 70: 1182-1190.

Frolov, I., Frolova, E. and Schlesinger, S. (1997). Sindbis virus replicons and Sindbis virus: assembly of chimeras and of particles deficient in virus RNA. *J. Virol.*, 71: 2819-2829.

Frolov, I., Agapov, E., Hoffman, T.A.Jr., Prágal, B.M., Lippa, M., Schlesinger, S. and Rice, C.M. (1999). Selection of RNA replicons capable of persistent noncytopathic replication in mammalian cells. *J. Virol.*, 73: 3854-3865.

Frolova, E., Frolov, I. and Schlesinger, S. (1997). Packaging signals in alphaviruses. *J. Virol.*, 71: 248-258.

Gates, M.C., Sheahan, B.J. and Atkins, G.J. (1984). The pathogenicity of the M9 mutant of Semliki Forest virus in immune-compromised mice. *J. Gen. Virol.*, 65: 73-80.

Giraud, A., Ataman-Onal, Y., Battail, N., Piga, N., Brand, D., Mandrand, B. and Verrier, B. (1999). Generation of monoclonal antibodies to native human immunodeficiency virus type 1 envelope glycoprotein by immunization of mice with naked RNA. *J. Virol. Methods*, 79: 75-84.

Glasgow, G.M., McGee, M.M., Sheahan, B.J. and Atkins, G.J. (1997). Death mechanisms in cultured cells infected by Semliki Forest virus. *J. Gen. Virol.*, 78: 1559-1563.

Glasgow, G.M., McGee, M.M., Tarbatt, C.J., Mooney, D.A., Sheahan, B.J. and Atkins, G.J. (1998). The Semliki Forest virus vector induces p53-independent apoptosis. *J. Gen. Virol.*, 79: 2405-2410.

Gleiser, C.A., Gochenour, W.S.Jr., Berge, T.O. and Tigert, W.D. (1962). The comparative pathology of experimental Venezuelan equine encephalitis infection in different animal hosts. *J. Infect. Dis.*, 110: 80-97.

Grandgirard, D., Studer, E., Monney, L., Belser, T., Fellay, I., Borner, C., and Michel, M.R. (1998). Alphaviruses induce apoptosis in Bcl-2-overexpressing Cells: evidence for a caspase-mediated, proteolytic inactivation of Bcl-2. *EMBO J.*, 17: 1268-1278.

Grieder, F.B., Davis, N.L., Aronson, J.F., Charles, P.C., Sellon, D.C., Suzuki, K. and Johnston, R.E. (1995). Specific restrictions in the progression of Venezuelan equine encephalitis virus-induced disease resulting from single amino acid changes in the glycoproteins. *Virology*, 206: 994-1006.

Griffin, D.E. and Hardwick, J.M. (1997). Regulators of apoptosis on the road to persistent alphavirus infection. *Ann. Review Microbiol.*, 51: 565-592.

Hahn, C.S., Hahn, Y.S., Braciale, T.J. and Rice, C.M. (1992). Infectious Sindbis virus transient expression vectors for studying antigen processing and presentation. *Proc. Natl. Acad. Sci. USA*, 89: 2679-83.

Haijou, M., Hill, K.R., Subramaniam, S.V., Hu, J.Y. and Raju, R. (1996). Nonhomologous RNA-RNA recombination events at the 3' nontranslated region of the Sindbis virus genome: hot spots and utilization of nonviral sequences. *J. Virol.*, 70: 5153-5164.

Hariharan, M.J., Driver, D.A., Townsend, K., Brumm, D., Polo, J.M., Belli, B.A., Catton, D.J., Hsu, D., Mittelstaedt, D., McCormack, J.E., Karavodin, L., Dubensky, T.W.Jr., Chang, S.M. and Banks, T.A. (1998). DNA immunization against herpes simplex virus: enhanced efficacy using a Sindbis virus-based vector. *J. Virol.*, 72: 950-958.

Heino, P., Dillner, J. and Schwartz, S. (1995). Human papillomavirus type 16 capsid proteins produced from recombinant Semliki Forest virus assemble into virus-like particles. *Virology*, 214: 349-359.

Herweijer, H., Latendresse, J.S., Williams, P., Zhang, G.F., Danko, I., Schlesinger, S. and Wolff, J.A. (1995). A plasmid-based self-amplifying Sindbis virus vector. *Hum. Gen. Ther.*, 6: 1161-1167.

Hevey, M., Negley, D., Pushko, P., Smith, J. and Schmaljohn, A. (1998). Marburg virus vaccines based upon alphavirus replicons protect guinea pigs and nonhuman primates. *Virology*, 251: 28-37.

Higgs, S., Olson, K.E., Klimowski, L., Powers, A.M., Carlson, J.O., Possee, R.D. and Beaty, B.J. (1995). Mosquito sensitivity to a scorpion neurotoxin expressed using an infectious Sindbis virus vector. *Insect Mol. Biol.*, 4: 97-103.

Higgs, S., Rayner, J.O., Olson, K.E., Davis, B.S., Beaty, B.J. and Blair, C.D. (1998). Engineered resistance in Aedes aegypti to a West African and a South American strain of yellow fever virus. *Am. J. Trop. Med. Hyg.*, 58: 663-670.

Hill, K.R., Hajjou, M., Hu, J.Y. and Raju, R. (1997). RNA-RNA recombination in Sindbis virus: roles of the 3' conserved motif, poly(A) tail, and nonviral sequences of template RNAs in polymerase recognition and template switching. *J. Virol.*, 71: 2693-2704.

Ijima, Y., Ohno, K., Ikeda, H., Sawai, K., Levin, B. and Meruelo, D. (1999). Cell-specific targeting of a thymidine kinase/ganciclovir gene therapy system using a recombinant Sindbis virus vector. *Int. J. Cancer,* 80: 110-118.

Ivanova, L., Schlesinger, S. and Olivo, P.D. (1999). Regulated expression of a Sindbis virus replicon by herpesvirus promoters. *J. Virol.*, 73: 1998-2005.

Jackson, A.C., SenGupta, S.K. and Smith, J.F. (1991). Pathogenesis of Venezuelan equine encephalitis virus infection in mice and hamsters. *Vet. Pathol.*, 28: 410-418.

Jain, S.K., DeCandido, S. and Kielian, M. (1991). Processing of the p62 envelope precursor protein of Semliki Forest virus. *J. Biol. Chem.*, 266: 5756-5761.

Jiang, W., Venugopal, K., and Gould, E. A. (1995). Intracellular interference of tick-borne flavivirus infection by using a single-chain antibody fragment delivered by recombinant Sindbis virus. *J. Virol.*, 69: 1044-1049.

Johanning, F.W., Conry, R.M., Lo Buglio, A.F., Wright, M., Sumerel, L.A., Pike, M.J. and Curiel, D.T. (1995). A Sindbis virus mRNA polynucleotide vector achieves prolonged and high level heterologous gene expression in vivo. *Nucl. Acids Res.*, 23: 1495-1501.

Kaluza, G., Lell, G., Reinacher, M., Stitz, L. and Willems, W.R. (1987). Neurogenic spread of Semliki Forest virus in mice. *Archiv. Virol.*, 93: 97-110.

Kamrud, K.I., Powers, A.M., Higgs, S., Olson, K.E., Blair, C.D., Carlson, J.O. and Beaty, B.J. (1995). The expression of chloramphenicol acetyltransferase in mosquitoes and mosquito cells using a packaged Sindbis replicon system. *Experimental Parasitology*, 81: 394-403.

Kamrud, K.I., Olson, K.E., Higgs, S., Carlson, J.O. and Beaty, B.J. (1998). Use of the Sindbis replicon system for expression of LaCrosse virus envelope proteins in mosquito cells. *Archiv. Virol.*, 143: 1365-1377.

Kohno, A., Emi, N., Kasai, M., Tanimoto, M. and Saito, H. (1998). Semliki Forest virus-based DNA expression vector transient protein production followed by cell death. *Gene Ther.*, 5: 415-418.

Kuhn, R.J., Niesters, H.G., Hong, Z. and Strauss, J.H. (1991). Infectious RNA transcripts from Ross River virus cDNA clones and the construction and characterization of defined chimeras with Sindbis virus. *Virology*, 182: 430-441.

Lebedeva, I., Fujita, K., Nihrane, A. and Silver, J. (1997). Infectious particles derived from Semliki Forest virus vectors encoding murine leukemia virus envelopes. *J. Virol.*, 71: 7061-7067.

Lee, A.H., Suh, Y.S., Sung, J.H., Yang, S.H. and Sung, Y.C. (1997). Comparison of various expression plasmids for the induction of immune response by DNA immunization. *Mol. Cells,* 7: 495-501.

Lee, S., Owen, K.E., Choi, H.K., Lee, H., Lu, G.G., Wengler, G., Brown, D.T., Rossmann, M.G. and Kuhn, R.J. (1996). Identification of a protein binding site on the surface of the alphavirus nucleocapsid and its implication in virus assembly. *Structure*, 4: 531-541.

Levine, B., Huang, Q., Isaacs, J.T., Reed, J.C., Griffin, D.E. and Hardwick, J.M. (1993). Conversion of lytic to persistent alphavirus infection by the bcl-2 cellular oncogene. *Nature*, 361: 739-742.

Levine, B., Goldman, J.E., Jiang, H.H., Griffin, D.E. and Hardwick, J.M. (1996). Bcl-2 protects mice against fatal alphavirus encephalitis. *Proc. Nat. Acad. Sci. USA*, 93: 4810-4815.

Li, K.J. and Garoff, H. (1996). Production of infectious recombinant Moloney murine leukemia virus particles in BHK cells using Semliki Forest virus-derived RNA expression vectors. *Proc. Natl. Acad. Sci. USA*, 93: 11658-11663.

Li, K.J. and Garoff, H. (1998). Packaging of intron containing genes into retrovirus vectors by alphavirus vectors. *Proc. Natl. Acad. Sci. USA*, 95: 3650-3654.

Liang, X.H., Jiang, H.H. and Levine, B. (1997). Expression of a biologically active antiviral antibody using a Sindbis virus vector system. *Mol. Immunol.*, 34: 907-917.

Liang, X.H., Kleeman, L.K., Jiang, H.H., Gordon, G., Goldman, J.E., Berry, G., Herman, B. and Levine, B. (1998). Protection against fatal Sindbis encephalitis by beclin, a novel Bcl-2-interacting protein. *J. Virol.*, 72: 8586-8596.

Liao, C.L., Lin, Y.L., Wang, J.J., Huang, Y.L., Yeh, C.T., Ma, S.H. and Chen, L.K. (1997). Effect of enforced expression of human bcl-2 on Japanese encephalitis virus-induced apoptosis in cultured cells. *J. Virol.*, 71: 5963-5971.

Liljeström, P. and Garoff, H. (1991). A new generation of animal cell expression vectors based on the Semliki Forest virus replicon. *Bio/Technology*, 9: 1356-1361.

Liljeström, P., Lusa, S., Huylebroeck, D. and Garoff, H. (1991). In vitro mutagenesis of a full-length cDNA clone of Semliki Forest virus: the 6,000-molecular-weight membrane protein modulates virus release. *J. Virol.*, 65: 4107-4113.

Lindenbach, B.D. and Rice, C.M. (1997). Trans-complementation of yellow fever virus NS1 reveals a role in early RNA replication. *J. Virol.*, 71: 9608-9617.

Lobigs, M. and Garoff, H. (1990). Fusion function of the Semliki Forest virus spike is activated by proteolytic cleavage of the envelope glycoprotein precursor p62. *J. Virol.*, 64: 1233-1240.

London, S.D., Schmaljohn, A.L., Dalrymple, J.M. and Rice, C.M. (1992). Infectious enveloped RNA virus antigenic chimeras. *Proc. Natl. Acad. Sci. USA*, 89: 207-211.

Lopez, S., Yao, J.S., Kuhn, R., Strauss, E. and Strauss, J. (1994). Nucleocapsid-glycoprotein interactions required for assembly of alphaviruses. *J. Virol.*, 68: 1316-1323.

Lundström, K. and Turpin, M.P. (1996). Proposed shizophrenia-related gene polymorphism: expression of the Ser9Gly mutant human dopamine D3 receptor with the Semliki Forest virus system. *Biochem. Biophys. Res. Commun.*, **225:** 1068-1072.

Lundström, K., Mills, A., Buell, G., Allet, E., Adami, N. and Liljeström, P. (1994). High-level expression of the human neurokinin-1 receptor in mammalian cell lines using the Semliki Forest virus expression system. *Eur. J. Biochem.*, 224: 917-921.

Lundström, K., Mills, A., Allet, E., Ceszkowski, K., Agudo, G., Chollet, A. and Liljeström, P. (1995a). High-level expression of G protein-coupled receptors with the aid of the Semliki Forest virus expression system. *J. Recept. Signal Transduct. Res.*, 15: 23-32.

Lundström, K., Vargas, A. and Allet, B. (1995b). Functional activity of a biotinylated human neurokinin 1 receptor fusion expressed in the Semliki Forest virus system. *Biochem. Biophys. Res. Comm.*, 208: 260-266.

Lundström, K., Michel, A., Blasey, H., Bernard, A.R., Hovius, R., Vogel, H. and Surprenant, A. (1997). Expression of ligand-gated ion channels with the Semliki Forest virus expression system. *J. Recept. Signal Transduct. Res.*, 17: 115-126.

Malone, J.G., Berglund, P., Liljeström, P., Rhodes, G.H. and Malone, R.W. (1997). Mucosal immune responses associated with polynucleotide vaccination. *Behring Inst. Mitt.*, 98: 63-72.

Marheineke, K., Lenhard, T., Haase, W., Beckers, T., Michel, H. and Reilander, H. (1998). Characterization of the human gonadotropin-releasing hormone receptor heterologously produced using the baculovirus/insect cell and the Semliki Forest virus systems. *Cell. Mol. Neurbiol.*, 18: 509-524.

McCue, L.A. and Anders, D.G. (1998). Soluble expression and complex formation of proteins required for HCMV DNA replication using the SFV expression system. *Protein Expr. Purif.*, 13: 301-312.

Monastyrskaia, K., Goepfert, F., Hochstrasser, R., Acuna, G., Leighton, J., Pink, J. R., and Lundström, K. (1999). Expression and intracellular localisation of odorant receptors in mammalian cell lines using Semliki Forest virus vectors. *J. Recept. Signal Transduct. Res.*, 19: 687-701.

Mor, G., Yamshchikov, G., Sedegah, M., Takeno, M., Wang, R., Houghten, R.A., Hoffman, S. and Klinman, D.M. (1996). Induction of neonatal tolerance by plasmid DNA vaccination of mice. *J. Clin. Invest.*, 98: 2700-2705.

Mossman, S., Bex, F., Berglund, P., Arthos, J., O'Neil, S.P., Riley, D., Maul, D.H., Bruck, C., Momin, P., Burny, A., Fultz, P. N., Mullins, J.I., Liljeström, P. and Hoover, E.A. (1996). Protection against lethal SIVsmmPBj14 disease by a recombinant Semliki Forest virus gp160 vaccine and by gp120 subunit vaccine. *J. Virol.*, 70: 1953-1960.

Nichols, W.W., Ledwith, B.J., Manam, S.V. and Troilo, P.J. (1995). Potential DNA vaccine integration into host cell genome. *Ann. N. Y. Acad. Sci.*, 772: 30-39.

Nilsson, M., von Bonsdorff, C.H., Weclewicz, K., Cohen, J. and Svensson, L. (1998). Assembly of viroplasm and virus-like particles of rotavirus by a Semliki Forest virus replicon. *Virology*, 242: 255-265.

Notka, F., Stahl-Hennig, C., Dittmer, U., Wolf, H. and Wagner, R. (1999). Construction and characterization of recombinant VLPs and Semliki Forest virus live vectors for comparative evaluation in the SHIV monkey model. *Biol. Chem.*, 380: 341-352.

Ohno, K., Sawai, K., Iijima, Y., Levin, B. and Meruelo, D. (1997). Cell-specific targeting of Sindbis virus vectors displaying IgG-binding domains of protein A. *Nat. Biotech.*, 15: 763-767.

Olivo, P.D., Frolov, I. and Schlesinger, S. (1994). A cell line that expresses a reporter gene in response to infection by Sindbis virus: a prototype for detection of positive strand RNA viruses. *Virology*, 198: 381-384.

Olivo, P.D., Collins, P.L., Peeples, M.E. and Schlesinger, S. (1998). Detection and quantitation of human respiratory syncytial virus (RSV) using minigenome cDNA and a Sindbis virus replicon: A prototype assay for negative-strand RNA viruses. *Virology*, 251: 198-205.

Olkkonen, V.M., Liljeström, P., Garoff, H., Simons, K. and Dotti, C.G. (1993). Expression of heterologous proteins in cultured rat hippocampal neurons using the Semliki Forest virus vector. *J. Neurosci. Res.*, 35: 445-451.

Olson, K.E., Carlson, J.O. and Beaty, B.J. (1992). Expression of the chloramphenicol acetyltransferase gene in Aedes albopictus (C6/36) cells using a non-infectious Sindbis virus expression vector. *Insect Mol. Biol.*, 1: 49-52.

Olson, K.E., Higgs, S., Hahn, C.S., Rice, C.M., Carlson, J.O. and Beaty, B.J. (1994). The expression of chloramphenicol acetyltransferase in Aedes albopictus (C6/36) cells and Aedes triseriatus mosquitoes using a double subgenomic recombinant Sindbis virus. *Insect Biochem. Mol. Biol.*, 24: 39-48.

Olson, K.E., Higgs, S., Gaines, P.J., Powers, A.M., Davis, B.S., Kamrud, K.I., Carlson, J.O., Blair, C.D. and Beaty, B.J. (1996). Genetically Engineered Resistance to Dengue-2 Virus Transmission In Mosquitoes. *Science*, 272: 884-886.

Opstelten, D.J., Wallin, M. and Garoff, H. (1998). Moloney murine leukemia virus envelope protein subunits, gp70 and Pr15E, form a stable disulfide-linked complex. *J. Virol.*, 72: 6537-6545.

Paul, N.L., Marsh, M., McKeating, J.A., Schulz, T.F., Liljeström, P., Garoff, H. and Weiss, R.A. (1993). Expression of HIV-1 envelope glycoproteins by Semliki Forest virus vectors. *AIDS Res. Hum. Retroviruses*, 9: 963-970.

Polo, J.M., Belli, B.A., Driver, D.A., Frolov, I., Sherrill, S., Hariharan, M.J., Townsend, K., Perri, S., Mento, S.J., Jolly, D.J., Chang, S.M.W., Schlesinger, S. and Dubensky, T.M. (1999). Stable alphavirus packaging cells lines for Sindbis virus and Semliki Forest virus-derived vectors. *Proc. Natl. Acad. Sci. USA*, 96: 4598-4603.

Powers, A.M., Olson, K.E., Higgs, S., Carlson, J.O. and Beaty, B.J. (1994). Intracellular immunization of mosquito cells to LaCrosse virus using a recombinant Sindbis virus vector. *Virus Res.*, 32: 57-67.

Pugachev, K.V., Mason, P.W. and Frey, T.K. (1995). Sindbis vectors suppress secretion of subviral particles of Japanese encephalitis virus from mammalian cells infected with SIN-JEV recombinants. *Virology*, 209: 155-166.

Pugachev, K.V., Mason, P.W., Shope, R.E. and Frey, T.K. (1995). Double-subgenomic Sindbis virus recombinants expressing immunogenic proteins of Japanese encephalitis virus induce significant protection in mice against lethal JEV infection. *Virology*, 212: 587-594.

Pushko, P., Parker, M., Ludwig, G.V., Davis, N.L., Johnston, R.E. and Smith, J.F. (1997). Replicon-helper systems from attenuated Venezuelan equine encephalitis virus - Expression of heterologous genes in vitro and immunization against heterologous pathogens in vivo. *Virology*, 239: 389-401.

Raju, R. and Huang, H.V. (1991). Analysis of Sindbis virus promoter recognition in vivo, using novel vectors with two subgenomic mRNA promoters. *J. Virol.*, 65: 2501-2510.

Raju, R., Subramaniam, S.V. and Haijou, M. (1995). Genesis of Sindbis virus by in vivo recombination of nonreplicative RNA precursors. *J. Virol.*, 69: 7391-7401.

Rayms-Keller, A., Powers, A.M., Higgs, S., Olson, K.E., Kamrud, K.I., Carlson, J.O., and Beaty, B.J. (1995). Replication and expression of a recombinant Sindbis virus in mosquitoes. *Insect Mol. Biol.*, 4: 245-251.

Rice, C.M., Levis, R., Strauss, J.H. and Huang, H.V. (1987). Production of infectious RNA transcripts from Sindbis virus cDNA clones: mapping of lethal mutations, rescue of a temperature-sensitive marker, and in vitro mutagenesis to generate defined mutants. *J. Virol.*, 61: 3809-3819.

Rolls, M.M., Webster, P., Balba, N.H. and Rose, J.K. (1994). Novel infectious particles generated by expression of the vesicular stomatitis virus glycoprotein from a self-replicating RNA. *Cell*, 79: 497-506.

Rolls, M.M., Haglund, K. and Rose, J.K. (1996). Expression of additional genes in a vector derived from a minimal RNA virus. *Virology*, 218: 406-411.

Ryan, M.D. and Drew, J. (1994). Foot-and-mouth disease virus 2A oligopeptide mediated cleavage of an artificial polyprotein. *EMBO J.*, 13: 928-933.

Ryan, M.D., King, A.M.Q. and Thomas, G.P. (1991). Cleavage of foot-and-mouth disease virus polyprotein is mediated by residues located within a 19 amino acid sequence. *J. Gen. Virol.*, 72: 2727-2732.

Salminen, A., Wahlberg, J.M., Lobigs, M., Liljeström, P. and Garoff, H. (1992). Membrane fusion process of Semliki Forest virus II: Cleavage dependent reorganization of the spike protein complex controls virus entry. *J. Cell Biol.*, 116: 349-357.

Sammin, D.J., Butler, D., Atkins, G.J. and Sheahan, B.J. (1999). Cell death mechanisms in the olfactory bulb of rats infected intranasally with Semliki Forest virus. *Neuropathol. Appl. Neurobiol.*, 25: 236-243.

Sawai, K. and Meruelo, D. (1998). Cell-specific transfection of choriocarcinoma cells by using Sindbis virus hCG expressing chimeric vector. *Biochem. Biophys. Res. Commun.*, 248: 315-323.

Sawai, K., Ohno, K., Ijima, Y., Levin, B. and Meruelo, D. (1998). A novel method of cell-specific mRNA transfection. *Mol. Genet. Metab.*, 64: 44-51.

Scallan, M.F., Allsopp, T.E. and Fazakerley, J.K. (1997). bcl-2 acts early to restrict Semliki Forest virus replication and delays virus-induced programmed cell death. *J. Virol.*, 71: 1583-1590.

Scheer, A., Bjorklof, K., Cotecchia, S. and Lundstrom, K. (1999). Expression of the alpha(1b)-adrenergic receptor and G protein subunits in mammalian cell lines using Semliki Forest virus expression system. *J. Recept. Signal Transduct. Res.*, 19: 369-378.

Schlesinger, S. and Weiss, B.G. (1994). Recombination between Sindbis virus RNAs. *Archiv. Virol.*, 9: 213-220.

Seabaugh, R.C., Olson, K.E., Higgs, S., Carlson, J.O. and Beaty, B.J. (1998). Development of a chimeric Sindbis virus with enhanced per Os infection of Aedes aegypti. *Virology*, 243: 99-112.

Sjöberg, E.M. and Garoff, H. (1996). The translation-enhancing region of the Semliki Forest virus subgenome is only functional in the virus-infected cell. *J. Gen. Virol.*, 77: 1323-1327.

Sjöberg, E.M., Suomalainen, M. and Garoff, H. (1994). A significantly improved Semliki Forest virus expression system based on translation enhancer segments from the viral capsid gene. *Bio/Technology*, 12: 1127-1131.

Skoging, U. and Liljeström, P. (1998). Role of the C-terminal tryptophan residue for the structure-function of alphavirus capsid protein. *J. Mol. Biol.*, 279: 865-872.

Skoging, U., Vihinen, M., Nilsson, L. and Liljeström, P. (1996). Aromatic interactions define the binding of the alphavirus spike to its nucleocapsid. *Structure*, 4: 519-529.

Smerdou, C. and Liljeström, P. (1999). Two-helper RNA system for production of recombinant Semliki forest virus particles. *J. Virol.*, 73: 1092-1098.

Smith, S.M., Maldarelli, F. and Jeang, K.T. (1997). Efficient expression by an alphavirus replicon of a functional ribozyme targeted to human immunodeficiency virus type 1. *J. Virol.*, 71: 9713-9721.

Strauss, J.H. and Strauss, E.G. (1994). The alphaviruses: Gene expression, replication, and evolution. *Microbiol. Rev.*, 58: 491-562.

Suomalainen, M., Liljeström, P. and Garoff, H. (1992). Spike protein-nucleocapsid interactions drive the budding of alphaviruses. *J. Virol.*, 66: 4737-4747.

Torresi, J., Meanger, J., Lambert, P., Li, F., Locarnini, S.A. and Anderson, D.A. (1997). High level expression of the capsid protein of hepatitis E virus in diverse eukaryotic cells using the Semliki Forest virus replicon. *J. Virol. Meth.*, 69: 81-91.

Torresi, J., Li, F., Locarnini, S.A. and Anderson, D.A. (1999). Only the non-glycosilated fraction of hepatitis E virus capsid (open reading frame 2) protein is stable in mammalian cells. *J. Gen. Virol.*, 80: 1185-1188.

Tsuji, M., Bergmann, C.C., Takita-Sonoda, Y., Murata, K., Rodrigues, E.G., Nussenzweig, R.S. and Zavala, F. (1998). Recombinant Sindbis viruses expressing a cytotoxic T-lymphocyte epitope of a malaria parasite or of influenza virus elicit protection against the correspondig pathogen in mice. *J. Virol.*, 72: 6907-6910.

Tubulekas, I. and Liljeström, P. (1998). Suppressors of cleavage-site mutations in the p62 envelope protein of Semliki Forest virus reveal dynamics in spike structure function. *J. Virol.*, 72: 2825-2831.

Verrier, F.C., Charneau, P., Altmeyer, R., Laurent, S., Borman, A.M. and Girard, M. (1997). Antibodies to several conformation-dependent epitopes of gp120/gp41 inhibit CCR-5-dependent cell-to-cell fusion mediated by the native envelope glycoprotein of a primary macrophage-tropic HIV-1 isolate. *Proc. Natl. Acad. Sci. USA*, 94: 9326-9331.

Wahlberg, J.M., Bron, R., Wilschut, J. and Garoff, H. (1992). Membrane fusion of Semliki Forest virus involves homotrimers of the fusion protein. *J. Virol.*, 66: 7309-7318

Wahlfors, J.J. and Morgan, R.A. (1999). Production of minigen-containing retroviral vectors using an alphavirus/retrovirus hybrid vector system. *Hum. Gene Ther.*, 10: 1197-1206.

Wahlfors, J.J., Xanthopoulos, K.G. and Morgan, R.A. (1997). Semliki Forest virus-mediated production of retroviral vector RNA in retroviral packaging cells. *Hum. Gene Ther.*, 8: 2031-2041.

Weiss, B.G. and Schlesinger, S. (1991). Recombination between Sindbis virus RNAs. *J.*

Virol., 65: 4017-4025.

Weiss, B., Rosenthal, R. and Schlesinger, S. (1980). Establishment and maintenance of persistent infection by Sindbis virus in BHK cells. *J. Virol.*, 33: 463-474.

White, C.L., Thomson, M. and Dimmock, N.J. (1998). Deletion analysis of a defective interfering Semliki Forest virus RNA genome defines a region in the nsP2 sequence that is required for efficient packaging of the genome into virus particles. *J. Virol.*, 72: 4320-4326.

Wolff, J.A., Ludtke, J.J., Acsadi, G., Williams, P. and Jani, A. (1992). Long-term persistence of plasmid DNA and foreign gene expression in mouse muscle. *Hum. Mol. Genet.*, 1: 363-369.

Xiong, C., Levis, R., Shen, P., Schlesinger, S., Rice, C.M., and Huang, H.V. (1989). Sindbis virus: an efficient, broad host range vector for gene expression in animal cells. *Science*, 243: 1188-1191.

Yadani, F.Z., Kohl, A., Prehaud, C., Billecocq, A. and Bouloy, M. (1999). The carboxy-terminal acidic domain of Rift Valley fever virus NSs protein is essential for the formation of filamentous structures but not for the nuclear localization of the protein. *J. Virol.*, 73: 5018-5025.

Zhang, J., Asselin-Paturel, C., Bex, F., Bernard, J., Chehimi, J., Willems, F., Caignard, A., Berglund, P., Liljeström, P., Burny, A. and Chouaib, S. (1997). Cloning of human Il-12 p40 and p35 DNA into the Semliki Forest virus vector: expression of Il-12 in human tumor cells. *Gene Ther.*, 4: 367-374.

Zhao, H., Lindqvist, B., Garoff, H., von Bonsdorff, C.H. and Liljeström, P. (1994). A tyrosine-based motif in the cytoplasmic domain of the alphavirus envelope protein is essential for budding. *EMBO J.*, 13: 4204-4211.

Zhou, X., Berglund, P., Rhodes, G., Parker, S.E., Jondal, M. and Liljestrom, P. (1994). Self-replicating Semliki Forest virus RNA as recombinant vaccine. *Vaccine*, 12: 1510-1514.

Zhou, X., Berglund, P., Zhao, H., Liljeström, P. and Jondal, M. (1995). Generation of cytotoxic and humoral immune responses by nonreplicative recombinant Semliki Forest virus. *Proc. Natl. Acad. Sci. USA*, 92: 3009-3013.

13. SUMMARY — Alphaviruses are enveloped viruses containing a single positive strand RNA molecule as genome. Several vectors derived from alphaviruses have been developed, which include Sindbis virus (SIN), Semliki Forest virus (SFV), and Venezuelan Equine Encephalitis virus (VEE). Alphavirus self-replicating RNA containing heterologous genes can be synthesized in vitro from plasmids having the recombinant alphavirus replicon sequences under the control of a prokaryotic promoter, such as SP6 or T7. High level expression of the heterologous proteins/RNA is obtained in cells transfected with this RNA. Although the system can be used for gene delivery directly as naked RNA, several packaging systems have been developed which allow the encapsidation of the alphaviral recombinant RNA into suicidal viral particles, increasing the efficiency of delivery of the recombinant genome into cells. A more recent variant of the system is based on a DNA/RNA layered alphaviral vector in which the recombinant replicon is transcribed from an RNA polymerase-II promoter, allowing direct delivery of DNA into the cells. Alphavirus vectors have been used to express a great number of proteins with many different purposes, including protein production and characterization, functional studies, vaccination, and gene therapy. Both the recombinant alphavirus particles and the alphavirus nucleic acid vectors have shown to be able to induce protective immune responses in different animal models. The possible application of these vectors in gene therapy faces, however, two limitations, which are the lack of specific targeting and the transient nature of the vector, due to the induction of apoptosis by the alphavirus replicon. Several strategies have been recently described to improve the targeting of alphavirus vectors and to develop noncytopathic vectors with potential use in gene therapy.

Key Words: Alphavirus; Semliki Forest virus; Sindbis virus; Venezuelan Equine Encephalitis virus; self-amplifying vector; suicide particles; protein production; nucleic acid vaccine; gene therapy

Progress in Gene Therapy: Basic and Clinical Frontiers, 433-453
R. Bertolotti *et al.* (Eds)

Anti-HIV ribozymes in the inhibition of HIV and AIDS

Greg Fanning, Lun Quan Sun and Geoff Symonds*

Johnson and Johnson Research, 1 Central Avenue, Australian Technology Park, Everleigh, Sydney, NSW 1430 Australia

Table of Contents

* Corresponding author. E-mail: gsymonds@MEDAU.JNJ.COM

1. INTRODUCTION

Human Immunodeficiency Virus (HIV) is the etiological agent of Acquired Immune Deficiency Syndrome (AIDS) (Gallo *et al.*, 1984). There is currently no cure for this disease and figures released by the World Health Organisation (May 1999) identify AIDS as the fourth leading cause of death worldwide and the number one cause in Africa. Highly Active Anti-Retroviral Therapy (HAART) is the favoured strategy for the treatment of AIDS and combines reverse transcriptase and protease inhibitors to reduce viral loads. These treatments do successfully reduce viral load but side effects, requirement for constant administration, HIV mutation, persistence of HIV reservoirs and lengthy time course for complete viral elimination (if at all possible, see below) are major limitations for the use of HAART as a stand-alone strategy to treat AIDS. Ribozymes are catalytic RNA molecules that can be designed to target HIV mRNA. This chapter discusses the potential use of anti-HIV ribozymes in the treatment of AIDS, including our own Phase I Clinical Trials using the anti-*Tat* ribozyme, Rz2.

Ribozymes as part of a AIDS treatment Regimen

Long-term HAART treatment has been analysed in the recent literature, highlighting a requirement for additional therapies (Furtado*et al.,* 1999; Zhang*et al.,* 1999). Zhang *et al.* (1999) and Furtado *et al.* (1999) demonstrated that HIV replication, as assessed by p24 levels, proviral DNA sequence analysis and estimates of levels of proviral DNA/RNA is compromised by HAART and drops to a plateau after 500 days of treatment (Furtado*et al.,* 1999). This plateau was significantly above zero, indicating that even at sites accessible to HAART, viral replication and HIV reservoirs persist. The side effects and toxicity of HAART are well documented and the estimate that this strategy would minimally take 10 - 60 years to eliminate HIV from both acute and latently infected cells make the search for alternative and additional treatments a priority (Saag and Kilby, 1999). It is apparent that at least a three pronged approach is required for AIDS management. Based on current knowledge, the three stages for HIV treatment could include: 1) Suppression of viral replication using HAART, 2) The mobilisation of HIV viral reservoirs, and 3) Restoration of the immune system and continued suppression of HIV replication. Stage one could be managed by HAART or a similar early aggressive treatment. Stage two has been approached recently with intermittent IL-2 administration (Chun*et al.,* 1999). Chun *et al.* (1999) described the treatment of HIV positive patients with IL-2 while receiving HAART. The aim was to stimulate resting T cell HIV reservoirs and thus render them vulnerable to HAART. A comparison between patients receiving intermittent interleukin-2 plus HAART and patients receiving HAART alone showed that in patients receiving IL-2 the pool of resting infected CD4+ T lymphocytes was significantly reduced (Chun*et al.,* 1999). The introduction of a third step in the treatment strategy could be envisaged involving strategies such as anti-RNA agents that aim to protect uninfected cells from infection and suppress the proviral gene product after viral gene integration.

2. HIV LIFE CYCLE AND TARGET POINTS FOR ANTI-RNA STRATEGIES

HIV belongs to the lentivirus family of complex retroviruses (Weiss, 1984). The life cycle of a retrovirus is mediated by encapsulated RNA that contains all information required for viral replication (Fig. 1). In brief, the life cycle of HIV begins by the binding of HIV receptors to the surface of a target cell followed by entry and uncoating of virus in the cytoplasm. This results in the integration of viral genes into the host genome via a DNA intermediate (termed provirus). The subsequent expression of replication, regulatory and accessory genes results in the propagation of new HIV viral particles (Fig. 2). The life cycle as described, occurs in CD4 positive T lymphocytes and other cell types such as macrophages. The subsequent development of AIDS is caused by the decline in the numbers of the CD4+ T lymphocyte cell population. As a result, the immune system is compromised and the host becomes susceptible to opportunistic infections. The other cells infected by HIV include: astrocytes, neurones and microvascular endothelial cells of the brain, epithelial cells, CD4 negative lymphocytes and thymocytes as well as cardiomyocytes (Speck*et al.,* 1999). Once infected, cells either continue the life cycle of HIV by producing infective particles or become latent viral reservoirs that ensure the persistence of the virus. It is not known whether these reservoirs contribute to specific AIDS-associated clinical syndromes but they are likely to have an impact on anti-HIV strategies as they have the capacity to continually re-emerge as virus-producing cells.

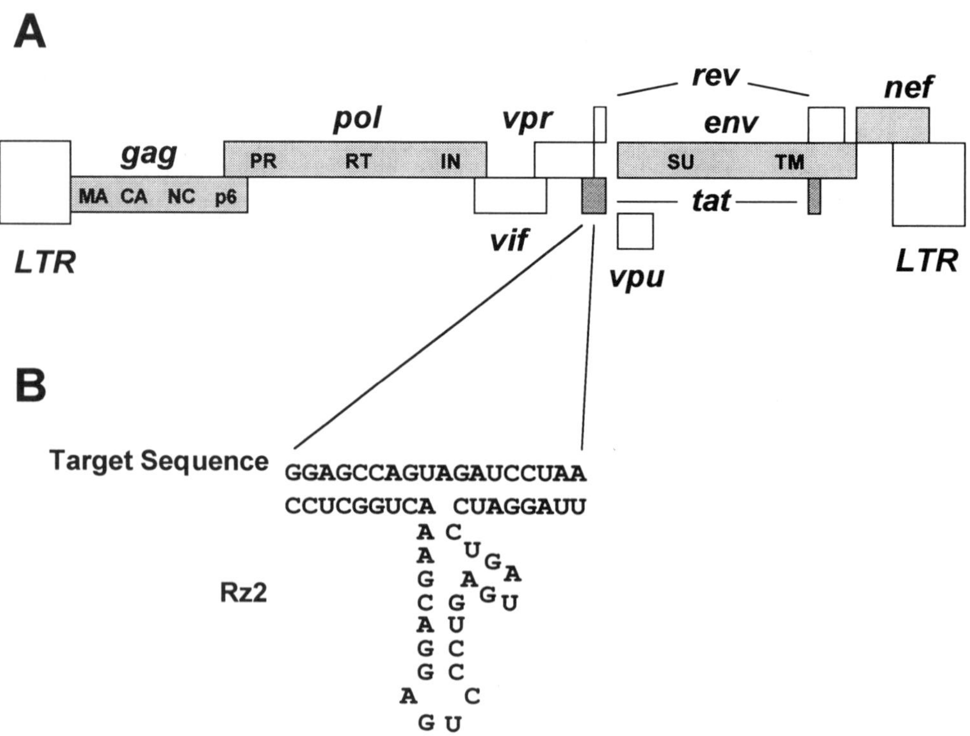

Figure 1. A. Genomic map of HIV-1. The HIV genome is approximately 9.8 kb in size and contains three replicative genes (*gag, pol, env*), five regulatory genes (*tat, rev, tev, vpr, nef*) and two accessory genes (*vpu and vif*). **B. Target site, sequence and secondary structure of the anti-*tat* ribozyme Rz2.**

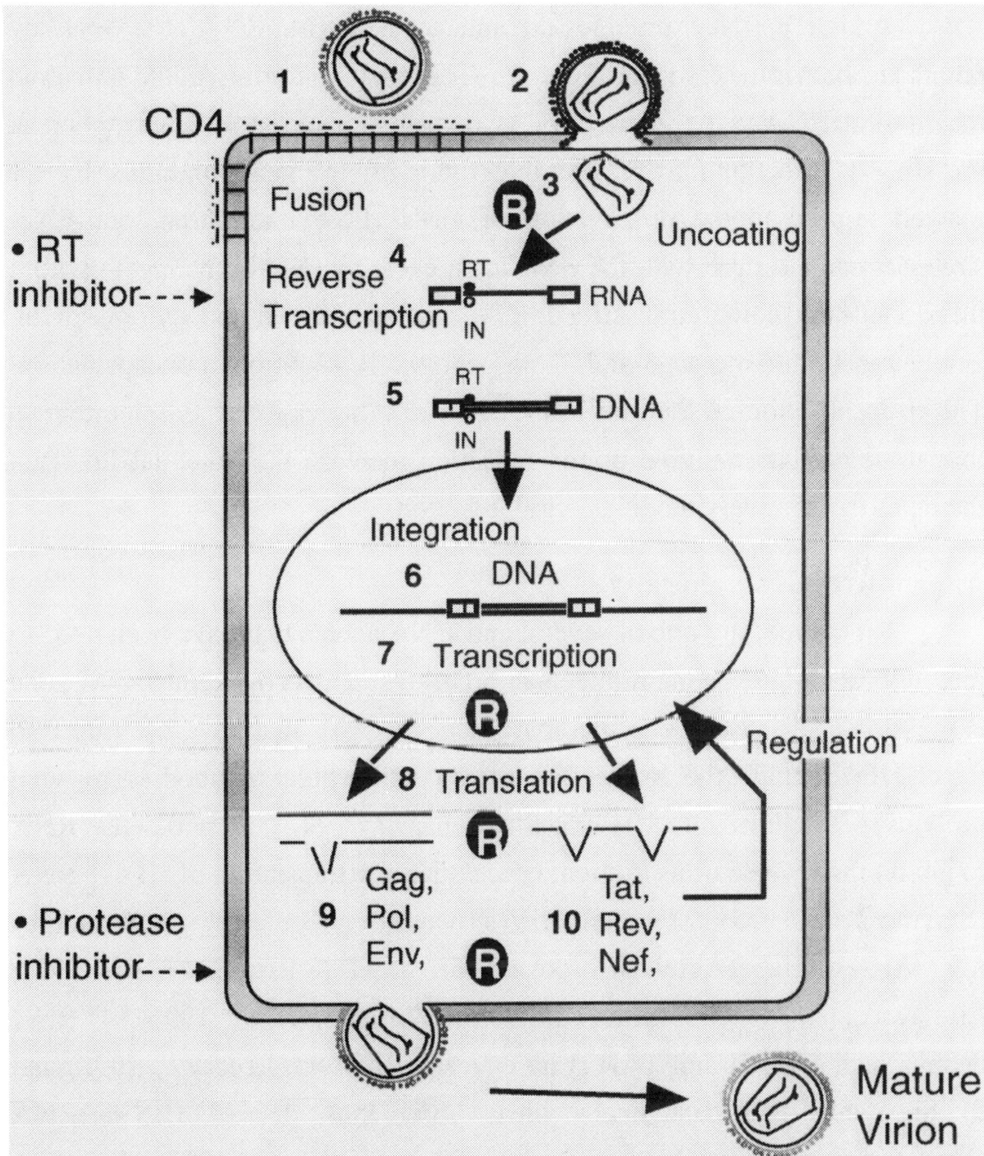

Figure 2. Infectious cycle of HIV-1 and potential sites of ribozyme action for the inhibition of HIV-1 replication. The virus enters the cell following binding to receptors on the cell membrane (**1**). The viral and host membrane fuse (**2**) and uncoating of the viral core ensues (**3**). The viral core contains viral RNA, reverse transcriptase and integrase enzymes (**4**). Viral RNA is converted to DNA by the reverse transcriptase and becomes double stranded in the cytoplasm (**5**) after which it is transported to the nucleus where it is integrated into the host cell genome by the action of integrase (**6**). Subsequent transcription (**7**) and translation (**8**) of packaging proteins (**9**) and regulatory proteins (**10**) facilitates the construction of mature viral particles which bud from the living cell. Potential sites for ribozyme action are indicated by R within a circle.

Anti-HIV RNA strategies can interact with viral RNA at several key points in the viral life cycle (Fig. 2). The first is during initial infection where viral RNA is exposed prior to reverse transcription. Interruption at this stage of infection may prevent integration of viral genes and thus, have a marked impact on the propagation of virus. Post integration, anti-RNA strategies can interfere with the replication cycle of HIV by acting 1) within the nucleus on newly transcribed HIV message, 2) in the cytoplasm on newly transported message and 3) on genomic RNA prior to encapsulation. The co-localisation of the anti-RNA agent and target RNA is important to this strategy. The multiple points of action increase the potential for such therapies to negatively modulate viral infection.

Alternative Gene therapy

Ribozymes are one of several anti-RNA agents that have been used to inhibit viral and host gene products in order to suppress the sequelae of gene expression of disease. Gene-therapy based strategies that have been used to target HIV include the expression of: 1) intracellular antibodies to viral proteins (Maciejewski*et al.,* 1995; Shaheen*et al.,* 1996), 2) anti-sense RNA to inhibit the reverse transcription, processing and translation of HIV-1 RNA (Smythe*et al.,* 1995), 3) mutant HIV structural or regulatory genes such as rev M10 or tat (Aguilar-Cordova*et al.,* 1995; Escaich*et al.,* 1995; Lisziewicz*et al.,* 1995; Caputo*et al.,* 1996; Woffendin*et al.,* 1996), 4) RNA decoys such as RRE and TAR (Lisziewicz*et al.,* 1993; Lee*et al.,* 1995) and 5) ribozymes that catalytically cleave and thus disable HIV-1 RNA (Heidenreich*et al.,* 1992; Rossi*et al.,* 1992; Crisell *et al.,* 1993; Homann*et al.,* 1993; Sun*et al.,* 1994; Sun 1995a). This is an expanding field and other anti-RNA technologies that could potentially be used against HIV include DNAzymes (Cairns*et al.,* 1999), minizymes (Kuwabara*et al.,* 1996), snorbozymes (Samarsky*et al.,* 1999), RNAi (Montgomery*et al.,* 1998), guide sequence directed Rnase-P cleavage (Yuan*et al.,* 1994), and gene targeting strategies such as homologous recombination. In general, the relative efficiencies of different anti-RNA approaches has yet to be elucidated but ribozymes that target and cleave specific mRNA sequences appear to have the advantage of being catalytic and therefore reusable.

3. RIBOZYMES

Ribozymes were first discovered in the protozoan *Tetrahymena* as part of what is now referred to as a group 1 self-splicing intron (Cech, 1988). Subsequent studies identified three further types of ribozymes from self cleaving infectious plant RNA, namely the 1) hammerhead ribozyme (Haseloff*et al.,* 1988) and 2) hairpin ribozyme (Hampel*et al.,* 1989), named on the basis of secondary structure and 3) delta ribozymes derived from hepatitis delta virus (Perrotta*et al.,* 1990). The derivation of minimum requirements and target sequences for these molecules led to the possibility (now being realised) for specific gene down regulation via site specific cleavage of mRNA. Hammerhead ribozymes have been the most widely investigated tool for RNA cleavage and the target site is generally GUX* (*indicates point of cleavage), although in certain cases NUX may be used (where N represents any nucleotide and X represents A, C or U) (Haseloff*et al.,* 1988). For hairpin ribozyme designs, *GUC is the sequence required (Hampel*et al.,* 1990).

a) Ribozymes can be targeted towards HIV

With a target defined, ribozymes can be designed against any mRNA containing the requisite sequence, including HIV mRNA. A basic search on Medline for "HIV and ribozymes" identified 220 references (1990-1999), of which approximately 37 papers appear to describe one or more new anti-HIV ribozymes. The ribozyme targets include *gag*, *pol*, *vif*, *ta*t, 5' LTR, TAR, U5, RRE, *rev*, *psi*, *env*, *nef*, *pro* and the host CCR5 gene. Three of these ribozymes targeting U5 and *Pol* (MY-2: Wong-Staal*et al.,* 1998), *Tat* and *Rev* (LTR/TAT-neo: Zaia*et al.,* 1998) and *Tat* (Rz2: Wang*et al.,* 1998; Cooper*et al.,* 1999; Amado et al 1999) have been used in Phase I Human

Clinical Trials. Several challenges confront anti-HIV ribozyme strategies. The two major challenges are 1) the realisation of levels of cell delivery that will have a clinical impact and 2) ensuring efficient *in vivo* action while limiting cell and host toxicity. The following discussion describes the steps and considerations for the design of ribozymes for clinical trials, using the development of Rz2 as an example.

b) Selection of *Tat* as a target

The first step in designing a ribozyme is determining a biologically important mRNA to target. Viruses carry by necessity few, if any genes that are superfluous to requirements and therefore possess many sites that are biologically suitable targets. The many sites in HIV that have already been targeted demonstrates this (see section above). The target mRNA must possess the triplet sequence recognised by the ribozyme and by inference this site must be accessible within its tertiary structure to ribozyme action (Sun*et al.,* 1994 and 1995b). The tertiary structure of RNA and potential protein interactions are difficult to predict but it is generally accepted that single-stranded regions and stem-loops of the RNA molecule are most accessible to ribozyme binding. However such considerations can only be approximated, this by using RNA secondary structure analysis software or chemical modification experiments (Campbell*et al.,* 1997).

In our work, the choice of site also considered two other factors relevant to HIV, namely mRNA splicing and mutation. Portions of the proviral sequence can be incorporated into differently spliced mRNAs and therefore to maximise the effectiveness of a ribozyme, it is important to consider sequences that are often found in a number of the sub-genomic mRNA species. The evolution of mutants is a feature of HIV infection and is a result of natural HIV replication, in particular the error rate of reverse transcriptase. This error rate combined with the selective pressure from the anti-retrovirals, reverse transcriptase and protease inhibitors leads to resistant variants. Based on these considerations, we have targeted the first exon of

Tat (Sun*et al.*, 1995a and 1995b). The protein product of the *Tat* gene enhances the rate of transcription initiation of the proviral sequence. *Tat* has been shown to be highly conserved in HIV clades and essential for the successful replication of HIV-1. Figure 1B shows the sequence and predicted hammerhead secondary structure of the anti-*Tat* ribozyme, Rz2.

c) Ribozyme synthesis and *in vitro* cleavage

Following selection of potential target sites, the corresponding ribozyme sequences can be synthesised and tested in *in vitro* cleavage reactions using standard conditions (50mM Tris HCl, pH 7.5, 10mM $MgCl_2$, at 37 °C) to assess their ability to produce the expected size cleavage products using polyacrylamide gel electrophoresis (Haseloff and Gerlach, 1988; Hampel*et al.*, 1993; Wang*et al.*, 1998). Whilst a correlation between *in vitro* cleavage efficiency and *in vivo* ribozyme activity has been found (Sun*et al.*, 1994), one is not always indicative of the other (Crisell*et al.*, 1993).

d) Pre-clinical assessment of anti-HIV activity of ribozymes

The next step in ribozyme development is to test its potential anti-HIV properties in a tissue culture system using a biological readout. For HIV, the standard method is to challenge ribozyme-containing T cells with laboratory adapted HIV strains and then assay for the production of p24. This requires that the ribozyme sequence be incorporated as a DNA copy into an expression vector. We have used constructs in which the ribozyme was cloned into the 3'-untranslated region of the neomycin resistance (*neo*) gene and driven either by the SV40 promoter in a plasmid-based expression vector or inserted into a Moloney Murine Leukemia virus-based retroviral vector. Ribozymes which targeted either the HIV *tat* gene or the ψ packaging site of HIV-1 (Sun*et al.*, 1994 and 1995a), protected T-lymphocyte cell lines from HIV-1 infection in terms of delay of HIV-1 replication and absolute virus levels (determined by syncytia formation and p24 ELISA antigen assay). In the T-cell line SupT 1, HIV replication of the virus was inhibited by 70-95% for the laboratory adapted HIV-1 isolates SF2 and IIIB and by 2 -4 logs for primary clinical isolates (Sun*et al.,* 1995b). In a second T lymphocyte cell culture system, CEM T4 cells were transduced with amphotropic retrovirus containing the same *neo*/ ribozyme expression cassette and conditions were established to allow a mixed pool of transduced cells to be challenged with HIV-1. As for the initial Sup T1 assay, the anti-HIV efficacy of constructs was determined by analysing p24 antigen production in comparison with cells transduced with the vector alone (Wang *et al.*, 1998).

In both systems, the most effective ribozyme construct, termed Rz2, was one directed to a site in the first coding exon of *tat* (Sun*et al.,* 1994, 1995a and 1995b; Wang*et al.,* 1998). When cloned into the MoMLV-based vector LNL6, this construct was designated RRz2 and formed the basis of our further work and Phase I Clinical Trials. This construct elicited a marked (70-90%) reduction of p24 antigen production when compared to LNL6 vector controls in a series of experiments. In addition, the RRz2 construct was demonstrated to offer protection against a range of clinical HIV-1 isolates, including AZT- and nevirapine-resistant strains (Wang*et al.,* 1998).

The results obtained in the T lymphocyte cell line studies were confirmed with primary cells using both total and CD4+ enriched non-HIV infected peripheral blood lymphocytes (PBLs). These PBLs were obtained from random normal blood donor packs by Ficoll/Hypaque separation and were CD4+ enriched using CD8+ MicroCELLector™ flasks (Applied Immune Systems/RPR Gencell) to deplete CD8+ T lymphocytes. Both total and CD4+ enriched PBLs were then transduced with LNL6 and RRz2 retrovirus, selected with an appropriate donor-specific dose of G418 and then challenged with HIV-1. RRz2, the retroviral construct shown to be effective in the pooled T lymphocyte assays was also effective in this assay giving inhibition of 70-90% when compared to the LNL6 control (Wang*et al.,* 1998).

e) Rz2 transduction of HIV positive PBLs

The RRz2 and LNL6 vectors were also used to transduce PBLs from HIV-1 infected patients and the results showed that cell viability was significantly higher in the ribozyme-transduced HIV-1 infected PBLs for a period of two weeks. No difference in viability (RRz2 versus LNL6-transduced) was observed in PBLs from non-infected donors (Wang*et al.*, 1998). This observation was the first evidence that the ribozyme could impact on the survival of HIV-1 infected patient derived PBLs in cell culture (Wang*et al.*, 1998) and implies that the ribozyme-expressing cells may have a growth viability advantage within HIV-1-infected patients.

f) Validation of specificity and *in vivo* action of Rz2

To validate the specificity of ribozyme action against HIV-1, we used another ribozyme, Rz1, which targets the 5' splicing region of the *tat* gene. This ribozyme was designed to cleave GUC N in which N is a G in HIV-1 IIIB and A in HIV-1 SF2. The data from both *in vitro* and *in vivo* studies with Rz1 showed that this ribozyme could protect cells only from those HIV isolates, the genomic sequence of which was cleavable *in vitro*. This study confirmed the importance of the first base pair distal to the NUX within helix I of the hammerhead structure for both *in vitro* and *in vivo* ribozyme activities (Sun*et al.*, 1995). The Rz2 ribozyme, in contrast to Rz1, is able to inhibit both strains of HIV-1, due to sequence conservation at the Rz2 target site (Sun*et al.*, 1995b; Wang*et al.*, 1998)

g) HIV Mutation

A feature of HIV replication is the frequent occurrence of mutations that can manifest as HIV strains/clades that are resistant to reverse transcriptase and protease inhibitors. *Tat* is highly conserved and mutants are less frequently encountered. But under selective pressure of an anti-*tat* ribozyme, mutant *tat* sequences may evolve. To test this, we used a multiple -passage assay to analyse HIV-1 sequence variation and viral replication dynamics in ribozyme-expressing cells. These studies demonstrated that Rz2 ribozyme expression in transduced human T cells did not provide selective conditions for the emergence of HIV resistant mutations over five sequential viral passages, and resulted in the rapid disappearance of certain quasi-species of HIV-1 (Wang*et al.,* 1998). Rapid disappearance of this "less fit" quasi-species in the Rz2 though not the LNL6 transduced cells argues for an *in vitro* selective pressure.

h) Target cells

Once a suitable ribozyme has been selected, the question becomes what target cells should be used. The two obvious candidates are peripheral blood CD4+ T lymphocytes that are specifically deficient in AIDS patients and the cells which ultimately differentiate into these T cells, namely the haematopoietic stem cell population. Protecting progenitor cells would appear a preferred choice for gene therapy, and current protocols focus on *ex vivo* manipulation of CD34 positive cells isolated from the periphery after mobilisation by systemically administered cytokines (Amado*et al.,* 1999). Future research investigating different cytokine combinations used in *ex vivo* manipulations and the inclusion of CD34 negative cells which are thought to represent a more primitive cell population may improve the ability of *ex vivo* manipulated cells to reconstitute the peripheral blood. Progenitor cells repopulate the periphery when there is "space" for new cells. The argument for this is that a rapid turnover of HIV-1 infected cells may (over time) yield reconstitution of the T lymphocyte population with gene therapeutic protected cells. Other approaches such as a suicide gene therapy approach that results in cell death upon infection (Caruso and Klatzmann, 1992; Caruso*et al.,* 1995; Chakrabarti*et al.,* 1996; Hamouda*et al.,* 1997; Kestler and Chakrabarti, 1997; Marcello and Giaretta, 1998) has been proposed and it may be that such an approach combined with protection of CD34 positive cells may act as an effective AIDS treatment. Other data, however, have shown that HIV infected patients have an impaired ability for T cell reconstitution, suggesting that a suicide based treatment may further compromise the immune system and have an adverse affect on recovery.

Alternative to a CD34 positive stem cell strategy is an approach to genetically modify T lymphocytes, and reinfuse these cells into the patient to produce a "protected" T lymphocyte population. The difficulty with such an approach may be in obtaining sufficient uninfected T lymphocytes from the peripheral blood of patients. However, even the treatment of already infected cells may be beneficial and have the effect of decreasing the rate of infection of other cells, thereby giving the immune system a slight but sufficient advantage. If the latter argument were true then a simpler approach to treating HIV infected patients would be the infusion of transduced CD4+ T lymphocytes. The genetic modification of these two cell populations, T lymphocytes and CD34 positive stem cells, has been used in our Phase I Clinical Trials and are discussed in the following sections.

4. PHASE I CLINICAL TRIALS OF RRz2

a) Study design

Two Phase I Clinical Trials have been initiated for the use of RRz2 to treat AIDS (Fig. 3 and 4). The first trial was conducted in identical twins discordant for HIV, in which healthy CD4+ lymphocytes from the uninfected twin were transduced with either LNL6 or RRz2 and then both cell populations transfused into the bloodstream of the HIV-positive twin (Wang*et al.,* 1998; Cooper*et al.,* 1999; Knop*et al.,* 1999). The second (non-twin based) trial isolated CD34+ cells from HIV infected patients for transduction with LNL6 and Rz2 (separately) followed by the re-infusion of ribozyme- and vector-containing CD34+ cells back into the same patient. The rationale for the latter trial being that the ribozyme containing CD34+ cells will differentiate *in vivo* and give rise to the various cell lineages (Amado*et al.,* 1999).

In each of the trials, the isolated cell populations were divided into two, and transduced with either the retrovirus vector containing the ribozyme or the vector alone, the latter as a control. Approximately equal

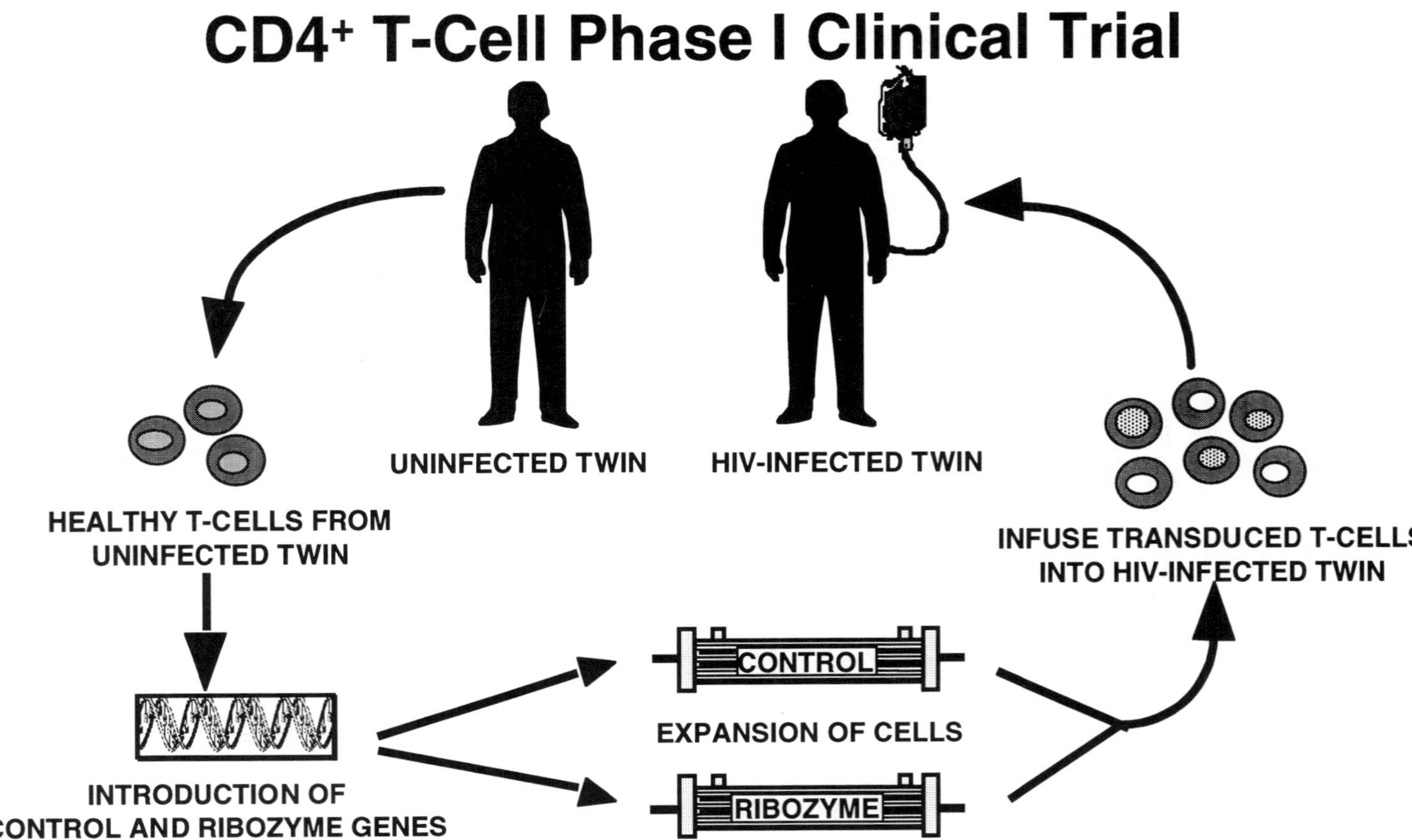

Figure 3. CD4 positive T lymphocyte based Clinical Trial. The trial involves pairs of identical twins discordant for HIV-1 infection. Healthy T -lymphocytes from the HIV-negative twin are separately transduced with ribozyme-containing and control retrovirus vectors, expanded *ex vivo* and then infused into the corresponding HIV-positive sibling.

numbers of these two transduced cell populations were then introduced into the patients. This allowed for monitoring the survival of the transduced cells by detecting ribozyme and control vector DNA sequences in peripheral blood using PCR (Knop*et al.,* 1999).

The overall objectives of the Phase I studies were to determine the safety, feasibility and ability to detect transduced cells following infusion of CD4+ and CD34+ cells into HIV-1 infected patients containing an anti-HIV ribozyme.

b) CD4+ Study

The CD4+ trial enrolled four HLA identical twins. These patients were infused with donor transduced CD4+. Approximately 1 x 10^8 white blood cells/kg were taken from each of the HIV negative twins by leukapheresis. CD4+ cells were isolated after separating lymphocytes from PBMCs by Ficoll/Hypaque density gradient and CD8+ T cell depletion using RPR Genecll Microcellector flasks. CD4+ cells were cultured within CELLMAX artificial capillary system in AIM-V medium containing 50 μg/L Streptomycin Sulfate, 10 μg/ml Gentamycin Sulfate and 5% heat-inactivated autologous plasma. CD4+ cells were stimulated with OKT3 antibody (day 0) and 100 U/ml IL-2 was present in the medium from day 2 onwards. Transduction was commenced when the cells reached log phase (day 3-5) in the 'CELLMAX' system.

c) CD34+ Study

In the CD34+ trial, seven patients were initially enrolled. Patients were treated with G-CSF (figrastim; Neupogen™) for four to six days. CD34+ cells were isolated from cells obtained from up to four separate apheresis procedures using the CellPro selection system. CD34+ cells were incubated with Stem Cell Factor and Megakaryocyte Growth and Development Factor prior to transduction with either the retroviral vector LNL6 or RRz2. Retrovirus-Transduced CD34+ cells were re-infused into the patient.

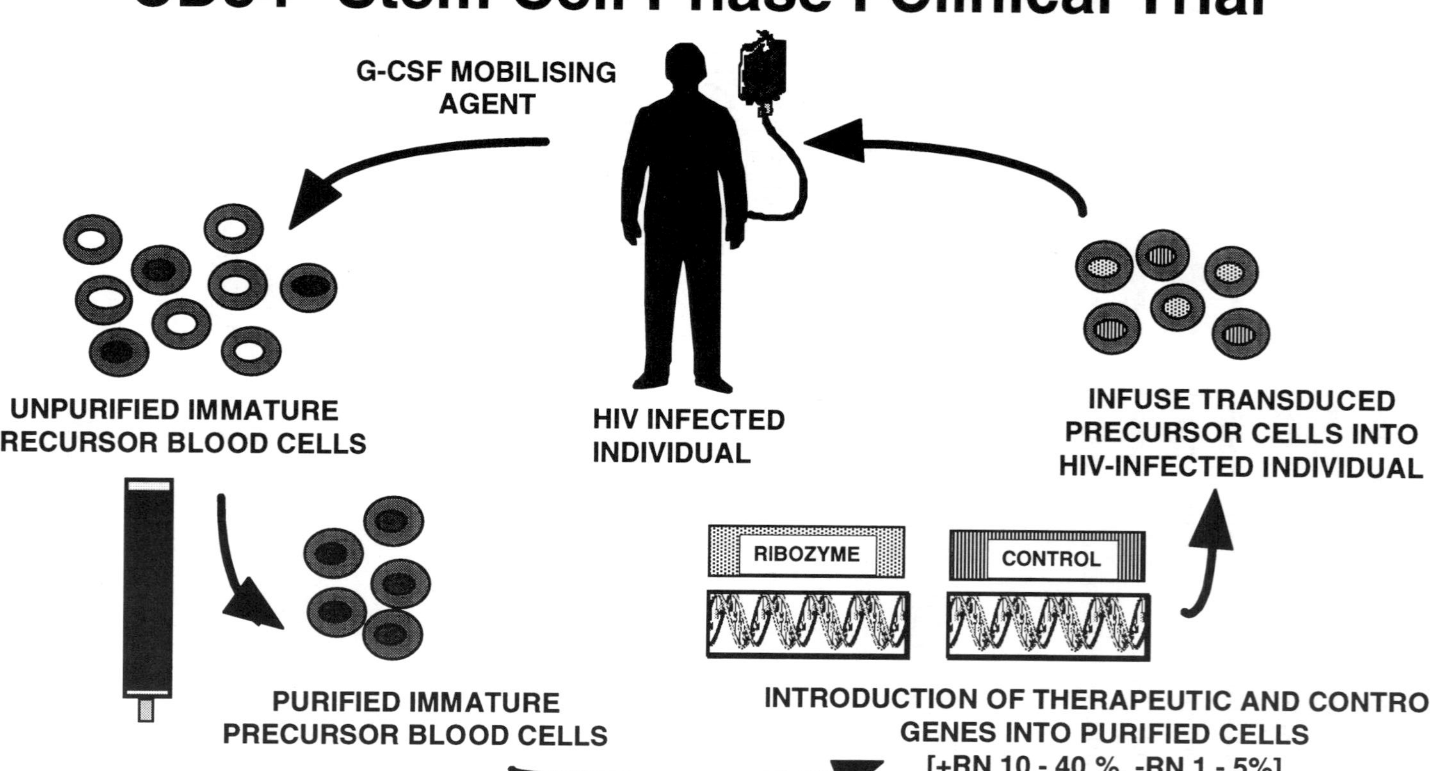

Figure 4. CD34 positive stem cell based Clinical Trial. Autologous pluripotent CD34 positive stem cells are collected from HIV-positive individuals after G-CSF mobilisation, further purified and then separately transduced with ribozyme-containing and control retroviral vectors before reinfusion. The transduced CD34 positive cells engraft into the bone marrow where they begin to differentiate into the component cells of the hematopoietic system.

5. CONCLUDING REMARKS

The aim of the Phase I Clinical Trials was to determine the feasibility and safety of introducing genetically modified CD4 positive T lymphocytes from a HLA identical twin or genetically modified autologous CD34 positive stem cells, into a HIV infected patient. In the post-transplant monitoring, the two main observations were: 1) there were no adverse effects from the treatment, and 2) cell marking was apparent in all patients (manuscripts in preparation).

AIDS is one of the most common causes of death in the world and represents a major therapeutic challenge. HIV pathogenesis is clearly complex and the multiple levels of control for proviral expression and the ability of the virus to mutate rapidly pose the greatest obstacles to current therapeutic regimens. Ribozymes are one of several possible gene therapy-based anti-HIV approaches; the cellular potential of these approaches relies on target specificity, multiple turnover, stable expression and lack of immunogenicity. Studies to date indicate that ribozymes can be effective in suppressing HIV-1 replication in tissue culture systems. Phase I Clinical Trials have determined that *ex vivo* transduction of CD4+ and CD34+ cells followed by reinfusion into the patient is an effective and safe mode of delivery for gene based therapies. The success of ribozymes in HIV patient care will ultimately depend on how they are integrated into a treatment regimen and improvements for *in vivo* action will depend upon developments in ribozyme design and delivery. It is conceivable that anti-HIV ribozymes will be used in conjunction with other treatments to facilitate recovery of the immune system and suppress viral replication. Such a gene therapy approach may have the major benefit of being effective for the life time of a patient. The clinical impact of the anti-HIV ribozyme Rz2 (and others) await the results of the current and future planned Phase II Clinical Trials.

6. REFERENCES

Aguilar-Cordova, E., Chinen, J., Donehower, L.A., Harper, J.W., Rice, A.P., Butel, J.S. and Belmont, J.W. (1995). Inhibition of HIV-1 by a double transdominant fusion gene. *Gene Ther.*, 2: 181-186.

Amado, R.G., Mitsuyasu, R.T., Symonds, G., Rosenblatt, J.D., Zack, J., Sun, L.Q, Miller, M., Ely, J. and Gerlach, W. (1999). A phase I trial of autologous CD34+ hematopoietic progenitor cells transduced with an anti-HIV ribozyme. *Hum. Gene Ther.,* 10: 2255-2270.

Cairns, M.J., Hopkins, T.M., Witherington, C., Wang, L. and Sun, L.Q. (1999). Target site selection for an RNA-cleaving catalytic DNA. *Nat. Biotechnol.,* 17: 480-486.

Campbell, T.B., McDonald, C.K. and Hagen, M. (1997). The effect of structure in a long target RNA on ribozyme cleavage efficiency. *Nucleic Acids Res.*, 25: 4985-4993.

Caputo, A., Grossi, M.P., Bozzini, R., Rossi, C., Betti, M., Marconi, P.C., Barbanti-Brodano, G. and Balboni, P.G. (1996). Inhibition of HIV-1 replication and reactivation from latency by tat transdominant negative mutants in the cysteine rich region. *Gene Ther.*, 3: 235-245.

Caruso, M. and Klatzmann, D. (1992). Selective killing of CD4+ cells harboring a human immunodeficiency virus-inducible suicide gene prevents viral spread in an infected cell population. *Proc. Natl. Acad. Sci. USA*, 89: 182-186.

Caruso, M., Salomon, B., Zhang, S., Brisson, E., Clavel, F., Lowy, I. and Klatzmann, D. (1995). Expression of a Tat-inducible herpes simplex virus-thymidine kinase gene protects acyclovir-treated CD4 cells from HIV-1 spread by conditional suicide and inhibition of reverse transcription. *Virology,* 206: 495-503.

Cech, T. R. (1988). Conserved sequences and structures of group I introns: building an active site for RNA catalysis--a review. *Gene*, 73: 259-271.

Chakrabarti, B.K., Maitra, R.K., Ma, X.Z. and Kestler, H.W. (1996). A candidate live inactivatable attenuated vaccine for AIDS. *Proc. Natl. Acad. Sci. USA*, **93**: 9810-9815

Chun, T. W., Engel, D., Mizell, S..B, Hallahan, C.W., Fischette, M., Park, S., Davey, R.T.Jr., Dybul, M., Kovacs, J.A., Metcalf, J.A., Mican, J.M., Berrey, M.M., Corey, L., Lane, H.C. and Fauci, A.S. (1999). Effect of interleukin-2 on the pool of latently infected, resting CD4+ T cells in HIV-1-infected patients receiving highly active anti-retroviral therapy. *Nat Med.*, 5: 651-655.

Cooper, D., Penny, R., Symonds, G., Carr, A., Gerlach, W., Sun, L.Q. and Ely, J. (1999). A marker study of therapeutically transduced CD4+ peripheral blood lymphocytes in HIV discordant identical twins. *Hum. Gene Ther.,* 10: 1401-1421.

Crisell, P., Thompson, S. and James, W. (1993). Inhibition of HIV-1 replication by ribozymes that show poor activity in vitro. *Nucleic-Acids Res.*, 21: 5251-5255.

Escaich, S., Kalfoglou, C., Plavec, I., Kaushal, S., Mosca, J.D. and Böhnlein, E. (1995). RevM10-mediated inhibition of HIV-1 replication in chronically infected T cells. *Hum. Gene Ther.*, 6: 625-634.

Furtado, M.R., Callaway, D.S., Phair, J.P., Kunstman, K.J., Stanton, J.L., Macken, C.A., Perelson, A.S. and Wolinsky, S.M. (1999). Persistence of HIV-1 transcription in peripheral-blood mononuclear cells in patients receiving potent antiretroviral therapy. *N. Engl. J. Med.*, 340: 1614-1622.

Gallo, R.C., Salahuddin, S.Z., Popovic, M., Shearer, G.M., Kaplan, M., Haynes, B.F., Palker, T.J., Redfield, R., Oleske, J., Safai, B. *et al.* (1984). Frequent detection and isolation of cytopathic retroviruses (HTLV-III) from patients with AIDS and at risk for AIDS. *Science*, 224: 500-503.

Hamouda, T., McPhee, R., Hsia, S.C., Read, G.S., Holland, T.C. and King, S.R. (1997). Inhibition of human immunodeficiency virus replication by the herpes simplex virus virion host shutoff protein. *J. Virol.*, **71**: 5521-5527.

Hampel, A. and Tritz, R. (1989). RNA catalytic properties of the minimum (-)sTRSV sequence. *Biochemistry*, 28: 4929-4933.

Hampel, A., Tritz, R., Hicks, M. and Cruz, P. (1990). 'Hairpin' catalytic RNA model: evidence for helices and sequence requirement for substrate RNA. *Nucleic Acids Res.*, 18: 299-304.

Hampel, A., Nesbitt, S., Tritz, R. and Altschuler, M. (1993). The hairpin ribozyme. In: *Antisense RNA and Ribozymes*, J. Rossi (Ed.), *Methods*, 5: 37-42.

Haseloff, J. and Gerlach, W.L. (1988). Simple RNA enzymes with new and highly specific endoribonuclease activities. *Nature*, 334: 585-591.

Heidenreich, O. and Eckstein, F. (1992). Hammerhead ribozyme-mediated cleavage of the long terminal repeat RNA of human immunodeficiency virus type 1. *J. Biol. Chem.*, 267: 1904-1909.

Homann, M., Tzortzakaki, S., Rittner, K., Sczakiel, G. and Tabler, M. (1993). Incorporation of the catalytic domain of a hammerhead ribozyme into antisense RNA enhances its inhibitory effect on the replication of HIV-1. *Nucleic Acids Res.*, 21: 2809-2814.

Kestler, H. W. and Chakrabarti, B. K. (1997). A live-virus "suicide" vaccine for human immunodeficiency virus. *Cleve. Clin. J. Med.*, 64: 269-274.

Knop, A.E., Arndt, A.J., Raponi, M., Boyd, M.P., Ely, J.A. and Symonds, G. (1999). Artificial capillary culture: Expansion and retroviral transduction of CD4+ T lymphocytes for clinical application. *Gene Ther.*, 6: 373-384.

Kuwabara, T., Amontov, S.V., Warashina, M., Ohkawa, J. and Taira, K. (1996). Characterization of several kinds of dimer minizyme: simultaneous cleavage at two sites in HIV-1 tat mRNA by dimer minizymes. *Nucleic Acids Res.*, 24: 2302-2310.

Lee, S.W., Gallardo, H.F., Gaspar, O., Smith C. and Gilboa, E. (1995). Inhibition of HIV-1 in CEM cells by a potent TAR decoy. *Gene Ther.*, **2**; 377-384.

Lisziewicz, J., Sun, D., Smythe, J., Lusso, P., Lori, F., Louie, A., Markham, P., Rossi, J., Reitz, M. and Gallo, R.C. (1993). Inhibition of human immunodeficiency virus type 1 replication by regulated expression of a polymeric Tat activation response RNA decoy as a strategy for gene therapy in AIDS. *Proc. Natl. Acad. Sci. USA*, 90: 8000-8004.

Lisziewicz, J., Sun, D., Lisziewicz, A. and Gallo, R.C. (1995). Antitat gene therapy: a candidate for late-stage AIDS patients. *Gene Ther.*, 2: 218-222.

Maciejewski, J.P., Weichold, F.F., Young, N.S., Cara, A., Zella, D., Reitz, M.S., Jr. and Gallo, R.C. (1995). Intracellular expression of antibody fragments directed against HIV reverse transcriptase prevents HIV infection in vitro. *Nat. Med.*, 1: 667-673.

Marcello, A. and Giaretta, I. (1998). Inducible expression of herpes simplex virus thymidine kinase from a bicistronic HIV1 vector. *Res. Virol.,* 149: 419-431.

Montgomery, M. K. and Fire, A. (1998). Double-stranded RNA as a mediator in sequence-specific genetic silencing and co-suppression. *Trends Genet* ., 14: 255-258.

Perrotta, A. T. and Been, M.D. (1990). The self-cleaving domain from the genomic RNA of hepatitis delta virus: sequence requirements and the effects of denaturant. *Nucleic Acids_Res.,* 18: 6821-6827.

Rossi, J.J., Elkins, D., Zaia, J.A. and Sullivan, S. (1992). Ribozymes as anti-HIV-1 therapeutic agents: principles, applications, and problems. *AIDS Res. Hum. Retroviruses*, 8: 183-189.

Saag, M. S. and Kilby, J.M. (1999). HIV-1 and HAART: a time to cure, a time to kill [news; comment]. *Nat. Med.,* 5: 609-611.

Samarsky, D.A., Ferbeyre, G., Bertrand, E., Singer, R.H., Cedergren, R. and Fournier, M.J. (1999). A small nucleolar RNA:ribozyme hybrid cleaves a nucleolar RNA target in vivo with near-perfect efficiency. *Proc. Natl. Acad. Sci. USA*, 96: 6609-6614.

Shaheen, F., Duan, L., Zhu, M., Bagasra, O. and Pomerantz, R.J. (1996). Targeting human immunodeficiency virus type 1 reverse transcriptase by intracellular expression of single-chain variable fragments to inhibit early stages of the viral life cycle. *J. Virol.*, 70: 3392-3400.

Smythe, J.A. and Symonds, G. (1995). Gene therapeutic agents: the use of ribozymes, antisense, and RNA decoys for HIV-1 infection - review. *Inflamm. Res.*, 44: 11-15.

Speck, R.F., Esser, U., Penn, M.L., Eckstein, D.A., Pulliam, L., Chan, S.Y. and Goldsmith, M.A. (1999). A trans-receptor mechanism for infection of CD4-negative cells by human immunodeficiency virus type 1. *Curr. Biol.*, 9: 547-550.

Sun, L. Q. and Symonds, G. (1994). Gene therapy to inhibit HIV replication. *Today's Life Sciences*, 7: 52-58.

Sun, L.Q., Warrilow, D., Wang, L., Witherington, C., Macpherson, J. and Symonds, G. (1994). Ribozyme-mediated suppression of Moloney murine leukemia virus and human immunodeficiency virus type I replication in permissive cell lines. *Proc. Natl. Acad. Sci. USA*, 91: 9715-9719.

Sun, L.Q., Pyati, J., Smythe, J., Wang, L., Macpherson, J., Gerlach, W. and Symonds, G. (1995a). Resistance to human immunodeficiency virus type 1 infection conferred by transduction of human peripheral blood lymphocytes with ribozyme, antisense or polymeric trans-activation response element constructs. *Proc. Natl. Acad. Sci. USA*, 92: 7272-7276.

Sun, L.Q., Wang, L., Gerlach, W.L. and Symonds, G. (1995b). Target sequence-specific inhibition of HIV-1 replication by ribozymes directed to *tat* RNA. *Nucleic Acids Res.*, 23: 2909-2913.

Sun, L.Q., Gerlach, W.L. and Symonds, G. (1998). The design, production and validation of an anti-HIV type1 ribozyme. In*: Therapeutic Application of Ribozymes,* Methods in Molecular Medicine, Vol. 11, K.J. Scanlon (Ed.), Humana Press, Totowa, pp. 51-64.

Wang, L., Witherington, C., King, A., Gerlach, W.L., Carr, A., Penny, R., Cooper, D., Symonds, G. and Sun, L.Q. (1998). Preclinical characterization of an anti-tat ribozyme for therapeutic application. *Hum. Gene Ther.*, 9: 1283-1291.

Weiss, R., Teich, N., Varmus, H. and Coffin, J. (1984). Taxonomy of Retroviruses. In: *RNA Tumor Viruses: Molecular Biology of Tumor Viruses,* R. Weiss, Teich, N., Varmus, H. and Coffin, J. (Eds.), Cold Spring Harbor Laboratory Press, Vol. 1, pp. 25 -208.

Woffendin, C., Ranga, U., Yang, Z.-Y., Xu, L. and Nabel, G.J. (1996). Expression of a protective gene prolongs survival of T-cells in human immunodeficiency virus-infected patients. *Proc. Natl. Acad. Sci. USA*, 93: 2889-2894.

Wong-Staal, F., Poeschla, E.M. and Looney, D.J. (1998). A controlled, Phase 1 clinical trial to evaluate the safety and effects in HIV-1 infected humans of autologous lymphocytes transduced with a ribozyme that cleaves HIV-1 RNA. *Hum. Gene Ther.*, 9: 2407-2425.

Yuan, Y. and Altman, S. (1994). Selection of guide sequences that direct efficient cleavage of mRNA by human ribonuclease P. *Science,* 263: 1269-1273.

Zaia, J.A., Rossi, J.J. *et al.* (1998). One year results after autologous stem cell transplantation using retrovirus transduced peripheral blood progenitor cells in HIV infected subjects. Abstract, American Society Hematology, *Blood.*

Zhang, L., Ramratnam, B., Tenner-Racz, K., He, Y., Vesanen, M., Lewin, S., Talal, A., Racz, P., Perelson, A.S., Korber, B.T., Markowitz, M. and Ho, D.D. (1999). Quantifying residual HIV-1 replication in patients receiving combination antiretroviral therapy. *N. Engl. J. Med.*, 340: 1605-1613.

6. SUMMARY — Human Immunodeficiency Virus (HIV) is a lentivirus, that integrates into the genomes of host cells as DNA copies. The RNA intermediates and RNA transcripts of integrated genes can be targeted by catalytic RNA molecules called ribozymes. The ability of ribozymes to target HIV RNA *in vitro* has been investigated by us and others, and several Phase I Clinical Trials have been initiated to determine the safety of expressing anti-HIV ribozymes in host cells via retroviral gene delivery. Ribozymes potentially form another arm in the treatment of AIDS patients, to assist immune recovery after viral load and has been targeted by contemporary treatments.

Key Words: HIV; Ribozyme; Gene Therapy; Clinical Trial

Progress in Gene Therapy: Basic and Clinical Frontiers, pp. 455-473
R. Bertolotti *et al.* (Eds)

Alteration of pre-mRNA splicing patterns by modified small nuclear RNAs

Linda Gorman, Daniel Schümperli[1] and Ryszard Kole*

Lineberger Comprehensive Cancer Center and Department of Pharmacology, University of North Carolina, Chapel Hill, NC 27599, USA, and [1]Abteilung für Entwicklungsbiologie, Zoologisches Institut der Universität Bern, Baltzerstrasse 4, CH-3012 Bern, Switzerland

Table of Contents

1. INTRODUCTION

A large number of genes undergo alternative splicing, a process which serves as one of the steps in post-transcriptional modulation of gene expression. Moreover, approximately 15% of point mutations that cause genetic disorders result in defective splicing. A considerable proportion of these mutations create new splice sites which compete with the existing ones to produce aberrantly spliced products. It follows that modification of

* Corresponding author. E-mail: kole@med.unc.edu

splicing should change gene expression, resulting in significant biological consequences. A small increase in the concentration of one splice variant, by definition, leads to a concomitant decrease in concentration of its counterpart and a major change in the ratio of the products. In comparison, transcriptional control of independent genes or their downregulation or upregulation in therapeutic protocols does not affect other genes that may constitute additional elements of a regulatory pathway.

We have demonstrated the feasibility of splicing modulation in the context of several models of genetic disorders such as thalassemia (Dominski and Kole, 1993; Gorman *et al.*, 1998; Sierakowska *et al.*, 1996), cystic fibrosis (Friedman *et al.*, 1999), and Duchenne muscular dystrophy (Wilton *et al.*, 1999). This was accomplished by targeting antisense oligonucleotides to splice sites or other sequence elements involved in splicing. Since this approach was recently reviewed (Kole, 1998; Mercatante and Kole, 1999), it will not be covered in this chapter.

Application of antisense oligonucleotides for downregulation of mRNA or splicing modification represents a pharmacological approach to gene regulation. Its advantages include precise control of dosing and route of application and the fact that the treatment may be terminated at will at any time. However, the temporary nature of the treatment is a drawback in the context of genetic disorders which may require lifelong administration of the drugs. An approach to circumvent this problem is to embed the antisense sequences in appropriate vectors that lead to permanent or at least long-term expression of antisense RNA. We have developed a series of constructs that provide intracellular expression of antisense sequences with small nuclear RNAs (snRNAs) as carriers. Our recent progress in these experiments, in the context of thalassemia, constitutes the main topic of this chapter.

2. THALASSEMIA

Thalassemia, a hereditary blood disorder, is one of the most common monogenic disorders in the world. A variety of defects in α- or β-globin genes, leading to α- or β-thalassemia, prevent formation of hemoglobin, the oxygen carrying component in red blood cells. Although, in principle, thalassemia could be treated by replacement of faulty genes, there have been difficulties developing vectors suitable for efficient delivery of large transgenes and for providing sustained expression of the transfected genes in a tissue-specific, properly regulated manner (Byun *et al.*, 1996; Shi *et al.*, 1997). However, regulated expression of α- and β-globin genes is difficult to achieve since they are controlled by large and complex control elements termed locus control regions (LCR). Vectors capable of accommodating sufficiently large fragments of DNA are not yet available, while truncated constructs, in spite of significant progress, do not provide the desired levels and specificity of expression (Ellis *et al.*, 1997; Sadelain *et al.*, 1995; Zhou *et al.*, 1996).

In addition to gene replacement, gene therapy for β-thalassemia may also be accomplished by manipulation of gene structure and expression. Chimeric RNA-DNA oligonucleotides can be used for site-specific removal of defects in the β-globin gene (Cole-Strauss *et al.*, 1996), although the efficiency of this elegant approach is questionable (Cole-Strauss *et al.*, 1999). Alteration of globin gene expression can also be achieved by treatment with hydroxyurea or butyric acid and its derivatives, drugs that induce expression of fetal hemoglobin genes, which can partially compensate for a lack of β-globin. This approach has been successful in clinical trials for sickle cell anemia (Charache *et al.*, 1996; Collins *et al.*, 1995; Perrine *et al.*, 1989; Sher *et al.*, 1995) but, unfortunately, not for thalassemia.

Among the close to two hundred mutations that cause β-thalassemia, those that lead to splicing defects are among the most important clinically. This affords an opportunity to modify splicing and repair the defective β-globin RNA transcripts. A promising feature of this approach is that in patients, the end result would be correctly spliced mRNA properly transcribed from the β-globin gene which remained in its natural chromosomal environment. This precludes the possibility of overexpression or inappropriate expression of β-globin mRNA and protein, an important consideration in the treatment of hemoglobinopathies.

In several forms of thalassemia (IVS2-654, IVS2-705, and IVS2-745), point mutations in intron 2 of the β-globin gene improve the match of the surrounding sequence to that of a consensus donor (5') splice site. In the transcribed pre-mRNA, the presence of the aberrant 5' splice site activates a 3' splice site upstream, resulting in incorrectly spliced β-globin mRNA containing a fragment of the intron (Fig. 1A). This fragment creates a premature stop codon resulting in a truncated β-globin polypeptide. Thus, in individuals homozygous for these mutations, the levels of the β-globin subunit of hemoglobin are drastically reduced (Schwartz and Benz, 1991).

3. MODIFIED U7 snRNA

As mentioned above, antisense oligonucleotides restored the production of correctly spliced β-globin mRNA and normal β-globin protein in cells expressing IVS2 mutations (Dominski and Kole, 1993; Sierakowska *et al.*, 1997 and 1996). Thus, an approach in which the antisense sequences, targeted to the IVS2-654, -705, and -745 splice sites, were expressed intracellularly as antisense RNAs seemed promising. This was accomplished by inserting antisense sequences into genes coding for snRNAs U7 and U1. Several naturally occurring snRNAs, for example U1, U2, U4, U6, and U7 contain antisense sequences of varying lengths, that bind to other RNAs (Reddy and Busch, 1988). This suggested that sequences antisense to the aberrant splice sites in the thalassemic β-globin pre-mRNA, when inserted into snRNAs to replace the native ones, would remain available to the target RNAs and would not destabilize the snRNA molecules. SnRNA genes also offer the advantage of having strong promoters, enabling the expression of multiple copies of antisense containing molecules to be expressed in a single cell. Furthermore, most snRNAs go through a biogenesis pathway that involves assembly of so-called Sm proteins on a U-rich sequence element,

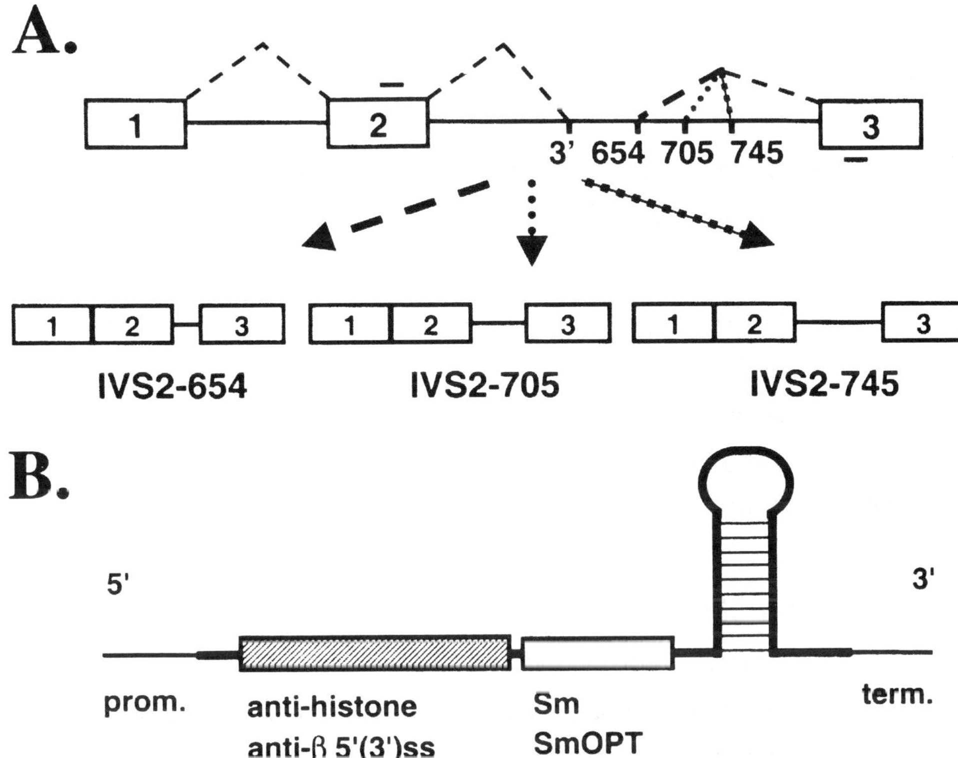

Figure 1. A) Splicing of mutant IVS2 β-globin pre-mRNAs. Boxes: exons; lines: introns; short bars above and below exons: primers used in RT-PCR analysis. The dashed lines represent correct and aberrant splicing pathways. The 3' cryptic splice site (3') and the aberrant 5' splice sites created by the IVS2-654, IVS2-705, and IVS2-745 mutations are indicated, as well as their respective splicing products. **B) Structure of U7 snRNA constructs.** Wild-type U7 snRNA (heavy line) includes a stem-loop structure, the U7-specific Sm sequence (open box) and a sequence antisense to the 3' end of histone pre-mRNA (stippled box). In anti-705 U7 snRNAs, the two sequences are replaced with the SmOPT sequence and with antisense sequences to the aberrant 3' or 5' splice sites in the β-globin gene, respectively. The promoter (prom.) and 3' end forming (term.) regions are indicated.

the Sm binding site, to form a stable ribonucleoprotein particle (snRNP), possibly including additional polypeptides. Most snRNAs acquire a 5' guanosine cap structure that becomes hypermethylated and additional nucleotides are also modified. The protein assembly and the cap modifications are responsible for nuclear import of snRNPs (Fischer *et al.*,

1991; Hamm *et al.*, 1990). For application of snRNPs in modification of splicing, their nuclear localization is an important characteristic since splicing takes place exclusively in the nucleus. In U7 snRNA, the first 18 nucleotides function as an antisense sequence by hybridizing with the downstream element of histone pre-mRNA during its 3' processing (Birchmeier *et al.*, 1984; Birnstiel and Schaufele, 1988; Bond *et al.*, 1991; Spycher *et al.*, 1994; Galli *et al.*, 1983). Since this was exactly the same length as that of the antisense oligonucleotides used for correction of splicing, it seemed likely that replacement of the anti-histone sequence with anti-globin sequence would produce U7 snRNA molecules that bind to β-globin pre-mRNA and correct aberrant splicing in a manner similar to antisense oligonucleotides.

In an effort to increase nuclear accumulation of U7 snRNA, its Sm binding site (AAUUUGUCUAG) was replaced with the consensus Sm binding sequence derived from the major spliceosomal snRNPs (SmOPT, AAUUUUUGGAG). This change had previously been shown to increase the efficiency of assembly and nuclear accumulation of the U7 snRNP (Grimm *et al.*, 1993), and render it functionally inactive in histone pre-mRNA processing (Grimm *et al.*, 1993; Stefanovic *et al.*, 1995). Most likely, the snRNPs assembled with U7SmOPT RNA contain only the abundant, major Sm proteins (Luhrmann *et al.*, 1990), but none of the possibly limiting U7-specific proteins (Smith *et al.*, 1991). This latter characteristic should prevent the U7 anti-β-globin constructs from being cytotoxic since they will not inhibit proper expression of histone genes. Based on these findings, the U7 gene with the SmOPT sequence was used to construct vectors expressing anti-β-globin U7 snRNAs (Fig. 1B).

Initial experiments indicated that the expression of U7 SmOPT snRNA, modified to hybridize to aberrant splice sites in IVS2-705 thalassemic human β-globin pre-mRNA, reduced the aberrant splicing of pre-mRNA and led to increased levels of the correctly spliced mRNA and β-globin protein (Gorman *et al.*, 1998). U7 constructs antisense to either the novel 5' splice site created by the 705 mutation (U7.5) or the cryptic 3' splice site activated in the aberrant splicing pathway (U7.3) were equally effective at restoring correct splicing in a dose dependent manner (Fig. 2). The sequence specificity of the observed antisense effects was confirmed since in cells transfected with the vector expressing anti-histone U7 snRNA (U7SmOPT), there was no correction of IVS2-705 pre-mRNA splicing (Fig. 2).

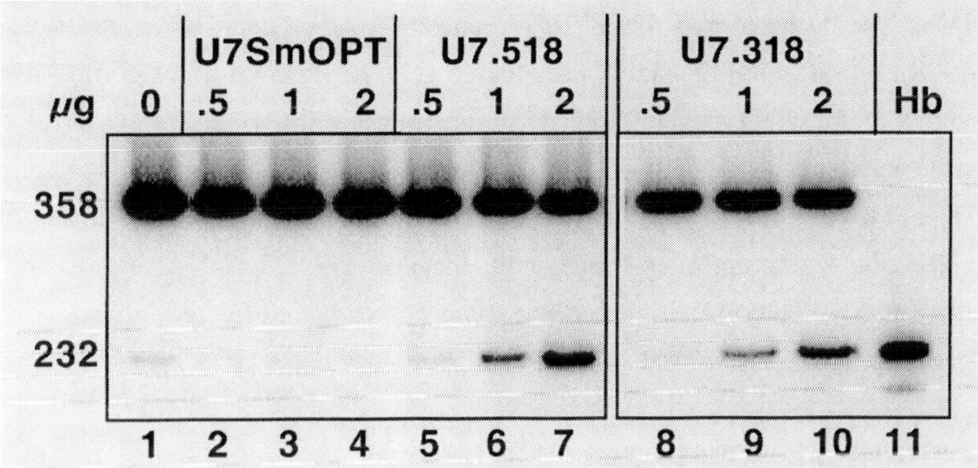

Figure 2. Correction of aberrant splicing by U7 snRNA (Gorman *et al.*, 1998). Total RNA from HeLa IVS2-705 cell line transiently transfected with 0.5, 1, and 2 μg of U7 snRNA constructs (shown at the top) was analyzed by RT PCR. Lane 1: untreated IVS2-705 cells. Lanes 2-4: cells transfected with U7SmOPT. Lanes 5-7, and 8-10: cells transfected with U7.5 and U7.3 constructs, respectively. Lane 11: RNA from human blood. The sizes (in nucleotides) of PCR bands representing aberrantly and correctly spliced mRNAs are indicated on the left.

Work then focused on further modifications of U7.3 to improve its ability to block splicing at the 3' splice site. The same cryptic 3' splice site is utilized by the splicing machinery in the three IVS2 thalassemic pre-mRNAs (Schwartz and Benz, 1991). Thus, modified U7 constructs directed to the 3' cryptic splice site should be useful for correction of splicing in cells expressing any of the three mutants. The antisense sequences were extended from 18 to 168 nucleotides with the assumption that a longer sequence would have higher affinity for the target site. The construct most efficient at correcting splicing contained 24 nucleotides of antisense sequence (U7.324), with levels of correction reaching 65% in transient transfections. Interestingly, there was no improvement in splicing correction by constructs with antisense sequences longer than 24 nucleotides.

The main advantage of the intracellularly expressed U7 vectors lies in their potential for permanent expression and concomitant correction of splicing. To test this possibility, stable HeLa cell lines that co-express the IVS2-705 β-globin and the U7.324 genes were generated (Gorman *et al.*, 1998). Importantly, the stably transfected cells had growth rates comparable to that of the wild type HeLa cells, suggesting that the modified U7 snRNA was not toxic to the cells. The cell lines permanently expressed correctly spliced β-globin mRNA, at levels that reached 55% (Fig. 3) and the levels of correction were commensurate with that of U7.324 RNA expression. Importantly, the correction of splicing was accompanied by production of β-globin protein (Fig. 3).

4. U7 snRNAs WITH DUAL SPECIFICITY

Since the cryptic 3' splice site is activated by the three thalassemic mutations, IVS2-654, -705, and -745, the U7.324 construct should, in principle, restore correct β-globin splicing in the context of all three IVS2 mutations. However, it was found by experiments with antisense oligonucleotides that the accessibility of splice sites in the three mutants differ, decreasing significantly from the IVS2-745 to -654 mutations (Sierakowska *et al.*, 1999). In agreement with these findings, the level of correctly spliced β-globin mRNA was close to 90% in HeLa IVS2-745 cells, but no more than 10% in HeLa IVS2-654 cells transfected with the U7.324 construct. The expression of U7.324 RNA was similar in all three cell lines, as tested by RNAse protection assays (Suter *et al.*, 1999).

Since an important prerequisite for modulating the splicing patterns of different genes is that the method should work efficiently in conjunction with splice sites of differing strength, we concentrated on optimising the

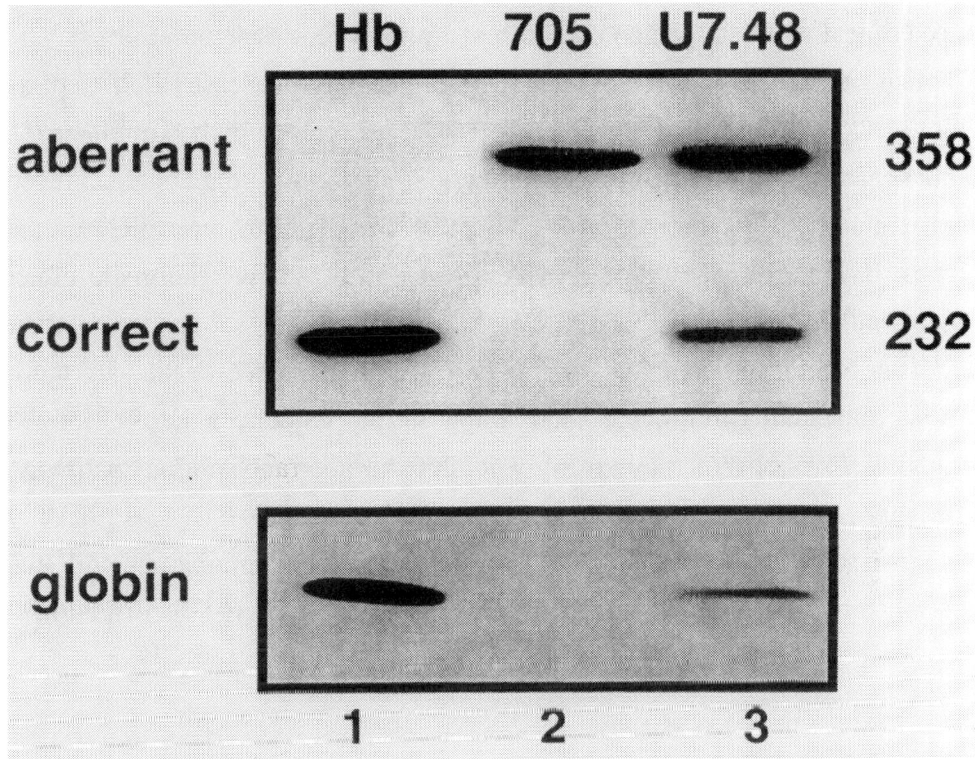

Figure 3. Correlation of β-globin protein and mRNA expression in stable cell line 705U7.324.48 (Gorman *et al.*, 1998). **Top)** Lane 1: RNA from human blood. Lane 2: RNA from IVS2-705 cell line. Lane 3: RNA from cell line 705U7.324.48 stably expressing U7.324 snRNA. **Bottom)** Western Blot of protein from cell line expressing full length β-globin protein (Lane 1). Lane 2: protein from IVS2-705 cell line. Lane 3: protein from cell line 705U7.324.48.

correction of splicing for the most resistant mutation, IVS2-654. The constructs targeted to the 5' splice site created by the IVS2-654 mutation (U7.5'654) or another U7 derivative, with a 21-nucleotide complementarity to two potential branch point adenosines upstream of the cryptic 3' ss (U7.BP) showed only borderline activity in Hela 654 cells. Interestingly, cotransfection of U7.318 and U7.5'654 led to a considerably higher splicing correction than that obtained with each construct alone. This result prompted us to test constructs in which two different antisense sequences were inserted into a single modified U7 SmOPT derivative (hereafter designated as a double target construct). Several combinations of antisense sequences directed against the 5' splice site created by the IVS2-654 mutation, cryptic 3' splice site, or the putative branch point region were at least as efficient as U7.324. More importantly, construct U7.BP+5'654, which combined the sequences antisense to the branch point and to the 654 5' splice site achieved approximately 30% correction of splicing of β-globin pre-mRNA, the highest of any constructs tested (Suter *et al.*, 1999). This synergistic effect reflects either cooperative binding of the two antisense sequences to their targets or the simultaneous binding of two distant sequence to the same pre-mRNA molecule forcing the intervening intron sequences into a looped structure (Fig. 4) that may strongly inhibit the aberrant splicing pathway. The latter possibility is supported by the fact that U7.BP+5'654 is not only the most active construct for IVS2-654, but also in combination with the other intron 2 mutations, IVS2-705 (85%) and IVS2-745 (95% correction).

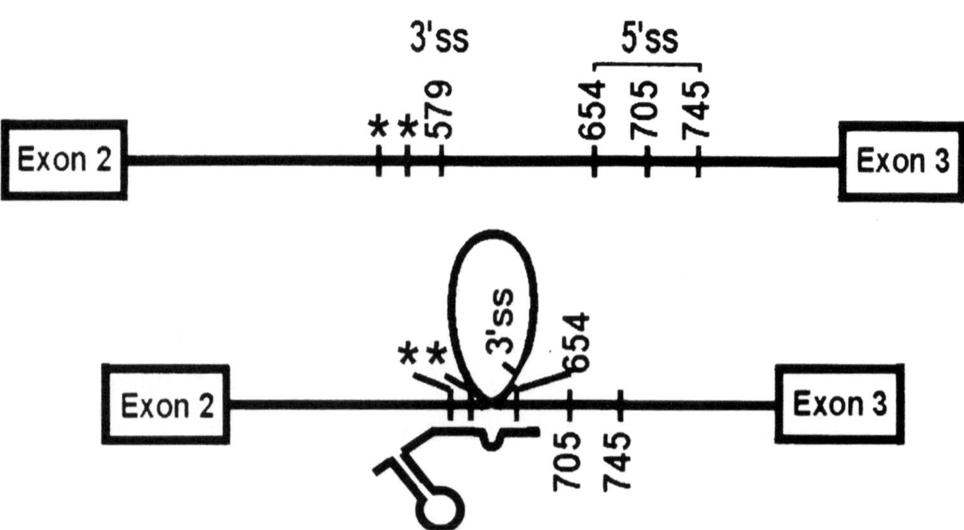

Figure 4. Model for the action of double target U7 snRNAs as exemplified by U7.BP+5'654. Intron 2 of the β-globin gene is proposed to be forced into a looped secondary structure that disfavors inclusion of the aberrant exon. Two putative branch points upstream of the cryptic 3' splice site are indicated by asterisks.

There, the sequence in β-globin pre-mRNA at position 654 is wild-type (leading to a C/A mismatch in the possible 20 bp hybrid) and not at the splice site but within the pseudo exon flanked by the activated aberrant 3' and 5' splice sites. Accordingly, the single target construct U7.5'654 is inactive against the IVS2-705 or IVS2-745 mutations. An additional double target construct combining sequences antisense to a region 60-80 nucleotides upstream of the cryptic 3' splice site and to the IVS2-654 5' splice site corrected β-globin mRNA splicing in the context of IVS2-705 and IVS2-745 mutants, *i.e.* when none of the two target sequences had any obvious involvement in β-globin pre-mRNA splicing. Taken together, these data strongly suggest that forcing the aberrant exon into a looped configuration as shown in Fig. 4 strongly disfavours the inclusion of this exon.

With this new double target strategy, the efficiency of splicing correction, even in the context of the highly resistant IVS2-654 mutation was considerably improved, both in transiently transfected cells and in permanently transformed HeLa cell lines (~50% correction). The *in vivo* applicability of this approach can now be tested in a transgenic mouse line that expresses the human IVS2-654 thalassemic β-globin gene in erythrocyte precursors (Lewis *et al.*, 1998). The double target strategy may also be effective at inducing skipping of natural exons in other genes that would be interesting for gene therapy or biotechnological applications.

5. U1 snRNA

Other snRNAs may also provide convenient delivery agents for antisense therapeutics. Both modified U1 and U6 snRNA have been used as antisense carriers to downregulate gene expression (Good *et al.*, 1997; Liu *et al.*, 1997; Montgomery and Dietz, 1997; Noonberg *et al.*, 1994; Yuo and Weiner, 1989). U1 snRNA was viewed as an attractive candidate for use against a pre-mRNA target since it is known to bind to 5' splice sites via a base pairing mechanism. In addition, experiments had shown that a modified, transiently transfected U1 snRNA could be efficiently transcribed, accounting for 25 to 30% of the total U1 RNA (Yuo and Weiner, 1989).

U1 snRNA contains the consensus Sm sequence (AAUUUGUGG), resulting in the accumulation of more U1 snRNA molecules in the nucleus than U7 snRNA. In fact, there are 1×10^6 copies of U1 per cell, while U7 snRNA is expressed at only 5×10^3 copies per cell (Baserga and Steitz, 1993). This difference in RNA levels is mostly due to a higher gene copy number (thirty for U1, one for U7) (Dahlberg and Lund, 1988; Gruber *et al.*, 1991), the Sm binding site that controls U1 snRNP assembly and nuclear accumulation, and to differences in the activity of the U1 promoter. As a result, production of U1 snRNP from each copy of the U1 gene is approximately 6 times more efficient than that of U7 snRNP. Thus, it appeared that U1 might provide a more effective antisense carrier than U7 snRNA.

Previous work had shown that modified U1 snRNA could bind to mutant pre-mRNAs to promote changes in splicing (Cohen *et al.*, 1993 and 1994). Therefore we replaced the U1 snRNA 9 nucleotide 5' splice site recognition sequence with sequences of different length antisense to the aberrant 5' or 3' splice sites in IVS2-705 pre-mRNA (Fig. 5). Both modified U1 snRNAs were expected to bind to the target sequence, but it was possible that the U1 targeted to the aberrant 5' splice site would retain its natural function and promote aberrant splicing instead of inhibiting it. In addition, the U1 snRNA targeted to the aberrant 5' splice site must compete with the large amounts of endogenous U1 snRNP that bind to that splice site during aberrant splicing.

Transient transfections with modified U1 snRNAs designed to bind to the cryptic 3' splice site in IVS2-705 thalassemic β-globin pre-mRNA resulted in varying levels of correctly spliced mRNA. As in the modified U7 constructs, elongation of the antisense sequences up to 24 nucleotides led to increased levels of correction. The U1.324 construct led to corrections comparable to those reached with its U7 counterpart.

Interestingly, modified U1 snRNA designed to bind to the IVS2-705 5' splice site (U1.524) did not correct aberrant β-globin splicing, even though it was produced at similar levels to U1.324 in transient transfections. Since experiments to prove that the modified 5' U1 snRNP was still functional in splicing were negative, we conclude that U1.524 was unable to compete with endogenous U1 for binding to the 5' splice site. The correction of splicing by U1.324 is consistent with this possibility, since it is directed to the 3' cryptic splice and does not have to compete with wild type U1 for binding.

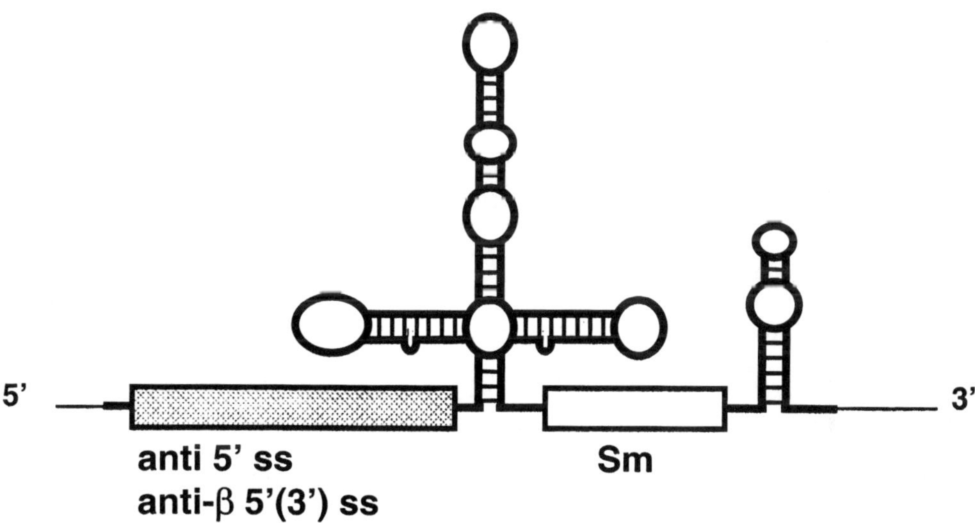

Figure 5. Structure of U1 snRNA constructs. Wild-type U1snRNA (heavy line) includes two stem-loops, the Sm sequence (open box) and a sequence antisense to the 5' splice sites of pre-mRNA (hatched box). In anti-705 U1 snRNAs, this sequence is replaced with antisense sequences to the aberrant 3' or 5' splice sites in the β-globin gene. The U1 promoter and 3' end forming regions are included (thin lines).

6. U1.U7 HYBRID snRNA

As another option for improvement of snRNAs as antisense vector, we have utilized the U1 promoter and terminator sequences to control production of a modified U7 gene (Fig. 6). The U7.324 snRNA gene was selected since among the single target U7 snRNAs it was the most effective in correcting splicing. As anticipated, the stronger promoter in the U1.U7.324 construct led to increased levels of U7 expression and the more effective correction of aberrant splicing of IVS2-705 β-globin pre-mRNA as compared to U7.324 and U1.324 snRNAs driven by their respective endogenous promoters. One can expect that the combination of this approach with a double target constructs would result in even more efficient vectors for antisense delivery.

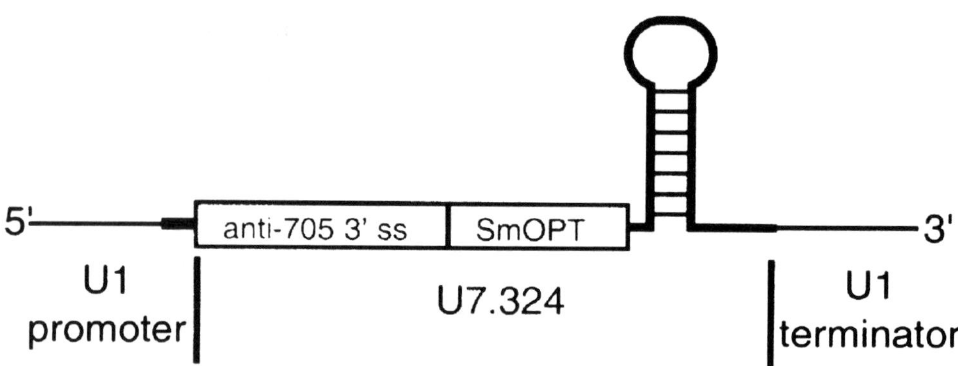

Figure 6. Structure of U1.U7.324 snRNA construct. U7.324 snRNA (black line) includes a stem-loop structure, the SmOPT sequence and 24 nucleotide long sequence antisense to the 3' cryptic splice site. The U1 promoter and 3' end forming regions are shown by thin gray lines. The first two nucleotides of U1 snRNA (thick gray line) were also included in the construct.

7. CONCLUSIONS

Modified U7 and U1 snRNAs were successful at correcting splicing for several reasons. For repair of a splicing mutation at the RNA level, high levels of expression of antisense RNA in the nucleus, where both expression of target pre-mRNAs and splicing occur, is desirable. Using U7 or U1 snRNAs as an antisense carrier guarantees nuclear localization, since these snRNAs are transported from the cytoplasm to the nucleus via Sm protein interactions. Due to their small size, secondary structure and tight interactions with common Sm and other snRNP-specific proteins (Luhrmann *et al.*, 1990), the snRNAs, or rather their snRNP complexes, are very stable. In clinical applications the above properties would reduce the frequency of patient treatment.

Restoration of β-globin and presumably hemoglobin to over 50% in thalassemic patients would be therapeutically significant since transfusion therapy raises the hemoglobin to even lower levels yet improves the clinical status of the affected individuals. Heterozygous individuals with 50% of the normal level of hemoglobin are essentially symptom free (Schwartz and Benz, 1991).

The ability to restore continuous production of a correct gene product in stable cell lines is highly encouraging, suggesting that gene therapy based on the antisense concept may be feasible. The patients' bone marrow could be transfected *ex vivo* with modified snRNAs vectors and reimplanted. The increased production of β-globin would reduce the imbalance between the α- and β-subunits of hemoglobin, improving the survival of erythroblasts and promoting the maturation of erythrocytes. The treated cells should persist in the blood stream for up to 120 days, the life span of erythrocytes (Eadie *et al.*, 1955). Therefore, even if the expression of the snRNA were short lived, due to lack of transfection of stem cells or promoter shut-off, both of which are common problems in the expression of transgenes (Palmer *et al.*, 1991; Page and Brownlee, 1997), the results may still be relatively long lasting.

The possibility of transgene overexpression is a serious concern in the treatment of β-thalassemia, as it could lead to a new imbalance between α- and β-globin subunits and, conceivably, to symptoms of α-thalassemia. Utilizing modified snRNAs to correct splicing offers an advantage, since the β-globin subunits may at best reach the wild type levels. Furthermore, the snRNAs should only affect cells that express β-globin pre-mRNA, *i.e.* nucleated erythroblasts.

The fact that we could accomplish splicing correction by different snRNA constructs and in three thalassemic mutations suggests that this approach may be applicable to other targets, be it splicing mutations in other genetic disorders or modification of normal splicing patterns of constitutively and alternatively spliced pre-mRNAs. Apart from the potential clinical applications, modified snRNAs may be used to permanently alter splicing patterns of specific pre-mRNAs in studies on the control of gene expression.

8. REFERENCES

Baserga, S.J. and Steitz, J.A. (1993). The Diverse World of Small Ribonucleoproteins. In: *The RNA World*, Gesteland, R.F. and Atkins, J.F. (eds.), Cold Spring Harbor Press, Plainview, NY, pp. 359-381.

Birchmeier, C., Schumperli, D., Sconzo, G. and Birnstiel, M.L. (1984). 3' editing of mRNAs: sequence requirements and involvement of a 60-nucleotide RNA in maturation of histone mRNA precursors. *Proc. Natl. Acad. Sci. USA*, 81: 1057-1061.

Birnstiel, M.L. and Schaufele, F. (1988). Structure and Function of Minor snRNPs. In: *Structure and Function of Major and Minor Small Nuclear Ribonucleoprotein Particles*, Birnstiel, M.L. (ed.), Springer, Berlin, pp. 155-182.

Bond, U.M., Yario, T.A. and Steitz, J.A. (1991). Multiple processing-defective mutations in a mammalian histone pre-mRNA are suppressed by compensatory changes in U7 RNA both in vivo and in vitro. *Genes Dev.*, 5: 1709-1722.

Byun, J., Kim, S.H., Kim, J.M., Yu, S.S., Robbins, P.D., Yim, J. and Kim, S. (1996). Analysis of the relative level of gene expression from different retroviral vectors used for gene therapy. *Gene Ther.,* 3, 780-788.

Charache, S., Barton, F.B., Moore, R.D., Terrin, M.L., Steinberg, M.H., Dover, G.J., Ballas, S.K., McMahon, R.P., Castro, O. and Orringer, E.P. (1996). Hydroxyurea and sickle cell anemia. Clinical utility of a myelosuppressive "switching" agent. The Multicenter Study of Hydroxyurea in Sickle Cell Anemia. *Medicine*, 75: 300-326.

Cohen, J.B., Broz, S.D. and Levinson, A.D. (1993). U1 small nuclear RNAs with altered specificity can be stably expressed in mammalian cells and promote permanent changes in pre-mRNA splicing. *Mol. Cell. Biol.*, 13: 2666-7266.

Cohen, J.B., Snow, J.E., Spencer, S.D. and Levinson, A.D. (1994). Suppression of mammalian 5' splice-site defects by U1 small nuclear RNAs from a distance. *Proc. Natl. Acad. Sci. USA,* 91: 10470-10474.

Cole-Strauss, A., Yoon, K., Xiang, Y., Byrne, B.C., Rice, M.C., Gryn, J., Holloman, W.K. and Kmiec, E.B. (1996). Correction of the mutation responsible for sickle cell anemia by an RNA-DNA oligonucleotide. *Science*, 273: 1386-1389.

Cole-Strauss, A., Gamper, H., Holloman, W.K., Munoz, M., Cheng, N. and Kmiec, E.B. (1999). Targeted gene repair directed by the chimeric RNA/DNA oligonucleotide in a mammalian cell-free extract. *Nucleic Acids Res.,* 27: 1323-30.

Collins, A.F., Pearson, H.A., Giardina, P., McDonagh, K.T., Brusilow, S.W. and Dover, G.J. (1995). Oral sodium phenylbutyrate therapy in homozygous beta thalassemia: a clinical trial. *Blood*, 85: 43-49.

Dahlberg, J.E. and Lund, E. (1988). *Structure and Function of Major and Minor Small Nuclear Ribonucleoprotein Particles*. Springer, Berlin.

Dominski, Z. and Kole, R. (1993). Restoration of correct splicing in thalassemic pre-mRNA by antisense oligonucleotides. *Proc. Natl. Acad. Sci. USA*, 90: 8673-8677.

Eadie, G.S., Brown, I.W. and Curtis, W.G. (1955). *J. Clin. Invest.*, 34: 629-637.

Ellis, J., Pasceri, P., Tan-Un, K.C., Wu, X., Harper, A., Fraser, P. and Grosveld, F. (1997). Evaluation of beta-globin gene therapy constructs in single copy transgenic mice. *Nucleic Acids Res.*, 25: 1296-1302.

Fischer, U., Darzynkiewicz, E., Tahara, S.M., Dathan, N.A., Luhrmann, R. and Mattaj, I.W. (1991). Diversity in the signals required for nuclear accumulation of U snRNPs and variety in the pathways of nuclear transport. *J. Cell Biol.*, 113: 705-714.

Friedman, K.J., Kole, J., Cohn, J.A., Knowles, M.R., Silverman, L.M., and Kole, R. (1999). Correction of aberrant splicing of CFTR gene by antisense oligonucleotides. *J. Biol. Chem.*, 274: 36193-36199.

Galli, G., Hofstetter, H., Stunnenberg, H.G. and Birnstiel, M.L. (1983). Biochemical complementation with RNA in the Xenopus oocyte: a small RNA is required for the generation of 3' histone mRNA termini. *Cell*, 34: 823-828.

Good, P.D., Krikos, A.J., Li, S.X., Bertrand, E., Lee, N.S., Giver, L., Ellington, A., Zaia, J.A., Rossi, J.J. and Engelke, D.R. (1997). Expression of small, therapeutic RNAs in human cell nuclei. *GeneTher.*, 4: 45-54.

Gorman, L., Suter, D., Emerick, V., Schumperli, D. and Kole, R. (1998). Stable alteration of pre-mRNA splicing patterns by modified U7 small nuclear RNAs. *Proc. Natl. Acad. Sci. USA*, 95: 4929-4934.

Grimm, C., Stefanovic, B. and Schumperli, D. (1993). The low abundance of U7 snRNA is partly determined by its Sm binding site. *EMBO J.*, 12: 1229-1238.

Gruber, A., Soldati, D., Burri, M. and Schumperli, D. (1991). Isolation of an active gene and of two pseudogenes for mouse U7 small nuclear RNA. *Biochim. Biophys. Acta,* 1088: 151-154.

Hamm, J., Darzynkiewicz, E., Tahara, S.M. and Mattaj, I.W. (1990). The trimethyl-guanosine cap structure of U1 snRNA is a component of a bipartite nuclear targeting signal. *Cell*, 62: 569-577.

Kole, R. (1998). Modification of pre-mRNA splicing by antisense oligonucleotides. In: *Applied Antisense Oligonucleotide Technology*, Stein, C.A. and Krieg, A.M. (eds.), Willey-Liss, Inc., New York, pp. 451-469.

Lewis, J., Yang, B., Kim, R., Sierakowska, H., Kole, R., Smithies, O. and Maeda, N. (1998). A common human beta globin splicing mutation modeled in mice. *Blood*, 91: 2152-2156.

Liu, D., Donegan, J., Nuovo, G., Mitra, D. and Laurence, J. (1997). Stable human immunodeficiency virus type 1 (HIV-1) resistance in transformed CD4+ monocytic cells treated with multitargeting HIV-1 antisense sequences incorporated into U1 snRNA. *J. Virol.*, 71: 4079-4085.

Luhrmann, R., Kastner, B. and Bach, M. (1990). Structure of spliceosomal snRNPs and their role in pre-mRNA splicing. *Biochim. Biophys. Acta*, 1087: 265-292.

Mercatante, D. and Kole, R. (1999). Modification of Alternative Splicing Pathways as a Potential Approach to Chemotherapy. *Pharmacology & Therapeutics*, In press.

Montgomery, R.A. and Dietz, H.C. (1997). Inhibition of fibrillin 1 expression using U1 snRNA as a vehicle for the presentation of antisense targeting sequence. *Hum. Mol. Genet.*, 6: 519-525.

Noonberg, S.B., Scott, G.K., Garovoy, M.R., Benz, C.C. and Hunt, C.A. (1994). In vivo generation of highly abundant sequence-specific oligonucleotides for antisense and triplex gene regulation. *Nucleic Acids Res.*, 22: 2830-2836.

Page, S.M. and Brownlee, G.G. (1997). An ex vivo keratinocyte model for gene therapy of hemophilia B. *J. Invest. Dermatol.*, 109: 139-145.

Palmer, T.D., Rosman, G.J., Osborne, W.R. and Miller, A.D. (1991). Genetically modified skin fibroblasts persist long after transplantation but gradually inactivate introduced genes. *Proc. Natl. Acad. Sci. USA*, 88: 1330-1334.

Perrine, S.P., Miller, B.A., Faller, D.V., Cohen, R.A., Vichinsky, E.P., Hurst, D., Lubin, B.H. and Papayannopoulou, T. (1989). Sodium butyrate enhances fetal globin gene expression in erythroid progenitors of patients with Hb SS and beta thalassemia. *Blood*, 74: 454-459.

Reddy, R. and Busch, H. (1988). Small Nuclear RNAs: RNA Sequences, Structure, and Modifications. In: *Structure and Function of Major and Minor Small Nuclear Ribonucleoprotein Particles*, Birnstiel, M.L. (ed.), Springer, Berlin, pp. 1-37.

Sadelain, M., Wang, C.H., Antoniou, M., Grosveld, F. and Mulligan, R.C. (1995). Generation of a high-titer retroviral vector capable of expressing high levels of the human beta-globin gene. *Proc. Natl. Acad. Sci. USA*, 92: 6728-6732.

Schwartz, E. and Benz, E.J. (1991). *Hematology: Basic Principles and Practices*. Churchill Livingstone, New York.

Sher, G.D., Ginder, G.D., Little, J., Yang, S., Dover, G.J. and Olivieri, N.F. (1995). Extended therapy with intravenous arginine butyrate in patients with beta-hemoglobinopathies [see comments]. *N. Engl. J. Med.*, 332: 1606-1610.

Shi, Q., Wang, Y. and Worton, R. (1997). Modulation of the specificity and activity of a cellular promoter in an adenoviral vector. *Human Gene Ther.*, 8: 403-410.

Sierakowska, H., Sambade, M.J., Agrawal, S. and Kole, R. (1996). Repair of thalassemic human beta-globin mRNA in mammalian cells by antisense oligonucleotides. *Proc. Natl. Acad. Sci. USA*, 93: 12840-12844.

Sierakowska, H., Montague, M., Agrawal, S. and Kole, R. (1997). Restoration of beta globin gene expression in mammalian cells by antisense oligonucleotides that modify the aberrant splicing patterns of thalassemic pre-mRNAs. *Nucleotides and Nucleosides*, 16: 1173-1182.

Sierakowska, H., Sambade, M.J., Schumperli, D. and Kole, R. (1999). Sensitivity of splice sites to antisense oligonucleotides in vivo. *RNA*, 5: 369-377.

Smith, H.O., Tabiti, K., Schaffner, G., Soldati, D., Albrecht, U. and Birnstiel, M.L. (1991). Two-step affinity purification of U7 small nuclear ribonucleoprotein particles using complementary biotinylated 2'-O-methyl oligoribonucleotides. *Proc. Natl. Acad. Sci. USA*, 88: 9784-9788.

Spycher, C., Streit, A., Stefanovic, B., Albrecht, D., Koning, T.H. and Schumperli, D. (1994). 3' end processing of mouse histone pre-mRNA: evidence for additional base-pairing between U7 snRNA and pre-mRNA. *Nucleic Acids Res.*, 22: 4023-4030.

Stefanovic, B., Hackl, W., Luhrmann, R. and Schumperli, D. (1995). Assembly, nuclear import and function of U7 snRNPs studied by microinjection of synthetic U7 RNA into Xenopus oocytes. *Nucleic Acids Res.*, 23: 3141-3151.

Suter, D., Tomasini, R., Reber, U., Gorman, L., Kole, R. and Schümperli, D. (1999). Double target antisense U7 snRNAs promote efficient skipping of an aberrant exon in three human ß-thalassemic mutations. *Hum. Mol. Genet.*, 8: 2415-2423.

Wilton, S.D., Lloyd, F., Carville, K., Fletcher, S., Honeyman, K., Agrawal, S. and Kole, R. (1999). Specific removal of the nonsense mutation from the mdx dystrophin mRNA using antisense oligonucleotides. *Neuromuscular Disorders*, 9: 330-338.

Yuo, C.Y. and Weiner, A.M. (1989). A U1 small nuclear ribonucleoprotein particle with altered specificity induces alternative splicing of an adenovirus E1A mRNA precursor. *Mol. Cell. Biol.*, 9: 3429-3437.

Zhou, S.Z., Li, Q., Stamatoyannopoulos, G. and Srivastava, A. (1996). Adeno-associated virus 2-mediated transduction and erythroid cell-specific expression of a human beta-globin gene. *Gene Ther.*, 3: 223-229.

9. SUMMARY — This chapter focuses on the alteration of gene expression through the use of modified U7 and U1 small nuclear RNAs (snRNAs). They were used primarily for restoration of correct splicing in thalassemic mutants of the human β-globin gene. Expression of modified U7 and U1 snRNAs blocked the utilization of aberrant splice sites present in the pre-mRNA of several thalassemic genes. Without access to the aberrant splice sites, the splicing machinery is forced to use the normal splice sites and correctly spliced β-globin mRNA is produced. The stable expression of modified snRNAs in mutant β-globin cell lines resulted in permanent correction of splicing. These results indicate the possibility of using this technology to permanently alter gene expression.

Key Words: snRNA; gene therapy; RNA splicing; β-globin; thalassemia

Progress in Gene Therapy: Basic and Clinical Frontiers, 475-512
R. Bertolotti *et al.* (Eds)

Single-base conversion of mammalian genes by an RNA-DNA oligonucleotide

Kyonggeun Yoon*

Department of Dermatology and Cutaneous Biology, and Department of Biochemistry and Molecular Pharmacology, Jefferson Institute of Molecular Medicine, Jefferson Medical College, Philadelphia, PA 19107, USA

Table of Contents

* Corresponding author. E-mail: kyonggeun.yoon@mail.tju.edu

1. INTRODUCTION

Gene targeting is currently the only procedure that can produce predefined alterations in the genome of eukaryotic cells. Such strategy is advantageous for gene therapy because transferred gene can be maintained, expressed, and regulated as the normal endogenous gene. The ability to find and pair homologous DNA molecules enables accurate gene targeting and is the central phenomenon underlying genetic recombination. The overall process can be thought of as a series of steps in a reaction pathway whereby DNA molecules are brought into homologous register, the four-stranded Holliday structure intermediate is formed, heteroduplex DNA is extended, and DNA strands are exchanged. Integration of transfected DNA upon homologous recombination with cognate chromosomal DNA has been reported to be infrequent in mammalian cells, ranging from one out of 10^5 to 10^7 events (Capecchi, 1989; Mansour *et al.*, 1988; Thomas and Capecchi, 1987; Yanez and Porter, 1998), and it occurs against a background of nonhomologous events that are more common. Homologous recombination between double-stranded DNA and chromosomes in higher eukaryotes has been exploited in numerous experimental systems where the aim is to inactivate or to replace a gene of interest (Capecchi, 1989; Mansour *et al.*, 1988; Thomas and Capecchi, 1987; Zheng *et al.*, 1991; Pennington and Wilson, 1991; Merrihew *et al.*, 1996; Wilson *et al.*, 1994). Several ingenious procedures have been devised to select or screen for the rare successful targeting products, but the low absolute frequency of specific recombination remains a serious limitation. Therefore, it will be important to develop strategies that increase the frequency of homologous recombination.

RNA has been shown to play an important role in DNA recombination and repair. Transcription stimulated homologous pairing by RecA (Kotani and Kmiec 1994a and 1994b). A rapid hybridization of RNA to single stranded DNA was promoted by RecA (Kirkpatrick and Radding, 1992). The RNA:DNA hybrids were more active in strand-exchange reaction than DNA duplexes (Kmiec *et al.*, 1994; Kotani *et al.*, 1996). Moreover, transcription of the switch region was required for recombination involved in class switching of immunoglobulin heavy chain genes (Daniels and Lieber, 1995a). The RNA transcript was suggested to induce a change in the DNA structure to stabilize a recombination intermediate (Daniels and Lieber, 1995b). Furthermore, DNA repair and transcription has been shown to be coupled by a preferential repair of the transcribed DNA strand (Hanawalt, 1994). A factor essential for transcription initiation, TFIIH, is also required for nucleotide excision repair throughout the genome and in expressed genes (Bootsma and Hoeijmakers, 1993). These observations led to a strategy for gene targeting in which vector-design would exploit the natural recombinogenicity of RNA-DNA hybrids and would feature hairpin capped ends avoiding destabilization or destruction by cellular helicases or exonucleases.

The feasibility of gene correction by the RNA-DNA oligonucleotide (RDO) was demonstrated first by correction of a point mutation in the plasmid encoding the alkaline phosphatase cDNA in CHO cells (Yoon *et al.*, 1996). Gene correction restored an enzymatic activity, thus enabling visualization of cells that have gained the alkaline phosphatase positive phenotype, as red cells (Yoon *et al.*, 1988; Hobbs and Yoon, 1994). Moreover, gene correction was confirmed by both functional enzyme activity and DNA sequence, without PCR analysis. This study provided the basis for subsequent applications of RDO for conversion of chromosomal genes. Since then, a site-specific chromosomal correction or mutation has been reported in several genes (Cole-Strauss *et al.*, 1996; Kren *et al.*, 1997; Xiang *et al.*, 1997; Kren *et al.*, 1998; Santana *et al.*, 1998). Recently, an RDO was shown to induce a sequence-specific mutation in the factor IX gene in cell culture and *in vivo* by intravenous injection (Kren *et al.*, 1998). These results demonstrated the generality for targeted sequence *in vitro* and *in vivo* by a double-stranded RDO. However, previous studies detected the gene conversion event by PCR amplification of genomic DNA without direct confirmation at the protein level (Cole-Strauss *et al.*, 1996; Kren *et al.*, 1997; Xiang *et al.*, 1997; Kren *et al.*, 1998 ; Santana *et al.*, 1998). Concerns have been raised about a potential artifact during PCR amplification, or by cross-contamination between cells, especially because no clonal analysis has been performed (Thomas and Capecchi, 1997; Stasiak *et al.,* 1997; Strauss, 1998).

Recently, we demonstrated for the first time a permanent and stable gene correction by an RDO at the level of genomic sequence, protein, enzymatic activity and phenotypic change by clonal analysis (Alexeev and Yoon, 1998). The corrected genotype was stable for several months, indicating a permanent and inheritable change has occurred in the genomic sequence. This study also characterized an allelic variation in gene conversion by RDO for the first time. All analyzed clones exhibited a correction of the point mutation in a single allele of the tyrosinase gene. The correction of both alleles is expected to be a rare event since it requires 2 separate targeting events by RDO. However, the frequency of gene correction appeared to vary among individual experiments (Alexeev and Yoon, 1998). Markedly different frequency of gene conversion was observed among several epithelial cells with varying degrees of transformation (Santana *et al.,* 1998). Cellular activities such as recombination and repair may vary according to the cell type, cycle and metabolism. It will be important to understand the mechanism involved in gene conversion by RDO to improve its efficiency and its general applicability.

2. STRUCTURE OF THE RDO

A hypothetical intermediate between RDO and a genomic target is diagrammed in Figure 1. Ten 2'-O-methyl RNA residues flank either side of a 5 residue DNA stretch, which contains the base change desired for the mutation. The sequence is complementary to the targeted DNA over a stretch of 25 residues that span the site of the mutation with the exception of a single base generating a base mismatch. The endogenous DNA repair systems would then recognize the mismatched base pair and change the base in either the RDO or the target sequence. Thus, gene conversion by RDO was suggested to involve both homologous strand pairing and DNA repair activities. In order to protect the 5'- and 3'-end of the oligonucleotide from exonucleolytic cleavage, an RDO is folded into a double-hairpin structure containing 4 T residues in each loop and a 5-base-pair GC clamp.

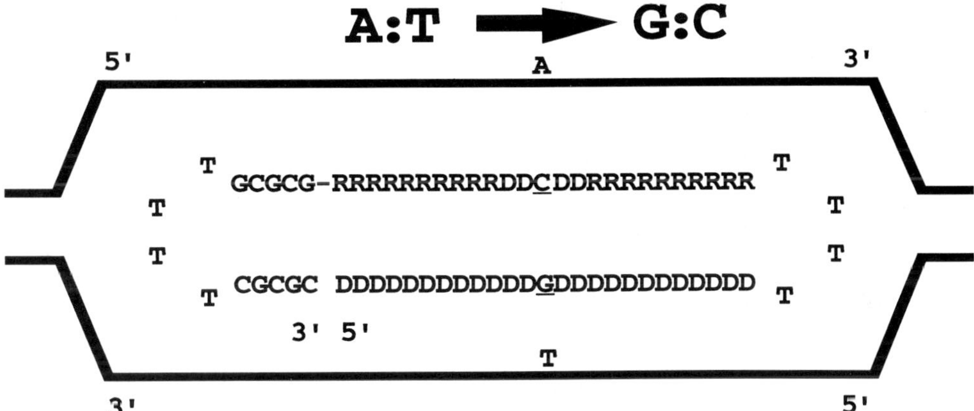

Figure 1. Hypothetical intermediate between an RDO and a genomic target. RDO consists of ten 2'-O-methyl RNA residues (R) flanking either side of a 5 residue DNA (D) stretch, which contains the base change desired for the mutation. The sequence of RNA is complementary to the targeted DNA over a stretch of 25 residues that span the site of the mutation. The top and the bottom strand of targeted DNA is designated as non-transcribed and transcribed-strand, respectively. A double-stranded genomic target is drawn to be paired with an RDO, generating a RNA:DNA and a DNA:DNA duplex. A strand pairing is suggested to be initiated by base paring of RNA residues in RDO and the genomic target. The RDO can create two DNA mismatches when it is paired with the targeted DNA duplex (A:C and G:T mismatches). Preferential DNA repair may occur at a mismatch residing in the DNA:DNA duplex (G:T), rather than a mismatch in the RNA:DNA duplex (A:C).

a) Incorporation of RNA: length of homology and strand-specificity of RNA

The RNA:DNA hybrid duplex was shown to be more active than DNA duplex in homologous recombination by RecA and Rec2 proteins (Kmiec *et al.*, 1994; Kotani *et al.*, 1996). The first design of RDO contained 20 RNA residues since a minimum homology of 15-20 bases of RNA was required in RecA strand exchange reaction (Kmiec *et al.*, 1994). Although the frequency of homologous recombination increases exponentially with the length of homology, RecA has been shown to pair DNA strands of relatively short homology, ranging 26-38 bp (Gonda and Radding, 1983; Hsich *et al.*, 1992). When the homologous region is actively transcribed, DNA paring was observed as little as a 15 bp homology by RecA (Kotani and Kmiec, 1994b). These observations suggest that the effect of RNA is to reduce the length of homology required for DNA pairing. The strand-pairing event has recently been shown to be mainly driven by the base pairing of RNA residues in the RDO and genomic target (Kumar R., personal communication). Longer and more complete homology of RNA to the target would likely improve conversion frequency.

An oligonucleotide containing RNA was highly effective in gene correction while a DNA oligonucleotide with an identical sequence exhibited no measurable gene correction (Yoon *et al.*, 1996). The importance of RNA in gene correction cannot be explained entirely by the thermodynamic stability of RNA:DNA duplex. The general conception that a RNA:DNA duplex is more stable than a DNA:DNA duplex is not always correct, because duplex stability is highly dependent on the base-composition and nearest-neighbor sequence (Lesnik and Freier, 1995; Sugimoto *et al.,* 1995). The stability of homopurine-homopyrimidine duplexes decreases in the order R(pu):D(py) > D(pu):D(py) > D(pu):R(py) (Roberts and Crothers, 1992; Riley and Maling, 1966; Martin and Tinoco, 1980). All RNA residues in the RDO were modified by 2'-O-methylation of the ribose sugar. 2'-O-methyl RNA binds to its RNA target with a higher affinity than unmodified RNA (Lesnik and Freier, 1998; Majlessi *et al.,* 1998). However, the stability of the 2'-O-methyl RNA:DNA duplex was similar to that of DNA:DNA (Majlessi *et al.,* 1998) or RNA:DNA duplexes (Wang and Kool, 1995). Thus, other factors are likely to cause the observed differences in gene correction between an RDO and all DNA containing oligonucleotide. The protein complex interacting with the RDO may be highly specific in the recognition of the unusual structure of the RNA:DNA duplex, distinguishing it from DNA:DNA duplex. The structure of heteroduplex was shown to be neither the A-form of typical RNA duplexes nor the B form of typical DNA duplexes (Bhattacharyya *et al.*, 1990; Roberts and Crothers, 1992; Tereshko *et al.*, 1998). The overall hybrid duplexes are closer to the A form than the B form, albeit with a narrower minor groove than RNA (Gyi *et al.*, 1996). This structure may play a role in an initial strand-pairing event and in discriminating between homology and non-homology during its interaction with a recombination complex.

Effect of the strand-specificity of RNA in gene correction by RDO was first examined in the episome system (Yoon *et al.*, 1996). The molecule containing RNA complementary to the non-transcribed-strand of the alkaline phosphatase cDNA was much more efficient in correction of the point mutation than the molecule containing RNA complementary to the transcribed-strand (Yoon *et al.*, 1996). However, 2 RDO with similar designs did not differ in causing the point mutation of the β-globin gene in $CD34^+$-selected bone marrow cells (Xiang *et al.*, 1997). A site-specific conversion of the factor IX gene occurred at an equal frequency, regardless of the polarity of the RNA strand in the primary hepatocytes (Kren *et al.*, 1998). The targeted-strand of the alkaline phosphatase cDNA has a high content of pyrimidine. When the RNA is complementary to the non-transcribed-strand of the alkaline phosphatase cDNA, a stable [R(pu):D(py)] duplex, which is close to A form, can be made during a strand-pairing event. On the other hand, the RDO containing the RNA strand complementary to the transcribed-strand may be involved in an unstable [D(pu):R(py)] duplex, which has an intermediate conformation between A and B forms. The different structure and stability of the [D(pu):R(py)] duplex may result in a low frequency of gene correction, in comparison to [R(pu):D(py)] duplex. The targeted-strand of the β-globin gene contains approximately equal proportions of purine and pyrimidine. Thus, a duplex with a similar stability is likely to be formed between the targeted DNA and an RDO containing a different polarity of RNA. The differential stability and structure of RNA:DNA duplex depending on the base composition of the targeted-strand may explain an observed strand-specificity of RNA among genes targeted (Yoon *et al.,* 1996; Xiang *et al.,* 1997; Kren *et al.,* 1998).

b) Incorporation of the DNA:DNA mismatch

Five DNA residues containing a single mismatched base pair between an RDO and the targeted sequence interrupted the stretch of RNA. We reasoned that DNA duplex might serve as a better substrate for enzymes involved in mismatch repair rather than RNA:DNA duplex. Since then, a mismatch in the RNA:DNA duplex was shown to be less efficiently repaired than a mismatch residing in the DNA:DNA duplex (Thaler *et al.,* 1996; Kamath-Loeb *et al.,* 1997). The original design of RDO can create two DNA mismatches when it is paired with the targeted DNA duplex: one with the transcribed-strand and the other with the non-transcribed-strand (Fig. 1). However, mechanism of selecting which strand is to be repaired has not yet been elucidated. In mammalian cells, the DNA mismatch repair was shown to occur in a strand-specific and a bi-directional manner, relative to a nick which is located 5' or 3' of the mismatch (Modrich and Lahue, 1996; Kat *et al.,* 1993; Fang and Modrich, 1993). Thus, it is also possible that RDO may not only enhance the strand-pairing activity but also facilitate the mismatch repair activity to the targeted strand preferentially. An RDO containing a single DNA mismatch between the non-transcribed-strand of the target DNA and the RNA-containing strand of RDO was shown to be much less effective in gene conversion (Kren *et al.,* 1997). This molecule contained no mismatch between the targeted DNA strand and the DNA-containing strand of RDO. This result suggests that DNA mismatch repair may preferentially occur at a mismatch residing in the DNA:DNA duplex, rather than a mismatch in the RNA:DNA duplex (see Fig. 1). All 8 base-base mismatches are subject to nick-directed correction (Modrich, 1997; Modrich and Lahue, 1996). However, some mismatches are better substrates than others and sequence effects have been observed. In general, Pu:Py and Pu:Pu mismatches are among the best substrates and Py:Py mismatches, such as

C:C mismatch, are processed with reduced efficiency (Modrich, 1997). Therefore, some mismatches are likely to be repaired better than others in the targeted repair by RDO.

c) Structural and chemical modifications of the RDO

For mammalian gene conversion experiments, it was necessary to make structural and chemical modifications of RDO. The RNA:DNA duplex was linked by a double-hairpin structure containing 4 T residues in each loop and a 5 base-pair GC clamp, to protect the 5'- and 3'-end of the oligonucleotide from exonucleolytic cleavage. The joint molecule formation proceeded efficiently even when the ends of the hybrid were capped with hairpin structures (Kmiec *et al.,* 1994; Kotani *et al.,* 1996). This modification also prevented tandem ligation, which has been known to inactivate transfected DNA in mammalian cells. Chemical modifications of the backbone, sugar and base, are common to all nucleic acid-based therapeutic approaches. The base or backbone modification should not appreciably alter thermodynamic or kinetic parameters for association of two strands. Modification should also improve the delivery of oligonucleotide to the cellular environment and increase the resistance to nuclease degradation. Moreover, modifications should not alter cellular functions responsible for biological activity. Although it is likely that chemical modification of RNA may affect mammalian recombination or repair activity, it was necessary to incorporate some modification to render the oligonucleotide resistant to the RNaseH. The first of such attempt was a 2'-O-methylation of the ribose sugar. We chose a substitution at the 2' position since the nature of the 2' substitution is the primary chemical difference between DNA and RNA. Moreover, the 2'-O-methyl RNA exhibited a similar affinity to the DNA target as unmodified RNA or DNA (Inoue *et al.*, 1987; Wang and Kool, 1995; Majlessi *et al.*, 1998). This modification also made the RDO resistant to RNaseH and other RNases (Monia *et al.*, 1993). Antisense oligonucleotides containing 2'-O-methyl modification targeted for H-ras and

HIV exhibited a higher potency in comparison to the non-modified oligonucleotide in tissue culture cells (Monia *et al.,* 1992; Shibahara *et al.,* 1989). Recently, RecA was shown to make a joint molecule between a single-stranded DNA and a hairpin composed of a 2'-O-methyl RNA and complementary DNA, albeit not as efficiently as a hairpin containing all DNA (Havre and Kmiec, 1998). This modification also made the synthesis of RDO much easier than regular RNA incorporation due to a protected 2'-hydroxyl group. However, it is important to note that the original RDO molecule was empirically designed. Thus, key parameters of the structure of RDO have yet to be determined for an efficient gene conversion.

3. SYNTHESIS, PURIFICATION AND CHARACTERIZATION OF RDO

RDO was synthesized in a 0.2 μmole scale using the 1000 A wide pore CPG on the ABI 392 DNA/RNA synthesizer by standard phosphoramidite procedures (Yoon *et al.,* 1996). Crude oligonucleotides were purified by polyacrylamide gel electrophoresis, containing 7M urea or 7M urea/40% formamide. We found no difference in efficiencies between the RDO purified by PAGE gel and by HPLC in gene conversion experiments. An active form of RDO is likely to be a double-hairpin structure (Fig. 1). Because of the symmetry of the molecule, RDO can also make multimeric concatemers at high concentration of oligonucleotide. In order to prevent multimerization, it is recommended that RDO be diluted at 100 nM in a low salt solution (10 mM Tris, 0.1mM EDTA), heated for 5 min at 95 °C, and quick-cooled to 4 °C to facilitate an unimolecular hairpin formation. These molecules are stable and can be stored at –20 °C for a long period of time. The melting temperature of RDO ranged from 78 to 90 °C, depending on salt concentration (R. Kumar, personal communication). Once a double-hairpin structure is formed, the melting temperature should be independent of concentration due to unimolecular unfolding of the molecule. Characterization of RDO can also be carried out using denatured polyacrylamide gel electrophoresis, containing 7M urea or 7M urea/40%

formamide. Although RDO is difficult to denature completely due to a large number of base-pairing, it migrates as a 68-mer when it is denatured prior to gel loading and run on a hot 7M urea/40% formamide-gel. Failure sequences migrated faster than the full-length RDO in a denaturing gel. Therefore, it is essential that any preparation of RDO be tested for its length and homogeneity.

4. DELIVERY, UPTAKE AND STABILITY OF THE RDO

The use of liposomes as delivery agents for DNA and RNA is a natural extension of their application as drug delivery agents. The efficiency of liposomes appears to depend on many variables: tissue type, lipid membrane composition, relative ratio of lipid to DNA, and the stability of the DNA-lipid complex. In recent years a series of cationic liposome agents have been introduced (Felgner and Ringfold, 1989). Their greater effectiveness is probably due to stronger DNA-lipid interactions and a greater level of association with net negative charge on the cell membrane. Recently, two different structures of the complex, lamellar and columnar conformation, has been determined by X-ray diffraction (Radler *et al.,* 1997; Koltover *et al.,* 1998). The lamellar structure consists of DNA condensing between the lipid bilayers of cationic liposome, with the cationic lipids aggregating where the DNA contacts the bilayer (Radler *et al.,* 1997). The columnar structure consists of DNA coated by cationic lipid monolayers and arranged on a two-dimensional hexagonal lattice (Koltover *et al.,* 1998). The columnar complex was shown to rapidly fuse to endosomes and to release the DNA molecule, resulting in higher transfection efficiency (Koltover *et al.,* 1998). Delivery of RDO has been carried out mostly by different formulation of liposomes, dendrimers or polyethyleneimines (Yoon *et al.,* 1996; Cole-Strauss *et al.,* 1996; Kren *et al.,* 1997; Xiang *et al.,* 1997; Kren *et al.*, 1998; Santana *et al.,* 1998; Alexeev and Yoon, 1998). It is important to test a variety of delivery methods, since the cellular uptake of RDO varied among different cells according to the mode of delivery. For each liposome, the molar ratio of the oligonucleotide to liposome needs to be optimized according to the uptake experiments.

a) Oligonucleotide uptake measurement using a radio-labeled RDO

Cellular uptake of oligonucleotide was measured by using a ^{32}P end-labeled RDO. At various times after transfection, cells were lysed and fractionated into cytoplasm and nucleus. A dose-dependent accumulation of oligonucleotide was observed in the nuclei and cytoplasms of human primary keratinocytes, HaCaT, and HeLa cells (Fig. 2). Uptake of the RDO was completed within 6 h since there was no appreciable increase in the intracellular concentration of oligonucleotide measured 24 h after transfection. The amount of oligonucleotide accumulated in the cytoplasm was higher than nucleus (Fig. 2A and 2B). Nuclear concentration of RDO was saturated, approaching 10^7-10^8 molecules per cell. Intracellular accumulation of antisense oligonucleotide was shown to be much lower than that of RDO (Yakubov *et al.*, 1989). This result indicates that a double-stranded RDO

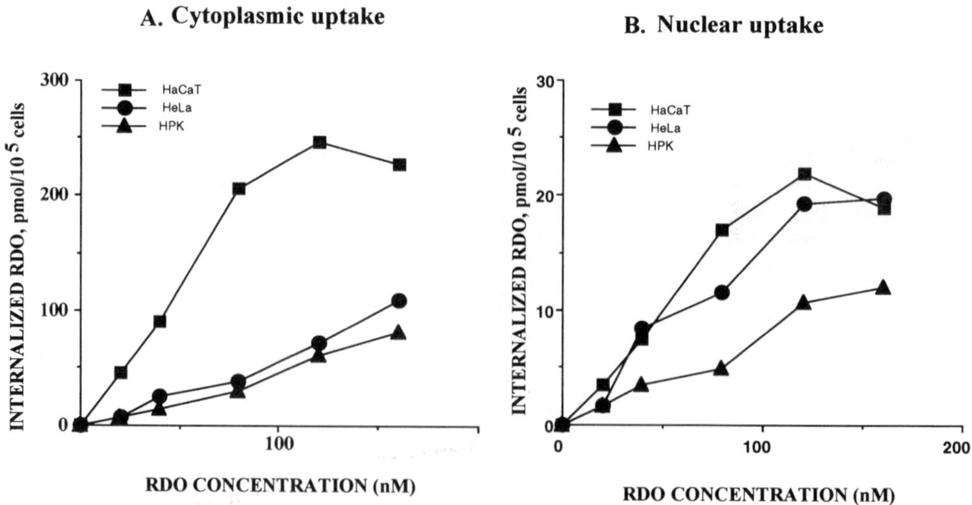

Figure 2. Cellular uptake of the RDO in human primary keratinocyte (HPK), HaCaT, and HeLa cells. Uptake is expressed in pmol of oligonucleotide per 100,000 cells.

may be transported to nucleus by a different pathway than a single-stranded oligonucleotide. Several nuclear proteins whose molecular weights ranged from 36 to 50 kDa were shown to be involved in the transport of oligonucleotide (Chin *et al.*, 1990; Leonetti *et al.*, 1991; Vlassov *et al.*, 1994). The stability of RDO can also contribute to an accumulation of high concentration of RDO.

b) Intracellular stability of the RDO

Intracellular stability of RDO was measured by a ^{32}P end-labeled oligonucleotide. The RDO isolated from nuclei contained approximately an equal amount of intact and of a distinct degraded product, which resulted from a cleavage at the T4-loop proximal to the 3'-end of the RDO (Fig. 3). Such a preferential cleavage suggests that the 5 GC bp may not be stable enough to protect the loop from exonuclease or endonuclease. Only degraded forms of RDO were detected in the cytoplasm (Fig. 3). The RDO

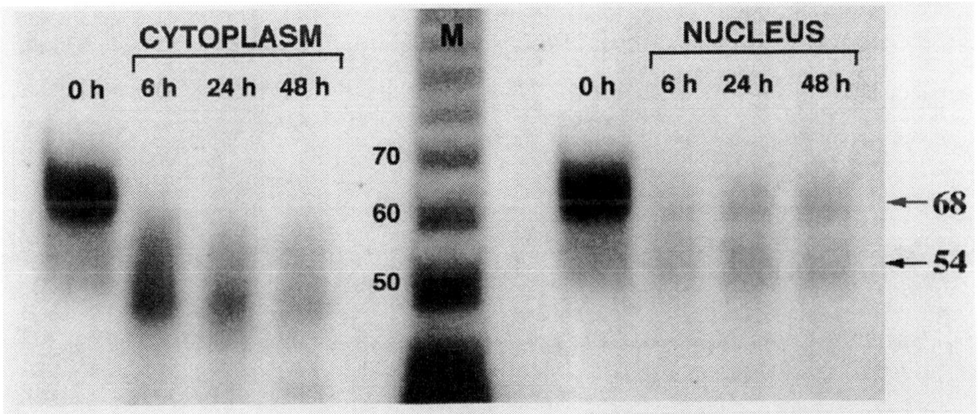

Figure 3. Stability of the RDO. ^{32}P end-labeled RDO was transfected into HeLa cells and extracted from cytoplasm and nucleus at various time intervals: 0, 6, 24 and 48 h. M shows ^{32}P end-labeled 10 bp marker. Full-length 68-mer RDO (arrow) migrates slightly faster than position 68 because the compact double-stranded structure of RDO is not completely denatured in 7M urea. The position of the degraded product is around 54 (arrow) suggesting a cleavage site at the 3' end junction between 2'-O-methyl RNA and DNA.

remained as a monomer inside the cells, and is thus not ligated into concatemers (Fig. 3). This observation also argues against the participation of RDO in nonhomologous recombination, in which end-to-end ligation of double-stranded DNA is often observed prior to integration.

c) Oligonucleotide uptake measurement using a fluorescein-conjugated RDO

Fluorescein-labeled RDO was transfected to measure its cellular uptake and localization in individual cells using confocal fluorescence microscopy. Over 90% of the cells were fluorescent in human primary keratinocytes, HaCaT, and HeLa cells, in both cytoplasm and nucleus (Fig. 4). The nucleus exhibited a distinct and punctuated accumulation of fluorescent oligonucleotide while cytoplasm showed a diffuse fluorescence. Human primary keratinocytes exhibited mostly nuclear fluorescence in comparison to HaCaT and HeLa cells. Similarly, over 90% of the cells were fluorescent in both the cytoplasm and nucleus of transfected albino melanocytes (Fig. 4). The delivery and stability of the RDO in the nucleus will be critical for gene conversion. Although we observed a consistently high uptake of fluorescein-conjugated RDO in the nucleus (Fig. 4), it is difficult to determine whether the RDO is accessible to the chromatin DNA.

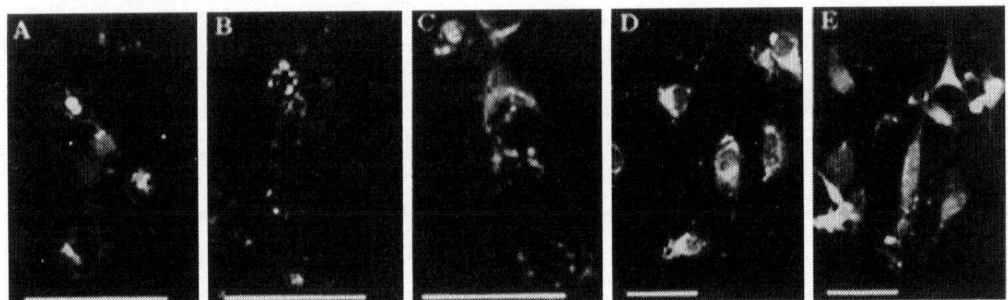

Figure 4. Confocal micrographs of primary keratinocyte (A), HaCaT (B), HeLa (C) and Melan-c (D-E) cells transfected with fluorescein-conjugated RDO. Cells were transfected with lipofectamine (A-C), superfectin (D) or cytofectin (E). Scale bar=50 μm.

5. SINGLE-BASE CONVERSION BY RDO IN MAMMALIAN CELLS

a) Correction of a targeted nucleotide by the RDO in an episomal gene

In order to test the feasibility of RDO-based conversion in mammalian cells, we chose an episomal target containing a single point mutation in human liver/bone/kidney alkaline phosphatase (AP) cDNA (Yoon *et al.,* 1996). This mutant plasmid has a missense mutation (G→A) at the position 711 of the cDNA, resulting in the loss of enzymatic activity (Fig. 5A). Several advantages exist in this system. By targeting episomal DNA, we reduced the target complexity considerably in comparison to the genomic DNA. A single cell that has gained the alkaline phosphatase positive phenotype can be visualized as red by histochemical staining, since the catalytic site of alkaline phosphatase is extracellular (Yoon *et al.,* 1988). Thus, targeted gene correction can be detected directly without selection or screening for the rare successful targeting events. An illustrative example of the design of an RDO is diagrammed in Fig. 5A. The AP1 RDO was designed to correct the missense mutation (A) to an active wild-type sequence (G). The AP2 has a design similar to AP1 but with a single base-pair change. Thus, AP2 has a complete homology to the mutant sequence and is not expected to correct the mutation. The AP3 molecule has the same sequence as AP1 but the RNA and DNA sequences are inverted with the RNA stretches complementary to the transcribed-strand of the alkaline phosphatase cDNA. The oligonucleotide AP4 contains the same sequence as AP1, but is comprised entirely of DNA residues. Various RDOs were introduced into CHO cells previously transfected with the mutant AP plasmid. Histochemical staining monitored the extent of the conversion to the wild-type phenotype, where red pigment was deposited on the cells expressing an active enzyme, shown in Fig. 5B. When cells with the mutant plasmid were transfected with the AP1, red cells appeared at a high frequency. In contrast, neither AP2 nor AP4 restored the enzymatic activity. Conversion to the wild type was

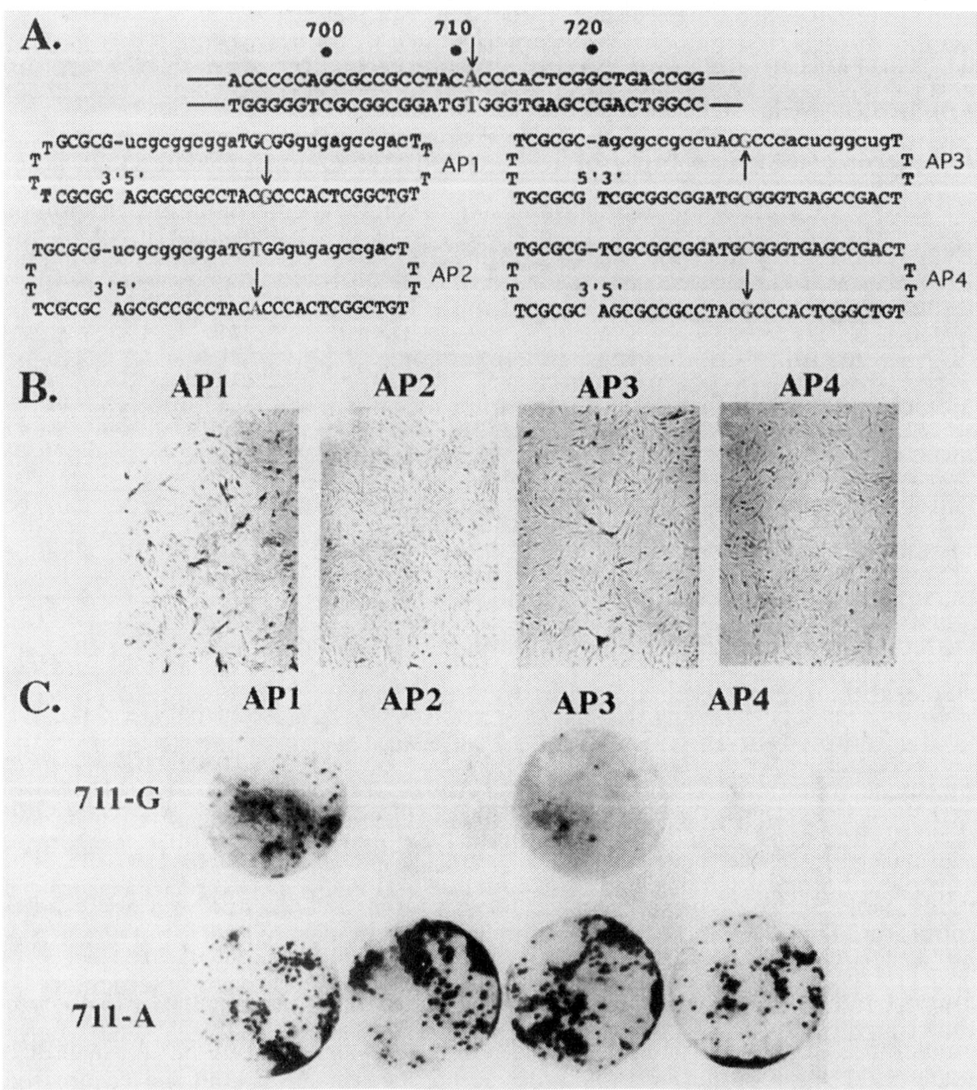

Figure 5. Correction of a point mutation of an episomal gene by RDO. **A)** Sequence of targeted DNA (top) and of corresponding RDOs (AP1-3) and DNA control (AP4). DNA residues are listed in upper case and 2'-O-methyl RNA residues in lower case. Arrows point to the targeted adenosine for correction (top) and to the corresponding position of each RDO. **B)** Histochemical staining of alkaline phosphatase activities of CHO cells transfected with 10 μg of the mutant alkaline phosphatase plasmid followed by AP1, AP2, AP3 and AP4, respectively, at 11 nM. **C)** Analysis of Hirt-DNA extracted from CHO cells transfected as indicated in B. Hirt-DNA was was transformed into bacteria and the resulting colonies were hybridized on duplicate filters with ^{32}P-end-labeled wild-type 711-G (top panel) and mutant-specific 711-A (bottom panel) probes, respectively.

observed at a low level when cells were transfected with AP3, a molecule in which the RNA stretches are complementary to the transcribed-strand. This difference could be attributed to the structure and thermodynamic stability of an intermediate, [D(pu):R(py)] versus [R(pu):D(py)] duplex, during the strand-pairing event.

In order to confirm the change at the DNA sequence level, a Hirt extract was made from the cells transfected with the mutant plasmid and various oligonucleotides. Hirt DNA was transformed into DH5α bacteria and transformants were screened for specific hybridization with a probe designed to distinguish between the point mutation (A) and the wild-type (G) sequence (711-A, 5'-CCGCCTAC**A**CCCACTCG-3' and 711-G, 5'-CCGCCTAC**G**CCCACTCG-3'). Approximately 70% of the colonies generated from the Hirt DNA made from cells transfected with the mutant plasmid and AP1 hybridized to the 711-A probe, while 30% of colonies exhibited hybridization to the 711-G probe, shown in Fig. 5C. Hybridization was specific and no cross-hybridization was observed between the two populations. All colonies from the Hirt extract prepared from cells transfected with the mutant plasmid and AP2 hybridized to the 711-A probe only, indicating a sequence-specificity of the RDO. Some colonies from the Hirt DNA isolated from cells transfected with the mutant plasmid and AP3 hybridized to the wild-type probe, but to a lesser extent than that of the AP1. The importance of RNA was demonstrated by the differential activities of oligonucleotides containing identical sequence with or without RNA stretches, AP1 and AP4, in eliciting corrections at biochemical and genetic levels. In this study, the gene correction was detected and confirmed at both functional enzyme activity and genetic level without PCR analysis. This study provided the basis for subsequent applications of RDO for chromosomal genes.

b) Conversion of a targeted nucleotide in chromosomal genes

Chromosomal correction of a point mutation in the β-globin gene by RDO was demonstrated in EBV-transformed lymphoblastoid cells from sickle cell anemia patients causing a single base (T→A) change (Cole-Strauss *et al.,* 1996). The RDO designed to convert a normal to sickle genotype (A→T) demonstrated a targeted gene conversion of $CD34^{+}$-enriched cell population from a normal donor with 5-11% conversion rate (Xiang *et al.,* 1997). A site-specific mutation in the alkaline phosphatase gene was achieved by the RDO designed to cause a point mutation (G→A) in HuH-7 cells (Kren *et al.,* 1997). The identical position was previously shown to be corrected (A→G) by an RDO which differed in a single base pair (Yoon *et al.,* 1996). Recently, an RDO was shown to induce a sequence-specific mutation in the factor IX gene in cell culture and *in vivo* by intravenous injection, resulting in a reduced factor IX activity and a marked prolongation of the activated partial thromboplastin time (Kren *et al.,* 1998).

(i) Frequencies of RDO-mediated gene conversion in different cells. In order to investigate whether the gene conversion strategy by RDO is generally applicable in mammalian cells, we measured the frequencies of gene conversion among different cells (Santana *et al.,* 1998). Highly transformed epithelial cells (HeLa) exhibited a dose-dependent conversion frequency of 5% by AP2 (Fig. 6C). In comparison, human primary keratinocytes or immortalized keratinocytes (HaCaT) did not show any detectable level of gene conversion (Fig. 6A and 6B). The extent of gene conversion events was analyzed utilizing PCR amplification of genomic DNA followed by restriction enzyme digestion of the 194 bp PCR product.

The wild type sequence of the bone/liver/kidney alkaline phosphatase (**G**CCCACTCGGC) is cleaved by the restriction enzyme *Bgl* I resulting in 114 and 80 bp fragments, while the mutated sequence (**A**CCCACTCGGC) is not cleaved. Sequencing of the 194 bp band confirmed a C→T change in the complementary strand indicating G→A in the coding strand. This eliminates the possibility that the 194 bp fragment could be generated by an incomplete *Bgl* I digestion. The concentration of oligonucleotide in the nuclei of HeLa cells was similar to that of HaCaT or human primary keratinocytes measured by a radiolabeled or a fluorescein-conjugated RDO (Fig. 2 and 4). Moreover, RDO exhibited a prolonged stability in the nucleus (Fig. 3). Thus, neither uptake nor nuclear stability of the RDO appears to be a limiting factor in gene conversion under these experimental conditions. These results indicate that the frequency of gene conversion varies among different cells, suggesting that cellular recombination and DNA repair activities may be important.

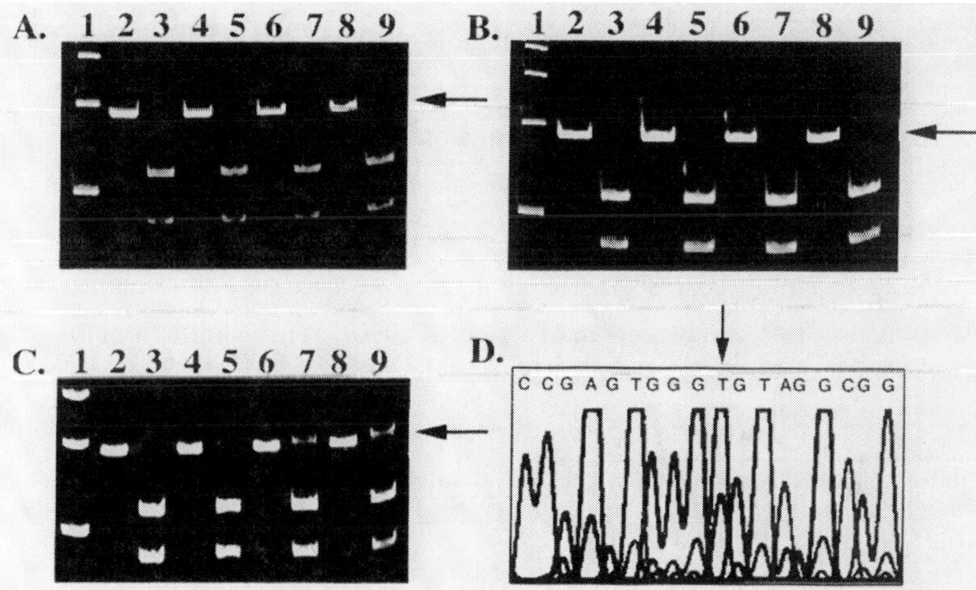

Figure 6. Variability of gene conversion by AP2. Gene conversion of wild type alkaline phosphatase was attempted on human primary keratinocyte (A), HaCaT (B) and HeLa (C) cells with increasing amount of AP2. Conversion to mutant sequence was monitored by *Bgl* I digestion of a 194 bp PCR product of the bone/liver/kidney alkaline phosphatase gene. The 194 bp product (arrow) is cleaved by the restriction enzyme *Bgl* I resulting in 114 and 80 bp fragments (lanes 3, 5 7 and 9) when it comprises the wild type sequence (**G**CCCACTCGGC) and is not cut (panel C, lanes 7 and 9) when it comprises the mutant one (**A**CCCACTCGGC). Lane 1 contains 100 bp marker. Lanes 2, 4, 6 and 8 show the migration of PCR products from cells treated with 0, 20, 110 and 160 nM AP2, respectively; lanes 3, 5, 7 and 9 correspond to *Bgl* I digestion of the PCR products from lanes 2, 4, 6 and 8, respectively. Direct sequencing of 194 bp PCR products demonstrates the anticipated single-base conversion (C→T) in the complementary strand (arrow, panel D) indicating a G→A change in the coding strand.

(ii) Stable and inheritable changes in genotype and phenotype of albino melanocytes. The need to develop a better experimental system to test this technology led us to explore the possibility of pigmentation change by an RDO in albino melanocytes (Alexeev and Yoon, 1998). This system offers a distinct advantage over other chromosomal gene conversion strategies. First, we can monitor live cells for pigmentation change by microscopy. Second, the black-pigmented cells can be easily visualized among albino cells with a minimum background, allowing detection of even a low frequency of genomic conversion. Third, we can clone converted melanocytes exhibiting the pigmentation change and characterize their genotype. Tyrosinase is an essential enzyme in melanin synthesis, produced by melanocytes, and sufficient for pigmentation change *in vitro* and *in vivo* (Halaban and Moellmann, 1993). Melanocytes derived from albino mice contain a homozygous point mutation (T**G**T→T**C**T) in the tyrosinase gene, resulting in amino acid sequence change from Cys to Ser (Halaban and Moellmann, 1993; Jackson and Bennett, 1990). This single amino acid change was shown to be responsible for the complete inactivation of tyrosinase and an absence of pigmentation. Thus, correction of the point mutation in the tyrosinase gene would result in the restoration of the enzyme activity and in change of the pigmentation of the cell.

The Tyr-A RDO was designed to correct a point mutation in the tyrosinase gene to an active wild-type sequence. Another RDO, Tyr-B, contains an identical sequence to the mutant tyrosinase and thus differs in a single base pair from Tyr-A (Fig. 7A). Melan-c cells containing this single

Figure 7. Variability in the frequency of gene correction in melanocytes. A) Tyr-RDO and targeted sequences in the tyrosinase gene. The target site is underlined. DNA and 2'-O-methyl RNA residues are in upper and lower case, respectively. The Tyr-A RDO contains 25 bp of sequence identical to the wild type tyrosinase whereas control Tyr-B corresponds to the albino mutant. **B)** Extent of pigmentation changes in melan-c cells. Micrographs of melan-c (A) and melan-c cells treated with 220 nM Tyr-B and 6 μg/ml of superfectin (B), 220 nM Tyr-A and 6 μg/mL of superfectin (C-E), 220 nM Tyr-A and 4 μg/ml of cytofectin (F) and 440 nM Tyr-A and 6 μg/ml of superfectin (G). C-G: different experiments; H and I: converted black-pigmented clones 5 and 20. Scale bar=100 μm.

A. INACTIVE TYROSINASE (ALBINO PHENOTYPE)

Phe Met Gly Phe Asn Cys Gly Asn Ser Lys Phe Gly Phe Gly Gly Pro

5'TTC ATG GGT TTC AAC TGC GGA AAC TCT AAG TTT GGA TTT GGG GGC CCA 3'

3'AAG TAC CCA AAG TTG ACG CCT TTG AGA TTC AAA CCT AAA CCC CCG GGT 5'

Tyr-A

T GCGCG ug acg ccu uuG ACA Tuc aaa ccu aa T
T T
T T
T CGCGC AC TGC GGA AAC TGT AAG TTT GGA TT T
3' 5'

Tyr-B

T GCGCG ug acg ccu uuG AGA Tuc aaa ccu aa T
T T
T T
T CGCGC AC TGC GGA AAC TCT AAG TTT GGA TT T
3' 5'

B.

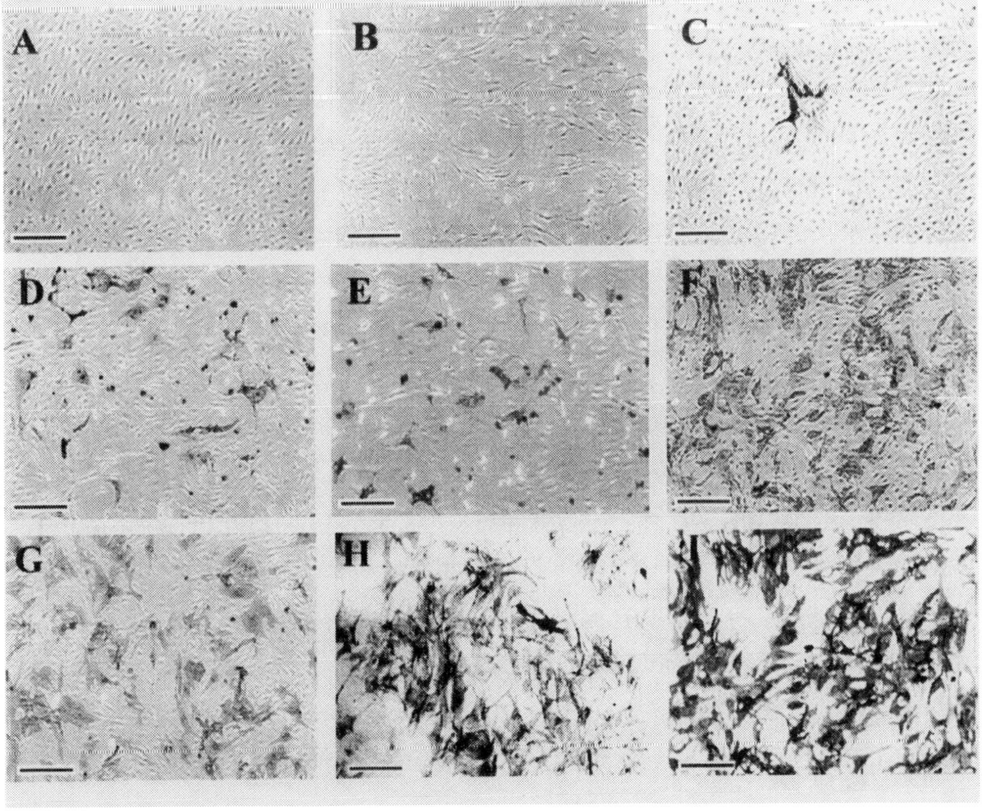

A B C

D E F

G H I

point mutation were transfected with Tyr-A at various concentrations ranging from 220 to 440 nM and monitored for a change in pigmentation for 2 weeks. Changes in pigmentation occurred 5-6 days after transfection. The extent of pigmentation change varied among at least 30 independent experiments we performed, ranging from 0.01% to 15%, shown in Fig. 7B. As a control, Tyr-B was transfected into melan-c cells and did not produce any pigmentation change in all 30 independent experiments. These results indicate a sequence-specific correction induced by the RDO.

This system also enabled us to monitor the pigmentation change in a living cell and to clone converted black-pigmented cells for characterization of genotype and protein expression. Cells were subcloned 5-10 times to insure isolation of a black-pigmented clone from single cell. All of the subclones maintained the pigmented phenotype for at least 3 months in tissue culture. We verified the genotype of converted black-pigmented cells by RFLP analysis and DNA sequencing. Genomic DNA was isolated and subjected to PCR amplification generating a 354 bp fragment. The gene conversion was measured by *Dde* I digestion of the PCR product . The PCR product from albino tyrosinase (**C**TAAG) was cleaved by the restriction enzyme *Dde* I, resulting in 144, 102, 73 and 35 bp fragments. In comparison, the PCR product from homozygous wild type tyrosinase (**G**TAAG) generates 179, 102 and 73 bp fragments upon *Dde* I digestion (Fig. 8A). Thus, a 179 bp and a 144 bp fragment are specific for the wild type and the mutant tyrosinase gene, respectively. *Dde* I digestion of the PCR product from all converted black-pigmented clones generated 179, 144, 102 and 73 bp fragments. The presence of both 179 and 144 bp fragments indicated a single-allele conversion in all black-pigmented clones. Sequencing of the 354 bp fragment from all these clones confirmed an equal mixture of C and G at the position indicated by an arrow (Fig. 8B). These results confirmed a targeted base change, C→G, in a single allele of melan-c cells by Tyr-A. No other sequence change was observed within a 400 bp region of the tyrosinase gene which has been sequenced in each clone.

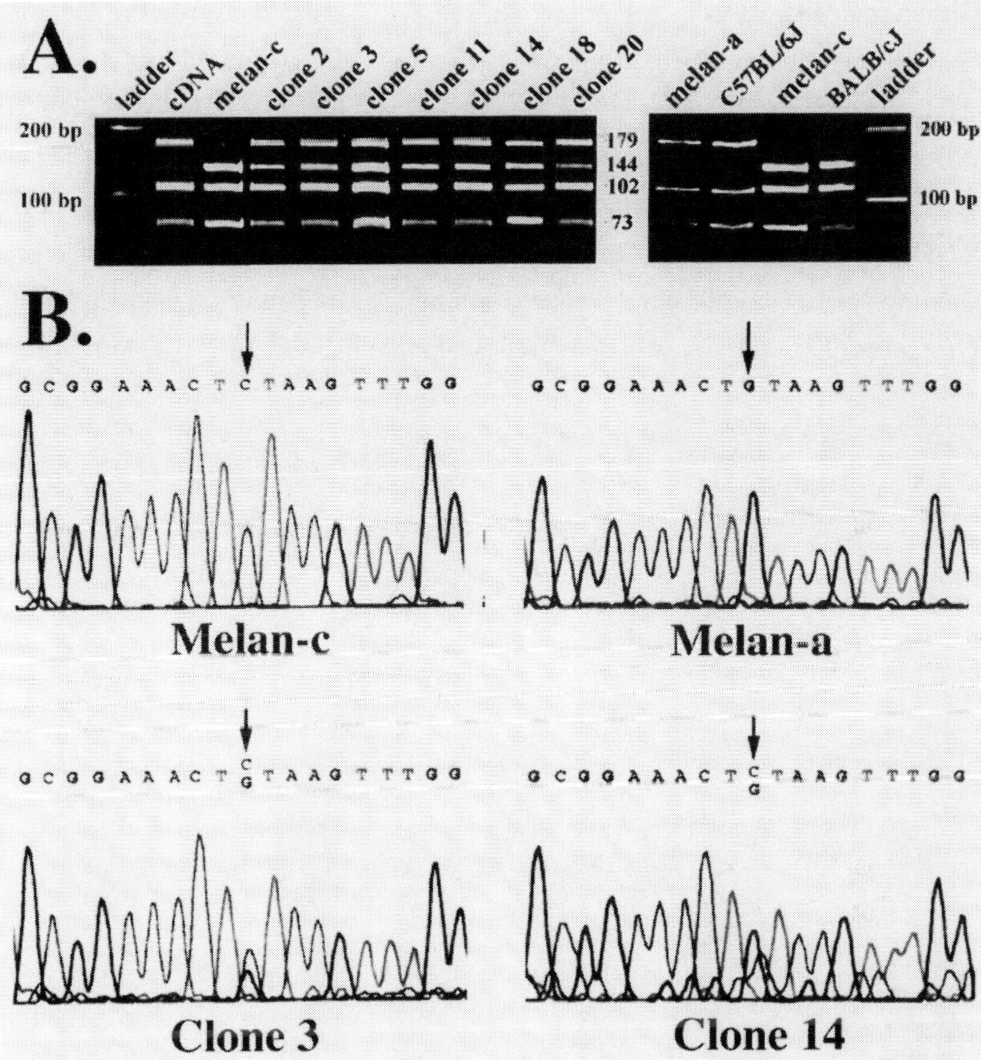

Figure 8. RFLP analysis of converted clones. A) Fragments generated by *Dde* I digestion of the PCR products from a wild type tyrosinase cDNA, melan-c cells and seven independently isolated black-pigmented clones. The RFLP analysis of genomic DNA isolated from melan-a (normal melanocytes), C57 BL/6J (black) mouse, melan-c and BALB/cJ (albino) mouse are also shown. **B)** DNA sequences of melan-c, melan-a, and of 2 converted black-pigmented clones. The arrow designates the targeted base for correction.

In order to verify the gene correction event at the protein level, cell lysates from individual clones were subjected to western blot analysis (Fig. 9) using the anti-tyrosinase peptide-specific polyclonal antibody αPEP7

(Jimenez *et al.,* 1991). Albino tyrosinase was detected as low-molecular-weight degradation products in Western blot due to proteolytic digestion (Halaban and Moellmann, 1993). In contrast, all corrected clones exhibited a full-length mature tyrosinase (Fig. 9A). Tyrosinase enzymatic activity can be detected in a non-denaturing gel, in which proteins are separated, upon incubation with L-DOPA. Oxidation of L-DOPA to melanin results in black staining of a single band corresponding to the molecular weight of tyrosinase. This activity was detected as a single band in all converted black-

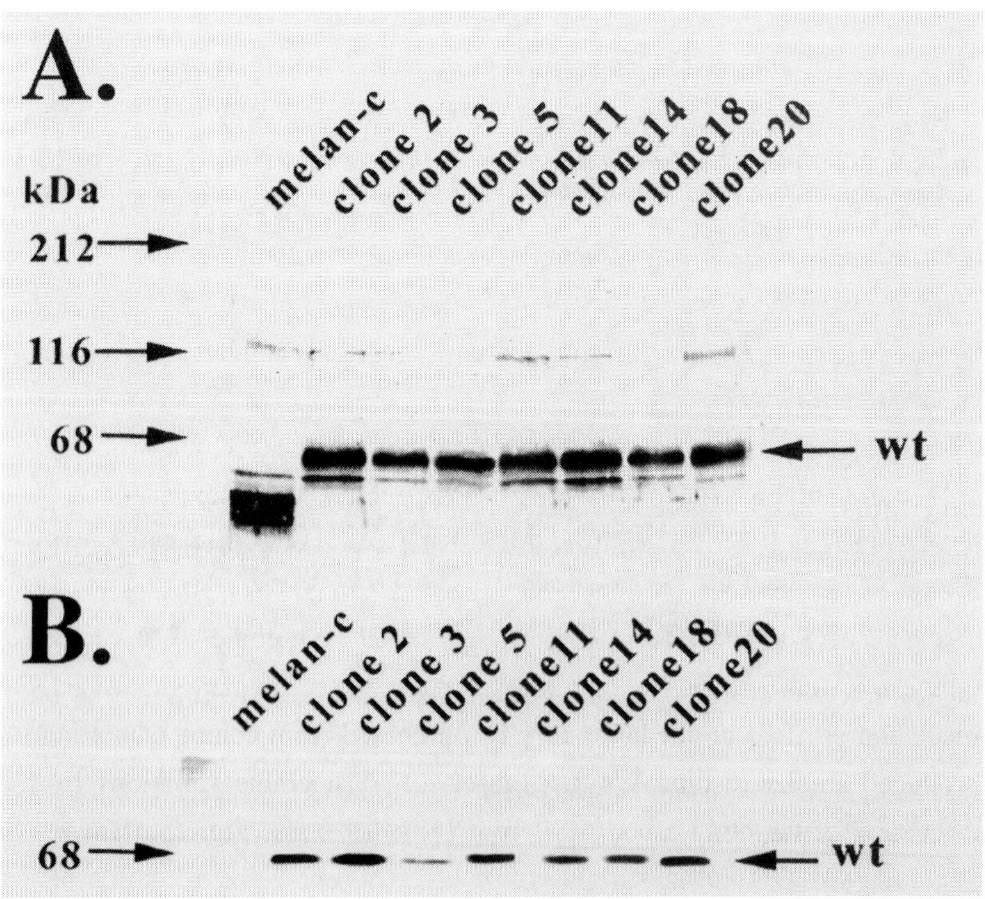

Figure 9. Immunoblot analysis (A) and L-DOPA zymogram (B) of melan-c cells and converted black-pigmented clones. All converted clones exhibited full-length tyrosinase (68 kDa) identified by antibody and by oxidation of L-DOPA. Tyrosinase from melan-c cells was detected as degraded form of the protein and lacked L-DOPA oxidation activity.

pigmented clones but not in melan-c cells (Fig. 9B). This study demonstrated for the first time a permanent and inheritable gene correction by RDO at the level of genomic sequence, protein and phenotypic change by clonal analysis.

6. CRITICAL ISSUES AND IMPROVEMENTS

a) Variability in the frequency of single-base conversion by the RDO

Because RDO was shown to work in several systems (Yoon *et al.*, 1996; Cole-Strauss *et al.*, 1996; Kren *et al.*, 1997; Xiang *et al.*, 1997; Kren *et al.*, 1998), many researchers have applied this technology to their gene of interest and most of them have failed so far (Strauss, 1998). Concerns have been raised on potential artifacts since high frequency conversion was determined by PCR amplification, especially without clonal analysis (Thomas and Capcccchi, 1997). Cross-contamination bctwccn β^A and β^S cclls has also been raised as an explanation of a high conversion frequency (Stasiak *et al.*, 1997).

We have repeated gene conversion experiments in β^A and β^S cells for at least 100 times with variable results. Except for a few experiments, we detected no gene conversion by the RFLP analysis. Only frequencies above 5% can be reproducibly detected by RFLP analysis. However, gene conversion was consistently detected by an allele-specific PCR analysis, indicating a low frequency of conversion. Different preparations of RDO, made and purified in our laboratory or purchased from commercial vendor, produced similar results. We also observed considerable variations in the frequency of targeting among different epithelial cells (Fig. 6). However, neither uptake nor nuclear stability of the RDO appears to be a limiting factor (Fig. 2, 3 and 4). We also experienced a large variation in the frequency of tyrosinase gene correction determined by cell pigmentation change among 30 independent experiments, ranging from 0.01% to 15% (Fig. 7B).

At this point, it is not clear just what causes so much experimental variations. Differences in the frequency of gene conversion by RDO could be attributed to the DNA recombination and repair activities among different cells. Epithelial cells with varying degrees of transformation, primary keratinocytes, immortalized keratinocytes and HeLa cells were chosen for comparison (Santana *et al.,* 1998). Because the DNA mismatch repair activity appears to be active in these cells (Lawrence and Benford, 1993; Diem and Runger, 1997; Fang and Modrich, 1993; Kat *et al.,* 1993), the observed differential frequency cannot be explained entirely by the DNA repair activity. Homologous recombination activity has been reported to vary a great deal among different cells and to be elevated in transformed cells (Mekeel *et al.,* 1997; Thyagarajan *et al.,* 1996). Tumor cells containing inactivated p53 recombined 100-fold higher than the isogenic primary cells (Mekeel *et al.,* 1997). Expression of the E6 protein (Meyn *et al.,* 1994) or SV40 large T antigen (Wiesmuller *et al*; 1996) increased the recombination frequency, presumably by inactivating p53 protein. Likewise, an interaction between Rad51 and p53 protein (Sturzbecher *et al.,* 1996) led to a high frequency of homologous recombination. Recently, other tumor suppressor proteins, BRCA1, BRCA2 and pRb were shown to interact with Rad 51 protein, suggesting participation of these proteins in DNA repair and recombination events (Scully *et al.,* 1997; Sharan *et al.,* 1997). Therefore, it is possible that recombination activity of these epithelial cells (human primary keratinocyte, HaCaT and HeLa) may differ according to the degree of transformation and/or the level of p53 protein. These inherent differences in recombination activity may partially account for different frequency in targeting by the RDO (Santana *et al.*, 1998).

It is plausible that the chromatin structure of highly transcribed or replicating region may be more accessible to strand-pairing by RDO. However, DNA replication appears not to be required, since RDO gene conversion was observed in quiescent G_0 hepatocytes and $CD34^+$-enriched cells (Xiang *et al.*, 1997; Kren *et al.*, 1998). Moreover, transcription may facilitate but may not be a necessary condition for gene conversion by RDO. A high level of gene correction was shown in episome system where alkaline phosphatase gene was transcribed (Yoon *et al.*, 1996). However, RDO was shown to be effective in gene conversion of a non-transcribed gene, exemplified by the β-globin gene in EBV-transformed lymphoblastoid cells (Cole-Strauss *et al.,* 1996), $CD34^+$-enriched cells (Xiang *et al.*, 1997), HuH-7 cells (Kren *et al.*, 1997) and HeLa cells (Santana *et al.*, 1998). Nevertheless, it is possible that the β-globin gene may be an exception because the origin of DNA replication has been shown to reside close to the region where gene correction is targeted (Aladjem *et al.,* 1995). The origin of replication was active in non-erythroid cells in which the β-globin gene was silent and DNase I resistant (Aladjem *et al.,* 1995). Proximity of the targeted region and the origin of replication may account for the efficient RDO correction in cells where the β-globin gene was not transcribed. A transient opening of the targeted chromosomal region may be sufficient to initiate strand pairing by RDO.

The variability in the frequency could also result from cell cycle, cell metabolism, and transfection condition. So far, gene conversion by RDO has been carried out in populations of unsynchronized cells. Although DNA replication may not be necessary for gene conversion, a cell synchronization study would likely shed some lights on experimental variability in mammalian cells. The repair of the transcribed-strand was efficient both in G_1 and G_2 to a similar extent, while non-transcribed-strands or inactive regions of DNA were not repaired (Petersen *et al.,* 1995). It is also possible that nuclear delivery of RDO may be dependent on cell cycle. The delivery and stability of RDO in the nucleus will be critical for successful targeting. Although we observed a consistently high uptake of fluorescein-conjugated RDO in the nucleus, it is difficult to determine whether the RDO is accessible to target the chromatin DNA. Each of these parameters is difficult to control and to measure in tissue culture condition. Endogenous gene expression caused by the RDO in a living cell, such as pigmentation change, will serve as a good model system to study the relationship between gene conversion and cellular activities.

Variability among different cells as well as in a single given system prompted us to develop a shuttle vector, which can measure the conversion frequency rapidly and reproducibly. A shuttle vector where a gene product is expressed in both mammalian and bacterial cells make it possible to transfer the conversion event by RDO in mammalian cells to bacteria for analysis. The advantage of this scheme is that, in bacteria, mutations or corrections can be rapidly and unambiguously scored. Extrachromosomal homologous recombination was shown to occur by a nonconservative single-strand annealing mechanism, which is different from chromosomal recombination (Maryon and Carroll, 1991; Lin *et al.*, 1990; Rouet *et al.*, 1994). Nevertheless, a shuttle system enables us to analyze events in mammalian cells using a convenient bacteria assay, to determine the feasibility of gene conversion in a given cell line, and to compare the frequency of targeting among different cells. We have developed a shuttle vector to assay targeted gene correction using *E. coli* β-galactosidase (Igoucheva *et al.*, 1999). This strategy utilizes convenient histochemical and spectrophotometric β-galactosidase assays in mammalian cells as well as a simple color selection (blue or white) in bacteria transformed by the Hirt DNA. We made the mutant β-galactosidase plasmid containing a single point mutation in the catalytic domain, resulting in a loss enzymatic activity. The frequency of gene correction by an RDO was measured by counting blue or white colonies generated by the Hirt DNA isolated from cells transfected with the mutant plasmid and an RDO. The first results indicate a dose-dependent gene correction, ranging from 0.1% to 1.4% (Igoucheva *et al.*, 1999).

b) Specificity of single-base conversion by the RDO

Most RDO conversion experiments were carried out by utilizing 2 RDOs which differ in a single base pair, one containing a mismatch and the other identical to the targeted sequence as a control: AP1 and AP2 in the alkaline phosphatase cDNA (Yoon *et al.,* 1996; Kren *et al.,* 1997), SC1 and SC2 in the β-globin gene (Cole-Strauss *et al.,* 1996; Xiang *et al.,* 1997), and Tyr-A and Try-B in the tyrosinase gene (Alexeev and Yoon, 1998). In all cases, RDO was shown to cause a sequence-specific change in genotype and phenotype. The specificity of RDO was further examined in the globin gene locus (Cole-Strauss *et al.,* 1996). The sequence of the β-globin gene exhibited over a 90% homology to that of the δ-globin gene (Poncz *et al.,* 1983). However, SC2 caused a specific mutation only in the β-globin but not in the δ-globin (Cole-Strauss *et al.,* 1996). SC2 contained 2 nucleotide mismatches within the RNA stretch to the δ-globin sequence but not to the β-globin sequence. Thus, it appears that 2 mismatches in the RNA region may be sufficient to disable gene conversion, indicating a stringent specificity of conversion.

Because RDO is a double-stranded oligonucleotide and is stable for prolonged period, we considered the possibility that RDO may cause nonhomologous recombination events leading to spurious mutations or random integration. DNA sequencing of several hundred bases surrounding the target base revealed no alteration in sequence in episomal or chromosomal targeting (Yoon *et al.*, 1996; Cole-Strauss *et al.*, 1996; Kren *et al.*, 1997; Xiang *et al.*, 1997; Kren *et al.*, 1998; Santana *et al.*, 1998; Alexeev and Yoon, 1998). The RDO was stable and remained as a monomer inside the cells, indicating that RDO molecules were not ligated into concatemers (Fig. 3). This observation also argues against the participation of RDO in nonhomologous recombination, in which end-to-end ligation of double-stranded DNA is often observed prior to integration. Moreover, the kinetics of conversion by RDO appears to be fast (<24 h), while most integration events of double-stranded DNA occur over much longer time periods (Capecchi, 1989; Mansour *et al.,* 1988; Thomas and Capecchi, 1987; Yanez and Porter, 1998). As a minimal screen to measure a potential mutagenicity of RDO, inactivation of the hypoxanthine-guanine phosphoribosyl transferase (HPRT) gene was compared between cells treated and untreated by RDO (Cole-Strauss *et al.,* 1996). There was no difference in the number of thioguanine resistant cells in the presence or absence of RDO, indicating that random mutagenesis by RDO is not likely to happen at a high frequency. Thus, the targeted conversion by RDO is highly sequence-specific and more frequent than random nonhomologous recombination.

c) Improvements in the design of the RDO

The key parameters of the structure of RDO will need to be optimized for an efficient gene conversion. Since the frequency of homologous recombination increases exponentially with the length of homology, a longer RDO would likely improve the targeting frequency. Ultimately, the length of homology will be limited by the synthetic capability. Targeting a few base pair insertion or deletion would likely to require a longer length of DNA. The original design of RDO can generate 2 DNA mismatches when it is paired with the targeted DNA duplex: one with each strand of its double-stranded target. It is likely that DNA mismatch repair may preferentially occur at a mismatch residing in the DNA:DNA duplex, rather than a mismatch in the RNA:DNA duplex. Nevertheless, a strand-specific repair will need to be investigated by comparing the conversion frequencies of RDOs containing a single mismatch between one strand of RDO, either in the RNA- or DNA-containing strand, and targeted DNA sequence. The strand-specificity of RNA in gene conversion will also need to be tested in various genes to draw conclusions on the polarity of RNA, either complementary to the transcribed- or non-transcribed-strand of the genomic sequence. The original design contained a double-hairpin to make RDO resistant to both 5' and 3' exonucleases. The size and composition of the loop contributes to the thermodynamic stability of the hairpin. Although a loop size of 4-6 bases was shown to be most stable (Gralla and Crothers, 1973), various sizes of the loop need to be tested for an efficient targeting. The base composition of the loop in the original design was thymidine. The tetra-loop sequences, GNRA and UNCG where R designates purine and N designates any base, present in ribosomal RNAs and hammerhead ribozymes, were shown to be most stable through internal G:A or U:G hydrogen bonding within the loop (Woese *et al.*, 1990). Therefore, the size and composition of the loop will need further optimization.

In addition, base, backbone and sugar modifications can be also incorporated to the future design of RDO to increase affinity to the target sequence and resistance to nucleases. Hydrophobic modifications at the 5-position of pyrimidines, such as 5-fluoro-2'-deoxy-uridine, 5-bromo-2'-deoxyuridine and 5-methyl-2'-deoxycytidine, enhance thermodynamic stability toward RNA or DNA target (Milligan *et al.*, 1993). Various backbone modifications, such as phosphorothioates, phosphoramidites and methylphosphonates, and those with nonphosphate internucleotide bonds, such as carbonates, carbamates, siloxanes, sulfonamides and polyamide nucleic acid have shown increased resistance to nucleases (Milligan *et al.*, 1993; Nielsen *et al.*, 1991). An RDO containing a sugar modification, 2'-O-methyl RNA, was the first attempt at chemical modification. Other sugar modifications, 2'-fluoro or 2'-methoxyethoxy, have been shown to increase thermodynamic stability of duplex and nuclease resistance (Freier and Altmann, 1997). These modifications need to be incorporated and tested for effectiveness in gene conversion. Above all, modifications should not alter cellular functions responsible for biological activity, in this case recombination and repair activity.

The highly sensitive and convenient assay utilizing *E. coli* β-galactosidase will be useful for optimization of RDO structure. This shuttle vector is also well suited for investigation of gene conversion in cells with various repair and recombination deficiencies. Moreover, this system can be utilized for biochemical characterization of intermediates and proteins involved in the process. Human cell extracts have been shown to repair efficiently a heteroduplex generated by a base mismatch in a site-specific and a strand-specific manner (Modrich and Lahue, 1996; Kat *et al.*, 1993; Fang and Modrich, 1993). Nuclear extracts competent for homologous recombination have been prepared from mammalian cells (Thyagarajan *et al.*, 1996; Thyagarajan and Campbell, 1997). Such nuclear or cell extracts can be also utilized to measure the frequency of RDO-mediated gene conversion (Cole-Strauss *et al.*, 1999; Igoucheva *et al.*, 1999), to optimize the structure of RDO, and to identify an intermediate involved in RDO gene conversion. As this technology is utilized more widely in the scientific community, it will be critical to establish a standard system where one can measure the frequency of gene conversion in a reproducible manner. This feasibility testing can be used to predict whether this technology may be applicable to a specific system, thus circumventing a prolonged experimen-tation in dark. Understanding of mechanism and improvement of the design of RDO will be essential in order to make this technology applicable for general gene conversion.

Acknowlegements

The author 1) thanks the members of her laboratory for their contributions during the development of the RDO technology: Drs. V. Alexeev, O. Igoucheva, A. Peritz and Ms. E. Santana, 2) acknowledges the seminal contribution of Dr. E. Kmiec in conception of RDO and members of his laboratory, in particular Ms. A. Cole-Strauss, 3) thanks Dr. R. Kumar and his group at Kimeragen Inc. for discussion and sharing unpublished results, Dr. S. Freier at ISIS Pharmaceuticals for comments, and Dr. J. Uitto at Jefferson Medical College for continued interests and support, 4) especially acknowledges Dr. I. Tinoco Jr. for teaching her not only oligonucleotide chemistry but also an attitude about doing science. This work was supported in part by NIH grants PO1-AR38923 and RO1-AR44350.

7. REFERENCES

Aladjem, M., Groudine, M., Brody, L., Dieken, E., Fournier, R., Wahl, G. and Epner, E. (1995). Participation of the human β-globin locus control region in initiation of DNA replication. *Science*, 270: 815-819.

Alexeev, V. and Yoon, K. (1998) Stable and inheritable changes in genotype and phenotype of albino melanocytes induced by an RNA-DNA oligonucleotide. *Nature Biotechnology*, 16: 1343-1346.

Bhattacharyya, A., Murchie, A.I. and Lilley, D.M. (1990). RNA bulges and the helical periodicity of double-stranded RNA. *Nature*, 343: 484-487.

Bootsma, D. and Hoeijmakers, J.H. (1993). DNA repair. Engagement with transcription. *Nature*, 363: 114-115.

Capecchi, M.R. (1989). Altering the genome by homologous recombination. *Science*, 244: 1288-1292.

Chin, D.J., Green, G.A., Zon, G., Szoka, F.C.Jr. and Straubinger, R.M. (1990) Rapid nuclear accumulation of injected oligodeoxyribonucleotides. New Biol., 2: 1091-1100.

Cole-Strauss, A., Yoon, K., Xiang, Y., Byrne, B.C., Rice, M.C., Gryn, J., Holloman, W.K. and Kmiec, E.B. (1996). Correction of the mutation responsible for sickle cell anemia by an RNA-DNA oligonucleotide *Science*, 273: 1386-1389.

Cole-Strauss, A., Gamper, H., Holloman, W.K., Munoz, M., Cheng, N. and Kmiec, E.B, (1999). Targeted gene repair directed by the chimeric RNA/DNA oligonucleotide in a mammalian cell-free extract. *Nucleic Acids Res.*, 27: 1323-1330.

Daniels, G.A. and Lieber, M.R. (1995a). Strand specificity in the transcriptional targeting of recombination at immunoglobulin switch sequences. *Proc. Natl. Acad. Sci. USA*, 92: 5625-5629.

Daniels, G.A. and Lieber, M.R. (1995b). RNA:DNA complex formation upon transcription of immunoglobulin switch regions: implications for the mechanism and regulation of class switch recombination. *Nucleic Acids Res.*, 23: 5006-5011.

Dash, P., Lotan, I., Knapp, M., Kandel, E.R. and Goelet, P. (1987). Selective elimination of mRNAs in vivo: complementary oligodeoxynucleotides promote RNA degradation by an RNase H-like activity. *Proc. Natl. Acad. Sci. USA*, 84, 7896-7900.

Diem, C. and Runger, T.M. (1997) Processing of three different types of DNA damage in cell lines of a cutaneous squamous cell carcinoma progression model. *Carcinogenesis*, 18: 657-662.

Fang, W.H. and Modrich, P. (1993). Human strand-specific mismatch repair occurs by a bidirectional mechanism similar to that of the bacterial reaction. *J. Biol. Chem.*, 268: 11838-11844.

Felgner, P.L. and Ringold, G.M. (1989) Cationic liposome-mediated transfection. *Nature*, 337: 387-388.

Freier, S.M. and Altmann, K.H. (1997). The ups and downs of nucleic acid duplex stability: structure-stability studies on chemically-modified DNA:RNA duplexes. *Nucleic Acids Res.*, 25: 4429-4443.

Gonda, D.K. and Radding, C.M. (1983). By searching processively RecA protein pairs DNA molecules that share a limited stretch of homology. *Cell*, 34: 647-654.

Gralla, J. and Crothers, D.M. (1973). Free energy of imperfect nucleic acid helices. II. Small hairpin loops. *J. Mol. Biol.*, 73: 497-511.

Gyi, J.I., Conn, G.L., Lane, A.N. and Brown, T. (1996). Comparison of the thermodynamic stabilities and solution conformations of DNA.RNA hybrids containing purine-rich and pyrimidine-rich strands with DNA and RNA duplexes. *Biochemistry*, 35: 12538-12548.

Halaban, R. and Moellmann, G. (1993). White mutants in mice shedding light on humans. *J. Invest. Dermatol.*, 100: 176S-185S.

Hanawalt, P.C. (1994). Transcription-coupled repair and human disease. *Science*, 266: 1957-1958.

Havre, P.A. and Kmiec, E.B. (1998) RecA-mediated joint molecule formation between 2'-O-methylated RNA/DNA hairpins and single-stranded targets. *Mol. Gen. Genet.*, 258: 580-586.

Hobbs, C.A. and Yoon K. (1994) Differential regulation of gene expression *in vivo* by triple helix-forming oligonucleotides as detected by a reporter enzyme. *Antisense Res. Dev.*, 4: 1-8.

Hsieh, P., Camerini-Otero, C.S. and Camerini-Otero, R.D. (1990). Pairing of homologous DNA sequences by proteins: evidence for three-stranded DNA. *Genes. Dev.*, 4: 1951-1963.

Hsieh, P., Camerini-Otero, C.S. and Camerini-Otero, R.D. (1992). The synapsis event in the homologous pairing of DNAs: RecA recognizes and pairs less than one helical repeat of DNA. *Proc. Natl. Acad. Sci. USA*, 89: 6492-6496.

Igoucheva, O., Peritz, A.E., Levy, D. and Yoon, K. (1999). A sequence-specific gene correction by an RNA-DNA oligonucleotide in mammalian cells characterized by transfection and nuclear extract using a lacZ shuttle system. *Gene Therapy*, 12: 1960-1971.

Inoue, H., Hayase, Y., Imura, A., Iwai, S., Miura, K., and Ohtsuka, E. (1987). Synthesis and hybridization studies on two complementary nona(2'-O- methyl)ribonucleotides. *Nucleic Acids Res.*, 15: 6131-6148.

Jackson, I.J. and Bennett, D. (1990). Identification of the albino mutation of mouse tyrosinase by analysis of an in vitro revertant. *Proc. Natl. Acad. Sci. USA*, 87: 7010-7014.

Jackson-Grusby, L., Laird, P., Magge, S., Moeller, B. and Jaenisch, R. (1997). Mutagenicity of 5-aza-2'-deo-xycytidine is mediated by mammalian DNA methyltransferase. *Proc. Natl. Acad. Sci. USA*, 94: 4681-4685.

Jimenez, M., Tsukamoto, K. and Hearing, V. (1991). Tyrosinases from 2 different loci are expressed by normal and by transformed melanocytes. *J. Biol. Chem.*, 266: 1147-1156.

Kamath-Loeb, A.S., Hizi, A., Tabone, J., Solomon, M.S. and Loeb, L.A. (1997) Inefficient repair of RNAxDNA hybrids. *Eur. J. Biochem.*, 250: 492-501.

Kat, A., Thilly, W., Fang, W., Longley, M.J., Li, G.M. and Modrich, P. (1993). An alkylation-tolerant, mutator human cell line is deficient in strand-specific mismatch repair. *Proc. Natl. Acad. Sci.USA*, 90: 6424-6428.

Kirkpatrick, D.P. and Radding, C.M. (1992). RecA protein promotes rapid RNA-DNA hybridization in heterogeneous RNA mixtures. *Nucleic Acids Res.*, 20: 4347-4353.

Kmiec, E.B., Cole, A., and Holloman, W.K. (1994). The REC2 gene encodes the homologous protein of Ustilago maydis. *Mol. Cell Biol.*, 14: 7163-7172.

Koltover, I., Salditt, T., Radler, J.O. and Safinya, C.R. (1998) An inverted hexagonal phase of cationic liposome-DNA complexes related to DNA release and delivery. *Science*, 281: 78-81.

Kotani, H. and Kmiec, E.B. (1994a). Transcription activates RecA-promoted homologous of nucleosomal DNA. *Mol. Cell Biol.*, 14: 1949-1955.

Kotani, H. and Kmiec, E.B. (1994b). A role for RNA synthesis in homologous events. *Mol. Cell Biol.*, 14: 6097-6106.

Kotani, H., Germann, M.W., Andrus, A., Vinayak, R., Mullah, B. and Kmiec, E.B. (1996). RNA facilitates RecA-mediated DNA and strand transfer between molecules bearing limited regions of homology. *Mol. Gen. Genet.*, 250: 626-634.

Kren, B.T., Bandyopadhyay, P. and Steer, C.J. (1998). In vivo site-directed mutagenesis of the factor IX gene by chimeric RNA/DNA oligonucleotides. *Nat. Med.* 4, 285-290.

Kren, B., Cole-Strauss, A., Kmiec, E. and Steer, C. (1997). Targeted nucleotide exchange in alkaline phosphatase gene of HuH-7 cells mediated by a chimeric RNA/DNA oligonucleotide. *Hepatology*, 25: 1462-1468.

Lawrence, J.N. and Benford, D.J. (1993) Chemically-induced unscheduled DNA synthesis in cultures of adult human epidermal keratinocytes. *Mut. Res.*, 291: 105-115.

Lee, S., Elenbaas, B., Levine, A. and Griffith, J. (1995). p53 and its 14 kDa C-terminal domain recognize primary DNA damage in the form of insertion/deletion mismatches. *Cell*, 81: 1013-1020.

Leonetti, J.P., Mechti, N., Degols, G., Gagnor, C. and Lebleu, B. (1991). Intracellular distribution of microinjected antisense oligonucleotides. *Proc. Natl. Acad Sci. USA*, 88: 2702-2706.

Lesnik, E. and Freier, S.M. (1995) Relative thermodynamic stability of DNA,RNA and DNA:RNA hybrid duplexes: Relationship with base composition and structure. *Biochemistry*, 34: 10807-10815.

Lesnik, E., Guinosso, C., Kawasaki, A.M., Sasmor, H., Zounes, M., Cummins, L., Ecker, D., Cook, D.P. and Freier, S.M. (1993). Oligodeoxynucleotides containing 2'-O-Methyl adenosine: Synthesis and effect on stability of DNA:RNA duplexes. *Biochemistry*, 32: 7832-7838.

Lesnik, E.A. and Freier, S.M. (1998). What affects the effect of 2'-alkoxy modifications? 1-Stabilization effect of 2'-methoxy substitutions in uniformly modified DNA oligonucleotides. *Biochemistry*, 37: 6991-6997.

Lin, F.M., Sperle, K. and Sternberg, N. (1990). Intermolecular recombination between DNAs introduced into mouse L cells is mediated by nonconservative pathway that leads to crossover products. *Mol. Cell. Biol.*, 10: 113-119.

Majlessi, M., Nelson, N.C. and Becker, M.M. (1998). Advantages of 2'-O-methyl oligoribonucleotide probes for detecting RNA targets. *Nucleic Acids Res.*, 26: 2224-2229.

Mansour, S., Thomas, K. and Capecchi, M.R. (1988). Disruption of the proto-oncogene int-2 in mouse embryo-derived stem cells: a general strategy for targeting mutations to non-selectable genes. *Nature*, 336: 348-352

Martin, F.H. and Tinoco, I.Jr. (1980). DNA-RNA hybrid duplexes containing oligo (dA:rU) sequences are exceptionally unstable and may facilitate termination of transcription. *Nucleic Acids Res.,* 8: 2295-2299.

Maryon, E. and Carroll, D. (1991) Characterization of recombination intermediates from DNA injected into Xenopus laevis oocytes: evidence for a nonconservative mechanism of homologous recombination. *Mol. Cell. Biol.*, 8: 3918-3928.

Mekeel, K.L., Tang, W., Kachnic, L.A., Luo, C.M., DeFrank, J.S. and Powell, S.N. (1997). Inactivation of p53 results in high rates of homologous recombination. *Oncogene*, 14: 1847-1857.

Merrihew, R., Marburger, K., Pennington, S., Roth, D. and Wilson, J. (1996). High-frequency illegitimate integration of transfected DNA at preintegrated target sites in mammalian genome. *Mol. Cell Biol.*, 16: 10-18.

Meyn, M.S., Strasfeld, L. and Allen, C. (1994) Testing the role of p53 in the expression of genetic instability and apoptosis in ataxia-telangiectasia. *Int. J. Radiat. Biol .*, 66: S141-S149.

Milligan, J.F., Matteucci, M.D. and Martin, J.C. (1993). Current concepts in antisense drug design. *J. Med Chem.*, 36: 1923-1937.

Modrich, P. (1997). Strand-specific mismatch repair in mammalian cells. *J. Biol. Chem.*, 272: 24727-24730.

Modrich, P. and Lahue, R. (1996). Mismatch repair in replication fidelity, genetic recombination, and cancer biology. *Annu. Rev. Biochem.*, 65: 101-133.

Monia, B.P, Johnston, J.F., Ecker, D.J., Zounces, M.A., Lima, W.F. and Freier, S.M. (1992). Selective inhibition of mutant Ha-ras mRNA expression by antisense oligonucleotides.*J. Biol. Chem.*, 267: 19954-19962.

Monia, B.P., Lesnik, E.A., Gonzalez, C., Lima, W.F., McGee, D., Guinosso, C.J., Kawasaki, A.M., Cook, P.D. and Freier, S.M. (1993). Evaluation of 2'-modified oligonucleotides containing 2'-deoxy gaps as antisense inhibitors of gene expression. *J. Biol. Chem.*, 268: 14514-14522.

Nielsen, P.E., Egholm, M., Berg, R.H. and Buchardt, O. (1991). Sequence-selective recognition of DNA by strand displacement with a thymine-substituted polyamide. *Science*, 254: 1497-1500.

Pennington, S.L. and Wilson, J.H. (1991) Gene targeting in chinese hamster ovary cells is conservative. *Proc. Natl. Acad. Sci. USA*, 88: 9498-9502.

Petersen, L.N., Orren, D.K and Bohr, V.A. (1995). Gene-specific and strand-specific DNA repair in the G1 and G2 phases of the cell cycle. *Mol. Cell Biol.*, 15: 3731-3737.

Poncz, M., Schwartz, E., Ballantine, M. and Surrey, S. (1983). Nucleotide sequence analysis of the β and δ-globin gene region in humans. *J. Biol. Chem.*, 258: 11599-11609.

Radler, J., Koltover, I., Salditt, T. and Sanfinya, C. (1997). Structure of DNA-cationic liposome comlexes: DNA intercalation in multilamella membranes in distinct interhelical packing regimes. *Science*, 275: 810-814.

Riley, M. and Maling, B. (1966). Physical and chemical characterization of two- and three-stranded adenine-thymine and adenine-uracil homopolymer complexes. *J. Mol. Biol.*, 20: 359-389.

Roberts, R.W. and Crothers, D.M. (1992). Stability and properties of double and triple helices: dramatic effects of RNA or DNA backbone composition. *Science,* 258: 1463-1466.

Rouer, P., Smih, F. and Jasin, M. (1994) Expression of a site-specific endonuclease stimulates homologous recombination in mammalian cells. *Proc. Natl. Acad. Sci. USA*, 91: 6064-6068.

Santana, E., Peritz, A.E., Iyer, S., Uitto, J. and Yoon, K. (1998). Different frequency of gene targeting events by the RNA-DNA oligonucleotide among epithelial cells. *J. Invest. Dermatology* (in the press).

Scully, R., Chen, J., Plug, A., Xiao, Y., Weaver, D., Feunteun, J., Ashley, T. and Livingston, D.M. (1997). Association of BRCA1 with Rad51 in mitotic and meiotic cells. *Cell*, 88: 265-275.

Sharan, S.K., Morimatsu, M., Albrecht, U., Lim, D.S., Regel, E., Dinh, C., Sands, A., Eichele, G., Hasty, P. and Bradley, A. (1997). Embryonic lethality and radiation hypersensitivity mediated by Rad51 in mice lacking Brca2 [see comments]. *Nature*, 386: 804-810.

Shibahara, S., Mukai, S., Morisawa, H., Nakashima, H., Kobayashi, S. and Yamamoto, N. (1989). Inhibition of human immunodeficiency virus (HIV-1) replication by synthetic oligo-RNA derivatives. *Nucleic Acids Res.*, 17: 239-252.

Stasiak, A., West, S.C. and Egelman, E.H. (1997). Sickle cell anemia research and a recombinant DNA technique [letter; comment]. *Science*, 277: 460-462.

Strauss, M. (1998). The site-specific correction of genetic defects [news; comment]. *Nat. Med.*, 4: 274-275.

Sturzbecher, H., Donzelmann, B., Henning, W., Knippschild, U. and Buchlop, S. (1996). p53 is linked directly to homologous recombination processes via RAD51/RecA protein interaction. *EMBO J.*, 15: 1992-2002.

Sugimoto, N., Nakano, S., Katoh, M., Nakamuta, H., Ohmichi, T., Yoneyama, M. and Sasaki, M. (1995). Thermodynamic parameters to predict stability of RNA/DNA hybrid duplexes. *Biochemistry*, 34: 11211-11216.

Tereshko, V., Portmann, S., Tay, E.C., Martin, P., Natt, F., Altmann, K-H and Egli, M. (1998). Correlating structure and stability of DNA duplexes with incorporated 2'-O-modified RNA analogs. *Biochemistry*, 37: 10626-10634.

Thaler, D.S., Liu, S. and Tombline, G. (1996). Extending the chemistry that supports genetic information transfer in vivo: phosphorothioate DNA, phosphorothioate RNA, 2'-O-methyl RNA, and methyl-phosphonate DNA. *Proc. Natl. Acad. Sci. USA*, 93: 1352-1356.

Thomas, K.R. and Capecchi, M.R. (1987). Site-directed mutagenesis by gene targeting in mouse embryo-derived stem cells. *Cell*, 51: 503-512.

Thomas, K.R. and Capecchi, M.R. (1997). Recombinant DNA technique and sickle cell anemia research [letter; comment]. *Science*, 275: 1404-1405.

Thyagarajan, B. and Campbell, C. (1997). Elevated homologous recombination acitivity in Fanconi Anemia Fibroblasts. *J. Biol. Chem.*, 272: 23328-23333.

Thyagarajan, B., McCormick-Graham, M., Romero, D. and Campbell, C. (1996). Characterization of homologous DNA recombination activity in normal/immortal mammalian cells. *Nucleic Acids Res.*, 24: 4084-4091.

Vlassov, V.V., Balakireva, L.A. and Yakubov, L.A. (1994) Transport of oligonucleotides across natural and model membranes. *Biochim. Biophys. Acta*, 1197: 95-108.

Wang, S. and Kool, E. (1995). Relative stabilities of triple helices composed of combination of DNA, RNA and 2'-O-methyl-RNA backbones: chimeric circular oligonucleotides as probes. *Nucleic Acids Res.*, 23: 1157-1164.

Wiesmuller, L., Cammenga, J. and Deppert, W.W. (1996). In vivo assay of p53 function in homologous recombination between simian virus 40 chromosomes. *J. Virol.*, 70: 737-744.

Wilson, J.H., Leung, W.Y., Bosco, G., Dieu, D. and Haber, J.E. (1994). The frequency of gene targeting in yeast depends on the number of target copies. *Proc. Natl. Acad. Sci. USA*, 91: 177-181.

Woese, C.R., Winker, S. and Gutell, R.R. (1990). Architecture of ribosomal RNA: constraints on the sequence of "tetra-loops". *Proc. Natl. Acad. Sci. USA*, 87: 8467-8471.

Xiang, Y., Cole-Strauss, A., Yoon, K., Gryn, J. and Kmiec, E. (1997) Targeted gene conversion in mammalian CD34$^+$-enriched cell population using a chimeric RNA/DNA oligonucleotide. *J. Mol. Med.*, 75: 829-835.

Yakubov, L., Deeva, E., Zarytova, V., Ivanova, E., Ryte, A., Yurchenko, L. and Vlassov, V. (1989). Mechanism of oligonucleotide uptake by cells: involvement of specific receptors? *Proc Natl Acad Sci USA*, 86: 6454-6458

Yanez, R.J. and Porter, A.C. (1998). Therapeutic gene targeting. *Gene Ther.*, 5: 149-159.

Yoon, K., Cole-Strauss, A. and Kmiec, E.B. (1996). Targeted gene correction of episomal DNA in mammalian cells mediated by a chimeric RNA/DNA oligonucleotide. *Proc. Natl. Acad. Sci. USA*,. 93: 2071-2076.

Yoon, K., Thiede, M.A. and Rodan, G.A. (1988) Alkaline phosphatase as reporter enzyme. *Gene*, 66: 11-17.

Zheng, H., Hasty, P., Brenneman, M.A., Grompe, M., Gibbs, R.A., Wilson, J.H. and Bradley, A. (1991). Fidelity of targeted recombination in human fibroblasts and murine embryonic stem cells. *Proc. Natl. Acad. Sci.USA*, 88: 8067-8071.

8. SUMMARY — An oligonucleotide composed of a contiguous stretch of RNA and DNA residues has been developed to facilitate correction of single-base mutations of episomal and chromosomal targets in mammalian cells. Design of the oligonucleotide exploited the highly recombinogenic RNA-DNA hybrids and featured hairpin capped ends avoiding destruction by cellular helicases or exonucleases. The RNA-DNA oligonucleotide (RDO) was shown to cause a site-specific chromosomal correction in several genes and to induce a sequence-specific mutation *in vivo* by intravenous injection. Moreover, a permanent and stable gene correction by the RDO was demonstrated by clonal analysis at the level of genomic sequence, protein and phenotypic change. The RDO might hold promise as a therapeutic method for the treatment of genetic diseases. However, the frequency of gene correction varies among different cells and individual experiments. Cellular activities such as recombination and repair may be important for gene conversion by the RDO. As this technology is more widely utilized in the scientific community, it will be important to understand the mechanism and to optimize the design of the RDO to improve its efficiency and its general applicability.

Key Words: gene repair, strand pairing, DNA repair, RNA-DNA hybrids

Progress in Gene Therapy: Basic and Clinical Frontiers, pp. 513-549
R. Bertolotti *et al.* (Eds)

Gene therapy: dsDNA-cored presynaptic filaments as vectors for gene repair and targeted integration of transgenes

Roger Bertolotti*

CNRS, Molecular Genetics, Department of Hepato-Gastroenterology, Faculty of Medicine, University of Nice Sophia Antipolis, 06107 Nice, France

Table of Contents

---

* E-mail: Roger.Bertolotti@unice.fr

1. INTRODUCTION

a) Toward gene repair and optimized minigene expression therapy

Current gene therapy relies on the transfer and expression of minigenes into patient's somatic cells (see introductory chapter of the book). This has been our first approach to gene therapy as a collaborative preliminary projet on the Gaucher disease with Dr. Edward Ginns at the National Institutes of Health in Bethesda (McKinney C., Mellouk S., LaMarca M., Bertolotti R. and Ginns E., 1990-1991, unpublished). The strategy is to enforce appropriate cells from a patient to synthesize an active protein or non-coding RNA upon introduction of the corresponding minigene, *i.e.* a basic artificial gene resulting from the insertion of DNA encoding a protein or a non-coding RNA into various expression cassettes (see introductory chapter of the book). Minigenes are used either to compensate the deficiency of their normal counterparts in inherited diseases or to provide therapeutic/prophylactic benefits for acquired disorders. Except for some viral vectors (*e.g.* alphaviruses or poxviruses: Berglund *et al.*, 1998; Moss, 1996), expression is effective when minigenes are in the nucleus of transfected cells. Minigenes are either free (transient expression) or integrated into host chromosomes (permanent expression). Yet, transient expression may also result from occasional unstable expression of integrated minigenes and from elimination of transfected cells. On the other hand, stable expression is eventually achieved with some virus-based shuttle plasmids that are maintained as self-replicating nuclear episomes (DiMaio *et al.*, 1982; Yates *et al*, 1985; Lutfalla *et al.*, 1989).

The use of these artificial genes either for transient or long-term expression holds great promises both for inherited diseases and acquired disorders; indeed, it is expected to revolutionize the practice of Medicine at the turn of the millenium (see introductory chapter of the book). However, such an approach does not yet provide for non-random (targeted) integration of the therapeutic minigene, a major concern for long-term expression (see below); in addition, it has limitations for inherited diseases where physiological genetic regulation is not easy to achieve with current minigenes (see introductory chapter of the book). Our new approach to gene therapy, initiated in december 1991, is aimed at solving both aforedescribed problems upon establishing an efficient technology for gene repair (inherited diseases) and targeted integration of therapeutic minigenes (acquired disorders).

b) Gene repair and targeted integration of therapeutic minigenes

The most obvious approach to tackle an inherited disease is to repair the dysfunctional gene in appropriate somatic cell targets. Reestablishing the wild type sequence of the mutant gene *in situ* requests efficient gene repair techniques that are under intensive investigations. They are essentially based on gene targeting, *i.e.* homologous recombination between chromosomal DNA and transfecting DNA (Capecchi, 1989). Such an exchange of DNA sequences provides the means to modify at will the sequence of the target gene. However, well-established gene targeting technology has such a poor efficiency that it is currently performed in conjunction with selectable markers in order to isolate rare valuable transfectant cells (Capecchi, 1989; Abuin and Bradley, 1996). Such protocols are fine to mutate pluripotent embryo-derived stem cells for the production of transgenic mice (Capecchi, 1989; Abuin and Bradley, 1996) and, possibly, for *ex vivo* gene therapy with some stem cells. They are however inappropriate for most gene therapy protocols where drastic selective growth of transfectant cells is impossible. This is why, in 1991, we devised a new approach to gene targeting aimed at increasing its efficiency to a level compatible with gene therapy protocols.

Our gene therapy approach is based on the transfer of premade presynaptic filaments, *i.e.* the active nucleoprotein complexes that mediate the key reaction from homologous recombination (Bertolotti R., Grant application to French MRT, 1991; Bertolotti, 1996a and 1996b). Such enzymatic complexes resulting from the *in vitro* polymerization of recombinase protein onto therapeutic DNA are expected to drive, upon transfection into target cells, efficient homologous DNA pairing and exchange with cognate chromosomal DNA (see below). Such a directed homologous recombination process should culminate in the efficient substitution of chromosomal DNA by homologous genomic DNA from wild type (gene repair) or mutant (gene inactivation) origin.

Importantly enough, we have invented chimeric presynaptic filaments with a double-stranded DNA (dsDNA) core that offer broad DNA exchange potentialities (Bertolotti, 1996b). These filaments may thus be able to mediate efficient targeted integration of therapeutic minigenes, *i.e.* non-random integration into appropriate chromosomal sites of target somatic cells (Bertolotti, 1998c and 1999a). Our recombinase-mediated gene therapy approach is thus discussed both in terms of gene repair for inherited diseases and of improved therapeutic minigene expression therapy for acquired disorders.

2. HOMOLOGOUS RECOMBINATION IN TRANSFECTANT CELLS

a) Transfecting DNA is subject to transient homologous recombination but integrates mainly upon illegitimate recombination

Transfected dividing cells give rise to stable transfectants when exogenous DNA is either integrated into host chromosomal DNA (Robins *et al.*, 1981; Folger *et al.*, 1982) or maintained as a self-replicating nuclear episome (see aforedescribed virus-based shuttle plasmids).

Under current transfection protocols, plasmid integration into host chromosomal DNA usually occurs as head-to-tail multimeric inserts (Fig. 1). As illustrated in figure 1, such multimeric structures are generated by homologous recombination before integration into host chromosomal DNA (Folger *et al.*, 1982). Indeed, in most transfected cells, exogenous DNA molecules are suject to an intense homologous recombination activity (Capecchi, 1989; Folger *et al.*, 1982; Miller and Temin, 1983; Lutfalla *et al.*, 1985). Such a homologous recombination activity coincides with the early transfection phase when newly introduced DNA is not yet packaged into chromatin (Capecchi 1989). It is transient and, in association with ligation and end-joining activities (Folger *et al.* 1982; Miller and Temin, 1983; Roth

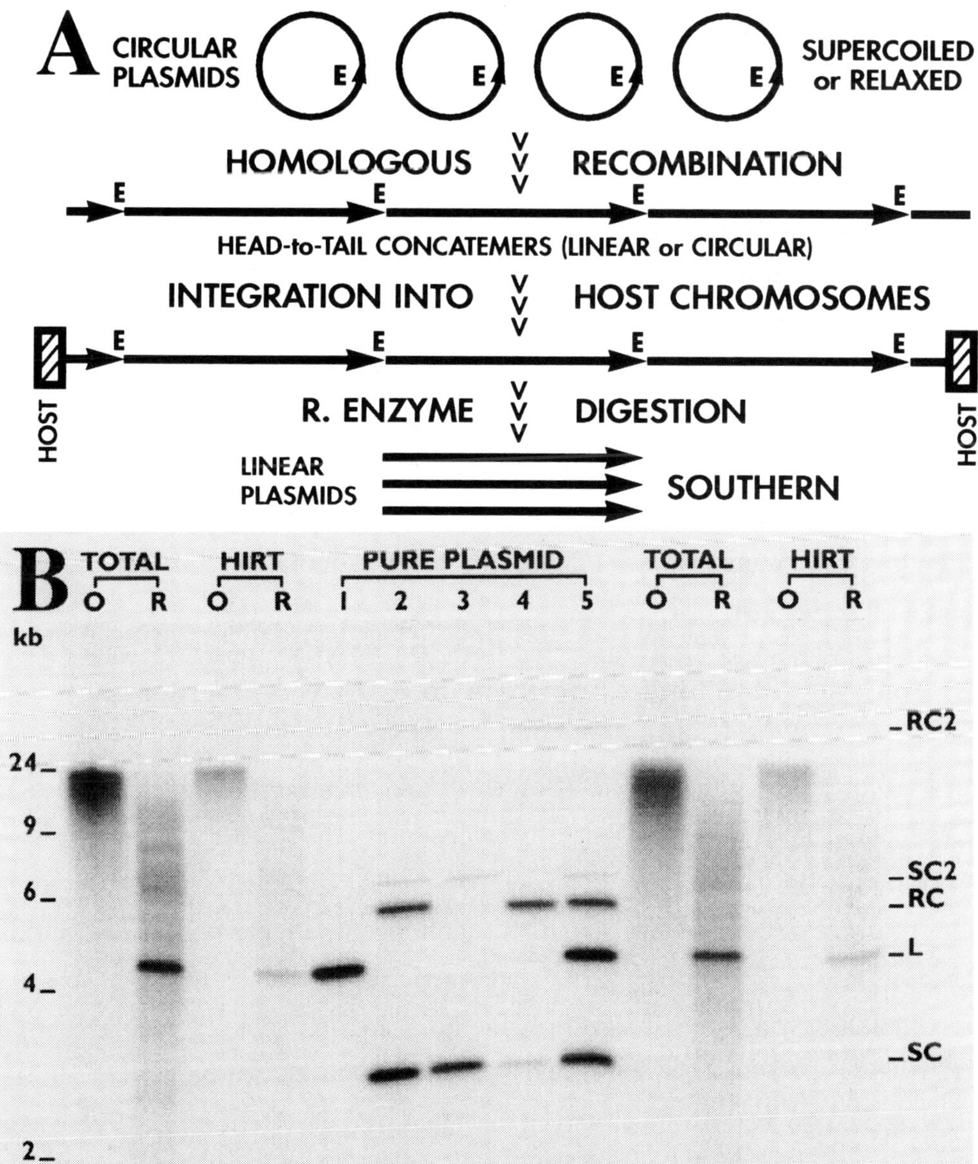

Figure 1. Formation and integration of head-to-tail concatemers in transfected mammalian cells (Bertolotti, 1997 and 1998a). **A.** Transfection scheme with circular plasmids. Digestion of circular plasmids or head-to-tail concatemers with restriction enzyme R that cuts the plasmid in a single point (E) gives rise to full-length linear plasmids. **B.** Southern blot analysis of the plasmid content of stable transfectants. Low-molecular weight (Hirt) and total DNA were extracted and digested either with enzyme O that does not cut the plasmid or with R. Plasmid-specific probes showed that plasmid DNA was associated to host chromosomal DNA (O lanes); no free plasmid either supercoiled (SC and dimer SC2) or relaxed circular (RC and dimer RC2) was ever found (see pure plasmid combinations). Digestion with R generated essentially full-length linear plasmids (L; R lanes) indicative of head-to-tail multimeric inserts (Lutfalla *et al.*, 1985).

et al., 1995; Critchlow and Jackson, 1998), generates concatemers (pure plasmids) or concatenates (plasmids and carrier DNA) that give rise to stable transformants upon integration into host chromosomes (Perucho *et al.*, 1980; Robins *et al.*, 1981). Unlike concatemer genesis, concatemer integration into chromatinized chromosomal DNA is usually not the result of homologous recombination. Even with isogenic DNA (Te Riele *et al.*, 1992; Thomas *et al.,* 1992), gene targeting, *i.e.* integration upon homologous recombination, is an unfrequent event; its frequency is only a few percent of random integration upon illegitimate recombination (Capecchi, 1989; Thomas *et al.*, 1986; Merrihew *et al.*, 1996; Yanez and Porter, 1998). However, although unfrequent, gene targeting in mammalian somatic cells is a highly significant process (Capecchi, 1989; Thomas *et al.*, 1992; Abuin and Bradley, 1996).

b) Somatic recombinases as RecA-like DNA repair enzymes

Such a transient homologous recombination activity of transfecting DNA molecules, either alone (Folger *et al.*, 1982; Lutfalla *et al.*, 1985; Capecchi, 1989) or with cognate chromosomal DNA (Thomas *et al.*, 1986; Capecchi, 1989), is in contrast with the absence of homologous chromosome recombination in most mammalian somatic cells and hybrids (Ephrussi, 1972; Panthier and Condamine 1991), indicating that somatic recombinases do not usually mediate homologous genetic recombination (Bertolotti, 1996a). Like the bacterial RecA protein (Roca and Cox, 1990), they appear to be mainly involved in DNA repair (Bertolotti, 1996a; see RAD51 below). Therefore, in december 1991, our new approach to gene therapy (Bertolotti R., Grant application to French MRT, 1991) started with the search for such a mammalian homolog of RecA (Borchiellini *et al.*, 1997) that could be used to produce *in vitro* nucleoprotein complexes analogous to the bacterial RecA presynaptic filament. The premade recombinase-DNA complexes should mimic, upon transfection, recombinational DNA repair that occurs in most dividing mammalian cell and thereby promote homologous recombination between transfecting DNA and cognate chromosomal DNA (Bertolotti, 1996a and 1996b).

3. RECOMBINASE-DNA NUCLEOPROTEIN VECTORS

a) Standard presynaptic filaments as gene conversion vectors

The eukaryotic RAD51 gene was cloned, sequenced and shown to encode a putative RecA-like protein in 1992 (Shinohara *et al.*, 1992; Aboussekra *et al.*, 1992; Basile *et al.*, 1992). However, the enzymatic activity of the RAD51 protein was demonstrated in 1994 only (Sung, 1994). In fact, the search for an eukaryotic recombinase has been an intensive disappointing task (see: Heyer, 1994; Borchiellini *et al.*, 1997) since RAD51 protein awaited an additional detailled enzymatic characterization (Sung and Robberson, 1995) before it could be established as a true recombinase. Indeed, like bacterial RecA (Kowalczykowski *et al.*, 1994; Roca and Cox, 1990), the eukaryotic RAD51 protein is able to mediate *in vitro* the key reaction from homologous recombination, *i.e.* ATP-dependent homologous DNA pairing-strand exchange with double-stranded DNA (Sung, 1994; Baumann *et al.*, 1996; Gutpa *et al.*, 1997). However, to be active under current experimental conditions, RAD51 requires the presence of replication protein A (RPA) (Sung, 1994) or eventually RAD52 (Benson *et al.*, 1998). Like in bacteria (Kowalczykowski *et al.*, 1994; Roca and Cox, 1990), the eukaryotic active molecule is a right-handed helical nucleoprotein (Fig. 2) formed, in the presence of ATP and RPA, upon recombinase protein polymerization onto single-stranded DNA (ssDNA) at a stochiometry of 3 nucleotides per monomer (Sung and Robberson, 1995). Adjunction of RAD52 protein may improve filament production (Sung, 1997a; Benson *et al.*, 1998; Shinohara and Ogawa, 1998; New *et al.*, 1998). Other proteins, such as RAD54 that stimulates homologous pairing activity of RAD51 *in vitro* (yeast; Petukhova *et al.*, 1998), appear to be involved in other ancillary functions. They may be part of a putative recombinosome that includes a number of proteins such as RAD51, RPA and RAD52 (Firmenich *et al.*, 1995; Hay *et al.*, 1995).

As shown in figure 2, the final product of the homologous DNA pairing-strand exchange reaction is an heteroduplex, *i.e.* a duplex DNA in which the complementary strands originate from different parental molecules. Pairing of DNA strands that differ in some points of their sequences results in mismatched bases. *In vivo*, such mismatches are the targets of cellular DNA repair systems (Friedberg *et al.*, 1995; Modrich and Lahue, 1996) that include strand-specific mismatch repair (Modrich, 1997). Elimination of a mismatch involves long-patch or short-patch mismatch repair mechanisms; it is achieved upon replacement of one of the mismatched bases by the correct match of the other base. This process results in gene conversion when the incoming base is maintained, *i.e.* the sequence of target dsDNA is changed (gene repair/inactivation).

Such a conversion of host genomic DNA may be achieved upon transfection of "naked[1]" ssDNA (Fujioka *et al.*, 1993; Yanez and Porter, 1998); its efficiency is however very poor with the potential exception of a new approach based on small PCR genomic fragments of an average length of about 500 nucleotides (Kunzelmann *et al.*, 1996; Goncz *et al.*, 1998). On the other hand, efficient correction of mismatched bases in non-replicating

Figure 2. Homologous DNA pairing and strand exchange reaction mediated by RAD51 protein *in vitro* (Bertolotti, 1999a). Under current experimental conditions, production of active filaments requires the presence of replication protein A (RPA) (Sung, 1994; Sung and Robberson, 1995; Baumann and West, 1997) and is facilitated by RAD52 (Sung, 1997a; Benson *et al.*, 1998; Shinohara and Ogawa, 1998; New *et al.*, 1998). RAD52 may obviate RPA requirement (Benson *et al.*, 1998). RAD54 enhances homologous pairing (yeast; Petukhova *et al.*, 1998) while RAD55-RAD57 heterodimers stimulate RAD51 activity (yeast; Sung, 1997b). Like with RecA (Howard-Flanders *et al.*, 1984; Roca and Cox, 1990; Radding, 1991; Stasiak and Egelman, 1994), the reaction occurs within the filament where DNA is underwound and stretched by a factor 1.5 compared to B-form dsDNA. Active human RAD51, RPA and RAD52 are produced in bacteria (Baumann *et al.*, 1996; Henricksen *et al.*, 1994; Benson *et al.*, 1998).

---

[1] Transfection protocols may include formation of complexes (Bertolotti *et al.*, 1998).

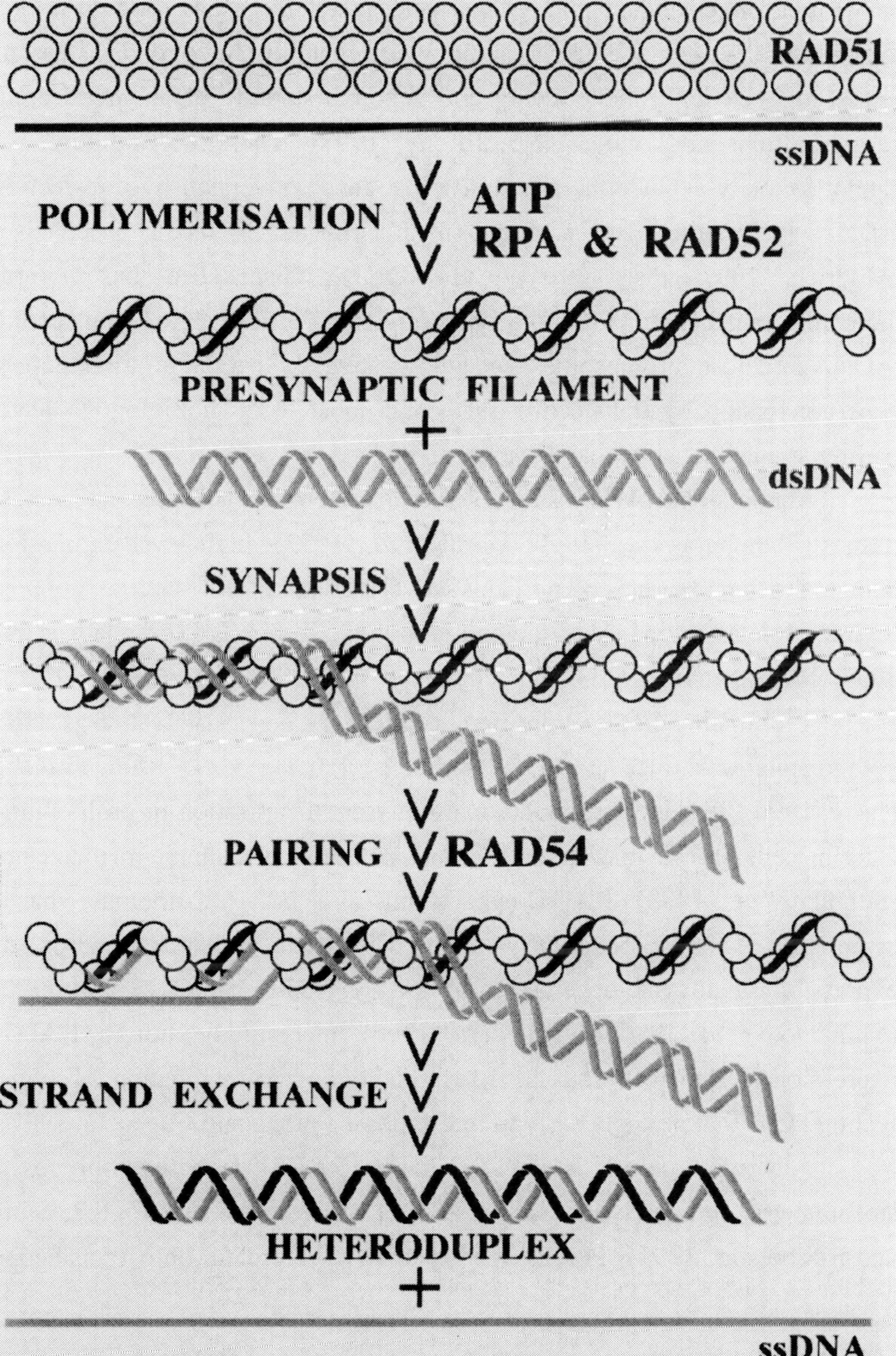
RAD51
ssDNA
POLYMERISATION
ATP
RPA & RAD52
PRESYNAPTIC FILAMENT
+
dsDNA
SYNAPSIS
PAIRING
RAD54
STRAND EXCHANGE
HETERODUPLEX
+
ssDNA

heteroduplex injected into cultured mammalian cell nuclei has been described long ago (Folger *et al.*, 1985); in addition, efficient mismatch repair is essensential to DNA replication fidelity and genomic stability of any replicating cell (Modrich and Lahue, 1996; Jiricny, 1998). Therefore, although we cannot formally exclude the possibility that gene conversion by transfecting ssDNA is impaired by a strand-specific mismatch repair mechanism (Modrich, 1997) that systematically remove mismatched bases from incoming transfecting DNA, the poor targeting[2] efficiency of transfecting dsDNA into host chromosomes (see below) suggests that the limiting step of gene conversion by transfecting ssDNA is most likely homologous DNA pairing-strand exchange that is catalyzed by RAD51 *in vitro*.

Mammalian RAD51 is an ubiquitous protein that is expressed in most tissues (Shinohara *et al.*, 1993; Morita *et al.*, 1993), high level expression being restricted to tissues comprising actively dividing cells such as gonads, thymus, spleen, uterus, intestine and embryonic materials (Shinohara *et al.*, 1993; Morita *et al.*, 1993; Yamamoto *et al.*, 1996). In fact, RAD51 is essential for mammalian cell proliferation: RAD51 gene inactivation is lethal both to embryonic mice and to cultured cells (Tsuzuki *et al.*, 1996; Lim and Hasty, 1996). In addition, conditional transgene inactivation in proliferating chicken cells shows that RAD51 is indispensable to genome maintenance (Sonoda *et al.*, 1998). RAD51 stands thus as a basic recombinase that is essential to recombinational DNA repair. Consistent with this function, its expression is cell cycle-dependent and is associated to late G1-S-G2 phases (Yamamoto *et al.*, 1996; Flygare *et al.*, 1996). Interestingly enough, RAD51 expression is also well correlated to homologous recombination of transfecting DNA that peaks in early to mid-S phase (Wong and Capecchi, 1987).

Although present in dividing cells, RAD51 is active only as a stochiometric complex with ssDNA and ATP (see above and Fig. 2; Sung and Robberson, 1995). Polymerization of RAD51 protein onto transfecting

---

[2] integration upon homologous recombination with host chromosomal DNA

ssDNA *in vivo* is likely to be less efficient than under optimized conditions *in vitro*. In addition, transfecting ssDNA is subjected to nuclease degradation both in the cytoplasm and the nucleus of recipient cells. Therefore, the recombinase sheath from premade presynaptic filaments should protect transfecting ssDNA and prevent the binding of proteins that may interfere with homologous recombination (Bertolotti, 1996b and 1998a). In conclusion, premade presynaptic filaments should thus provide transfecting ssDNA with protection, with RAD51 nuclear tagging[3] and with synapsing ability for chromosomal dsDNA (Bertolotti, 1996b and 1998a). Importantly enough, with or without the intermediate formation of recombinosomes (see above), RAD51 coating should provide for putative protein-protein interaction sites which are absent in bacterial RecA protein and which are supposed to be essential for an efficient spooling and processing of exogenous ssDNA into chromatinized mammalian chromosomal dsDNA (Bertolotti, 1996a, 1996b and 1998a).

Upon transfection into human cells, human recombinase presynaptic filaments should therefore be able of carrying out genome-wide search on chromatinized human DNA (Bertolotti, 1996a, 1996b and 1998a) in the same way *E. coli* RecA filaments do it *in vitro* on naked human DNA (Ferrin and Camerini-Otero, 1991). The ensuing directed homologous recombination should thus generate specific alterations to endogenous dsDNA sequences resulting in efficient gene inactivation or repair (Bertolotti, 1996a, 1996b and 1998a).

b) Chimeric dsDNA-cored filaments as gene targeting vectors

One of the obvious limitations of standard presynaptic filaments is the level of base pair heterology between transfecting ssDNA and host target DNA. RecA protein-mediated exchange of DNA strands can accomodate quite important DNA lesions and mismatched base-pair (Kowalczykovwski *et al.*, 1994). Such information is missing for RAD51 *in vitro*. However,

---

[3] RAD51 is a nuclear protein (Haaf *et al.*, 1995).

homologous recombination between transfecting dsDNA and cognate chromosomal DNA has been shown to be strongly inhibited by a few percents of base pair heterology (Deng and Capecchi, 1992; Te Riele *et al.*, 1992). In addition, many mutations result from deletions or insertions of multiple base sequences (see Fig. 5 and EMBL/GeneBank file No. M26434 for mutations of the human HPRT gene). This is why, upon publication of the enzymatic properties of RAD51 (Sung and Robberson, 1995), we devised a more sophisticated approach and invented dsDNA-cored filaments (Bertolotti, 1996b). As illustrated in Fig. 3, the idea is to eliminate mismatch problems by utilizing dsDNA as a therapeutic cassette and drive its targeted integration by using ssDNA tails that are 100% identical to host chromosomal dsDNA. We thus moved from gene conversion to "true" gene targeting, *i.e.* integration of transfecting dsDNA into host chromosomal DNA upon homologous recombination (Capecchi, 1989).

The point is that, unlike bacterial RecA protein, RAD51 has rougthly the same affinity for dsDNA as for ssDNA and forms helical RecA-like filaments both with ssDNA and dsDNA (Ogawa *et al.*, 1993; Benson *et al.*, 1994; Sung and Robberson, 1995). However, dsDNA-nucleoprotein filaments are inactive (Sung and Robberson, 1995). Importantly enough, coating dsDNA with RAD51 prevents duplex DNA from participating in the

Figure 3. Targeting recombination-locked dsDNA cassettes to chromosomal DNA with dsDNA-cored RAD51 filaments: gene repair/inactivation or targeted integration of minigenes (chromatinian structures are not shown; Bertolotti, 1997, 1998a and 1999a). The method consists in 1) producing *in vitro* dsDNA-cored presynatic filaments by polymerizing RAD51 protein onto appropriate DNA backbones (Fig. 5), and 2) transferring them (transfection) into recipient cells where recombination-activated ssDNA tails can drive homologous recombination with chromosomal cognate DNA. Substitution of an endogenous dsDNA segment by an exogenous recombination-free dsDNA cassette (targeted integration) should thus result from a double crossing-over mechanism involving both ssDNA tails from the targeting vector. The driving forces are therefore 1) the RAD51-coated ssDNA tails that are 100% identical to their respective targets and 2) the no-mismatch heteroduplexes resulting from homologous pairing of these activated ssDNA tails with cognate chromosomal strands. Importantly enough, adjunction of RAD52 protein to ATP and RPA may improve *in vitro* filament production (Sung, 1997a; Benson *et al.*, 1998; Shinohara and Ogawa, 1998; New *et al.*, 1998).

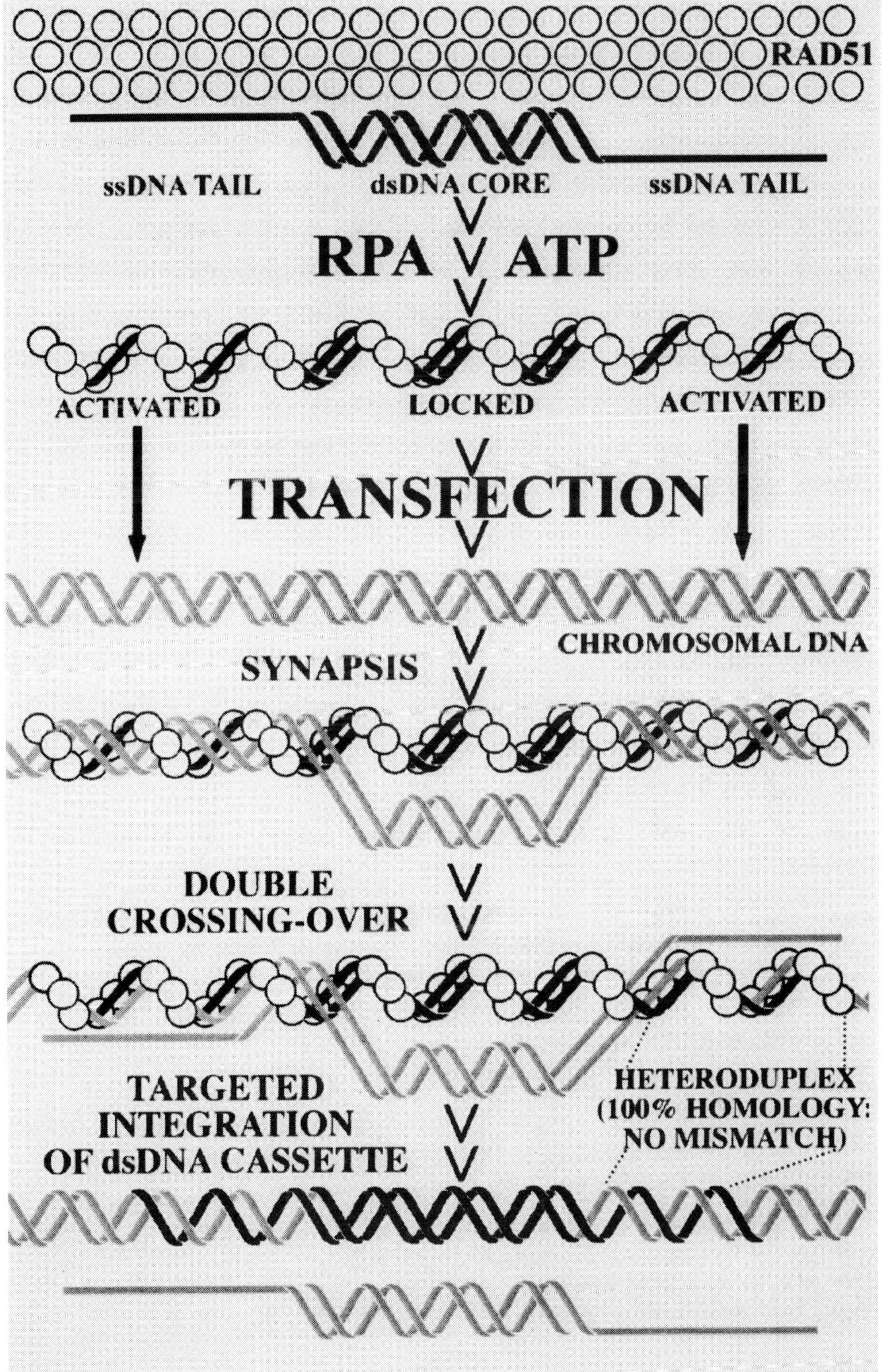
RAD51
ssDNA TAIL
dsDNA CORE
ssDNA TAIL
RPA
ATP
ACTIVATED
LOCKED
ACTIVATED
TRANSFECTION
CHROMOSOMAL DNA
SYNAPSIS
DOUBLE
CROSSING-OVER
TARGETED
INTEGRATION
OF dsDNA CASSETTE
HETERODUPLEX
(100% HOMOLOGY:
NO MISMATCH)

pairing reaction but does not affect the ability of uncoated concurrent dsDNA to react with RAD51-ssDNA filaments (Sung and Robberson, 1995). The idea is thus to devise chimeric targeting filaments in which therapeutic heterologous bases are comprised in a dsDNA core (Fig. 3). Indeed, coating such DNA backbones with RAD51 should generate recombination-activated ssDNA tails and recombination-locked dsDNA cores. Having a ssDNA tail at each end, these filaments should initiate homologous pairing-strand exchange reactions on both flanking sides of the dsDNA core, resulting, like in a double crosing-over scheme, in the substitution of an endogenous dsDNA segment by an exogenous recombination-free dsDNA cassette (Fig. 3). Such a reaction will be all the more efficient as the sequences of the ssDNA tails are 100% identical to their respective cognate host targets. Perfect identity between these ssDNA tails and their respective targets eliminates mismatch problems (see Fig. 3) and optimizes homologous DNA pairing-strand exchange reactions on both sides of the therapeutic dsDNA cassette. The dsDNA cassette is hermetic to homologous recombination; therefore, heterologous therapeutic bases, comprised in such a dsDNA cassette, should not interfere with homologous recombination.

Table 1. Putative effects of recombinase-coating on integration of transfecting DNA into mammalian host chromosomes (Bertolotti, 1997)

"Naked"* DNA	
-Gene Targeting (homologous recombination):	rare
-Random Integration ("illegitimate" recombination):	frequent
DNA-Recombinase Nucleoprotein Filaments	
-Putative direct effect:	increase of gene targeting
-Putative indirect effects:	reduction/suppression of random integration
	reduction/suppression of concatemer formation

(recombinase coating should prevent homologous recombination between transfecting DNA molecules and should protect ssDNA and dsDNA-cores from a variety of DNA-binding proteins such as end-joining ones**)

*Transfection protocols may include formation of precipitates with calcium-phosphate (Graham and Van der Eb, 1973) or of various complexes (Bertolotti *et al.*, 1998).
**In addition, end-to-end ligation described for linear dsDNA (Baumann *et al.*, 1996) should be impossible with standard or dsDNA-cored ssDNA filaments.

Another important feature of our dsDNA-cored presynaptic filaments is the opportunity to obviate unspecific targeting to repetitive chromosomal elements. Indeed, as illustrated with the human HPRT gene (Fig. 4 and 5), mammalian genomic DNA mainly comprises introns with repetitive elements such as *Alu* and LINE. Therefore, with our dsDNA-cored filaments, we should avoid unspecific targeting by excluding such ubiquitous intervening sequences from the ssDNA tails and inserting them in the dsDNA cassette when necessary (Fig. 5; Bertolotti, 1996b).

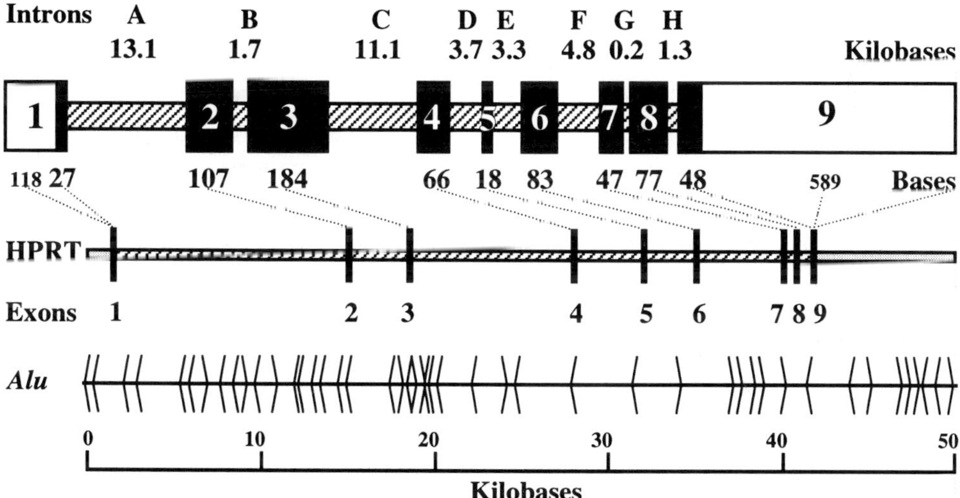

Figure 4. Structure of the human hypoxanthine-guanine phosphoribosyl transferase (HPRT) gene (based on Edwards *et al.*, 1990, and EMBL/GeneBank M26434) (Bertolotti, 1997 and 1998a). The HPRT gene comprises 9 exons (boxes) and 8 introns (hatched areas). Introns and flanking sequences (stippled areas) contain repetitive elements such as *Alu* (46 repeats, positioned with orientation) and LINE (not shown). Enlarged at the upper part of the figure, exons, not in scale with introns, span about 1364 bases out of a ~40 kb genomic sequence. The coding sequence is 657 base-long (filled boxes). Exons 1 and 9 comprise 5' and 3' non-coding regions (empty boxes), respectively.

c) DNA backbones for gene repair/inactivation

A series of DNA-backbones for nucleoprotein filaments aimed at inactivating then repairing the HPRT gene from human cells were designed (Bertolotti, 1996b) using EMBL / GenBank Data file No. M26434. This file comprises the nucleotide sequence from the human HPRT gene, its structure and its molecular alterations in a series of Lesch-Nyhan mutants (Edwards *et al.*, 1990). Some DNA backbones corresponding to two such mutants dealing with exon 3 are shown both for standard presynaptic filaments and dsDNA-cored ones (Fig. 5). Other selected mutations involve either introns or exons.

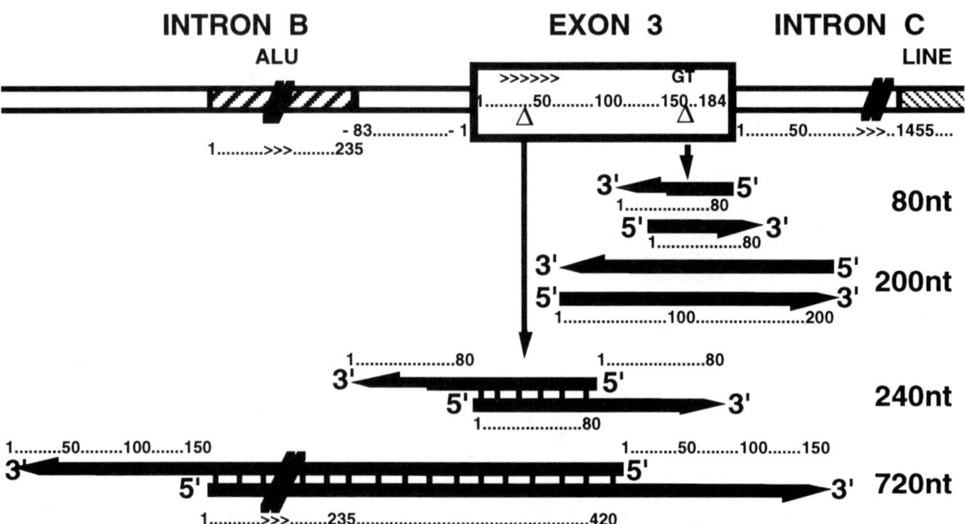

Figure 5. Designing HPRT DNA-backbones for standard and dsDNA-cored presynaptic filaments (Bertolotti, 1996b). Mutations Lesch-Nyhan RJK 2108 (>>>>) and 1332 (GT) are deletions (Δ) of either 40 base-pairs (bp) or 2 bp (GT/CA) from exon 3. Flanking exon 3 (184 bp), the 3' end of intron B (1.7 kb) and the 5' end of intron C (11.1 kb) are shown with their respective terminal *Alu* and proximal LINE repeat sequences. Single-stranded DNA stretches (ssDNA; thick horizontal arrows) are either derived from mutant (inactivating filaments) or wild-type DNA (reconstructing filaments). Backbones for chimeric presynaptic filaments comprise a double-stranded DNA (dsDNA) core for therapeutic heterologous bases and repetitive elements. Both ssDNA tails of dsDNA-cored backbones are 3' overhanging ones. Small arabic numbers refer to nucleotide (nt) length.

DNA backbones shown in figure 5 have been previously discussed (Bertolotti, 1996b and 1998a). Backbones for standard presynaptic filaments are illustrated for a mutation resulting from a deletion of two base-pairs (bp) in exon 3. Two sets of sens and antisens strands designed for the GT / CA deletion of mutant Lesch-Nyhan RJK 1332 are thus shown on the right part of figure 5. Short oligonucleotide-based filaments should be the best choice both for production and cell-internalization. However, although *in vitro* pairing can involve less than one helical repeat of DNA with RecA (Hsieh *et al.*, 1992), targeting efficiency upon transfection of RAD51-ssDNA filaments into mammalian cells may require oligonucleotide stretches longer than 25-50 nucleotides (nt). Aside from strand length, targeting efficiency may depend on strand type (sens or antisens) and concentration, on the position of heterologous bases on the strand and on properties inherent to recipient cell types. In addition, DNA methylation and discrete chromatinian structures may also interfere.

Both standard and dsDNA-cored presynaptic filaments may be used for point-mutations or very short lesions. However, illustrations for dsDNA -cored filaments have been restricted to backbones designed for a larger DNA lesion, *i.e.* a 40 bp deletion (>>>>>>; Fig. 5) in exon 3. The DNA backbones of two dsDNA-cored filaments corresponding to mutant Lesch-Nyhan RJK 2108 are thus presented in figure 5. The first one is 240 nt long. It comprises an 80 bp dsDNA core including a 40 bp cluster corresponding to deletion RJK 2108 and two 80 nt ssDNA tails. The second one is 720 nt long. It comprises a 420 bp dsDNA core including a 235 bp *Alu* repeat and the aforementioned 40 bp cluster; its tails are 150 nt long. In these cases, targeting efficiency may depend on the total length of the nucleoprotein filament as well as on the length of each component, *i.e.* tails and core. As above, potential parameters dealing with ssDNA tails include methylation of target host DNA, chromatinian structures and cell type. However, the new and main parameter is the single-stranded tail type, *i.e.* 3' or 5' overhang. Importantly enough, actual designs are based on RAD51 protein that has been shown to mediate *in vitro* the ATP-dependent homologous DNA pairing-strand exchange reaction with a polarity opposite to that of RecA (Sung and Robberson, 1995; Baumann and West, 1997). Therefore, 3' overhanging ssDNA tails are shown in figure 5 for dsDNA-cored filaments that will be used in conjunction with control filaments of the converse 5' overhang design (Bertolotti, 1996b). Such controls are all the more important as recent experiments have described a bidirectional strand exchange polarity for yeast RAD51 (Namsaraev and Berg, 1998).

d) DNA backbones for targeted integration of minigenes

Chimeric dsDNA-cored filaments offer broad DNA exchange potentialities and may therefore accomodate minigenes. In this case, the minigene of interest is inserted into the dsDNA core of appropriate recombinase-DNA nucleoprotein filaments. Such a new design should fit both sequence replacement vectors and sequence insertion vectors (Capecchi, 1989). Targeted integration of a minigene into a specific host chromosome is thus driven by genomic ssDNA tails. These tails are 100% identical to their respective chromosomal targets; they specify the integration site of the minigene.

3. DISCUSSION

a) Targeting with current vectors or with dsDNA-cored RAD51 nucleoprotein filaments

Although central to homologous recombination, homologous pairing and DNA strand exchange require additional activities in order to have the exchange process achieved. As shown in figure 3, outgoing DNA must be cleaved from chromosomal DNA and replaced upon ligation of incoming DNA. Such reactions are typical from gene targeting with replacement vectors (Thomas *et al.*, 1986; Thomas and Capecchi, 1987; Thomas *et al.*, 1992) where crossovers often occur near the ends of the vectors (Deng *et al.*, 1993). The enzymatic machinery able to carry out the double crossingover reaction is thus present in somatic cells. In addition, gene targeting can be performed with similar efficiencies in a variety of cell types from different species although there is a widespread inconsistent feeling that it is more proficient in embryonic stem cells than in primary or immortalized somatic cells (Yanez and Porter, 1998).

The propensity to integrate exogenous dsDNA at a specific chromosomal location upon homologous recombination is thus an ubiquitous property of mammalian dividing cells. However, under current procedures, genc targeting is achieved with very low frequencies even with isogenic DNA (Te Riele *et al.*, 1992; Deng and Capecchi,1992; Yanez and Porter, 1998). Consequently, gene targeting is mainly performed in conjunction with selectable markers in order to isolate rare valuable transfectant cells (Capecchi, 1989; Thomas *et al.*, 1992; Te Riele *et al.*, 1992; Abuin and Bradley, 1996). Such protocols cannot be used for most gene therapy protocols where drastic selective growth of transfectant cells is impossible.

Current gene targeting vectors are plain dsDNA (Capecchi, 1989; Thomas *et al.*, 1992; Te Riele *et al.*, 1992; Abuin and Bradley, 1996). By contrast to our dsDNA-cored presynaptic filaments, their design is not optimized for homologous recombination. First, they have to be processed in transfected cells to yield ssDNA-tailed molecules similar to our dsDNA-cored DNA-backbones (Fig. 5). *In vivo*, such a processing is accomplished by unwinding of the dsDNA by a helicase, or by degradation of one strand of the duplex by a strand-specific exonuclease, or by a combination of these activities (Eggleston and West, 1996; Bianco *et al.*, 1998). In bacteria, a 3-subunit enzyme that possesses DNA helicase, and ss- and dsDNA nuclease activities (RecBCD; Eggleston and West, 1996) is tailored for this purpose. It involves a specific DNA sequence (Chi; Myers and Stahl, 1994). "Chi-like" motifs have been recently identified in mammalian cells (Olson P. and Dornburg R., 1996, personal communication; Pan *et al.*, 1998; Olson and Dornburg, 1999), suggesting that a mammalian homolog of the RecBCD enzyme may be in charge of providing DNA substrate for homologous recombination in mammalian cells. This hypothesis is consistent with our early observation that transfecting dsDNA plasmids require mammalian genomic DNA for efficient transient homologous recombination (Lutfalla *et al.*, 1985). In eukaryotic cells, such a RecBCD-like processing might rely on

Mre11 complexes where 3' ssDNA tail genesis may result from versatile Mre11 nuclease activities: ssDNA endonuclease, 3'-to-5' ssDNA exonuclease and 3'-to-5' dsDNA exonuclease (Usui *et al.*, 1998; Trujillo *et al.*, 1998). Indeed, like bacterial RecBCD, yeast and human Mre11 complexes comprise a putative DNA helicase and have ss- and dsDNA nuclease activities (Paull and Gellert, 1998; Trujillo *et al.*, 1998; Usui *et al.*, 1998); they can possibly produce 3'-overhanging strands eventhough the dsDNA exonuclease activity of Mre11 is a 3'-to-5' one (Trujillo *et al.*, 1998; Usui *et al.*, 1998). The potential involvement of Chi-like sequences remains to be solved. Whether sequence-specific or not, such a processing of dsDNA may thus be a serious limitating factor to efficient gene targeting. In addition, this processing is in competition with unspecific nuclease degradation and with neutralizing effects of other DNA-binding proteins (Bertolotti, 1998a) such as the end-joining ones (Critchlow and Jackson, 1998). Moreover, as discussed above, *in vivo* polymerization of RAD51 on ssDNA overhangs is not likely to be as efficient as under optimized *in vitro* conditions.

With our premade dsDNA-cored filaments where DNA is protected by a RAD51 protein sheath and ssDNA tails ready for synapsis, we thus escape the initial degradative step of dsDNA processing and ssDNA-recombinase nucleation. In addition, we can possibly achieve optimized efficiency with fairly short ssDNA tails. Indeed, current gene targeting vectors require long target-homologous dsDNA ends for high-fidelity gene targeting (Thomas *et al.*, 1992; Deng and Capecchi, 1992). Requirement on both sides of the vector for such a long stretch (1 to 7 kb) of dsDNA homologous to cognate host dsDNA may result from the aforedescribed ssDNA-tail processing mechanism based on Chi-like sequences. Such sequences may not be ubiquitous enough to provide short ssDNA tails at random. This hypothesis is consistent with the long 3' overhanging ssDNA tails of double-strand break products that have been described in yeast (Cao *et al.*, 1990; Sun *et al.*, 1991). In addition, in our dsDNA-cored filaments, ssDNA tails are 100% identical to their host DNA targets which is not the case with current dsDNA vectors even isogenic DNA ones. Moreover, unlike most current dsDNA vectors, the ssDNA tails from our vectors do not comprise repetitive elements that may hamper targeting efficiency. We thus believe that our premade dsDNA-cored presynaptic filaments may be effective with fairly short ssDNA tails. With ssDNA tails ranging from 50 to 150 nucleotides, and dsDNA cores long enough to avoid potential steric interferences between the upstream recombination complex and the downstream one, we may be able to generate a broad spectrum of active RAD51 nucleoprotein filaments.

b) RAD51, DMC1 and other mammalian RecA-like proteins

RAD51 stands as the somatic recombinase we have been looking for in the early phase of our program (Bertolotti R., Grant application to French MRT, 1991; Borchiellini *et al.*, 1997). Although there are other somatic RecA-like proteins (Rice *et al.*, 1997; Albala *et al.*, 1997; Dosanjh *et al.*, 1998; Cartwright *et al.*, 1998a and 1998b; Pittman *et al.*, 1998; Liu *et al.*, 1998; Kawabata and Saeki, 1998), they seem to be unable to carry the key reaction from homologous recombination, *i.e.* homologous DNA pairing-strand exchange with double-stranded DNA. They appear to be involved in ancillary functions like yeast RAD55 and RAD57 that stimulate RAD51 activity upon heterodimerisation (Sung, 1997b) or XRCC2 and XRCC3 that may be involved in assembly or stabilisation of RAD51 complexes *in vivo* (Liu *et al.*, 1998; Bishop *et al.*, 1998); such ancillary functions may be associated to cell-cycle regulation as illustrated for REC2/RAD51L1 on cell cultures and transgenic mice (Havre *et al.*, 1998; Shu *et al.*, 1999). However, although RAD51 is also present in meiotic cells where it associates with meiotic prophase chromatin (Bishop, 1994; Haaf *et al.*, 1995; Teresawa *et al.*, 1995; Ashley *et al.*, 1995), there is a meiosis-specific RecA-like protein (DMC1; Bishop *et al.*, 1992; Kobayashi *et al.*, 1993; Habu *et al.*, 1996) that carries the homologous DNA pairing-strand exchange reaction *in vitro* (Li *et al*, 1997). Unlike RAD51, DMC1 activity has a strong dependence on DNA base composition suggesting that its function may be restricted to special sites in chromosomal DNA (Li *et al.*, 1997). Therefore, unlike RAD51, DMC1 appears not to meet the universal sequence requirement of our recombinase-mediated gene therapy approach. In addition, its meiotic cell-specific distribution suggests that it may request other germ-specific proteins to be active in somatic cells.

c) Oligonucleotide-mediated correction of point mutations

Interestingly enough, chimeric RNA-DNA oligonucleotides with a hybrid duplex structure were recently devised that appear now as a potential partial alternate to our recombinase-DNA nucleoprotein filaments. The active molecule is a 68 base-long RNA-DNA oligonucleotide (RDO) that is folded into a hairpin-capped duplex (Yoon *et al.*, 1996). Although its empirical design is based on the observation that RNA residues stimulate DNA pairing mediated *in vitro* by bacterial RecA or fungal Rec2 proteins (Yoon *et al.*, 1996), this RDO promotes single-base conversion of target mammalian genes (Cole-Strauss *et al.*, 1996; Kren *et al.*, 1997, 1998 and 1999; Santana *et al.*, 1998; Alexeev and Yoon, 1998; Bandyopadhyay *et al.*, 1999) by a mechanism that is still a matter of debate (Cole-Strauss *et al.*, 1999; Igoucheva *et al.*, 1999). The requirement for functional hMSH2 mismatch repair protein in a cell-free extract assay for RDO-mediated single-base substitution (Cole-Strauss *et al.*, 1999) is consistent with the conversion mechanism discussed above for standard ssDNA-RAD51 presynaptic filaments. However, the DNA pairing enzyme(s) or complexe(s) involved in the putative initial pairing of the RDO with target chromosomal DNA remains to be identified (Cole-Strauss *et al.*, 1999; Igoucheva *et al.*, 1999).

In any case, RDOs have been shown to be effective at genomic targeting *in vitro* (Cole-Strauss *et al.*, 1996) and *in vivo* (Kren *et al.,* 1998). As expected, the resulting single-base conversion is stable through clonal propagation (Alexeev and Yoon, 1998). Importantly enough, RDOs appear to be active both on proliferating and resting cells (Kren *et al.*, 1998). Therefore, although there are still questions about the universality (genes and cell-types; Yoon, 1999 and this book) and suggested efficiency (Strauss, 1998; Yoon, 1999 and this book) of the method, RDO-mediated correction of point mutations stands as a promissing gene repair technique that is complementary to our recombinase-mediated one. On the other hand, single-stranded bifunctional oligonucleotides have also been recently described that provide attractive correction of chromosomal point mutations in human cultured cells (Culver *et al.*, 1999). They comprise a third strand-forming domain for synergistic targeting of a single-base conversion repair domain. Therefore, from single-base conversion mediated by the aforedescribed oligonucleotides to broad DNA exchange potentialities relying on our dsDNA-cored presynaptic filaments, gene repair technology is now emerging as a promising approach for gene therapy of inherited diseases.

4. CONCLUSION: RECOMBINASE-MEDIATED GENE THERAPY

a) Selectable single-copy genes as target/control set models

Our model system that has been described for gene repair/inactivation (Bertolotti, 1996a, 1996b and 1998a) is also perfectly fitted for experimenting targeted integration of minigenes (Bertolotti, 1999a). We can monitor the targeted integration of a minigene into the target gene (targeted disruption; Capecchi, 1989) in the same way we do for gene repair/ inactivation. Indeed, in both cases, our approach is a two facetted one: on the one hand, it is aimed at increasing targeting efficiency to a level compatible with gene therapy protocols, and on the other hand, we need to reduce or suppress random integration of transfecting DNA. The latter point is a major safety concern since random integration of exogenous DNA into host chromosomes is a true mutagenic event. Such an insertional mutagenesis process may hit a cancer-prone gene (*e.g.* oncogene, tumor suppressor or DNA repair gene) thereby increasing the propensity of the cell to become cancerous. This problem is all the more important as, under current transfection protocols, exogenous DNA is essentially integrated at random (see above and introductory chapter of the book). Therefore, in order to monitor both types of events, we use a set of two selectable "single-copy" genes where one is the target and the other is the control for random insertional mutagenesis. Complementary PCR amplification and sequencing of integrated DNA and flanking host sequences provide for direct demonstration of targeted integration, either for gene/repair inactivation or minigene expression therapy.

Our original model system relies on the human HPRT and ornithine transcarbamoylase (OTC) genes (Bertolotti, 1996a and 1996b). Both the HPRT and OTC genes provide for positive and negative selections of cultured cells (see Bertolotti, 1998a and 1998b). In addition, they are "single-copy" X-linked genes, *i.e.* they are hemizygous in males and there is only a single functional copy in adult female cells (X chromosome inactivation; Lyon, 1996). They are therefore an ideal set for gene targeting experiments where each successful event can be readily scored upon positive or negative selection of transfectant cells either in HPRT- or OTC-specific growth media (table 2). Thus, upon transfecting human cells with RAD51-HPRT nucleoprotein filaments or with control RAD51-OTC ones, we can evaluate both the targeting efficiencies and the rates of unspecific insertional mutagenesis (Table 2; Bertolotti, 1996b). On the other hand, random integration is more accurately evaluated in experiments dealing with minigenes since, in this case, transfection efficiency (random + targeted integrations) can be directly determined upon selection in a minigene-specific growth medium (*e.g.* G418 medium for the neomycine minigene; Southern and Berg, 1982; Bertolotti *et al.*, 1995) and then compared to gene targeting efficiency (selection in a medium specific to the disrupted target gene, *e.g.* counter-selection for the HPRT gene), thereby providing a direct estimation of the suppressive effect of recombinase coating on integration upon non-homologous recombination (see: Table 1; Bertolotti, 1999a and 1999b).

Table 2. HPRT and OTC genes (Bertolotti, 1997)

Features

-"Single" copy (X-linked)
-Similar structure*: large (~40 and 70 kb)
9 and 10 exons
~1.5 kb mRNA
-Well-characterized mutations
-Selection and counter-selection for both genes**

Strategy

-Target:	HPRT (Targeting efficiencies)
-Control:	OTC (Rates of unspecific insertional mutagenesis)

*Edwards *et al.*, 1990; Horwich *et al.*, 1984; Hata *et al.*, 1988.
**see Armbruster *et al.*, 1992; Bertolotti, 1996b and 1998b.

The HPRT/OTC set of "single-copy" target genes is fitted to our hepatic cell model comprising well-differentiated human HepG2 clones that have been isolated in liver-specific growth media (Lutfalla *et al.*, 1989; Armbruster *et al.*, 1992). Ornithine transcarbamoylase is a tissue-specific function that is restricted to the liver, kidney and small intestine. Therefore, concurrent experiments on other human cell types require to associate the ubiquitously expressed HPRT gene to a non-hepatic one, *i.e.* a gene that is either involved in a household function or specific of the differentiated state of the target cell type. The current choice of human genes providing both for positive and negative selection is very limited; it includes autosomal genes with which single-copy-like targeting conditions require to use either heterozygous or homozygous mutant cells.

b) Toward *ex vivo* and *in vivo* recombinase-mediated gene therapy

From targeting selectable single-copy genes *in vitro* toward targeting non-selectable ones in specific somatic cells *in vivo*, our aim is to develop a recombinase-mediated gene targeting technology the efficiency of which would be compatible with a selection-free processing of gene-modified cells.

Although non-selectable genes can occasionally confer an *in vivo* selective survival to gene-modified cells that may accumulate to therapeutic levels (Kohn *et al.*, 1998; Bunting *et al.*, 1998; Overturf *et al.*, 1996), such a situation is exceptional. Gene targeting efficiency is therefore the cornerstone of the approach both for gene repair and minigene expression therapy. For many genetic diseases, the minimal number of gene-corrected cells for therapeutic achievement is estimated to be around 10% of the appropriate target cell population.

Like for other approaches, the ability to deliver therapeutic DNA to the correct cells is thus an essential feature for recombinase-mediated gene therapy. For our long-term gene therapy approach, progenitor/stem cells stand as the choice targets since their use obviates cell replacement problems with non-modified patient's cells (see: Verma and Somia, 1997; Bertolotti, introductory chapter of the book). Importantly enough, the *in vitro* manipulation and multiplication of a growing number of stem/progenitor cells provide an attractive *ex vivo* strategy for gene therapy in which cells are returned to the patient after genetic modification *in vitro* (see introductory chapter). The converse approach, *i.e.* tissue-specific delivery of therapeutic DNA *in vivo*, is under intensive investigations (see introductory chapter of the book) and has been previously discussed for our recombinase-DNA complexes (Bertolotti, 1998a and 1998b). Current *in vivo* protocols for clinical trials are restricted to topical applications where tissue-specific targeting is replaced by restricted local diffusion of the delivery vehicle (Anderson, 1998).

c) Promises and prospects of recombinase-mediated gene therapy

Gene repair and efficient targeted integration of therapeutic minigenes hold great promises both for inherited diseases and acquired disorders. In addition, they open new and safer avenues for future potential tantalizing applications for non-disease conditions when long-term safety concerns will be completely cleared.

Gene repair therapy for inherited diseases. The interest of gene repair is obvious for gene therapy of genetic diseases where timely correction of the mutated gene should prevent or reverse the genetic disorder. Recovery of normal genetic regulation is granted since the target gene is maintained in its original genomic background. Importantly enough, the timing of gene therapy has to be carrefully adjusted to the disease course. As illustrated with the Lesch-Nyhan syndrome (Bertolotti, 1998b), it is an essential therapeutic consideration because complete reversion of a disease is impossible after the onset of critical pathological stages.

Optimized long-term minigene expression therapy. For approaches involving minigenes, targeted integration should eliminate carcinogenic hazards (see above) and should provide for better control of transgene expression. Improved control of transgene expression should result both from the choice of the integration site and from single-copy integration of the minigene (Bertolotti, 1998c and 1999a). Indeed, as pinpointed in table 1, recombinase-coating should prevent both homologous recombination (Folger *et al.*, 1982) and end-joining (Roth *et al.*, 1995; Critchlow and Jackson, 1998) activities between transfecting DNA molecules, thereby avoiding concatemer formation and ensuing multicopy integration of the minigene at a single site (see Fig. 1; Folger *et al.*, 1982; Miller and Temin, 1983; Lutfalla *et al.*, 1985). On the other hand, the targeting sequences of our dsDNA-cored presynaptic filaments are free of repetitive elements (see above) that could promote the potential, but unlikely, multisite integration of a minigene in the host cell.

Functional enhancement and ageing prevention. Another potential interest of our gene targeting technology deals with modifications of endogenous gene expression patterns upon targeted insertion/alteration of *cis*-acting (see Bertolotti, introductory chapter of the book) or post-transcriptional regulatory (Atwater *et al.*, 1990; Siomi and Dreyfuss, 1997) elements. The main applications of such an approach are likely to be tantalazing functional enhancement and ageing prevention. In this respect, they are complementary to minigene expression therapy that has already been proposed for ageing prevention (Martinez-Serrano *et al.*, 1996) and for the protection of injury prone athletes (Gerich *et al.*, 1996) or for alopecia and loss of hair-color (Li and Hoffman, 1995; Sato *et al.*, 1999). Interestingly enough, production of "improved" proteins (Gerich *et al.*, 1996) is one aspect of functional enhancement where gene targeting can provide either for modification of the endogenous DNA coding sequence or for non-random integration of the corresponding minigene.

5. REFERENCES

Aboussekhra, A., Chanet, R., Adjiri, A. and Fabre, F. (1992). Semidominant suppressors of Srs2 helicase mutations of *S. cerevisiae* map in the *RAD51* gene, whose sequence predicts a protein with similarities to procaryotic RecA proteins. *Mol. Cell. Biol.*, 12: 3224-3234.

Abuin, A. and Bradley, A. (1996). Recycling selectable markers in mouse embryonic stem cells. *Mol. Cell. Biol.*, 16: 1851-1856.

Albala, J.S., Thelen, M.P., Prange, C., Fan, W., Christensen, M., Thompson, L.H. and Lennon, G.G. (1997). Identification of a novel human RAD51 homolog, RAD51B. *Genomics*, 46: 476-479.

Alexeev, V. and Yoon, K. (1998). Stable and inheritable changes in genotype and phenotype of albino melanocytes induced by an RNA-DNA oligonucleotide. *Nature Biotechnology*, 16: 1343-1346.

Anderson, F.W. (1998). Human gene therapy. *Nature*, 392 suppl. : 25-30.

Armbruster, L., Cavard, C., Briand, P. and Bertolotti, R. (1992). Selection of variant hepatoma cells in liver-specific growth media: regulation at the mRNA level. *Differentiation*, 50: 25-33.

Ashley, T., Plug, A. W., Xu, J., Solari, A. J., Reddy, G., Golub, E. and Ward, D. C. (1995). Dynamic changes in RAD51 distribution on chromatin during meiosis in male and female vertebrates. *Chromosoma*, 104: 19-28 (1995).

Atwater, J.A., Wisdom, R. and Verma, I.M. (1990). Regulated mRNA stability. *Annu. Rev. Genet.*, 24: 519-541.

Bandyopadhyay, P., Ma, X., Linehan-Stieers, C., Kren, B.T. and Steer, C.J. (1999). Nucleotide exchange in genomic DNA of rat hepatocytes using RNA/DNA oligonucleotides. Targeted delivery of liposomes and polyethyleneimine to the asialoglycoprotein receptor. *J. Biol. Chem.*, 274: 10163-10172.

Basile, G., Aker, M. and Mortimer, R. K. (1992). Nucleotide sequence and transcriptional regulation of the yeast recombinational repair gene *RAD51*. *Mol. Cell. Biol.*, 12: 3235-3246.

Baumann, P. and West, S.C. (1997). The human Rad51 protein: polarity of strand transfer and stimulation by hRP-A. *EMBO J.*, 17: 5198-5206.

Baumann, P., Benson, F. and West, S. C. (1996). Human Rad51 protein promotes ATP-dependent homologous pairing and strand transfer reactions *in vitro*. *Cell*, 87: 757-766.

Benson, F. E., Stasiak, A. and West, S. C. (1994). Purification and characterization of the human Rad51 protein, an analogue of *E. coli* RecA. *EMBO J.*, 13: 5764-5771.

Benson, F.E., Baumann, P. and West, S.C. (1998). Synergistic actions of Rad51 and Rad52 in recombination and DNA repair. *Nature*, 391: 401-404.

Berglund, P., Smerdou, C., Fleeton, M.N., Tubulekas, I. and Liljestrom P. (1998). Enhancing immune responses using suicidal DNA vaccines. *Nature Biotechnol.*, 16: 562-565.

Bertolotti, R. (1996a). Recombinase-mediated gene therapy: targeting single stranded DNA to chromosomal DNA with human RAD51 presynaptic fibers. *Hepatology* , 24: 484A.

Bertolotti, R. (1996b). Recombinase-mediated gene therapy: strategies based on Lesch-Nyhan mutants for gene repair/inactivation using human RAD51 nucleoprotein filaments. *Biogenic Amines*, 12: 487-498.

Bertolotti, R. (1997). Recombinase-mediated gene therapy: strategies for gene repair/ inactivation using human RAD51 nucleoprotein filaments. In: *Advances in Molecular Toxicology*, C. Reiss and H. Parvez (Eds), VSP, Utrecht, N.L., pp. 445-466.

Bertolotti, R. (1998a). Recombinase-DNA nucleoprotein filaments as Gene Therapy vectors. *Biogenic Amines*, 14: 41-65.

Bertolotti, R. (1998b). Prospects of recombinase-mediated HPRT Gene Therapy for Lesch-Nyhan disease. In: *Neurochemical Markers of Degenerative Nervous Diseases and Drug Addiction*, G.A. Qureshi, H. Parvez, P. Caudy and S. Parvez (Eds), VSP, Utrecht, N.L., pp. 133-170.

Bertolotti, R. (1998c). Gene therapy 1998: transient or stable minigene expression and gene repair/inactivation. *Biogenic Amines*, 14: 389-406.

Bertolotti, R. (1999a). Recombinase-DNA nucleoprotein filaments as vectors for gene repair /inactivation and targeted integration of minigenes. *Biogenic Amines*, 15: 169-195.

Bertolotti, R. (1999b). Recombinase-mediated gene therapy: gene repair/inactivation and targeted integration of minigenes. in: *Progress in Hypertension, Vol. 4, Genetic, immune and molecular predisposition to hypertension,* P.M. Frossard, H. Parvez, M. Minami, S. Parvez and H. Saito (Eds.), VSP, Utrecht, N.L., pp. 217-263.

Bertolotti, R. and Lutfalla, G. (1983). Genes programming cell differentiation. *J. Cell. Biochem.*, 7A: 149.

Bertolotti, R., Armbruster-Hilbert, L. and Okayama, H. (1995). Liver fructose-1,6-bisphosphatase cDNA: trans -complementation of fission yeast and characterization of two human transcripts. *Differentiation*, 59: 51-60.

Bertolotti, R., Parvez, S.H. and Nagatsu, T. (Eds) (1998). Gene Therapy 1998, Part 1. *Biogenic Amines*, 14: 387-572.

Bertolotti, R., Parvez, S.H. and Nagatsu, T. (Eds) (1999). Gene Therapy 1998, Part 2. *Biogenic Amines*, 15: 1-195.

Bianco, P.R., Tracy, R.B. and Kowalczykowski, S.C. (1998). DNA strand exchange proteins: a biochemical and physical comparison. *Front. Biosci.*, 3: D570-603.

Bishop, D. K. (1994). RecA homologs DMC1 and RAD51 interact to form multiple nuclear complexes prior to meiotic chromosome synapsis. *Cell*, 79: 1081-1092.

Bishop, D. K., Park, D., Xu, L. and Kleckner, N. (1992). DMC1: a meiosis-specific yeast homolog of *E. coli recA* required for recombination, synaptonemal complex formation, and cell cycle progression. *Cell*, 69: 439-456.

Bishop, D.K., Ear, U., Bhattacharyya, A., Calderone, C., Beckett, M., Weichselbaum, R.R. and Shinohara, A. (1998). XRCC3 is required for assembly of RAD51 complexes *in vivo*. *J. Biol. Chem.*, 273: 21482-21488.

Borchiellini, P., Angulo, J. F. and Bertolotti, R. (1997). Genes encoding mammalian recombinases: cloning approach with anti-RecA antibodies. *Biogenic Amines*, 13: 195-215.

Bunting, K.D., Sangster, M.Y., Ihle, J.N. and Sorrentino, B.P. (1998). Restoration of lymphocyte function in Janus kinase 3-deficient mice by retroviral-mediated gene transfer. *Nature Med.*, 4: 58-64.

Cao, L., Alani, E. and Kleckner, N. (1990). A pathway for generation and processing of double-strand breaks during meiotic recombination in S. cerevisiae. *Cell*, 61: 1089-1101.

Capecchi, M. (1989). The new mouse genetics: altering the genome by gene targeting. *Trends Genet.*, 5: 70-76.

Cartwright, R., Dunn, A.M., Simpson, P.J., Tambini, C. and Thacker, J. (1998a). Isolation of a novel human and mouse genes of the recA/RAD51 recombination-repair gene family. *Nucleic Acids Res.*, 26: 1653-1659.

Cartwright, R., Tambini, C., Simpson, P.J. and Thacker, J. (1998b). The XRCC2 DNA repair gene from human and mouse encodes a novel member of the recA/RAD51 family. *Nucleic Acids Res.*, 26: 3084-3089.

Cole-Strauss, A., Yoon, K., Xiang, Y., Byrne, B., Rice, M., Gryn, J., Holloman, W. and Kmiec, E. (1996). Correction of the mutation responsible for sickle cell anemia by an RNA-DNA oligonucleotide. *Science*, 273: 1386-1389.

Cole-Strauss, A., Gamper, H., Holloman, W.K., Munoz, M., Cheng, N. and Kmiec, E.B. (1999). Targeted gene repair directed by the chimeric RNA/DNA oligonucleotide in a mammalian cell-free extract. *Nucleic Acids Res.*, 27: 1323-1330.

Critchlow, S.E. and Jackson, S.P. (1998). DNA end-joining: from yeast to man. *Trends Biochem. Sci.*, 23: 394-398.

Culver, K.W., Hsieh, W.T., Huyen, Y., Chen, V., Liu, J., Khripine, Y. and Khorlin, A. (1999). Correction of chromosomal point mutations in human cells with bifunctional oligonucleotides. *Nat. Biotechnol.*, 17: 989-993.

Deng, C. and Capecchi, M. (1992). Reexamination of gene targeting frequency as a function of the extent of homology between the targeting vector and the target locus. *Mol. Cell. Biol.*, 12: 3365-3371.

Deng, C., Thomas, K. and Capecchi, M. (1993). Location of crossovers during gene targeting with insertion and replacement vectors. *Mol. Cell. Biol.*, 13: 2134-2140.

DiMaio, D., Treisman, R. and Maniatis, T. (1982). Bovine papilloma-virus vector that propagates as a plasmid in both mouse and bacterial cells. *Proc. Natl. Acad. Sci. U.S.A.*, 79: 4030-4034.

Dosanjh, M.K., Collins, D.W., Fan, W., Lennon, G.G., Albala, J.S., Shen, Z. and Schild, D. (1998). Isolation and characterization of RAD51C, a new human member of the RAD51 family of related genes. *Nucleic Acids Res.*, 26: 1179-1184.

Edwards, A., Voss, H., Rice, P., Civitello, A., Stegemann, J., Schwager, C., Zimmermann, J., Erfle, H., Caskey, C. and Ansorge, W. (1990). Automated DNA sequencing of the human *hprt* locus. *Genomics*, 6: 593-608.

Eggleston, A. and West, S. (1996). Exchanging partners: recombination in *E. coli*. *Trends Genet.*, 12: 20-26.

Ephrussi, B. (1972). *Hybridization of somatic cells*. Princeton University Press, N.J.

Ferrin, L. and Camerini-Otero, R. D. (1991). Selective cleavage of human DNA: RecA-assisted restriction endonuclease (RARE) cleavage. *Science*, 254: 1494-1497.

Firmenich, A.A., Elias-Arnanz, M. and Berg P. (1995). A novel allele of Saccharomyces cerevisiae RFA1 that is deficient in recombination and repair and suppressible by RAD52. *Mol. Cell. Biol.*, 15: 1620-1631

Flygare, J., Benson, F. and Hellgren, D. (1996). Expression of the human RAD51 gene during the cell cycle in primary human peripheral blood lymphocytes. *Biochim. Biophys. Acta*, 1312: 231-236.

Folger, K., Wong, E., Wahl, G. and Capecchi, M. (1982). Patterns of integration of DNA microinjected into cultured mammalian cells: evidence for homologous recombination between injected plasmid DNA molecules. *Mol. Cell. Biol.*, 2: 1372-1387.

Folger, K.R., Thomas, K. and Capecchi, M.R. (1985). Efficient correction of mismatched bases in plasmid heteroduplex injected into cultured mammalian cell nuclei. *Mol. Cell. Biol.*, 5: 70-74.

Friedberg, E. C., Walker, G. C. and Siede, W. (1995). *DNA repair and mutagenesis*. American Society for Microbiology, Washington D.C.

Fujioka, K., Aratani, Y., Kusano, K. and Koyama, H. (1993). Targeted recombination with single-stranded DNA vectors in mammalian cells. *Nucleic Acids Res.*, 21: 407-412.

Gerich, T. G., Fu, F.H., Robbins, P.D. and Evans, C.H. (1996). Prospects for gene therapy in sports medicine. *Knee Surg. Sports Traumatol. Arthrosc.*, 4: 180-187.

Goncz, K., Kunzelmann, K., Xu, Z. and Gruenert, D.C. (1998). Targeted replacement of normal and mutant CFTR sequences in human airway epithelial cells using DNA fragments. *Hum Mol Genet.*, 7: 1913-1919.

Graham, F. and Van der Eb, J. (1973). A new technique for the assay of infectivity of human adenovirus 5 DNA. *Virology*, 52: 456-467.

Gupta, R., Bazemore, L. R., Golub, E. and Radding, C. (1997). Activities of human recombination protein RAD51. *Proc. Natl. Acad. Sci. USA*, 94: 463-468.

Haaf, T., Golub, E., Reddy, G., Radding, C. and Ward, D. (1995). Nuclear foci of RAD51 recombination protein in somatic cells after DNA damage and its localization in synaptonemal complexes. *Proc. Natl. Acad. Sci. U.S.A.*, 92: 2298-2302.

Habu, T., Taki, T., West, A., Nishimune, Y and Morita, T. (1996). The mouse and human homologs of DMC1, the yeast meiosis-specific homologous recombination gene, have a common unique form of exon-skipped transcript in meiosis. *Nucleic Acids Res.*, 24: 470-477.

Hata, A., Tsuzuki, T., Shimada, K., Takiguchi, M., Mori, M. and Matsuda, I. (1988). Structure of the human ornithine transcarbamylase gene. *J. Biochem.*, 103: 302-308.

Havre, P.A., Rice, M.C., Noe, M. and Kmiec, E.B. (1998). The human REC2/RAD51B gene acts as a DNA damage sensor by inducing G1 delay and hypersensitivity to ultraviolet irradiation. Cancer Res., 58: 4733-4739.

Hays, S., Firmenich, A. and Berg, P. (1995). Complex formation in yeast double-strand break repair: participation of Rad51, Rad52, Rad55 and Rad57 proteins. *Proc. Natl. Acad. Sci. U.S.A.*, 92: 6925-6929.

Henricksen, L., Umbricht, C. and Wold, M. (1994). Recombinant replication protein A: expression, complex formation and functional characterization. *J. Biol. Chem.*, 269: 11121-11132.

Heyer, W.D. (1994). The search for the right partner: homologous pairing and DNA strand exchange proteins in eukaryotes. *Experientia*, 50: 223-233.

Horwich, A. L., Fenton, W., Williams, K., Kalousek, F., Kraus, J. P., Doolittle, R., Konigsberg, W. and Rosenberg, L. (1984). Structure and expression of a complementary DNA for the nuclear coded precursor of human mitochondrial ornithine transcarbamylase. *Science*, 224: 1068-1074.

Howard-Flanders, P., West, S. C. and Stasiak, A. (1984). Role of RecA protein spiral filaments in genetic recombination. *Nature*, 309: 215-220.

Hsieh, P., Camerini-Otero, C. and Camerini-Otero, R. D. (1992). The synapsis event in the homologous pairing of DNAs: RecA recognizes and pairs less than one helical repeat of DNA. *Proc. Natl. Acad. Sci. U.S.A.*, 89: 6492-6496.

Igoucheva, O., Peritz, A.E., Levy, D. and Yoon, K. (1999). A sequence-specific gene correction by an RNA-DNA oligonucleotide in mammalian cells characterized by transfection and nuclear extract using a lacZ shuttle system. *Gene Ther.*, 6: 1960-1971.

Jiricny, J. (1998). Replication errors: challenging the genome. *EMBO J.*, 17: 6427-6436.

Kawabata, M. and Saeki, K. (1998). Sequence analysis and expression of a novel mouse homolog of escherichia coli recA gene 1. *Biochim. Biophys. Acta*, 1398: 353-358.

Kobayashi, T., Hotta, Y. and Tabata, S. (1993). Isolation and characterization of a yeast gene which is homologous with a meiosis specific cDNA from a plant. *Mol. Gen. Genet.*, 237: 225-232.

Kohn, D.B. and Parkman, R. (1997). Gene therapy for newborns. *FASEB J.*, 11: 635-639

Kohn, D.B., Hershfield, M.S., Carbonaro, D., Shigeoka, A., Brooks, J., Smogorzewska, E.M., Barsky, L.W., Chan, R., Burotto, F., Annett, G., Nolta, J.A., Crooks, G., Kapoor, N., Elder, M., Wara, D., Bowen, T., Madsen, E., Snyder, F.F., Bastian, J., Muul, L., Blaese, R.M., Weinberg, K. and Parkman, R. (1998). T lymphocytes with a normal ADA gene accumulate after transplantation of transduced autologous umbilical cord blood CD34+ cells in ADA-deficient SCID neonates. *Nature Med.*, 4: 775-780.

Kowalczykowski, S., Dixon, D., Eggleston, A., Lauder, S. and Rehrauer, W. (1994).

Biochemistry of homologous recombination in *E. coli*. *Microbiol. Rev.*, 58: 401-465.

Kren, B.T., Cole-Strauss, A., Kmiec, E.B. and Steer, C.J. (1997). Targeted nucleotide exchange in the alkaline phosphatase gene of HuH-7 cells mediated by a chimeric RNA/DNA oligonucleotide. *Hepatology*, 25: 1462-1468.

Kren, B.T., Bandyopadhyay, P. and Steer, C.J. (1998). *In vivo* site-directed mutagenesis of the *factor IX* gene by chimeric RNA/DNA oligonucleotides. *Nature Med.*, 4: 285-290.

Kren, B.T., Parashar, B., Bandyopadhyay, P., Chowdhury, N.R., Chowdhury, J.R. and Steer, C.J. (1999). Correction of the UDP-glucuronosyltransferase gene defect in the gunn rat model of crigler-najjar syndrome type I with a chimeric oligonucleotide. *Proc. Natl. Acad. Sci. USA*, 96: 10349-10354.

Kunzelmann, K., Legendre, J-Y, Knoell, D.L., Escobar, L.C., Xu, Z. and Gruenert, D.C. (1996). Gene targeting of CFTR DNA in CF epithelial cells. *Gene Ther.*, 3: 859-867.

Leff, S., Rosenfeld, M. and Evans, R. (1986). Complex transcriptional units: diversity in gene expression by alternative RNA processing. *Ann. Rev. Biochem.*, 55: 1091-1117.

Li, L. and Hoffman, R.M. (1995). The feasability of targeted selective gene therapy of the hair follicule. *Nature Med.*, 1: 705-706.

Li, Z., Golub, E.I., Gupta, R. and Radding, C.M. (1997). Recombination activities of HsDmc1 protein, the meiotic human homolog of RecA protein. *Proc. Natl. Acad. Sci. U.S.A.*, 94: 11221-11226.

Lim, D.S. and Hasty, P. (1996). A mutation in mouse RAD51 results in an early embryonic lethal that is suppressed by a mutation in p53. *Mol. Cell. Biol.*, 16: 7133-7143.

Liu, N., Lamerdin, J.E., Tebbs, R.S., Schild, D., Tucker, J.D., Shen, M.R., Brookman, K.W., Siciliano, M.J., Walter, C.A., Fan, W., Narayana, L.S., Zhou, Z.Q., Adamson, A.W., Sorensen, K.J., Chen, D.J., Jones, N.J. and Thompson, L.H. (1998). XRCC2 and XRCC3, new human RAD51-family members, promote chromosome stability and protect against DNA cross-links and other damages. *Mol. Cell*, 1: 783-793.

Lutfalla, G., Blanc, H. and Bertolotti, R. (1985). Shuttling of integrated vectors from mammalian cells to *E. coli* is mediated by head-to-tail multimeric inserts. *Somat. Cell Mol. Genet.*, 11: 223-238.

Lutfalla, G., Armbruster, L., Dequin, S. and Bertolotti, R. (1989). Construction of an EBNA-producing line of well-differentiated human hepatoma cells and of appropriate Epstein-Barr virus-based shuttle vectors. *Gene*, 76: 27-39.

Lyon, M. (1996). X-chromosome inactivation: pinpointing the centre. *Nature*, 379: 116-117

Martinez-Serrano, A., Fischer, W., Soderstrom, S., Ebendal, T. and Bjorklund A. (1996). Long-term functional recovery from age-induced spatial memory impairments by nerve growth factor gene transfer to the rat basal forebrain. Proc. Natl. Acad. Sci. USA, 93: 6355-6360.

Merrihew, R., Marburger, K., Pennington, S., Roth, D. and Wilson, J. (1996). High-frequency illegitimate integration of transfected DNA at preintegrated target sites in a mammalian genome. *Mol. Cell. Biol.*, 16: 10-18

Miller, C. and Temin, H. (1983). High-efficiency ligation and recombination of DNA fragments by vertebrate cells. *Science*, 220: 606-609.

Modrich, P. (1997). Strand-specific mismatch repair in mammalian cells. *J. Biol. Chem.*, 272: 24727-24730.

Modrich, P. and Lahue, R. (1996). Mismatch repair in replication fidelity, genetic recombination, and cancer biology. *Annu. Rev. Biochem.*, 65: 101-133.

Morita, T., Yoshimura, Y., Yamamoto, A., Murata, K., Mori, M., Yamamoto, H. and Matsushiro, A. (1993). A mouse homolog of the Escherichia coli recA and Saccharomyces cerevisiae RAD51 genes. *Proc. Natl. Acad. Sci. USA*, 90: 6577-6580.

Moss, B. (1996). Genetically engineered poxviruses for recombinant gene expression, vaccination and safety. *Proc. Natl. Acad. Sci. USA*, 93: 11341-11348.

Myers, R. and Stahl, F. (1994). Chi and the RecBCD enzyme of *Escherichia coli. Ann. Rev. Genet.*, 28: 49-70.

Namsaraev, E.A. and Berg, P. (1998). Branch migration during Rad51-promoted strand exchange proceeds in either direction. *Proc. Natl. Acad. Sci. USA*, 95: 10477-10481.

New, J., Sugiyama, T., Zaitseva, E. and Kowalczykowski, S. (1998). Rad52 protein stimulates DNA strand exchange by Rad51 and replication protein A. *Nature*, 391: 407-410.

Ogawa, T., Yu, X., Shinohara, A. and Egelman, E. H. (1993). Similarity of the yeast RAD51 filament to the bacterial RecA filament. *Science*, 259: 1896-1899.

Olson, P. and Dornburg, R. (1999). Capture of a recombination activating sequence from mammalian cells. *Gene Ther.*, 6: 1819-1825.

Overturf, K., Al-Dhalimy, M., Tanguay, R., Brantly, M., Ou, C.N., Finegold, M. and Grompe, M. (1996). Hepatocytes corrected by gene therapy are selected *in vivo* in a murine model of hereditary tyrosinaemia type I. *Nature Genet.*, 12: 266-273.

Pan, Q., Rabbani, H. and Hammarstrom, L. (1998). Characterization of human γ 4 switch region polymorphisms suggests a meiotic recombinational hot spot within the Ig locus: influence of S region length on IgG4 production. *J. Immunol.*, 161: 3520-3526.

Panthier, J. and Condamine, H. (1991). Mitotic recombination in mammals. *BioEssays*, 13: 351-356.

Paull, T. and Gellert, M. (1998). The 3' to 5' exonuclease activity of Mre11 facilitates repair of DNA double-strand breaks. *Mol. Cell*, 1: 969-979.

Perucho, M., Hanahan, D. and Wigler, M. (1980). Genetic and physical linkage of exogenous sequences in transformed cells. *Cell*, 22: 309-317.

Petukhova, G., Stratton, S. and Sung, P. (1998). Catalysis of homologous DNA pairing by yeast RAD51 and Rad54 proteins. Nature, 393: 91-94.

Pittman, D.L., Weinberg, L.R. and Schimenti, J.C. (1998). Identification, characterization, and genetic mapping of RAD51D, a new mouse and human RAD51/RecA-related gene. *Genomics*, 49: 103-111.

Radding, C.M. (1991). Helical interactions in homologous pairing and strand exchange driven by RecA protein. *J. Biol. Chem.*, 266: 5355-5358.

Rice, M.C., Smith, S.T., Bullrich, F., Havre, P. and Kmiec, E.B. (1997). Isolation of human and mouse genes based on homology to REC2, a recombinational repair gene from the fungus Ustilago maydis. *Proc. Natl. Acad. Sci. USA*, 94: 7417-7422.

Robins, D., Ripley, S., Henderson, A. and Axel, R. (1981). Transforming DNA integrates into the host chromosome. *Cell*, 23: 29-39.

Roca, A. and Cox, M. (1990). The RecA protein: structure and function. *Crit. Rev. Biochem. Mol. Biol.*, 25: 415-455.

Roth, D. B., Lindahl, T. and Gellert, M. (1995). How to make ends meet. *Curr. Biol.*, 5: 496-499 (1995).

Santana, E., Peritz, A.E., Iyer, S., Uitto, J. and Yoon, K. (1998). Different frequency of gene targeting events by the RNA-DNA oligonucleotide among epithelial cells. *J. Invest. Dermatol.*, 111: 1172-1177.

Sato, N., Leopold, P.L. and Crystal, R.G. (1999). Induction of the hair growth phase in postnatal mice by localized transient expression of Sonic hedgehog. *J. Clin. Invest.*, 104: 855-864.

Shinohara, A. and Ogawa, T. (1998). Stimulation by Rad52 of yeast Rad51-mediated recombination. *Nature*, 391: 404-407.

Shinohara, A., Ogawa, H. and Ogawa, T. (1992). Rad51 protein involved in repair and recombination in *S. cerevisiae* is a RecA-like protein. *Cell*, 69: 457-470.

Shinohara, A., Ogawa, H., Matsuda, Y., Ushio, N., Ikeo, K. and Ogawa, T. (1993). Cloning of human, mouse and fission yeast recombination genes homologous to RAD51 and recA. *Nature Genetics*, 4: 239-243.

Shu, Z., Smith, S., Wang, L., Rice, M.C. and Kmiec, E.B. (1999). Disruption of muREC2/RAD51L1 in mice results in early embryonic lethality which can Be partially rescued in a p53(-/-) background. *Mol. Cell. Biol.*, 19: 8686-8693.

Siomi, H. and Dreyfuss, G. (1997). RNA-binding proteins as regulators of gene expression. *Curr. Opin. Genet.* Dev., 7: 345-353.

Southern, P.J. and Berg, P. (1982). Transformation of mammalian cells to antibiotic resistance with a bacterial gene under control of the SV40 early region promoter. *J. Mol. Appl. Genet.*, 1: 327-341.

Sonada, E., Sasaki, M.S., Bluerstedde, J-M., Bezzubova, O., Shinohara, A., Ogawa, H., Takata, M., Yamaguchi-Iwai, Y. and Takeda, S. (1998). Rad51-deficient vertebrate cells accumulate chromosomal breaks prior to cell death. *EMBO J.*, 17: 598-608.

Stasiak, A. and Egelman, È.H. (1994). Structure and function of RecA-DNA complexes. *Experientia*, 50: 192-203.

Strauss, M. (1998). The site-specific correction of genetic defects. *Nat. Med.*, 4: 274-275.

Sun, H., Treco, D. and Szostak, J.W. (1991). Extensive 3'-overhanging, single-stranded DNA associated with the meiosis-specific double-strand breaks at the ARG4 recombination initiation site. *Cell*, 64: 1155-1161.

Sung, P. (1994). Catalysis of ATP-dependent homologous DNA pairing and strand exchange by yeast RAD51 protein. *Science*, 265: 1241-1243.

Sung, P. (1997a). Function of yeast Rad52 protein as a mediator between replication protein A and the Rad51 recombinase. *J. Biol. Chem.*, 272: 28194-28197.

Sung, P. (1997b). Yeast Rad55 and Rad57 proteins form a heterodimer that functions with replication protein A to promote DNA strand exchange by Rad51 recombinase. *Genes Dev.*, 11: 1111-1121.

Sung, P. and Robberson, D. (1995). DNA strand exchange mediated by a RAD51-ssDNA nucleoprotein filament with polarity opposite to that of RecA. *Cell*, 82: 453-461.

Teresawa, M., Shinohara, A., Hotta, Y., Ogawa, H. and Ogawa, T. (1995). Localization of RecA-like recombination proteins on chromosomes of the lily at various meiotic stages. *Genes Dev.*, 9: 925-934.

Te Riele, H., Maandag, E. and Berns, A. (1992). Highly efficient gene targeting in embryonic stem cells through homologous recombination with isogenic DNA constructs. *Proc. Natl. Acad. Sci. USA*, 89: 5128-5132.

Thomas, K. and Capecchi, M. (1987). Site-directed mutagenesis by gene targeting in mouse embryo-derived stem cells. *Cell*, 51: 503-512.

Thomas, K., Folger, K. and Capecchi, M. (1986). High frequency targeting of genes to specific sites in the mammalian genome. *Cell*, 44: 419-428.

Thomas, K., Deng, C. and Capecchi, M. (1992). High-fidelity gene targeting in embryonic stem cells by using sequence replacement vectors. *Mol. Cell. Biol.*, 12: 2919-2923.

Trujillo, K.M., Yuan, S-S. F., Lee, E.Y-H. and Sung, P. (1998). Nuclease activities in a complex of human recombination and DNA repair factors Rad50, Mre11 and p95. *J. Biol. Chem.*, 273: 21447-21450.

Tsuzuki, T., Fujii, Y., Sakumi, K., Tominaga, Y., Nakao, K., Sekiguchi, M., Matsushiro, A., Yoshimura, Y. and Morita, T. (1996). Targeted disruption of the RAD51 gene leads to lethality in embryonic mice. *Proc. Natl. Acad. Sci. USA*, 93: 6236-6240.

Usui, T., Ohta, T., Oshiumi, H., Tomizawa, J-I., Ogawa, H. and Ogawa, T. (1998). Complex formation and functional versatility of Mre11 of budding yeast in recombination. *Cell*, 95: 705-716.

Verma, I. M. and Somia, N. (1997). Gene therapy -promises, problems and prospects. *Nature*, 389: 239-242.

Wong, E.A. and Capecchi, M.R. (1987). Homologous recombination between coinjected DNA sequences peaks in early to mid-S phase. *Mol Cell Biol.*, 7: 2294-2295.

Yamamoto, A., Taki, T., Yagi, H., Habu, T., Yoshida, K., Yoshimura, Y., Yamamoto, K., Matsushiro, A., Nishimune, Y. and Morita, T. (1996). Cell cycle-dependent expression of the mouse RAD51 gene in proliferating cells. *Mol. Gen. Genet.*, 251: 1-12.

Yanez, R.J. and Porter, A.C. (1998). Therapeutic gene targeting. *Gene Ther.*, 5: 149-159.

Yates, J., Warren, N. and Sugden, B. (1985). Stable replication of plasmids derived from Epstein-Barr virus in various mammalian cells. *Nature*, 313: 812-815.

Yoon, K. (1999). Single-base conversion of mammalian genes by an RNA-DNA oligonucleotide. *Biogenic Amines*, 15: 137-167.

Yoon, K., Cole-Strauss, A. and Kmiec, E. (1996). Targeted gene correction of episomal DNA in mammalian cells mediated by a chimeric RNA-DNA oligonucleotide. *Proc. Natl. Acad. Sci. USA*, 93: 2071-2076.

Yoshimura, Y., Morita, T., Yamamoto, A. and Matsushiro, A. (1993). Cloning and sequence of the human RecA-like gene cDNA. *Nucleic Acids Res.*, 21: 1665-1665.

7. SUMMARY — The most obvious approach to tackle an inherited disease is to repair the dysfunctional gene in appropriate somatic cell targets. Reestablishing the wild type sequence of the mutant gene *in situ* requests efficient gene repair techniques that are under intensive investigations. Gene targeting, *i.e.* homologous recombination between chromosomal DNA and transfecting DNA, mediates DNA exchanges that provide the means to modify at will the sequence of the target gene. However, well-established gene targeting technology has such a poor efficiency that it cannot apply to gene therapy protocols. This is why, in 1991, we devised a new approach to gene targeting based on premade presynaptic filaments, *i.e.* active recombinase-DNA nucleoprotein complexes that mediate the key reaction from homologous recombination. Aimed at repairing mutant genes (inherited diseases) and at inactivating viral/deleterious genes (AIDS, hepatitides, cancer, ...), the original idea was to 1) generate *in vitro* standard presynaptic filaments upon polymerizing human recombinase protein onto single-stranded DNA (ssDNA) and then 2) mimic, upon transfection, recombinational DNA repair that occurs in most dividing mammalian cells. Upon publication of the enzymatic properties of recombinase RAD51, we devised a more sophisticated approach and invented chimeric presynaptic filaments with a double-stranded DNA (dsDNA) core. In these dsDNA-cored filaments, therapeutic heterologous bases and repetitive elements are comprised in recombination-locked dsDNA cores. In transfected cells, homologous recombination should be driven by both recombination-activated ssDNA tails that are 100% identical to their respective chromosomal targets. Such a process should result in the substitution of host chromosomal DNA segments by exogenous recombination-free therapeutic dsDNA cassettes. Importantly enough, in addition to genomic DNA from wild type (gene repair) or mutant (gene inactivation) origin, these dsDNA cassettes may also accomodate therapeutic minigenes for efficient targeted (non-random) integration into host chromosomes. Targeted integration is essential to optimized long-term minigene expression therapy because 1) it should eliminate carcinogenic hazards associated to random insertional mutagenesis, and 2) it should provide for better control of transgene expression. Therefore, unlike complementary emerging single-base substitution approaches based on RNA-DNA oligonucleotides, our dsDNA-cored presynaptic filaments are discussed in terms of broad DNA exchange potentialities and of universal vectors for gene repair/inactivation (inherited diseases) and for targeted integration of therapeutic minigenes (acquired disorders).

Key Words: gene therapy; gene repair/inactivation; targeted integration of minigenes; homologous recombination; human RAD51; dsDNA-cored presynaptic filaments

List of Contributors

JEFFREY S. BARTLETT
Gene Therapy Center, CF Pulmonary Research Center, and Department of Medicine, University of North Carolina at Chapel Hill, Chapel Hill, NC 27599, USA

ROGER BERTOLOTTI
CNRS, Molecular Genetics, Department of Hepato-Gastroenterology, Faculty of Medicine, University of Nice-Sophia Antipolis, 06107 Nice, France

SYLVIA BRUNNER
Institute of Biochemistry, Vienna University Biocenter, Dr. Bohr-Gasse 9/3, A-1030 Vienna, Austria

SANDRA A. CALAROTA
Microbiology and Tumorbiology Center, Karolinska Institute, and Swedish Institute for Infectious Disease Control, Stockholm, Sweden

C. THOMAS CASKEY
Merck Research Laboratories, Department of Human Genetics, WP26A-3000, Sumneytown Pike, West Point, PA 19486, USA

KLAUS CICHUTEK
Department of Medical Biotechnology, Paul-Ehrlich-Institut, Paul-Ehrlich-Strasse 51-59, D-63225 Langen, Germany

DONG-SHEN FAN
Division of Genetic Therapeutics, Center for Molecular Medicine, and Department of Neurology, Jichi Medical School, Tochigi 329-0498, Japan

GREG FANNING
Johnson and Johnson Research, 1 Central Avenue, Australian Technology Park, Everleigh, Sydney, NSW 1430 Australia

ANTONIA FOLLENZI
Laboratory for Gene Transfer and Therapy, IRCC, Institute for Cancer Research, University of Torino Medical School, Strada Provinciale 142, 10060 Candiolo (Torino), Italy

KEN-ICHI FUJIMOTO
Department of Neurology, Jichi Medical School, Tochigi 329-0498, Japan

TOSHIKO FUKUMA
Division of Gastroenterology-Hepatology, Department of Medicine, University of Connecticut Health Center, 263 Farmington Avenue, Farmington, CT 06030-1845, USA

LINDA GORMAN
Lineberger Comprehensive Cancer Center, and Department of Pharmacology, University of North Carolina, Chapel Hill, NC 27599, USA

YUTAKA HANAZONO
Division of Genetic Therapeutics, Center for Molecular Medicine, Jichi Medical School, Tochigi 329-0498, Japan

MICHELLE HAUCHECORNE
INSERM U.458, Hôpital Robert Debré, 48 Bd Sérurier, 75019 Paris, France

KUNIHIKO IKEGUCHI
Department of Neurology, Jichi Medical School, Tochigi 329-0498, Japan

YASUFUMI KANEDA
Division of Gene Therapy Science, Graduate School of Medicine, Osaka University, Suita, Osaka 565-0871, Japan

RALF KIRCHEIS
Boehringer Ingelheim Austria, Dr. Boehringer-Gasse 5-11, A-1121 Vienna, Austria

ANNE KJERRSTRÖM
Microbiology and Tumorbiology Center, Karolinska Institute, and Swedish Institute for Infectious Disease Control, Stockholm, Sweden

KAZUTO KOBAYASHI
Department of Molecular Genetics, Institute of Biomedical Sciences, Fukushima Medical University School of Medicine, Fukushima 960-1295, Japan

DONALD B. KOHN
Division of Research Immunology/Bone Marrow Transplantation, Childrens Hospital Los Angeles, Departments of Pediatrics and Microbiology, University of Southern California School of Medicine, Los Angeles, CA 90027, USA

RYSZARD KOLE
Lineberger Comprehensive Cancer Center and Department of Pharmacology, University of North Carolina, Chapel Hill, NC 27599, USA

AKIHIRO KUME
Division of Genetic Therapeutics, Center for Molecular Medicine, Jichi Medical School, Tochigi 329-0498, Japan

MARGARETHA KURSA
Boehringer Ingelheim Austria, Dr. Boehringer-Gasse 5-11, A-1121 Vienna, Austria

DAVID S. LATCHMAN
Department of Molecular Pathology, Windeyer Institute of Medical Sciences, University College London Medical School, Windeyer Bldg., Cleveland Street, London W1P 6DB, UK

JEAN-MARIE LEHN
Laboratoire de Chimie des Interactions Moléculaires, Collège de France, 11 Place Marcelin Berthelot, 75005 Paris, France

PIERRE LEHN
INSERM U.458, Hôpital Robert Debré, 48 Bd Sérurier, 75019 Paris, France

PETER LILJESTRÖM
Microbiology and Tumor Biology Center, Karolinska Institutet, Box 280, S-17177, Stockholm, and Department of Vaccine Research, Swedish Institute for Infectious Disease Control, S-17182 Solna, Sweden

HEIKE MERGET-MILLITZER
Department of Medical Biotechnology, Paul-Ehrlich-Institut, Paul-Ehrlich-Strasse 51-59, D-63225 Langen, Germany

MARC MESNIL
Unit of Multistage Carcinogenesis, International Agency for Research on Cancer, F-69372 Lyon cedex 08, France

HIROAKI MIZUKAMI
Division of Genetic Therapeutics, Center for Molecular Medicine, Jichi Medical School, Tochigi 329-0498, Japan

RYUICHI MORISHITA
Division of Gene Therapy Science, Graduate School of Medicine, Osaka University, Suita, Osaka 565-0871, Japan

MANAL A. MORSY
Merck Research Laboratories, Department of Human Genetics, WP26A-3000, Sumneytown Pike, West Point, PA 19486, USA

PETER MULLER
Department of Medical Biotechnology, Paul-Ehrlich-Institut, Paul-Ehrlich-Strasse 51-59, D-63225 Langen, Germany

SHIN-ICHI MURAMATSU
Department of Neurology, Jichi Medical School, Tochigi 329-0498, Japan

TOSHIHARU NAGATSU
Division of Molecular Genetics, Institute for Comprehensive Medical Science, Graduate School of Medicine, Fujita Health University, Toyoake 470-0192, Japan

IMAHARU NAKANO
Department of Neurology, Jichi Medical School, Tochigi 329-0498, Japan

HIROSHI NAKASIMA
Division of Gene Therapy Science, Graduate School of Medicine, Osaka University, Suita, Osaka 565-0871, Japan

LUIGI NALDINI
Laboratory for Gene Transfer and Therapy, IRCC, Institute for Cancer Research, University of Torino Medical School, Strada Provinciale 142, 10060 Candiolo (Torino), Italy

JEAN NAVARRO
INSERM U.458, Hôpital Robert Debré, 48 Bd Sérurier, 75019 Paris, France

MATHIEU H. M. NOTEBORN
Leadd BV, and Department of Molecular Cell Biology, Leiden University, PO. Box 9503, 2300 RA, Leiden, The Netherlands

MATSUO OGAWA
Department of Neurology, Jichi Medical School, Tochigi 329-0498, Japan

MANFRED OGRIS
Institute of Biochemistry, Vienna University Biocenter, Dr. Bohr-Gasse 9/3, A-1030 Vienna, Austria

NOUFISSA OUDRHIRI
INSERM U.458, Hôpital Robert Debré, 48 Bd Sérurier, 75019 Paris, France

KEIYA OZAWA
Division of Genetic Therapeutics, Center for Molecular Medicine, Jichi Medical School, Kawachi-gun, Tochigi 329-0498, Japan

MICHEL PEUCHMAUR
INSERM U.458, Hôpital Robert Debré, 48 Bd Sérurier, 75019 Paris, France

RAJEN RAMASAWMY
INSERM U.458, Hôpital Robert Debré, 48 Bd Sérurier, 75019 Paris, France

S. RIQUIER
Laboratoire de Chimie des Interactions Moléculaires, Collège de France, 11 Place Marcelin Berthelot, 75005 Paris, France

RICHARD JUDE SAMULSKI
Gene Therapy Center, and Department of Pharmacology, University of North Carolina, Chapel Hill, NC 27599, USA

DIRK SCHADENDORF
Clinical Cooperation Unit for Dermato-oncology (DKFZ), Clinics Mannheim, University of Heidelberg, Theodor kutzer Ufer 1, 68135 Mannhein, Germany

SUSANNE SCHÜLLER
Institute of Biochemistry, Vienna University Biocenter, Dr. Bohr-Gasse 9/3, A-1030 Vienna, Austria

DANIEL SCHÜMPERLI
Abteilung für Entwicklungsbiologie, Zoologisches Institut der Universität Bern, Baltzerstraße 4, CH-3012 Bern, Switzerland

YANG SHEN
Division of Genetic Therapeutics, Center for Molecular Medicine, and Department of Neurology, Jichi Medical School, Tochigi 329-0498, Japan

CRISTIAN SMERDOU
Microbiology and Tumor Biology Center, Karolinska Institutet, Box 280, S-17177, Stockholm, Sweden

JÖRN STITZ
Department of Medical Biotechnology, Paul-Ehrlich-Institut, Paul-Ehrlich-Strasse 51-59, D-63225 Langen, Germany

CANDACE SUMMERFORD
Gene Therapy Center and Department of Pharmacology, University of North Carolina, Chapel Hill, NC 27599, USA

LUN QUAN SUN
Johnson and Johnson Research, 1 Central Avenue, Australian Technology Park, Everleigh, Sydney, NSW 1430 Australia

YUANSHENG SUN
Clinical Cooperation Unit for Dermato-oncology (DKFZ), Clinics Mannheim, University of Heidelberg, Theodor kutzer Ufer 1, 68135 Mannhein, Germany

GEOFF SYMONDS
Johnson and Johnson Research, 1 Central Avenue, Australian Technology Park, Everleigh, Sydney, NSW 1430 Australia

RENEE TOURY
INSERM U.458, Hôpital Robert Debré, 48 Bd Sérurier, 75019 Paris, France

NAOKI TSUBONIWA
Division of Gene Therapy Science, Graduate School of Medicine, Osaka University, Suita, Osaka 565-0871, Japan

MASASHI URABE
Division of Genetic Therapeutics, Center for Molecular Medicine, Jichi Medical School, Tochigi 329-0498, Japan

ALEX J. VAN DER EB
Department of Molecular Cell Biology, Leiden University, P.O. Box 9503, 2300 RA, Leiden, The Netherlands

JEAN-PIERRE VIGNERON
Laboratoire de Chimie des Interactions Moléculaires, Collège de France, 11 Place Marcelin Berthelot, 75005 Paris, France

ERIC VIVIEN
INSERM U.458, Hôpital Robert Debré, 48 Bd Sérurier, 75019 Paris, France

ERNST WAGNER
Boehringer Ingelheim Austria, Dr. Boehringer-Gasse 5-11, A-1121 Vienna, Austria, and Institute of Biochemistry, Vienna University Biocenter, Dr. Bohr-Gasse 9/3, A-1030 Vienna, Austria

BRITTA WAHREN
Microbiology and Tumorbiology Center, Karolinska Institute, and Swedish Institute for Infectious Disease Control, Stockholm, Sweden

LIONEL WIGHTMAN
Boehringer Ingelheim Austria, Dr. Boehringer-Gasse 5-11, A-1121 Vienna, Austria

CATHERINE H. WU
Division of Gastroenterology-Hepatology, Department of Medicine, University of Connecticut Health Center, 263 Farmington Avenue, Farmington, CT 06030-1845, USA

George Y. WU
Division of Gastroenterology-Hepatology, Department of Medicine, University of Connecticut Health Center, 263 Farmington Avenue, Farmington, CT 06030-1845, USA

HIROSHI YAMASAKI
Unit of Multistage Carcinogenesis, International Agency for Research on Cancer, 150 Cours Albert Thomas, F-69372 Lyon cedex 08, France

KYONGGEUN YOON
Department of Dermatology & Cutaneous Biology, and Department of Biochemistry and Molecular Pharmacology, Jefferson Institute of Molecular Medicine, Jefferson Medical College, Philadelphia, PA 19107, USA

Subject Index